Automotive
Quality Systems
Handbook

Automotive
Quality Systems
Handbook

Second Edition

ISO/TS 16949:2002 Edition

David Hoyle

ELSEVIER
BUTTERWORTH
HEINEMANN

AMSTERDAM • BOSTON • HEIDELBERG • LONDON • OXFORD • NEW YORK •
PARIS • SAN DIEGO • SAN FRANCISCO • SINGAPORE • SYDNEY • TOKYO

Elsevier Butterworth-Heinemann
Linacre House, Jordan Hill, Oxford OX2 8DP
30 Corporate Road, Burlington, MA 01803

First published 2000
Second edition 2005

British Library Cataloguing in Publication Data
A catalogue record for this book is available from the British Library

Library of Congress Cataloguing in Publication Data
Library of Congress Control Number: 2005924355

ISBN 0 7506 6663 3

For information on all Elsevier Butterworth-Heinemann publications
visit our website at www.books.elsevier.com

Working together to grow libraries in developing countries

www.elsevier.com | www.bookaid.org | www.sabre.org

ELSEVIER   BOOK AID
International   Sabre Foundation

Typeset by Charon Tec Pvt. Ltd, Chennai, India
www.charontec.com
Printed and bound in Great Britain

Contents

Preface

The first edition of this handbook was published in response to ISO/TS 16949:1999 just prior to the total revision of the ISO 9000 family of standards. Three years later ISO/TS 16949 was revised to take account of the changes in ISO 9000. As ISO/TS 16949 embodies the complete text of ISO 9001, this second edition of the *Automotive Quality Systems Handbook* is based on the fourth edition of the *ISO 9000 Quality System Handbook* which was produced to align with ISO 9001:2000. The three previous editions addressed ISO 9000:1987 and 1994.

Approach to the requirements

Although the handbook required a complete rewrite, I have maintained its basic purpose, i.e. that of providing the reader with an understanding of each requirement of ISO/TS 16949 through explanation, examples, lists, tables and diagrams. As there are over 400 requirements in ISO/TS 16949, this has led to a book of considerable size; it was not intended as a quick read! It was and remains a source of reference. I have adhered to following the structure of the standard, adding clause numbers to the headings to make it user-friendly. The handbook is therefore laid out so as to follow the section numbers of ISO/TS 16949:2002.

In the previous edition the approach taken was to identify a requirement and then explain its meaning followed by some examples to illustrate how it could be implemented. In this new edition I have followed a more structured approach and for each requirement answered three basic questions: *What does it mean? Why is it important? How is it implemented?* This is so that the requirements may be perceived as reflecting concepts valid outside the context of ISO/TS 16949.

I make no apologies for the layout. I have chosen to address the requirements of the standard in the sequence they are presented in the standard itself. Although it is claimed that the standard is based on a process approach, as will become apparent the requirements are not presented as a sequence of actions in a process neither are the sections wholly representative of business processes but it depends on what one regards as a process. The summaries at the beginning of Chapters 4–8 do attempt to link together the requirements in a cycle but they are hampered by the headings ascribed to the clauses in the standard.

At the end of each chapter there is a section titled Summary of requirements that will act as a checklist for system developers and auditors. However, it is not intended that the list be used by auditors during a system audit but used afterwards in order to test the completeness of the system. I have included a section titled Food for thought. This is intended to cause the reader to reflect on the previous chapter, perhaps even change perceptions but mostly confirm understanding.

ISO/TS 16949:2002 contains the full text of ISO 9001:2000 supplemented by additional automotive requirements. In some cases the supplemental and additional automotive requirements merely clarify an existing ISO 9001 requirement, in others they simply duplicate requirements that would be included in a contract but in other cases they add new requirements that one could derive from existing ISO 9001 requirements, but not without some insight.

Where the additional automotive requirements merely complement, clarify or particularize an existing ISO 9001 requirement, the automotive requirement is addressed under the ISO 9001 heading and either the ISO 9001 requirement reworded or expanded.

Where an automotive requirement is not so obviously derived from an ISO 9001 requirement or addressing it with an ISO 9001 requirement would make it difficult to find in this book, it is addressed under a new heading.

The interpretations are those of the author and should not be deemed to be those of the International Automotive Task Force, International Organization for Standardization, any National Standards Body or any Certification Body.

How to use this handbook

As stated previously this is a reference book for users of the ISO/TS 16949 and as such it contains concepts as well as details.

If you are new to the field of quality management you might start by reading Chapter 1 in order to gain an appreciation of the fundamental principles and the broader perspective of quality management. You could skip Chapter 2 because you won't have any preconceived ideas about the standard and move onto Chapter 3. To obtain an overview of

ISO/TS 16949, you could then read the Requirement Summaries at the start of each chapter from Chapter 4 through 8 and then seek out particular requirements of interest.

If you have experience of using QS-9000 or ISO/TS 16949 and are charged with improving your management system you might start with Chapter 2 where you will gain an insight into how the standard has been used, misused and misunderstood. If you then dip into Chapter 1 especially the last sections on Process management and Developing an effective quality management system, you might begin changing your perceptions so that when you begin browsing Chapters 4 through 8 which deal with the requirements you will see them within an entirely different framework. Finally you might use the Requirement lists at the end of Chapters 4 through 8 to help check that your system addresses all relevant requirements.

If you have a problem with interpreting a particular requirement, simply go to the heading carrying the Clause number about which you are interested.

If you are auditing management systems against ISO/TS 16949 you certainly need to read Chapter 2 to gain an appreciation of the problems that have been caused by its inappropriate use. You should then study Chapter 1 to appreciate the concepts and principles, especially the sections on quality management principles and process management. In preparing for the audit you might consult the appropriate sections in Chapters 4 through 8 in order to gain an insight into what evidence you should be looking for when you examine the organization's processes. After completing the audit you might use the Requirement lists at the end of Chapters 4 through 8 to help check that you have covered all relevant requirements.

Whatever your purpose you would benefit from studying the Glossary of terms because the meaning given might well differ from that which you may have assumed the term to mean and thus it will affect your judgement.

Structure and style of the handbook

The first three chapters provide background information with the subsequent five chapters dealing with the sections of ISO/TS 16949 that contain the requirements. In this way the chapter numbers of the book mirror the section headings of ISO/TS 16949.

In view of the differing perceptions, when the term ISO/TS 16949 is used in this book it means the standard and not its attendant infrastructure. Comment on any aspect of the infrastructure will be referred to it by its usual name – auditing, consulting, certification, training or accreditation.

I have retained the direct style of writing referring to the reader as 'you'. You may be a manager, an auditor, a consultant, an instructor, a member of staff, a student or simply an interested reader. You may not have the power to do what is recommended in this book but may know of someone who does whom you can influence. There will be readers who feel I have laboured topics too much but it never ceases to amaze me how many different ways a certain word, phase or requirement might be interpreted.

The type of organizations that might seek ISO/TS 16949 certification are by no means all Tier 1 suppliers to the big vehicle makers. As a result, the interpretations in the book are given with respect to as wide a readership as practical. I have not assumed that the only people reading the book will be from organizations supplying customer-specified products; and the interpretations also relate to organizations supplying customer-selected products.

This book is written for those who want to improve the performance of their organization, and whether or not certification is a goal or indeed required, I hope it will continue to provide a source of inspiration in the years ahead.

David Hoyle
Monmouth
E-mail: hoyle@transition-support.com

Acknowledgements

In December 1998, I was fortunate to be invited to participate in the development of the ISO/TS 16949 certification scheme and later to represent the UK Society of Motor Manufacturers and Traders (SMMT) on the IATF Training Council set up to design the original auditor qualification programme. I am indebted to Bob Lawrie and Robin Lock of the Society of Motor Manufacturers and Traders (SMMT), the members of the IATF Training Council including Robert Frank and Markus Stang of the VDA, Antonio Ciancio of ANFIA, Jacques Letheicq of Euro-Symbiose and Al Peterson and Jodi Shorma of Plexus-Training who with the other members of the Training Council created an environment rich in ideas, knowledge and experience during those early days of the programme. I am grateful to the many clients and associates I have talked with during the transition from QS-9000 to the publication of ISO/TS 16949:2002 on related consultancy and training assignments; to my business partner John Thompson whose ideas and insight provided the clarity needed to explain the requirements in the wider context of business management and for his teachings on process management; to my wife Angela for her constructive comment and editing of the manuscript and to Jonathan Simpson of Elsevier who commissioned the work.

The teachings of P. F. Drucker have been a constant inspiration particularly in clarifying issues on strategic management. The teachings of W. E. Deming have been particularly useful for this edition in clarifying the theory of variation and confirming my ideas on systems theory. The teachings of J. M. Juran have also been a constant inspiration particularly concerning breakthrough and control principles and quality planning. The treatment of *competence* in Chapter 6 would not have been possible without the teachings of Shirley Fletcher and contributions from John Thompson.

Extracts from ISO 9001:2000 are reproduced with the permission of the British Standards Institution on behalf of ISO under licence number 2003SK/135. BSI and ISO publications can be obtained from BSI Customer Services, 389 Chiswick High Road, London W4 4AL. United Kingdom. (Tel: +44 (0) 20 8996 9001). Email: cservices@bsi-global.com

Chapter 1

Basic concepts

What is important is not the system but the creativity of human beings who select
and interpret the information.
Taiichi Ohno, Former Executive Vice President, Toyota Motor Company

Introduction

The automotive industry is the biggest industry in the world and constantly changing. Over 8 million people working for 50 manufacturers produced over 60 million vehicles in 2003 with production rising by 6% by mid-2004. As revealed by Table 1.1,[1] 80% of the vehicles were produced by just 12 manufacturers and over the next two decades there is speculation that this may reduce to just six. There has been modest growth since 1999 and apart from some changes in position, the top 15 vehicle manufacturers have remained the same over the same period.

Our lives revolve around the automobile. We use it to transport goods and people, for work, for sport, for war and for pleasure, and in many ways our world would not be the same without it. With oil prices rising since the start of the second Gulf War, today's consumers are interested in more fuel efficient, smarter and safer automobiles, and consumer groups are pressing for sustainability thus impacting not only the selection of materials to build automobiles but also the efficiency and effectiveness of the production processes.

Until the oil crisis of 1973 and subsequent recession, it was an industry that pushed products onto the market in quantities derived from sales forecasts. This mass production strategy depended on sustaining consumer demand but when oil prices soared, the post-World War II gas guzzling cars were no longer wanted. Faced with having to cut costs while producing a small number of many types of cars, in 1945 Toyota began to

Table 1.1 *Worldwide motor vehicle production*

Rank	Vehicle manufacturer	Vehicle production (millions) includes passenger cars, light commercial vehicles, heavy trucks, buses and coaches					
		1999	2000	2001	2002	2003	2003 Pareto
1	General Motors (including Opel-Vauxhall)	8.42	8.13	7.58	8.33	8.19	13.52%
2	Ford (including Jaguar-Volvo cars)	6.64	7.32	6.68	6.73	6.57	24.36%
3	Toyota	5.46	5.95	6.05	6.63	6.24	34.66%
4	Volkswagen Group (including Audi, Seat and Skoda)	4.78	5.11	5.11	5.02	5.02	42.96%
5	Daimler-Chrysler (with Evobus)	4.83	4.67	4.36	4.46	4.23	49.95%
6	PSA Peugeot Citroën	2.53	2.88	3.10	3.26	3.31	55.41%
7	Nissan	2.46	2.63	2.56	2.72	2.94	60.27%
8	Honda	2.43	2.51	2.67	2.90	2.92	65.10%
9	Hyundai-Kia	1.31	2.49	2.52	2.64	2.70	69.55%
10	Renault-Dacia-Samsung	2.35	2.51	2.38	2.33	2.39	73.49%
11	Fiat-Iveco	2.63	2.64	2.41	2.19	2.08	76.92%
12	Suzuki-Maruti	1.52	1.46	1.54	1.70	1.81	79.92%
13	Mitsubishi	1.56	1.83	1.65	1.82	1.58	82.53%
14	Mazda	0.97	0.93	0.96	1.04	1.15	84.43%
15	BMW including Rover*	1.15*	0.83	0.95	1.09	1.12	86.28%
	Total (top 15)	49.04	51.89	50.52	52.86	52.25	
	World production	**56.01**	**58.39**	**56.33**	**58.84**	**60.56**	
	% change since 1999		+4.26%	+0.56%	+5.06%	+8.12%	

develop a different strategy – one in which the consumer pulls the goods they need thus eliminating overproduction. However, while both Toyota and Ford used the same workflow system, Toyota managed to eliminate the warehouse.[2] The Henry Ford quip "that you can have any colour as long as it is black" characterized by mass production but it was also characterized by excessive waste. The economies of scale tied the manu-facturer to providing a limited choice, but an order-driven production strategy would turn this around causing parts to be produced only on demand and so the consumer could have within reason, any colour, any combination of features and engines, i.e. a car to their specification. In the late 1980s and early 1990s as companies became aware of the Toyota Production System, they too started to adopt what is being called a "lean production" philosophy.[3]

The industry is now global. Cars designed in one country are assembled in another with parts made in a third, fourth or fifth country. Organizations operating in the automotive sector continually have to improve product quality and delivery, reduce material and labour costs, waste and vulnerability in the supply chain and hence attract some of the best engineering talent.

It is the management systems of organizations that need to be responsive to these changes because it is these systems that produce the 60 million or so vehicles every year. It is therefore vital that the management systems in the automotive industry are designed to enable organizations to satisfy the needs of customers and other stakeholders, not sim-ply short-term profits for shareholders. It is against this background that pressure for a common set of quality management system requirements emerged. Generic international quality management system requirements came in 1987 with ISO 9000 and, from an amalgam of company-specific requirements in the early 1990s, the first international automotive specification ISO/TS 16949 emerged in 1999 followed by a second edition in 2002. This standard will remain current until at least 2008 when ISO 9001:2000 will complete its review cycle.

Poor-quality components entering the supply chain create massive problems for the vehicle manufacturers thus making product quality key to survival and so we begin by looking at the meaning of the word "quality" and the principles and practices that have emerged to determine and manage it.

Principles or prescription

One of the great problems in our age is to impart understanding in the minds of those who have the ability and opportunity to make decisions that affect our lives. There is no shortage of information – in fact, there is too much now we can search a world of information from the comfort of our armchair. We are bombarded with information but it is not knowledge – it does not necessarily lead to understanding. With so many con-flicting messages from so many people, it is difficult to determine the right thing to do. There are those whose only need is a set of principles from which they are able to determine the right things to do. There are countless others who need a set of rules derived from principles that they can apply to what they do and indeed others who need a detailed prescription derived from the rules for a particular task. In the translation from principles to prescription, inconsistencies arise. Those translating the principles into rules or requirements are often not the same as those translating the rules into a detailed prescription. The principles in the field of quality management have not arisen out of academia but from life in the workplace. Observations from the workplace have been taken into academia, analysed, synthesized and refined to emerge as universal principles. These principles have been expressed in many ways and in their constant refreshment the language is modernized and simplified, but the essence is hardly changed.

Without a set of principles, achieving a common understanding in the field of quality management would be impossible. Since Juran, Deming and Feigenbaum wrote about quality management in the 1950s there has been considerable energy put into codifying the field of quality management and a set of principles from which we can derive useful rules, regulations and requirements has emerged. This chapter addresses these principles in a way that is intended to impart understanding not only in the minds of those who prefer principles to prescription, but also in the minds of those who prefer prescriptions. There is nothing intrinsically wrong with wanting a prescription. It saves time, it's repeatable, it's economic and it's the fastest way to get things done but it has to be right. The receivers of prescriptions need enough understanding to know whether what they are being asked to do is appropriate to the circumstances they are facing.

The concepts expressed in this chapter embody universal principles and have been selected and structured in a manner that is considered suitable for users of ISO/TS 16949. It is not intended as a comprehensive guide to quality management – some further reading is given as footnotes appropriately. ISO/TS 16949 also contains concepts some of which are questionable but these will be dealt with as they arise. The aim is to give the reader a balanced view and present a logical argument that is hoped will lead to greater understanding. As ISO/TS 16949 is supposed to be about the achievement of quality, there is no better place to start than with an explanation of the word *quality*.

Quality

We all have needs, wants, requirements and expectations. Needs are essential for life, to maintain certain standards, or essential for products and services, to fulfil the purpose for which they have been acquired. According to Maslow,[4] man is a wanting being; there is always some need he or she wants to satisfy. Once this is accomplished, that particular need no longer motivates him and he or she turns to another, again seeking satisfaction. Everyone has basic physiological needs that are necessary to sustain life (food, water, clothing and shelter). Maslow's research showed that once the physiological needs are fulfilled, the need for safety emerges. After safety come social needs followed by the need for esteem and finally the need for self-actualization or the need to realize ones

Figure 1.1 *Hierarchy of needs*

full potential. Satisfaction of physiological needs is usually associated with money – not money itself but what it can buy. The hierarchy of needs is shown in Figure 1.1.

These needs are fulfilled by the individual purchasing, renting or leasing products or services. Corporate needs are not too dissimilar. The physiological needs of organizations are those necessary to sustain survival. Often profit comes first because no organization can sustain a loss for too long but functionality is paramount – the product or service must do the job for which it is intended regardless of it being obtained cheaply. Corporate safety comes next in terms of the safety of employees, and the safety and security of assets followed by social needs in the form of a concern for the environment and the community as well as forming links with other organizations and developing contacts. Esteem is represented in the corporate context by organizations purchasing luxury cars, winning awards, superior offices and infrastructures, and possessing those things that give it power in the marketplace and government. Self-actualization is represented by an organization's preoccupation with growth, becoming bigger rather than better, seeking challenges and taking risks. However, it is not the specific product or service that is needed but the benefits that possession brings that is important. This concept of benefits is the most important and key to the achievement of quality. Unfortunately, ISO/TS 16949 or ISO 9001 certification falls into the category of physiological needs simply because for organizations in the automotive supply chain, it has become a necessity for survival in this industry sector.

Requirements are what we request of others and may encompass our needs but often we don't fully realize what we need until after we have made our request. For example, now that we own a mobile phone we discover, we really need hands-free operation when using the telephone while driving a vehicle. Our requirements at the moment of sale may or may not therefore express all our needs. By focusing on benefits resulting from products and services, needs can be converted into wants such that a need for food may be converted into a want for a particular brand of chocolate. Sometimes the *want* is not essential but the higher up the hierarchy of needs we go, the more a *want* becomes essential to maintain our social standing, esteem or to realize our personal goals. Our requirements may therefore include such wants what we would like to have but are not essential for survival.

In growing their business organizations, create a demand for their products and services but far from the demand arising from a want that is essential to maintain our social standing, it is based on an image created for us by media advertising. We don't need spring vegetables in the winter but because industry has created the organization to supply them, a demand is created that becomes an expectation. Spring vegetables have been available in the winter now for so long that we expect them to be available in the shops and will go elsewhere if they are not. But they are not essential to survival, to safety, to esteem or to realize our potential, and their consumption may in fact harm our health because we are no longer absorbing the right chemicals to help us survive the cold winters. We might want it, even need it but it does us harm and there are plenty of organization ready to supply us products that will harm us.

Expectations are *implied needs or requirements*. They have not been requested because we take them for granted, we regard them to be understood within our particular society as the accepted norm. They may be things to which we are accustomed, based on fashion, style, trends or previous experience. One therefore expects sales staff to be polite and courteous, electronic products to be safe and reliable, policemen to be honest, coffee to be hot, etc. One would like politicians to be honest but in some countries we have come to expect them to be corruptible, dishonest or, at least, economical with the truth! On the other hand, we expect businessmen to be dishonest, unfair, corruptible and selfish, and it comes as no surprise to read about long-drawn-out court cases involving fraud and deceit.

Quality
The degree to which a set of inherent characteristics fulfils a need or expectation that is stated, generally implied or obligatory.

In supplying products or services there are three fundamental parameters that determine their saleability. They are price, quality and delivery. Customers require products and services of a given quality to be delivered by or be available by a given time and to be of a price that reflects value for money. These are the requirements of customers. An organization will survive only if it creates and retains satisfied customers and this will only be achieved if it offers for sale products or services that respond to customer needs and expectations as well as requirements. While price is a function of cost, profit margin and market forces, and delivery is a function of the organization's efficiency and effectiveness. Quality is determined by the extent to which a product or service successfully serves the purposes of the user during usage (not just at the point of sale). Price and delivery are both transient features, whereas the impact of quality is sustained long after the attraction or the pain of price and delivery has subsided.

The word *quality* has many meanings:

- a degree of excellence,
- conformance with requirements,
- the totality of characteristics of an entity that bear on its ability to satisfy stated or implied needs,
- fitness for use,
- fitness for purpose,
- freedom from defects imperfections or contamination,
- delighting customers.

These are just a few meanings; however, the meaning used in the context of ISO 9000 was concerned with the totality of characteristics that satisfy needs but in the 2000 version this has changed. Quality in ISO 9000:2000 is defined as the degree to which a set of inherent characteristics fulfils the requirements. The former definition focused on an entity that was described as a product or service but with this new definition, the implication is that quality is relative to what something should be and what it is. The something may be a product, service, decision, document, piece of information or any output from a process. In describing an output, we express it in terms of its characteristics. To comment on the quality of anything we need a measure of its characteristics and a basis for comparison. By combining the definition of the terms *quality* and *requirement* in ISO 9000:2000, quality can be expressed as *the degree to which a set of inherent characteristics fulfils a need or expectation that is stated, generally implied or obligatory.*

This means that when we talk of anything using the word quality it simply implies that we are referring to the extent or degree to which a requirement is met. Therefore, environmental, safety, security and health problems are in fact quality problems because an expectation or a requirement has not been met. Even if it had there would be no problem.

Having made the comparison we can still assess whether the output is "fit for use". In this sense the output may be of poor quality but remain fit for use. The specification is often an imperfect definition of what a customer needs; because some needs can be difficult to express clearly and it doesn't mean that by not conforming, the product or service is unfit for use. It is also possible that a product that conforms to requirements may be totally useless. It all depends on whose requirements are being met. For example, if a company sets its own standards and these do not meet customer needs, its claim to producing quality products is bogus. On the other hand, if the standards are well in excess of what the customer requires, the price tag may well be too high for what customers are prepared to pay; there probably isn't a market for a gold-plated mouse-trap, for instance, except as an ornament perhaps!

The interested parties

<table>
<tr><td>

Interested party

Person or group having a benevolent interest in the performance or success of an organization, which includes customers, owners, employees, contractors, suppliers, investors, unions, partners or society.

</td><td>

Organizations depend on customers because without them there is no business but in order to satisfy these customers, organizations also depend on a number of other parties that provide it with resources and sanction its operations. There are parties other than the customer that have an interest or stake in the organization and what it does but may not receive a product. The term *quality* is not defined relative to customers but to requirements and these interested parties do have requirements. ISO 9000:2000 defines an interested party as *a person or group having an interest in the performance or success of an organization*. But the organization may not have an interest in all of them and on examination perhaps the word *interest* is not quite appropriate. Consider for instance, competitors, criminals and terrorists. None of these has put anything into the organization

</td></tr>
</table>

and their interest is more likely to be malevolent than benevolent so in these cases the organization fights off their interests rather than satisfying them. A better word to *interest* would be *stake* but in some cultures this translates as a bet that you place on a horse to win a race. However, it is common in the west for the term *stakeholder* to be used in preference to the term interested party, as it implies benevolence unlike the term interested parties which might be benevolent or malevolent.

Parties with a benevolent interest (stakeholders) are customers, owners, employees, contractors, suppliers, investors, unions, partners or society. They all expect something in return for their stake and can withdraw it should the expected benefits not be returned. When you produce products you are producing them within the intent that all these parties benefit but particularly for the benefit of customers. The other parties are not particularly interested in the products and services themselves but may be interested in their effects on their investment, their well-being and the environment.

The customer

<table>
<tr><td>

Customer

Organization that receives a product or service, which includes purchaser, consumer, client, end user, retailer or beneficiary.

</td><td>

A product that possesses features that satisfy customer needs is a quality product. Likewise, one that possesses features that dissatisfy customers is not a quality product. So the final arbiter on quality is the customer. The customer is the only one who can decide whether the quality of the products and services you supply is satisfactory and you will be conscious of this either by direct feedback or by loss of sales, reduction in market share and, ultimately, loss of business. This brings us back to benefits. The customer acquires a product for the benefits that possession will bring. Therefore, if the product fails to

</td></tr>
</table>

deliver the expected benefits it will be considered by the customer to be of poor quality. So when making judgements about quality, the requirement should be expressed in terms of benefits not a set of derived characteristics. In the foregoing it was convenient to use the term customer but the definition of *quality* does not only relate to customers. Dissatisfy you customers than they withdraw their stake and take their business elsewhere.

Clearly the customer is the only stakeholder that brings in revenue and therefore meeting their needs and expectations is paramount but not at the expense of the other stakeholders. The secret is to satisfy customers in a way that will satisfy the needs of other stakeholders (see the section Quality management principles later in this chapter).

> **Stakeholder**
>
> A person or organization that has freedom to provide something to or withdraw something from an enterprise.

The internal customer

We tend to think of products and services being supplied to customers and in the wake of total quality management (TQM), we also think of internal and external customers but in reality there is no such thing as an internal customer. A customer is a stakeholder; they have entered into a commitment in return for some benefits that possession of a product or experience of a service may bring. The internal receivers of products are not stakeholders therefore they are not customers. ISO 9000:2000 defines the customer as an organization or person that receives a product. It is implied that the organization and person referred to is external to the organization supplying the product because to interpret the term customer as either internal or external would make nonsense of requirements in ISO/TS 16949 where the term customer is used.

The supply chain

The transaction between the customer and the supplier is often a complex one. There may be a supply chain from original producer through to the end user. At each transaction within this supply chain, the receiving party needs to be satisfied. It is not sufficient to simply satisfy the first receiver of the product or service. All parties in the supply chain need to be satisfied before you can claim to have supplied a quality product. Admittedly, once the product leaves your premises you may lose control and therefore cannot be held accountable for any damage that may become the product, but the inherent characteristics are your responsibility.

Society

Society is a stakeholder because it can withdraw its support for an organization. It can protest or invoke legal action. Society is represented by the regulators and regardless of whether or not a customer specifies applicable regulations *you* are under an obligation to comply with those that apply. The regulator is not interested in whether you satisfy

Regulator
A legal body authorized to enforce compliance with the laws and statutes of a national government.

your customers, your employees or your investors – the regulator couldn't care less if you go bankrupt! Its primary concern is the protection of society. The regulators take their authority from the law that should have been designed to protect the innocent. Regulators are certainly stakeholders because they can withdraw their approval.

Employees

Employees may not be interested in the products and services, but are interested in the conditions in which they are required to work. Employees are stakeholders because they can withdraw their labour.

Suppliers

Suppliers are interested in the success of the organization because it may in turn lead to their success. However, suppliers are also stakeholders because they can withdraw their patronage. They can choose their customers. If you treat your suppliers badly such as delaying payment of invoices for trivial mistakes, you may find they terminate the supply at the first opportunity possibly putting your organization into a difficult position relative to its customer's commitments.

Investors

Often the most common type of stakeholder, owners, investors including banks and shareholders are interested in protecting their stake in the business. They will withdraw their stake if the organization fails to perform. Poorly conceived products and poorly managed processes and resources will not yield the expected return and the action of investors can directly affect the supply chain – although they are not customers, they are feeding the supply chain with much needed resources. In the event that this supply of resource is terminated, the organization ceases to have the capability to serve its customers.

The success of any organization therefore depends on understanding the needs and expectations of all the stakeholders, not just its customers and on managing the organization in a manner that leads to the continued satisfaction of all parties. Table 1.2[5] shows the criteria used by different interested parties. It tends to suggest that for an organization to be successful it needs to balance (not trade-off) the needs of the stakeholders such that all are satisfied. There are those who believe that a focus on customers alone will result in the other parties being satisfied. There are those who believe that a focus on shareholder value will result in all other parties being satisfied. The

Table 1.2 *Criteria used by interested parties to judge organization effectiveness or success*

Interested party	Success criteria
Owner	Financial return
Employees	Job satisfaction, pay and conditions, and quality of leadership
Customers	Quality of products and services
Community	Contribution to the community (jobs, support for other traders in the community), care for the local environment
Suppliers	Satisfactory mutual trading
Investors	Value of shares
Government	Compliance with legislation

problem is that the interested party is motivated by self-interest and may not be willing to compromise.

The characteristics of quality

Classification of products and services

If we group products and services (entities) by type, category, class and grade we can use the subdivision to make comparisons on an equitable basis. But when we compare entities we must be careful not to claim one is of better quality than the other unless they are of the same grade. Entities of the same type have at least one attribute in common. Entities of the same grade have been designed for the same functional use and therefore comparisons are valid. Comparisons on quality between entities of different grades, classes, categories or types are invalid because they have been designed for a different use or purpose.

Let us look at some examples to illustrate the point. Food is a type of entity. Transport is another entity. Putting aside the fact that in the food industry the terms *class* and *grade* are used to denote the condition of post-production product, comparisons between *types* is like comparing fruit and trucks, i.e. there are no common attributes. Comparisons between *categories* are like comparing fruit and vegetables. Comparisons between *classes* are like comparing apples and oranges. Comparisons between *grades* is like comparing eating apples and cooking apples.

Now let us take another example. Transport is a type of entity. There are different categories of transport such as airliners, ships, automobiles and trains; they are all modes of transport but each has many different attributes. Differences between categories of transport are therefore differences in *modes* of transport. Within each category there are differences in class. For manufactured products, differences between classes imply differences in *purpose*. Luxury cars, large family cars, small family cars, vans, trucks, four-wheel drive vehicles, etc. fall within the same category of transport but each was designed for a different purpose. Family cars are in a different class to luxury cars; they were not designed for the same purpose. It is therefore inappropriate to compare a Cadillac with a Chevrolet or a Rolls Royce Silver Shadow with a Ford Mondeo. Entities designed for the same purpose but having different specifications are of different grades. A Ford Mondeo GTX is a different grade to a Mondeo LX. They were both designed for the same purpose but differ in their performance and features and hence comparisons on quality are invalid.

You can legitimately compare the quality of entities if comparing entities of the same grade. If a low-grade product or service meets the needs for which it was designed, it is of the requisite quality. If a high-grade product or service fails to meet the requirements for which it was designed, it is of poor quality, regardless of it still meeting the requirements for the lower grade. There is a market for such differences in products and services but should customer's expectations change then what was once acceptable for a particular grade may no longer be acceptable and regrading may have to occur.

Where manufacturing processes are prone to uncontrollable variation it is not uncommon to grade products as a method of selection. The product that is free of imperfections would be the highest grade and would therefore command the highest price. Any product with imperfections would be downgraded and sold at a correspondingly lower price. Examples of such practice arise in the fruit and vegetables trade, and the ceramics, glass and textile industries. In the electronic component industry, grading is a common practice to select devices that operate between certain temperature ranges. In ideal conditions, all devices would meet the higher specification but due to variations in the raw material or in the manufacturing process only a few may actually reach full performance. The remainder of the devices has a degraded performance but still offer all the functions of the top-grade component at lower temperatures. To say that these differences are not differences in *quality* would be misleading, because the products were all designed to fulfil the higher specification. As there is a market for such products it is expedient to exploit it. There is a range over which product quality can vary and still create satisfied customers. Outside the lower end of this range the product is considered to be of poor quality.

Quality and price

Most of us are attracted to certain products and services by their price. If the price is outside our reach we don't even consider the product or service, whatever its quality, except perhaps to form an opinion about it. We also rely on price as a comparison, hoping that we can obtain the same characteristics at a lower price. In the luxury goods market, a high price is often a mark of quality but occasionally it is a confidence trick aimed at making more profit for the supplier. When certain products and services are rare the price tends to be high and when plentiful the price is low, regardless of their quality. One can purchase the same item in different stores at different prices, some as much as 50% less, many at 10% less than the highest price. You can also receive a discount for buying in bulk, buying on customer credit card and being a trade customer rather than a retail customer. Travellers know that goods are more expensive at an airport than from a country craft shop. However, in the country craft shop, defective goods or "seconds" may well be on sale, whereas at the airport the supplier will as a rule, want to display only the best examples. Often an increase in the price of a product may indicate a better after-sales service, such as free on-site maintenance, free delivery and free telephone support line. The discount shops may not offer such benefits.

The price label on any product or service should be for a product or service free of defects. If there are defects the label should say as much, otherwise the supplier may well be in breach of national laws and statutes. Price is therefore not an inherent feature or characteristic of the product. It is not permanent and as shown above varies without any change to the inherent characteristics of the product. Price is a feature of the service associated with the sale of the product. Price is negotiable for the same quality of product. Some may argue that quality is expensive but in reality, the saving you make on buying low-priced goods could well be eroded by inferior service or differences in the cost of ownership.

Quality and cost

Philip Crosby published his book *Quality Is Free* in 1979 and caused a lot of raised eyebrows among executives because they always believed the removal of defects was an in-built cost in running any business. To get quality you had to pay for inspectors to detect the errors! What Crosby told us was that if we could eliminate all the errors and reach zero defects, we would not only reduce our costs but also increase the level of customer satisfaction by several orders of magnitude. In fact there is the cost of doing the right things right first time and the cost of *not* doing the right things right first time. The latter are often referred to as *quality costs* or the cost incurred because failure is possible. Using this definition, if failure of a product, a process or a service is not possible, there would be no *quality costs*. It is rather misleading to refer to the cost incurred because failure is possible as *quality costs* because we could classify the costs as avoidable costs and unavoidable costs. We have to pay for labour, materials, facilities, machines, transport, etc. These costs are unavoidable but we are also paying in addition some cost to cover the prevention, detection and removal of errors. Should customers have to pay for the errors made by others? There is a basic cost if failure is not possible and an additional cost in preventing and detecting failures and correcting errors because our prevention and detection programmes are ineffective. However, there is variation in all processes but it is only the variation that exceeds the tolerable limits that incurs a penalty. If you reduce complexity and install failure-prevention measures you will be spending less on failure detection and correction. There is an initial investment to be paid, but in the long term you can meet your customer's requirements at a cost far less than you were spending previously. Some customers are now forcing their suppliers to reduce internal costs so that they can offer the same products at lower prices. This has the negative effect of forcing suppliers out of business. While the motive is laudable the method is damaging to industry. There are inefficiencies in industry that need to be reduced but imposing requirements will not solve the problem. Co-operation between customer and supplier would be a better solution and when neither party can identify any further savings the target has been reached. Customers do not benefit by forcing suppliers out of business. So is *quality free*?

High quality and low quality; poor quality and good quality

When a product or service satisfies our needs we are likely to say it is of good quality and likewise when we are dissatisfied we say the product or service is of poor quality. When the product or service exceeds our needs we will probably say it is of high quality and likewise if it falls well below our expectations we say it is of low quality.

These measures of quality are all subjective. What is good to one may be poor to another. In the under-developed countries, any product, no matter what the quality, is welcomed. When you have nothing, even the poorest of goods is better than none. A product may not need to possess defects for it to be regarded as poor quality, which means it may not possess the features that we would expect, such as access for maintenance. These are design features that give a product its saleability. Products and services that conform to customer requirements are considered to be products of acceptable quality. However, we need to express our relative satisfaction with products and services, and as a consequence use subjective terms, such as high, low, good or poor quality. If a product that meets customer requirements is of acceptable quality, what do we call one that does not quite meet the requirements, or perhaps exceeds the requirements? An otherwise acceptable product has a blemish, i.e. is it now unacceptable? Perhaps not because it may still be far superior to other competing products in its acceptable features and characteristics.

While not measurable, these subjective terms enable customers to rate products and services according to the extent to which they satisfy their requirements. However, to the company supplying products and services, a more precise means of measuring quality is needed. To the supplier, a quality product is one that meets in full the perceived customer requirements.

Quality characteristics

Any feature or characteristic of a product or service that is needed to satisfy customer needs or achieve fitness for use is a *quality characteristic*. When dealing with products the characteristics are almost always technical characteristics, whereas service quality characteristics have a human dimension. Some typical quality characteristics are given below:

Product characteristics

Accessibility	Disposability	Odour	Security	Testability
Availability	Emittance	Operability	Size	Traceability
Appearance	Flammability	Portability	Susceptibility	Toxicity
Adaptability	Flexibility	Producibility	Storability	Transportability
Cleanliness	Functionality	Reliability	Strength	Vulnerability
Consumption	Interchangeability	Reparability	Taste	Weight
Durability	Maintainability	Safety		

Service quality characteristics

Accessibility	Comfort	Dependability	Flexibility	Responsiveness
Accuracy	Competence	Efficiency	Honesty	Reliability
Courtesy	Credibility	Effectiveness	Promptness	Security

These are the characteristics that need to be specified and their achievement controlled, assured, improved, managed and demonstrated. These are the characteristics that form the subject matter of the product requirements referred to in ISO/TS 16949. When the value of these characteristics is quantified or qualified they are termed *product requirements*. We used to use the term *quality requirements* but this caused a division in thinking that resulted in people regarding quality requirements as the domain of the quality personnel and technical requirements being the domain of the technical personnel. All requirements are *quality* requirements – they express needs or expectations that are intended to be fulfilled by a process output that possesses inherent characteristics. We can therefore drop the word *quality*. If a modifying word is needed in front of the word requirements it should be a word that signifies the subject of the requirements. Transportation system requirements would be requirements for a transportation system, audio speaker design requirements would be requirements for the design of an audio speaker, component test requirements would be requirements for testing components and management training requirements would be requirement for training managers. The requirements of ISO 9000 and its derivatives such as ISO/TS 16949 are often referred to as *quality requirements* as distinct from other types of requirements but this is misleading. ISO 9000 is no more a quality requirement than is ISO 1000 on SI units, ISO 2365 for Ammonium nitrate or ISO 246 for Rolling Bearings. The requirements of ISO 9001 are quality management system requirements – requirements for a quality management system and the requirements of ISO/TS 16949 are requirements for the application of ISO 9001 for automotive production and relevant service part organizations (as displayed on its cover).

Quality, reliability and safety

There is a school of thought that distinguishes between quality and reliability and quality and safety. Quality is thought to be a non-time-dependent characteristic and reliability a time-dependent characteristic. Quality is thought of as conformance to specification regardless of whether the specification actually meets the needs of the customer or society. If a product or service is unreliable, it is clearly unfit for use and therefore of poor quality. If a product is reliable but emits toxic fumes, is too heavy or not transportable when required to be, it is of poor quality. Similarly, if a product is unsafe it is of poor quality even though it may meet its specification in other ways. In such a case the specification is not a true reflection of customer needs. A nuclear plant may meet all the specified safety requirements but if society demands greater safety standards, the plant is not meeting the requirements of society, even though it meets the immediate customer requirements. You therefore need to identify the interested parties in order to determine the characteristics that need to be satisfied. The needs of all these parties have to be satisfied in order for *quality* to be achieved. But, you can say, "This is a quality product as far as my customer is concerned". Figure 1.2 shows some of the characteristics of product quality – others have been identified previously.

Quality parameters

Differences in design can be denoted by grade or class but can also be the result of poor attention to customer needs. It is not enough to produce products that conform to the specifications or supply services that meet management's requirements. Quality is a composite of three *parameters*: quality of design, quality of conformance and quality of use, which are summarized below:

1 *Quality of design* is the extent to which the design reflects a product or service that satisfies customer needs and expectations. All the necessary characteristics should be designed into the product or service at the outset.

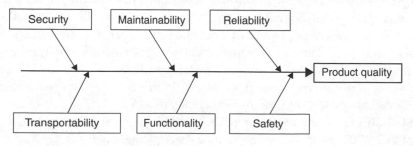

Figure 1.2 *Determinants of product quality*

2   *Quality of conformance* is the extent to which the product or service conforms to the design standard. The design has to be faithfully reproduced in the product or service.
3   *Quality of use* is the extent by which the user is able to secure continuity of use from the product or service. Products need to have a low cost of ownership, be safe and reliable, maintainable in use and easy to use.

Products or services that do not possess the right features and characteristics either by design or by construction are products of poor quality. Those that fail to give customer satisfaction by being uneconomic to use are also products of poor quality, regardless of their conformance to specifications. Often people might claim that a product is of good quality but of poor design, or that a product is of good quality but it has a high maintenance cost. These notions result from a misunderstanding because product quality is always a composite of the quality of design, conformance and use.

Dimensions of quality

In addition to quality parameters there are three *dimensions of quality* which extend the perception beyond the concepts outlined previously:

*   *The business quality dimension*: This is the extent to which the business serves the needs of all interested parties and is the outward facing view of the organization. The interested parties are not only interested in the quality of particular products and services but judge organizations by their potential to create wealth, the continuity of operations, the sustainability of supply, care of the environment, and adherence to health, safety and legal regulations. Changes in business strategy, direction or policies might yield improvement in business quality.
*   *The product quality dimension*: This is the extent to which the products and services provided meet the needs of specific customers. Enhancement to product features to satisfy more customers might yield improvement in product quality.
*   *The organization quality dimension*: This is the extent to which the organization maximizes its efficiency and effectiveness and is the inward facing view of the organization. Efficiency is linked with productivity which itself is linked with the motivation of personnel and the capability and utilization of resources. Effectiveness is linked with the utilization of knowledge focusing on the right things to do. Seeking best practice might yield improvement in organizational quality. This directly affects all aspects of quality.

Many organizations only concentrate on the product quality dimension, but the three are interrelated and interdependent. Deterioration in one leads to a deterioration in the others, perhaps not immediately but eventually.

As mentioned previously, it is quite possible for an organization to satisfy the customers for its products and services, and fail to satisfy the other interested parties. Some may argue that the producers of pornographic literature, nuclear power, non-essential drugs, weapons, etc. harm society and so regardless of these products and services being of acceptable quality to their customers, they are not regarded by society as benefiting the quality of life. Within an organization, the working environment may be oppressive – there may be political infighting – the source of revenue may be so secure that no effort is made to reduce waste. In such situations organizations may produce products and services that satisfy their customers. We must separate the three concepts above to avoid confusion. When addressing quality, it is necessary to be specific about the object of our discussion. Is it the quality of products or services, or the quality of organization in which we work, or the business as a whole, about which we are talking? If we only intend that our remarks apply to the quality of products, we should say so.

Achieving sustaining and improving quality

Several methods have evolved to *achieve, sustain and improve quality*; they are quality control, quality improvement and quality assurance, which are collectively known as *quality management*. Techniques such as quality planning, quality costs, "just-in-time" and statistical process control are all elements of these three methods.

Quality management

There are two schools of thought on quality management. One views quality management as the management of success and the other the elimination of failure. They are both valid. Each approaches the subject from a different angle.

The "success" school is characterized by five questions:[6]

1   What are you trying to do?
2   How do you make it happen?
3   How do you know it's right?
4   How do you know it's the best way of doing it?
5   How do you know it's the right thing to do?

The "failure elimination" school is characterized by five different questions:

1   How do you know what is needed?
2   What could affect your ability to do it right?
3   What checks are made to verify achievement?
4   How do you ensure the integrity of these checks?
5   What action is taken to prevent a recurrence of failure?

In an ideal world, if we could design products, services and processes that could not fail we would have achieved the ultimate goal. Success means not only that products, services and processes fulfil their function but also that the function is what customers' desire.

Failure means not only that products, services and processes would fail to fulfil their function but also that their function was not what customers desired. A gold-plated mousetrap that does not fail is not a success if no one needs a gold-plated mousetrap!

The introductory clause of ISO 9001:1994 contained a statement that the aim of the requirements is to achieve customer satisfaction by prevention of nonconformities. (This was indicative of the failure school of thought.) It was as though the elimination of nonconformity would by itself lead to satisfied customers. We know better now and this is recognized in the introductory clause of ISO 9001:2000 which contains the following statement: "This International Standard ... aims to enhance customer satisfaction through the effective application of the system ... and the assurance of conformity to customer and applicable regulatory requirements." (This is indicative of the success school of thought.)

In reality you cannot be successful unless you know of the risks you are taking and plan to eliminate, reduce or control them. A unification of these approaches is what is therefore needed for organizations to achieve, sustain and improve quality. You therefore need to approach the achievement of quality from two different angles and answer two questions: What do we need to do to succeed and what do we need to do to prevent failure? Of course if your answer to the first question included those things you need to do prevent failure, the second question is superfluous.

Quality does not appear by chance, or if it does it may not be repeated. One has to design quality into the products and services. It has often been said that one cannot inspect quality into a product. A product remains the same after inspection as it did before, so no amount of inspection will change the quality of the product. However, what inspection does is, measure the quality in a way that allows us to make decisions on whether or not to release a piece of work. Work that passes inspection should be quality work but inspection unfortunately is not 100% reliable. Most inspection relies on human judgement and this can be affected by many factors, some of which are outside our control (such as the private life, health or mood of the inspector). We may also fail to predict the effect that our decisions have on others. Sometimes we go to great lengths in preparing organization changes and find to our surprise that we neglected something or underestimated the effect of something. We therefore need other means to deliver quality products; i.e. we have to adopt practices that enable us to achieve our objectives while preventing failures from occurring.

Quality management principles

As explained at the beginning of this chapter, we need principles to help us determine the right things to do, and understand why we do and what we do. The more prescription we have the more we get immersed in the detail and lose sight of our objectives – our purpose – our reason for doing what we do. Once we have lost sight of our purpose, our actions and decisions follow the mood of the moment. They are swayed by the political climate or fear of reprisals. We can so easily forget our purpose when in heated discussion when it's not who you are, but what you say and to whom you say it that is deemed important. Those people who live by a set of principles often find themselves cast out of the club for saying what they believe. However, with presence of mind and recollection of the reasons why the principles are important for survival, they could just redeem themselves and be regarded as an important contributor.

A quality management principle is defined by ISO/TC 176 as *a comprehensive and fundamental rule or belief, for leading and operating an organization, aimed at continually improving performance over the long term by focusing on customers while addressing the needs of all other interested parties.* Eight principles (see Figure 1.3) have emerged as fundamental to the management of quality.

All the requirements of ISO/TS 16949 are related to one or more of these principles. These principles provide the reasons for the requirements and are thus very important. Each of these is addressed below. Further guidance on the application of these principles is provided in Hoyle and Thomson (2000)[7].

Customer focus

This principle is expressed as follows:

Organizations depend on their customers and therefore should understand current and future customer needs, meet customer requirements and strive to exceed customer expectations.

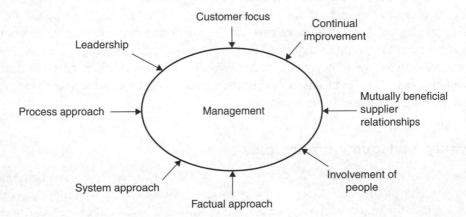

Figure 1.3 *The eight quality management principles*

Customers are the lifeblood of every organization. All organizations provide something to others – they do not exist in isolation. We should remember that customers are not simply purchasers but any person or organization that receives a product or service from the organization. Not-for-profit organizations therefore have customers. Customer focus means putting your energy into satisfying customers and understanding that profitability or avoidance of loss comes from satisfying customers. Profit is not the reason for an organization's existence. Profit is needed in order to grow the organization so that it may satisfy more customers. A profit focus is an inward seeking focus, a customer focus is an outward seeking focus. Customer focus means organizing work as a process that converts customers needs into satisfied customers. It means that all processes have a customer focus.

The principle means that everyone in the organization needs to be customer focused – not simply the top management or the sales personnel. If people were to ask themselves before making a decision, 'what does the customer need?', the organization would begin to move its focus firmly in the direction of its customers. Customer focus is also about satisfying needs rather than wants. A customer may want ISO/TS 16949 certification but in reality, it is business improvement that may be needed. While an ISO/TS 16949 certificate may appear to give satisfaction initially, this may be short lived as the customer slowly realizes that possession of the certificate did not result in the growth of business that was expected.

> **The transition**
>
> Inward seeking focus to outward seeking focus.

The customer-focus principle is reflected in ISO/TS 16949 through the requirements addressing:

- communication with the customer,
- care for customer property,
- the determination of customer needs and expectations,
- appointment of a management representative,
- management commitment.

Leadership

This principle is expressed as follows:

> *Leaders establish unity of purpose and direction for the organization. They should create and maintain the internal environment in which people can become fully involved in achieving the organization's objectives.*

Leaders exist at all levels in an organization – they are not simply the ones at the top. Within every team there needs to be a leader – one who provides a role-model consistent with the values of the organization. It is the behaviour of leaders (our role models) that influence our lives – not just in the business world but also in our family and leisure activities. People naturally concentrate on what they are measured against. It is therefore

<table>
<tr><td>

The transition

Aggravation
to
motivation

</td></tr>
</table>

vital that leaders measure the right things. Without a good leader an organization will go where the tide takes it, and as is so predictable with tides, they will be cast on the shoreline like the flotsam and jetsam of our society. Strong leadership will drive an organization in its chosen direction – away from disasters but towards success. But leadership alone will not bring the right success. It needs to be in combination with all the other principles. Leadership without customer focus will drive organizations towards profit for its own sake. Leadership without involving people will leave behind those who do not share the same vision – hence the second part of the principle. *Leaders are responsible for the internal environment*. If the workforce is unhappy, de-motivated and dissatisfied, it is the fault of the leaders. The culture, vision, values, beliefs and motivation in an organization arise from leadership. Good leadership strives to bring about a set of shared values – a shared vision so that everyone knows what the organization is trying to do and where it is going. A lack of vision and a disparate mix of values create conflict. A happy ship comes about by having good leadership. Regardless of Captain Blyth's orders and subsequent exoneration by the Royal Navy, the crew's mutiny on the Bounty in 1789 was down to a failure in leadership – a failure to create the conditions that motivated people to meet the organization's objectives.

The leadership principle is reflected in ISO/TS 16949 through the requirements addressing:

- the setting of objectives and policies,
- planning,
- internal communication,
- employee motivation and empowerment,
- creating an effective work environment.

Involvement of people

This principle is expressed as follows:

> *People at all levels are the essence of an organization and their full involvement enables their abilities to be used for the organization's benefit.*

It is not uncommon for those affected by decisions to be absent from the discussions with decision-makers. Decisions that stand the test of time are more likely to be made when those affected by them have been involved. Employees cannot employ a part of a person – they take the whole person or none at all. Every person has knowledge and experience beyond the job when he or she has been assigned to perform. Some are leaders in the community, some are architects of social events, building projects and expeditions. No one is limited in knowledge and experience to the current job they do. This principle means that management should tap this source of knowledge, encourage personnel to make a contribution and utilize their personal experience. It also means that management should be open – not hide its discussions unless national or business security could be threatened. Closed-door management leads to distrust among the workforce. Managers should be seen to operate with integrity and this means involving the people.

> **The transition**
>
> Operate
> to
> co-operate

The involvement of people principle is reflected in ISO/TS 16949 through the requirements addressing:

- participation in design reviews;
- defining objectives, responsibilities and authority;
- creating an environment in which people are motivated;
- internal communication;
- identifying competence needs.

Process approach

This principle is expressed as follows:

> *A desired result is achieved more efficiently when related resources and activities are managed as a process.*

All work is a process because it uses resources to perform actions that produce results. In the organizational sense, such processes add value to the input. Processes are therefore dynamic – they cause things to happen. An effective process would be one where the results were those that were required to fulfil the purpose of the organization. Every job involves people or machines equipped with resources performing a series of tasks to produce an output. No matter how simple the task, there is always an objective or a reason for doing it, the consumption of resources and expenditure of energy, a sequence of actions, decisions concerning their correctness, a judgement of completeness and an output which should be that which was expected. The organization exists to create and satisfy customers and other interested parties therefore the organization's processes must serve the needs of these interested parties (see also the section Process management principles later in this chapter).

> **The transition**
>
> Procedure approach
> to
> process approach

A process is as capable of producing rubbish as a procedure is capable of wasting resources – therefore processes need to be managed effectively for the required results to be produced. The process approach to management is therefore not simply converting inputs into outputs that meet requirements but about managing processes that:

- have a clearly defined purpose and objective that is based on the needs of the interested parties;
- are designed to achieve these objectives through tasks that use capable human, physical and financial resources and information;

- produce outputs that satisfy the interested parties;
- measure, review and continually improve process efficiency and effectiveness.

The process approach principle is reflected in ISO/TS 16949 through the requirements addressing:

- the identity of processes;
- defining process inputs and outputs;
- providing the infrastructure, information and resources for processes to function.

System approach to management

This principle is expressed as follows:

> *Identifying, understanding and managing interrelated processes as a system contributes to the organization's effectiveness and efficiency in achieving its objectives.*

A system is an ordered set of ideas, principles and theories or a chain of operations that produce specific results. To be a chain of operations, the operations need to work together in a regular relationship. Taking a systems approach to management means managing the organization as a system of processes so that all the processes fit together, the inputs and outputs are connected, resources feed the processes, performance is monitored and sensors transmit information which cause changes in performance and all parts work together to achieve the organization's objectives.

> **The transition**
>
> Functional approach to systems approach

This view of a system clearly implies a system is dynamic and not static. The system is not a random collection of elements, procedures and tasks but a set of interconnected processes. The systems approach recognizes that the behaviour of any part of a system has some effect on the behaviour of the system as a whole. Even if the individual processes are performing well, the system as a whole is not necessarily performing equally well. For example, assembling the best electronic components regardless of specification may not result in a world-class computer or even one that will run, because the components may not fit together. It is the interaction between parts and in the case of a management system, between processes, and not the actions of any single part or process that determines how well a system performs. Systems developed to meet QS-9000 were often based on a functional approach to management, i.e. the systems were collection of functions or departments – not processes.

The system approach principle is reflected in ISO/TS 16949 through the requirements addressing:

- establishing, implementing and maintaining the management system;
- interconnection, interrelation and sequence of processes;

- the links between processes;
- establishing measurement processes.

Continual improvement

This principle is expressed as follows:

> Continual improvement of the organization's overall performance should be a permanent objective of the organization.

This means that everyone in the organization should be continually questioning its performance and seeking ways to reduce variation, continually questioning their methods and seeking better ways of doing things, continually questioning their targets and seeking new targets that enhance the organization's capability. Performance, methods and targets are the three key areas where improvement is necessary for organizations to achieve and sustain success.

> **The transition**
>
> Error correction
> to
> course correction

ISO 9000:2000 defines continual improvement as a recurring activity to increase the ability to fulfil requirements. Improvement is therefore relative to a timescale. If the improvement recurs once a week, once a month, once a year or once every 5 years, it can be considered as "recurring". The scale of the improvement is also relative. Improvement can be targeted at specific characteristics, specific activities, specific products, specific processes or specific organizations. When targeted at a specific characteristic it may involve reducing variation in the measured characteristic. When targeted at specific products it may involve major modification – product upgrade. When targeted at the organization it may involve major re-organization or re-engineering of processes. To appreciate the scope of meaning you need to perceive requirements as a hierarchy of needs. At the lowest level are the needs of the task, passing through to the needs of the process, the needs of the system and ultimately the needs of the organization. At each level continual improvement is about improving efficiency and improving effectiveness.

It has become fashionable in certain sectors to use the term "Continuous Improvement" rather than "Continual Improvement". Continuous means without breaks or interruption such as continuous stationery. "Continual" means repeated regularly and frequently – a term that fits the concept of improvement rather better and is used in ISO/TS 16949.

The continual improvement principle is reflected in ISO/TS 16949 through the requirements addressing:

- improvement processes,
- identifying improvements,
- reviewing documents and processes for opportunities for improvement.

Factual approach to decision-making

This principle is expressed as follows:

Effective decisions are based on the analysis of data and information.

Facts are obtained from observations performed by qualified personnel using devices, the integrity of which is known. The factual approach to decision-making leads us to take certain actions. To make decisions on the basis of facts we need reliable mechanisms for collecting facts such as measurement systems. We need valid methods for interpreting the facts and producing information in a form that enables sound decisions to be made. The factual approach leads us to control activities based on fact rather than opinion or emotion. It means using statistical techniques to reveal information about a process, rather than reacting to variation that is an inherent characteristic of the system. However, used in isolation this principle can be dangerous.

The transition

Subjective
to
objective

An obsession with numbers tends to drive managers into setting targets for things that the individual is powerless to control. A manager may count the number of designs that an engineer completes over a period. The number is a fact, but to make a decision about that person's performance on the basis of this fact is foolish – the engineer has no control over the number of designs completed and even if she did, what does it tell us about the quality of the designs? Nothing! Each design is different so the time to complete each one varies. Each customer is different so the time taken to establish customer needs varies. Setting a target for the number of designs to be completed over a period might lead to the engineer rushing them, injecting errors in order to fulfil a meaningless target. It is therefore necessary to approach the decision in a different way. Firstly decide what decision you want to make and then determine what facts you need in order to make the decision. When you know what facts you need, determine how such facts will be obtained and what methods need to be used to obtain them. Assess the risks of the information being bogus or invalid and put in place measures to ensure its integrity. Work backwards from the decision you need to make to the information you require, not forward from the information to a decision you might make with it. This gives data collection a purpose; for without purpose, data collection is a waste of resources. Don't collect data for the sake of it, just because you can on the pretext that it might come in useful.

The factual approach principle is reflected in ISO/TS 16949 through the requirements addressing:

* reviews, measurements and monitoring to obtain facts;
* control of measuring devices;
* analysis to obtain facts from information;

- records for documenting the facts;
- approvals based on facts.

Mutually beneficial supplier relationships

This principle is expressed as follows:

> *An organization and its suppliers are interdependent and a mutually beneficial relationship enhances the ability of both to create value.*

The customer-focus principle drew our attention to the fact that organizations depend on their customers. It is also valid to state that organizations depend on their suppliers. Suppliers provide the materials, resources and often many services that were once provided by internal functions. The organizations of the 21st century are more dependent on their suppliers than ever before. The quest for lower and lower costs with higher and higher performance has caused many organizations to consider the economics of continuing to operate their own support services. There has been a recognition that organizations were trying to be good at everything rather than being good at their core business. This has led to single-function organizations serving many customers where there is entirely mutual dependency. However, there is another reason that has led to stronger supplier relationships.

The transition

Adversarial approach
to
alliance approach

Over the last 100 years the market for goods and services has changed dramatically. Prior to the 1920s most firms focused on production in the belief that a quality product will sell itself. From the 1920s to the 1950s, many firms focused on selling what they could make regardless of whether the customer actually needed it. From the 1950s to the 1990s the market turned around from a seller's market to a buyer's market as customers became more discerning and firms began to focus on identifying customer needs, and producing products and services that satisfied these needs. During the last 10 years, customer orientation has been taken one step further by focusing on establishing and maintaining relationships with both the customers and the suppliers.[8] From a simple exchange between buyer and seller, there evolved strategic alliances and partnerships that cut inventory, packaging and most importantly cut the costs of acquiring new customers and suppliers. There is a net benefit to both parties. For the customer, the supplier is more inclined to keep its promises because the relationship secures future orders. There is more empathy – the customer sees the supplier's point of view and vice versa. There is more give and take that binds the two organizations closer together and ultimately there is trust that holds the partnership together. Absent will be adversarial relationships and one-off transactions when either party can walk away from the deal. The partnerships will also encourage better after-sales care and more customer focus throughout the organization (everyone knows their customers because there are fewer of them).

The mutually beneficial supplier relationships principle is reflected in ISO/TS 16949 through the requirements addressing:

* control of suppliers,
* evaluation of suppliers,
* analysis and review of supplier data.

Using the principles

The principles can be used in validating the design of processes, in validating decisions, in auditing system and processes. You look at a process and ask:

* Where is the customer focus in this process?
* Where in this process is there leadership, guiding policies, measurable objectives and the environment that motivates the workforce to achieve these objectives?
* Where in this process is the involvement of people in the design of the process, the making of decisions, the monitoring and measurement of performance and the improvement of performance?
* Where is the process approach to the accomplishment of these objectives?
* Where is the systems approach to the management of these processes, the optimization of performance, the elimination of bottlenecks?
* Where in the process are the facts collected and transmitted to the decision-makers?
* Where is there continual improvement in performance, efficiency and effectiveness of this process?
* Where is there a mutually beneficial relationship with suppliers in this process?

Alignment of principles with the Business Excellence Model

There is a very close match between the eight quality management principles and the Business Excellence Model used by British Quality Foundation and the EFQM. The Excellence Model consists of nine criteria that are used to assess the overall strengths of an organization and measure its progress towards "best in class". The model is based on the premise that **Customer Satisfaction**, **People** (employee) **Satisfaction** and **Impact on Society** are achieved through **Leadership** driving **Policy and Strategy**, **People Management**, **Partnerships and Resources** and **Processes** leading ultimately to excellence in **Business Results**. These are the nine criteria and supporting these are eight fundamental concepts that have some similarity with the eight quality management principles as shown in Table 1.3.

The differences between the Business Excellence Model and the Quality Management Principles are small enough to be neglected if you take a pragmatic approach. If you take a pedantic approach you can find many gaps but there are more benefits to be gained by looking for synergy rather than for conflict.

Table 1.3 *Comparison between business excellence principles and quality management principles*

Business excellence concepts	Quality management principles
Customer focus The customer is the final arbiter of product and service quality and customer loyalty, retention and market share gain are best optimized through a clear focus on the needs of current and potential customers.	*Customer focus* Organizations depend on their customers and therefore should understand current and future customer needs, meet customer requirements and strive to exceed customer expectations.
Leadership and constancy of purpose The behaviour of an organization's leaders creates a clarity and unity of purpose within the organization and an environment in which the organization and its people can excel.	*Leadership* Leaders establish unity of purpose and direction for the organization. They should create and maintain the internal environment in which people can become fully involved in achieving the organization's objectives.
People development and involvement The full potential of an organization's people is best released through shared values and a culture of trust and empowerment, which encourages the involvement of everyone.	*Involvement of people* People at all levels are the essence of an organization and their full involvement enables their abilities to be used for the organization's benefit.
Management by processes and facts Organizations perform more effectively when all interrelated activities are understood and systematically managed and decisions concerning current operations and planned improvements are made using reliable information that includes stakeholder perceptions.	*Process approach* A desired result is achieved more efficiently when related resources and activities are managed as a process. *Factual approach to decision-making* Effective decisions are based on the analysis of data and information. *Systems approach* Identifying, understanding and managing interrelated processes as a system contributes to the organization's effectiveness and efficiency in achieving its objectives.
Continuous learning, innovation and improvement Organizational performance is maximized when it is based on the management and sharing of knowledge within a culture of continuous learning, innovation and improvement	*Continual improvement* Continual improvement of the organization's overall performance should be a permanent objective of the organization.

Table 1.3 *(continued)*

Business excellence concepts	Quality management principles
Partnership development An organization works more effectively when it has mutually beneficial relationships, built on trust, sharing of knowledge and integration with its partners.	*Mutually beneficial supplier relationships* An organization and its suppliers are interdependent and a mutually beneficial relationship enhances the ability of both to create value.
Public responsibility The long-term interest of the organization and its people are best served by adopting an ethical approach and exceeding the expectations and regulations of the community at large.	*Public responsibility* There is no equivalent principle in ISO 9000; however, ISO 9004 Clause 5.2.2 stresses that the success of the organization depends on understanding and considering current and future needs and expectations of the interested parties.
Results orientation Excellence is dependent on balancing and satisfying the needs of all relevant stakeholders.	*Results orientation* There is no equivalent principle in ISO 9000; however, ISO 9004 Clause 5.2.2 does recommend that an organization should identify its interested parties and maintain a balanced response to their needs and expectations.

Quality control

The ISO 9000 definition states that *quality control* (commonly abbreviated as QC) is part of quality management focused on fulfilling requirements. What the definition fails to tell us is that controls regulate performance. Control is sometimes perceived as undesirable as it removes freedom, but if everyone were free to do just as they liked there would be chaos. Controls prevent change and when applied to quality they regulate performance and prevent undesirable changes being present in the quality of the product or service being supplied. When operations are under control they are predictable and predictability is a factor that is vital for any organization to be successful. If you cannot predict what might happen when a process is initiated, you are relying on chance. The quality of products and services cannot be left to chance.

The simplest form of quality control is illustrated in Figure 1.4. Quality control can be applied to particular products, to processes that produce the products or to the output of the whole organization by measuring the overall performance of the organization.

Quality control is often regarded as a post-event activity: i.e. a means of detecting whether quality has been achieved and taking action to correct any deficiencies.

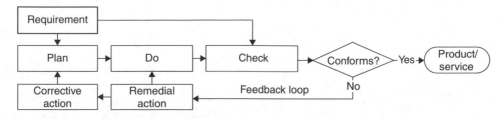

Figure 1.4 *Generic control model*

However, one can control results by installing sensors before, during or after the results are created. It all depends on where you install the sensor, what you measure and the consequences of failure.

The progressive development of controls from having no control of quality to installing controls at all key stages from the beginning to the end of the product cycle is illustrated in Figure 1.5. As can be seen, if you have no controls, quality products are produced by chance and not by design. The more controls you install the more certain you are of producing products of consistent quality but more control does not mean more inspection.

Control before the event

Some failures cannot be allowed to occur and so must be prevented from happening through rigorous planning and design. One example is the use of reliability prediction performed before the design is complete to predict whether product reliability will meet the specification. Another is the use of competence-based assessment techniques where personnel are under close supervision until they demonstrate competences following which supervisory controls are removed. This allows you to remove output checks because you know that if you were to inspect the work you would find it to be correct. Instead of checking every product produced, you check competency periodically and assign responsibility to personnel for checking their own work. Another method is the use of prevention-based error-proofing mechanisms that sense an abnormality that is about to happen, and then signal the occurrence or halt processing, depending on the severity, frequency or downstream consequences.

Control during the event

Some failures must be corrected immediately using automatic controls or error proofing. By continuous monitoring of parameters in a processing plant the temperature, pressure, quantities, etc. are adjusted to maintain output within specified limits. Electronic components are designed so that they can only be inserted in the correct orientation. Computer programs are designed so that routines will not run unless the correct type of data is entered in every field.

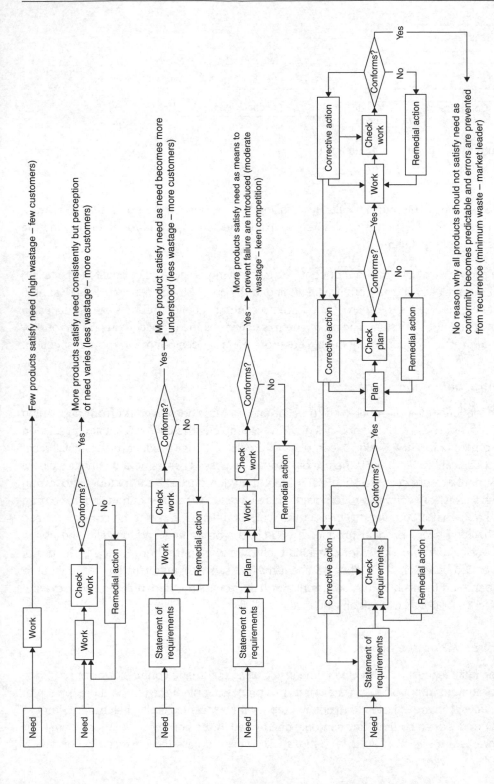

Figure 1.5 Development of controls

Control after the event

Where the consequences of failure are less severe or where other types of sensors are not practical or possible, output verification can be used as a means of detecting failure early in the process and preventing the subject product passing through subsequent stages in the process and increasing the cost of rectification. Although errors have occurred, measures taken to contain them or remove the defective products from the production stream automatically are another form of error-proofing methods.

Product inspection and test is *control after the event* because it occurs after the product is produced. Where failure cannot be measured without observing trends over longer periods, you can use information controls. They do not stop immediate operations but may well be used to stop further operations when limits are exceeded.

It is often deemed that quality assurance serves prevention and quality control detection, but a control installed to detect failure before it occurs serves prevention, such as reducing the tolerance band to well within the specification limits. So quality control can prevent failure. Assurance is the result of an examination whereas control produces a result. *Quality assurance* does not change the product, but *quality control* does.

Quality control as a label

"Quality control" is also the term used as the name of a department. In most cases Quality Control departments perform inspection and test activities and the name derives from the authority that such departments have been given. They sort good products from bad products and authorize the release of the good products. It is also common to find that quality control departments perform supplier control activities, which are called *supplier quality assurance* or vendor control. In this respect they are authorized to release products from suppliers into the organization either from the supplier's premises or on receipt in the organization.

To control anything requires the ability to effect change; therefore, the title quality control department is a misuse of the term, because such departments do not in fact change the quality of the product they inspect. They do act as a regulator if given the authority to stop release of product, but this is control of supply and not of control of quality. Authority to change product usually remains in the hands of the producing departments. It is interesting to note that similar activities within a design department are not called "quality control" but "design assurance" or some similar term. "Quality control" has for decades been a term applied primarily in the manufacturing areas of an organization and it is therefore difficult to change people's perceptions after so many years of the term's incorrect use.

In manufacturing, inspection and test activities have been transferred into the production departments of organizations, sometimes retaining the labels and sometimes reverting to the inspection and test labels. However, the term quality control is used less frequently in the west probably because of the decline of manufacturing. It has not been widely used in the service sector. A reason for this could be that it is considered more of a concept than a function.

Universal sequence of steps

The following steps can accomplish control of quality, or anything else for that matter:[9]

1 Determine the subject of control, i.e. what is to be regulated.
2 Define a unit of measure, i.e. express the control subject in measurable terms such as quantities, ratios, indices, rating, etc.
3 Establish a standard level of performance, i.e. a target to aim for.
4 Produce plans for controls that specify the means by which the characteristics will be achieved, and variation detected and removed.
5 Organize resources to implement the plans for quality control.
6 Select a sensor to sense variance from specification.
7 Install the sensor at the stage in the process appropriate whether you need to control before, during or after results are produced.
8 Collect and transmit data to a place for analysis.
9 Verify the results and establish whether the variance is within the range expected for a stable process (the *status quo*).
10 Diagnose the cause of any variance beyond the expected range.
11 Propose remedies and decide on the action needed to restore the *status quo*.
12 Take the agreed action and check that process stability has been restored.

Variation

Variation is present in all systems. Nothing is absolutely stable. If you monitor the difference between the measured value and the required value of a characteristic and plot it on a horizontal timescale in the order the products were produced, you would notice that there is variation over time. There does not have to be a required value to spot variation. If you monitor any parameter over time (duration, resource consumption, strength, weight, etc.) you will see a pattern of variation that with an appropriate scale will show up significant deviations from the average. If you plot the values as a histogram you will observe that there is a distribution of results around the average. As you repeat the plot for a new set of measurements of the same characteristic, you will notice that there is variation between this second set and the first. In studying the results you will observe variation in:

- the location of the average for each plot,
- the spread of the values,
- the shape of the distribution.

The factors causing these variations are referred to as "assignable" or "special causes".

Special cause

The cause of variations in the location, spread and shape of a distribution are considered special or assignable because the cause can be assigned to a specific or special condition that does not apply to other events. They are causes that are not always present. Wrong material, inaccurate measuring device, worn-out tool, sick employee, weather conditions, accident, stage omitted are all one-off events that cannot be predicted. When they occur they make the shape, spread or location of the average change. The process is not predictable while special cause variation is present. Eliminating the special causes is part of quality control, see Steps 9–11 above.

Once all the special causes of variation have been eliminated the shape and spread of the distribution and the location of the average become stable, the process is under control – the results are predictable. However, it may not be producing conforming product. You may be able to predict that the process could produce one defective product in every 10 produced. There may still be considerable variation but it is random. A stable process is one with no indication of a special cause of variation and can be said to be in *statistical control*. Special cause variation is not random, and it is unpredictable. It occurs because something has happened that should not have happened so you should search for the cause immediately and eliminate it. The person running the process should be responsible for removing special causes unless these causes originate in another area when the source should be isolated and eliminated.

Common cause

Once the special cause of variation has been removed, the variation present is left to chance, it is random or what is referred to as common causes. This does not mean that no action should be taken but to treat each deviation from the average as a special cause will only lead to more problems. The random variation is caused by factors that are inherent in the system. The operator has done all she can to remove the special causes, the rest are down to management. This variation could be caused by poor design, working environment, equipment maintenance or inadequacy of information. Some of these events may be common to all processes, all machines, all materials of a particular type, all work performed in a particular location or environment, or all work performed using a particular method. This chain of events is illustrated in Figure 1.6.

This shows that by removing special causes, the process settles down and although non-conformities remain, performance becomes more predictable. Further improvement will not happen until the common causes are reduced and this requires action by management. However, the action management takes should not be to look for a scapegoat – the person whom they believe caused the error, but to look for the root cause – the inherent weakness in the system that causes this variation.

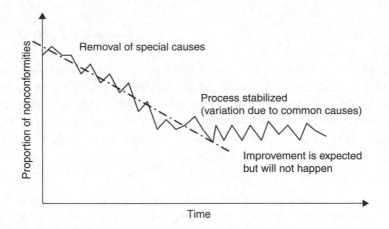

Figure 1.6 *Stabilizing processes*

Common cause variation is random and therefore adjusting a process on detection of a common cause will destabilize the process. The cause has to be removed, not the process adjusted. When dealing with either common cause or special cause problems the search for the root cause will indicate whether the cause is random and likely to occur again or a one off event. If it is random, only action on the system will eliminate it. If it is a one-off event, no action on the system will prevent its recurrence – it just has to be fixed. Imposing rules will not prevent a nonconformity caused by a worn out tool that someone forgot to replace. A good treatment of common cause and special cause variation is given in Deming (1982)[10].

With a stable process the spread of common cause variation will be within certain limits. These limits are not the specification limits but are limits of natural variability of the process. These limits can be calculated and are referred to as the upper and lower control limits (UCL and LCL, respectively). The control limits may be outside the upper and lower specification limits to start with but as common causes are eliminated, they close in and eventually the spread of variation is all within the specification limits. Any variation outside the control limits will be rare and will signal the need for corrective action. This is illustrated in Figure 1.7.

Keeping the process under control is process control. Keeping the process within the limits of the customer specification is quality control. The action needed to make the transition from process control to quality control is an improvement action and this is dealt with next.

Quality improvement
Firstly, we need to put *quality improvement* (abbreviated as QI) in context because it is minefield of terms and concepts that overlap one another. There are three things

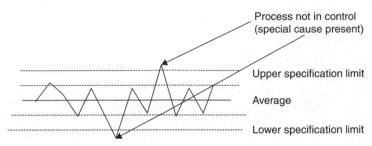

Figure 1.7 *Control and specification limits*

that are certain in this life, death, taxes and change! We cannot improve anything unless we know its present condition and this requires measurement and analysis to tell us whether improvement is both desirable and feasible. Improvement is always relative. Change is improvement if it is beneficial and a retrograde step if it is undesirable but there is a middle ground where change is neither desirable nor undesirable – it is inevitable and there is nothing we could or should do about it. Change is a constant. It exists in everything and is caused by physical, social or economic forces. Its effects can be desirable, tolerable or undesirable. Desirable change is change that brings positive benefits to the organization. Tolerable change is change that is inevitable and yields no benefit or may have undesirable effects when improperly controlled. The challenge is to cause desirable change and to eliminate, reduce or control undesirable change so that it becomes tolerable change. Juran writes on improvement thus "Putting out fires is not improvement of the process – neither is discovery and removal of a special cause detected by a point out of control. This only puts the process back to where it should have been in the first place".[11] This we call *restoring the status quo*. If eliminating special causes is not improvement but maintaining the *status quo*, that leaves two areas where the improvement is desirable – the reduction of common cause variation and the raising of standards.

Figure 1.8 illustrates the continuing cycle of events between periods of maintaining performance and periods of change. The transition from one target to another may be gradual on one scale but considered a breakthrough on another scale. The variation around the target value is due to common causes that are inherent in the system. This represents the expected performance of the process. The spike outside the average variation is due to a special cause – a one-off event that can be eliminated. These can be regarded as fires and is commonly called *fire fighting*. Once removed the process continues with the average variation due to common causes.

When considering improvement by raising standards, there are two types of standards – one for results achieved and another for the manner in which the results are achieved. We could improve on the standards we aim for, the level of performance, the target or the goal but use the same methods. There may come a point when the existing methods

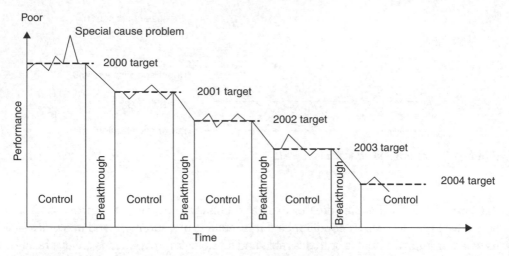

Figure 1.8 *Continual improvement*

won't allow us to achieve the standard, when we need to devise a new method, a more efficient or effective method or due to the constraints on us, we may choose to improve our methods simply to meet existing standards.

This leads us to ask four key questions:

1   Are we doing it right?
2   Can we keep on doing it right?
3   Are we doing it in the best way?
4   Is it the right thing to do?

The ISO 9000 definition of *quality improvement* states that it is that part of quality management focused on increasing the ability to fulfil quality requirements. If we want to reduce the common cause variation we have to act on the system. If we want to improve efficiency and effectiveness we also have to act on the system and both are not concerned with correcting errors but concerned with doing things better and doing different things.

There is a second dimension to improvement – it is the rate of change. We could improve "gradually" or by a "step change". Gradual change is also referred to as incremental improvement, continual improvement or *kaizen*. "Step change" is also referred to as "breakthrough" or a "quantum leap". Gradual change arises out of refining the existing methods, modifying processes to yield more and more by consuming less and less. Breakthroughs often require innovation, new methods, techniques, technologies and new processes.

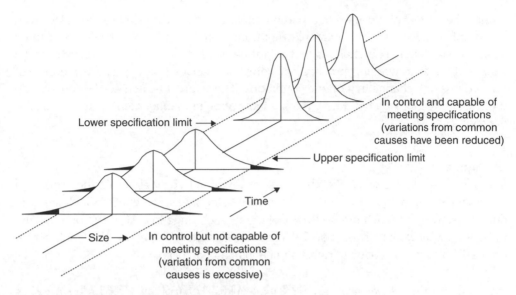

Lower specification limit

In control and capable of meeting specifications (variations from common causes have been reduced)

Upper specification limit

Time

Size

In control but not capable of meeting specifications (variation from common causes is excessive)

Figure 1.9 *From process control to capable process*

Are we doing it right?
- *Would the answer be this?* No, we are not because every time we do it we get it wrong and have to do it again.
- *Or would it be this?* Yes, we are because every time we do it we get it right – we never have to do it over again.

Quality improvement in this context is for better control and is about improving the rate at which an agreed standard is achieved. It is therefore a process for reducing the spread of common cause variation so that all products meet agreed standards. This is illustrated in Figure 1.9. It is not about removing special cause variation – this requires the corrective action process.

This type of improvement is only about reducing variation about a mean value or closing the gap between actual performance and the target. This is improvement by better control. The target remains static and the organization gets better and better until all output meets the target or falls between the acceptance limits.

When a process is stable the variation present is only due to random causes. There may still be unpredictable excursions beyond the target due to a change in the process but this is special cause variation. Investigating the symptoms of failure, determining the root cause and taking action to prevent recurrence can eliminate the special cause and reduce random causes. A typical quality improvement of this type might be to reduce the spread of variation in a parameter so that the average value coincides with the nominal value. Another example might be to reduce the defect rate from three sigma to six sigma. The changes that might be needed to meet this objective can be simple changes in working practices or complex changes that demand a redesign of the process or a change in working conditions. These might be achieved using existing methods or technology but it may require innovation in management or technology to accomplish.

Six sigma

In a perfect world, we would like the range of variation to be well within the upper and lower specification limits for the characteristics being measured but invariably we produce defectives. If there were an 80% yield from each stage in a 10-stage process, the resultant output would be less than 11% and as indicated in Table 1.4. We would obtain only 107,374 good products from an initial batch of 1 million.

Even if the process stage yield were 99% we would still only obtain 95,617 less products than we started with. It is therefore essential that multiple stage processes have a process stage yield well in excess of 99% and it is from this perspective that the concept of six sigma emerges.

The Greek letter "σ" called sigma is the symbol used to represent standard deviation – a measure of the spread of frequency distributions. It is the root-mean-square deviation

Table 1.4 *Ten-stage process yield*

Stage	Yield/stage	Total yield (%)	Initial population per million
1	0.80	80	800,000
2	0.80	64	640,000
3	0.80	51.2	512,000
4	0.80	41	409,600
5	0.80	32.8	327,680
6	0.80	26.2	262,144
7	0.80	21	209,715
8	0.80	16.8	167,772
9	0.80	13.4	134,218
10	0.80	10.7	107,374

of the readings in a series from their average value. Table 1.5 shows the number of products meeting requirements and the equivalent defects per million products for a range of standard deviations.

Assuming a normal distribution of results at the six sigma level you would expect 0.002 parts per million or ppm but when expressing performance in ppm, it is common practice to assume that the process mean can drift 1.5 sigma in either direction. The area of a normal distribution beyond 4.5 sigma from the mean is 3.4 ppm. As control charts will detect any process shift of this magnitude in a single sample, the 3.4 ppm represents a very conservative upper boundary on the nonconformance rate.[12]

Although the concept of six sigma can be applied to non-manufacturing processes you cannot assume as was done in Table 1.4 that the nonconformities in a stage output are rejected as unusable by the following stages. A person may pass through 10 stages in a hospital but you cannot aggregate the errors to produce a process yield based on stage errors. Patients don't drop out of the process simply because they were kept waiting longer than the specified maximum. You have to take the whole process and count the number of serious errors per 1 million patients.

Can we keep on doing it right?
- *Would the answer be this?* No, we can't because the supply of resource is unpredictable, the equipment is wearing out and we can't afford to replace it.
- *Or would it be this?* Yes, we can because we have secured a continual supply of resources and have in place measures that will provide early warning of impending changes.

Table 1.5 *Process yield at various sigma values*

Sigma	Product meeting requirements (%)	Number of errors per million products	
		Assuming normal distribution	Assuming 1.5 sigma drift (ppm)
1	68.26	317,400	697,672.15
2	95.45	45,500	308,770.21
3	99.73	2700	66,810.63
4	99.9937	63	6209.70
5	99.999943	0.57	232.67
6	99.9999998	0.002	3.4

This question is about continuity or sustainability. It is not enough to do it right first time once – you have to keep on doing it right and this is where a further question helps to clarify the issue.

What affects our ability to maintain this performance?
It could be resources as in the example, but to maintain the *status quo* might mean innovative marketing in order to keep the flow of customer orders of the type that the process can handle. Regulations change, staff leave, emergencies do happen – can you keep on doing it right under these conditions?

Are we doing it in the best way?
- *Would the answer be this?* We have always done it this way and if it isn't broke why fix it?
- *Or would it be this?* Yes, we think so because we have compared our performance with the best in class and we are as good as they are.

One might argue that any target can be met providing we remove the constraints and throw lots of money at it. Although the targets may be achieved, the achievement may consume too much resource; time and materials may be wasted – in other words, there may be a better way of doing it. By finding a better way you release resources to be used more productively. Over 14 years since the introduction of ISO 9000, it is strange that more organizations did not question if there was a better way than writing all those procedures, filling in all those forms, insisting on all those signatures. However, ISO 9000 did not require these things – there was more than one way of interpreting the requirement.

The search for a better way is often more effective when in the hands of those doing the job and you must therefore embrace the "leadership" and "involvement of people" principles in conjunction with continual improvement.

Is it the right thing to do?
- *Would the answer be this?* I don't know – we always measure customer satisfaction by the number of complaints.
- *Or would it be this?* Yes, I believe it is because these targets relate very well to the organizations objectives.

Quality improvement in this context is accomplished by raising standards and is about setting a new level of performance, a new target that brings additional benefits for the interested parties. These targets are performance targets for products, processes and the system. They are not targets established for the level of errors, such as nonconformities, scrap and customer complaints.

One needs to question whether the targets are still valid. These new targets have to be planned targets as exceeding targets sporadically is a symptom of out-of-control situations. Targets need to be derived from the organization's goals but as these change the targets may become disconnected. Targets that were once suitable are now obsolete – they are not the right things to do any longer. Functions are often measured by their performance against budget. We need to ask whether this is the right thing to do: Does it lead to optimizing organizational performance? You may have been desensitized to the level of nonconformities or customer complaints – they have become the norm – is this the right level of performance to maintain or should there be an improvement programme to reach much a lower level of rejects.

New standards are created through a process that starts at a feasibility stage and progresses through research and development to result in a new standard, proven for repeatable applications. Such standards result from innovations in technology, marketing and management. A typical quality improvement of this type might be to redesign a range of products to increase the achieved reliability from 1 failure every 5,000 hours to 1 failure every 100,000 hours. Another example might be to improve the efficiency of the service organization so as to reduce the guaranteed call-out time from the specified 36 to 12 hours or improve the throughput of a process from 1000 to 10,000 components/week. Once again, the changes needed may be simple or complex and might be achieved using existing technology but it may require innovation in technology to accomplish.

The transition between where quality improvement stops and quality control begins is where the level has been set and the mechanisms are in place to keep quality on or above the set level. In simple terms, if quality improvement reduces quality costs from 25% of turnover to 10% of turnover, the objective of quality control is to prevent the quality costs rising above 10% of turnover. This is illustrated in Figure 1.10.

Improving quality by better control or raising standards can be accomplished by the following steps:[13]

1   Determine the objective to be achieved, e.g. new markets, products or technologies, or new levels of organizational efficiency or managerial effectiveness, new national standards or government legislation. These provide the reasons for needing change.
2   Determine the policies needed for improvement, i.e. the broad guidelines to enable management to cause or stimulate the improvement.
3   Conduct a feasibility study. This should discover whether accomplishment of the objective is feasible and propose several strategies or conceptual solutions for consideration. If feasible, approval to proceed should be secured.
4   Produce plans for the improvement that specifies the means by which the objective will be achieved.
5   Organize the resources to implement the plan.

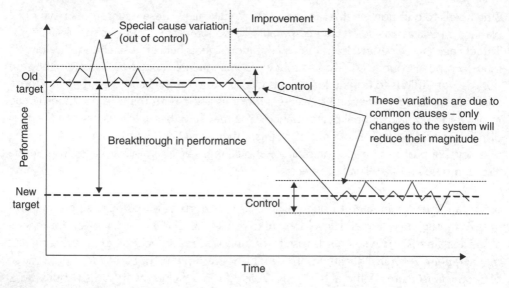

Figure 1.10 *Breakthrough and control*

6 Carry out research, analysis and design to define a possible solution and credible alternatives.
7 Model and develop the best solution and carry out tests to prove it fulfils the objective.
8 Identify and overcome any resistance to the change in standards.
9 Implement the change, i.e. put new products into production and new services into operation.
10 Put in place the controls to hold the new level of performance.

This improvement process will require controls to keep improvement projects on course towards their objectives. The controls applied should be designed in the manner described previously.

Quality assurance
The ISO 9000 definition states that *quality assurance* (commonly abbreviated as QA) is part of quality management focused on providing confidence that quality requirements will be fulfilled. Both the customers and the managers have a need for quality assurance because they are not in a position to oversee operations for themselves. They need to place trust in the producing operations, thus avoiding constant intervention.

Customers and managers need:

1 Knowledge of what is to be supplied. (This may be gained from the sales literature, contract or agreement.)

2   Knowledge of how the product or service is intended to be supplied. (This may be gained from the supplier's proposal or offer.)

3   Knowledge that the declared intentions will satisfy customer requirements if met. (This may be gained from personal assessment or reliance on independent certifications.)

4   Knowledge that the declared intentions are actually being followed. (This may be gained by personal assessment or reliance on independent audits.)

5   Knowledge that the products and services meet the specified requirements. (This may be gained by personal assessment or reliance on independent audits.)

You can gain an assurance of quality by testing the product or service against prescribed standards to establish its capability to meet them. However, this only gives confidence in the specific product or service purchased and not in its continuity or consistency during subsequent supply. Another way is to assess the organization that supplies the products or services against prescribed standards to establish its capability to produce products of a certain standard. This approach may provide assurance of continuity and consistency of supply.

Quality assurance activities do not control quality, they establish the extent to which quality will be, is being or has been controlled. All quality assurance activities are post-event activities and off-line, and serve to build confidence in results, in claims, in predictions, etc. If a person tells you they will do a certain job for a certain price in a certain time, can you trust them or will they be late, overspent and under specification? The only way to find out is to gain confidence in their operations and that is what quality assurance activities are designed to do. Quite often, the means to provide the assurance need to be built into the process, such as creating records, documenting plans, documenting specifications, reporting reviews, etc. Such documents and activities also serve to control quality as well as assure it. ISO/TS 16949 provides a basis for obtaining an assurance of quality, if you are the customer, and a basis for controlling quality, if you are the supplier.

Quality assurance is often perceived as the means to prevent problems but this is not consistent with the definition in ISO 9000. In one case the misconception arises due to people limiting their perception of quality control to control after the event; not appreciating that you can control an outcome before the event by installing mechanisms to prevent failure, such as automation, error-proofing and failure prediction.

In another case, the misconception arises due to the label attached to the ISO 9000 series of standards. They are sometimes known as the quality assurance standards when in fact, as a family of standards, they are quality management system standards. The requirements within the standards do aim to prevent problems, and consequently the standard is associated with the term *quality assurance*. ISO/TS 16949 is designed for use in assuring customers that suppliers have the capability of meeting their requirements. It is true that by installing a quality management system, you will gain an assurance of quality, but assurance comes about through knowledge of what will be, is being or has been done, rather than by doing something. Assurance is not an action but a result. It results from obtaining reliable information that testifies to the accuracy or validity of some event or product.

Labelling the prevention activities as quality assurance activities may have a negative effect, particularly if you have a Quality Assurance Department. It could send out signals with the aim of the Quality Assurance Department is to prevent things from happening! Such a label could unintentionally give the department a law enforcement role.

Quality Assurance Departments are often formed to provide both customer and management with confidence that quality will be, is being and has been achieved. However, another way of looking on Quality Assurance Departments is as Corporate Quality Control. Instead of measuring the quality of products, they are measuring the quality of the business and by doing so are able to assure management and customers of the quality of products and services. The following steps can obtain an assurance of quality:

1   Acquire the documents that declare the organization's plans for achieving quality.
2   Produce a plan that defines how an assurance of quality will be obtained, i.e. a quality assurance plan.
3   Organize the resources to implement the plans for quality assurance.
4   Establish whether the organization's proposed product or service possesses characteristics that will satisfy customer needs.
5   Assess operations, products and services of the organization, and determine where and what the quality risks are.
6   Establish whether the organization's plans make adequate provision for the control, elimination or reduction of the identified risks.
7   Determine the extent to which the organization's plans are being implemented and risks contained.
8   Establish whether the product or service being supplied has the prescribed characteristics.

In judging the adequacy of provisions you will need to apply the relevant standards, legislation, codes of practice and other agreed measures for the type of operation, application and business. These activities are quality assurance activities and may be subdivided into design assurance, procurement assurance, manufacturing assurance, etc. Auditing, planning, analysis, inspection and test are some of the techniques that may be used.

Level of attention to quality

In the first section of the Introduction to ISO/TS 16949:2002 is a statement that might appear progressive but depending on how it is interpreted could be regressive. The statement is: "The adoption of a quality management system should be a strategic decision of an organization". What would top management be doing if they did this? Would they be agreeing to manage the organization as a system of interconnected processes in order to deliver stakeholder satisfaction, or would they be simply agreeing to document the approach they take to the management of product quality and to subsequently do what they have documented? It all comes down to their understanding of the word *quality* and this is what will determine the level of attention to quality.

Whilst the decision to make the *management of quality* a strategic issue will be an executive decision, the attention it is given at each level in the organization will have a bearing on the degree of success attained. There are three primary organization levels: the *enterprise level, the business level* and the *operations level*;[14] between each level there are barriers.

At the enterprise level, the executive management responds to the 'voice' of ownership and is primarily concerned with profit, return on capital employed, market share, etc. At the business level, the managers are concerned with products and services and so respond to the 'voice' of the customer. At the operational level, the middle managers, supervisors, operators, etc. focus on processes that produce products and services and so respond to the 'voice' of the processes carried out within their own function.

In reality, these levels overlap, particularly in small organizations. The Chief Executive Officer (CEO) of a small company will be involved at all three levels whereas in the large multinational, the CEO spends all of the time at the enterprise level, barely touching the business level, except when major deals with potential customers are being negotiated. Once the contract is won, the CEO of the multinational may confine his or her involvement to monitoring performance through metrics and goals.

Quality should be a strategic issue that involves the owners because it delivers fiscal performance. Low quality will cause fiscal performance ultimately to decline.

The typical focus for a quality management system is at the operations level. ISO/TS 16949 is seen as an initiative for work process improvement. The documentation is often developed at the work process level and focused on functions. Much of the effort is focused on the processes within the functions rather than across the functions and only involves the business level at the customer interface, as illustrated in Table 1.6. For the application of ISO/TS 16949 to be successful, quality has to be a strategic issue with *every* function of the organization embraced by the management system that is focused on satisfying the needs of all interested parties.

Quality management systems
Philosophy
A system is an ordered set of ideas, principles and theories or a chain of operations that produce specific results and to be a chain of operations, the operations need to work together in a regular relationship. Shannon defined a system as a group or set of objects

Table 1.6 *Attention levels*

Organizational level	Principle process focus	Basic team structure	Performance issue focus	Typical quality system focus	Ideal quality system focus
Enterprise	Strategic	Cross business	Ownership	Market	Strategic
Business	Business	Cross functional	Customer	Administrative	Business process
Operations	Work	Departmental	Process	Task process	Work process

united by some form or regular interaction or interdependence to perform a specified function.[15] Deming defines a system as a series of functions or activities within an organization that work together for the aim of the organization. These three definitions appear to be consistent although worded differently.

A quality management system is not a random collection of procedures, tasks or documents (which many quality systems are). Quality management systems are like air-conditioning systems – they need to be designed. All the components need to fit together, the inputs and outputs need to be connected, sensors need to feed information to processes which cause changes in performance and all parts need to work together to achieve a common purpose.

ISO 9000 defines a quality management system as a set of interrelated or interacting processes that achieve the quality policy and quality objective. But the word quality gets in the way of our thinking. It makes us think that quality management systems operate alongside environmental management systems, safety management systems, and financial management systems. In Clause 3.11 of ISO 9000 it is stated that the quality management system is "that part of the organization's management system that focuses on the achievement of outputs in relation to the quality objectives"; therefore, the quality management system must exist to achieve the organization's quality objectives. This concept was unclear in QS-9000 with the result that many quality systems were focused on procedures for their own sake rather than on serving objectives. It would appear therefore that other parts of the management system are intended to serve the achievement of specific objectives. For example, we could establish:

- safety systems to serve safety objectives,
- environmental systems to serve environmental objectives,
- security systems to serve security objectives,

- human resource systems to serve human resource objectives,
- marketing systems to serve marketing objectives,
- innovation systems to serve innovation objectives,
- financial systems to serve financial objectives.

Many organizations have appointed specific managers to achieve each of these objectives so that we have for instance, an Environmental Manager, fulfilling Environmental Objectives through an Environmental Management System and a Quality Manager fulfilling Quality Objectives through a Quality System. Do the same for the others and you would have multiple management systems. This is what functional management produces and as a result puts the managers in potential conflict with each other as each tries to achieve their objectives independently of the others.

However, several questions arise: "Are quality objectives, objectives of the same kind as the other objectives or are these other objectives a subset of quality objectives?" and "Is the quality management system just one of a series of systems or is it the parent system of which the others are a part?"

> The quality management system is not part of the management system, it is the management system.

To find the answer it is necessary to go back a step. Which comes first an objective or a need? We don't set financial objectives because we think its a good idea, there is a need that has its origins in the organization's mission statement. The mission statement tells us what our goal is – where are we going. Without customers there is no business therefore the basic purpose of a business is to satisfy a particular want in society and so create a customer. Its mission is related to these wants and is expressed in specific terms. To be effective, a mission statement should always look outside the business not inside.[16] For example, a mission that is focused on increasing market share is an inwardly seeking mission whereas a mission that is focused on bringing cheap digital communication to the people is an outwardly seeking mission statement. From the mission statement we can ask, "What affects our ability to accomplish our goal?" The answers we get become our critical success factors and it is these factors that shape our objectives if our success depends on:

- the safety of our products, we need safety objectives;
- securing the integrity of information entrusted to us by our customers, then we need security objectives;
- the impact our operations have on the environment, we need environmental objectives;
- capital investment in modern plant and machinery, we need financial objectives.

> Allowing a constraint to become the overriding objective deflects attention away from the true purpose for which the organization was formed.

This list is incomplete, but if we were to continue, would we find a reason for having quality objectives? Business will only

Outputs and outcomes

Outputs tend to imply products and services whereas outcomes include impacts of the business on its surroundings, its employees, etc.

create customers if they satisfy their needs; therefore, success in all businesses depends on fulfilling customer needs and expectations. All the objectives only arise as a result of the organization seeking to create and satisfy customers. There is no environmental objective, impact or anything else if the organization does not have customers. Objectives for the environment, safety, security, finance, human resources, etc. only have meaning when taken in the context of what the business is trying to do (which is to create and satisfy customers). Therefore in reality, satisfying customers is the only true objective; all others are *constraints* that affect the manner in which the organization satisfies its customers.

Quality is defined in ISO 9000 as the degree to which a set of inherent characteristics fulfils requirements. Note that the definition is not limited to customer requirements and the inherent characteristics are not limited to products. It could apply to any set of requirements – internal or external, technical or non-technical including health, safety and environmental requirements. It could also apply to any process outcome – products, services, decisions, information, impacts, etc. It extends to all those with an interest in the business. Quality is therefore a term that describes the condition of business outcomes. Everything a business does must directly or indirectly affect the condition of its outcomes and therefore all business objectives are quality objectives. Therefore, we do not need quality objectives in addition to all the other objectives because all objectives are quality objectives and the quality management system is not part of the management system – it *is* the management system. We can therefore describe the relationship between the management system and the organization diagrammatically as shown in Figure 1.11.

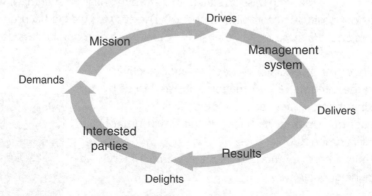

Figure 1.11 *The business management cycle (a specialists' view)*

The management system is the way the organization operates, the way it carries out its business, the way thing are. Its purpose is to enable the organization to accomplish its mission, it purpose, its goals and its objectives. All organizations possess a management system. Some are formal and some are informal. Even in a one-person business, that person will have a way of working – a way of achieving his or her objectives. That way is the system and comprises the behaviours, processes and resources employed to achieve those objectives. The system comprises everything that affects the results. It only has to be formalized when the relationships grow too large for one person to manage by relying on memory.

It is unlikely that you will be able to produce and sustain the required quality unless you organize yourselves to do so. Quality does not happen by chance, it has to be managed. No human endeavour has ever been successful without having been planned, organized and controlled in some way.

Scope of the system

As the quality management system is the means by which the organization achieves its objectives, it follows that the scope of the system (what it covers) is every function and activity of the organization that contributes to these objectives. This should leave no function or activity outside the system. The system must also include suppliers because the organization depends on its suppliers to achieve its objectives. The chain of processes from the customer interface and back again includes the suppliers.

Including every function and activity within the system should not be interpreted as compelling every function and activity to certification to ISO/TS 16949 – far from it. The scope of the system does not need to be the same as the scope of certification as is explained in Chapter 2 and addressed further in Chapter 3.

Design of the system

Imagine you are designing an air-conditioning system. You would commence by establishing the system requirements, then design a system that meets the requirements, document the design and build a prototype. You would then test it and when satisfied it functions under the anticipated operating conditions, launch into production. If problems are detected during production, solutions would be developed and the design documentation changed before recommencing production. If problems were experienced during maintenance, the design documentation would be consulted to aid in the search for the fault. If improvements are to be made, once again the design documentation would be consulted and design changes made and the documentation revised before implementation in production. This traditional cycle for products therefore has some redeeming features:

- Design does not commence without a specification of requirements; if it does, the wrong product is likely to be designed.
- Designs are documented before product is manufactured; if they are not documented, it is likely that the product cannot be manufactured or will not fit together or function as intended.
- Designs are proven before launching into production; if production commences before design proving, the product will probably fail on test or in service.
- Design documentation is changed before changes are implemented in production; if documentation is not changed before implementation, the product will be different each time it is made; solved problems will recur and no two installations will be alike.

If we apply the same logic to the design and implementation of a management system, we would:

- define the requirements before commencing management system design, i.e. we would establish the objectives the system is required to achieve;
- document the management system design before implementation;
- verify that the management system meets the requirements before commitment to full operation;
- document changes to the management system before implementation in practice.

But what often happens is:

- the management system development commences without a specification of requirements or a clear idea of the objectives is need to achieve; Often the system exists only to meet ISO 9000, ISO/TS 16949 or some other standard;
- the management system is documented before it has been designed;
- the management system is made fully operational before being verified it meets the requirements;
- changes are made to practices before they are documented;
- improvements are made to the management system without consulting the documentation because it is often out of date.

As the management system is the means by which the organization achieves its objectives, the management system delivers the organization's products. (This includes hardware, software, services and processed material including information products.)

If we analyse the factors on which the quality of these products depend we would deduce they include:

- the style of management (autocratic, democratic, participative, directive, etc.);
- the attitude and behaviour of the people (positive, negative, etc.);

- the capability of the available resources (capacity, responsiveness, technology);
- the quantity and quality of the available resources (materials, equipment, finance, people);
- the condition and capability of the facilities, plant and machinery;
- the physical environment in which people work (heat, noise, cleanliness, etc.);
- the human environment in which people work (freedom, empowerment, health and safety).

It follows therefore that a management system consists of the processes required to deliver the organization's products and services as well as the resources, behaviours and environment on which they depend. It is therefore not advisable to even contemplate a management system simply as a set of documents or if we do go someway towards ISO 9000:2000, a set of processes that simply converts inputs into outputs. Three out of the seven factors above relate to the human element – we therefore cannot afford to ignore it.

Following the argument above, if the management system is a collection of processes, we can think of the organization as a system of interconnected processes and therefore change Figure 1.12 so that it reflects reality.

Functional approach versus process approach

Organizations are usually structured functionally where a function is a collection of activities that make a common and unique contribution to the purpose and mission of the business.[17] This results in marketing, engineering, production, purchasing, quality, maintenance and accounts departments, etc. However, the combined expertise of design, quality, purchasing and production are needed to fulfil a customers requirement. It is rare to find one department that fulfils an organizational objective without the support

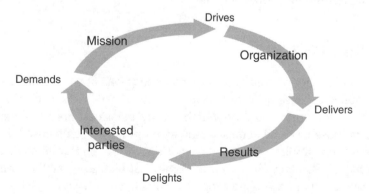

Figure 1.12 *The business management cycle (a pragmatic view)*

Table 1.7 *Function versus process*

Attribute	Functional approach	Process approach
Objectives focus	Satisfying departmental ambitions	Satisfying stakeholder needs
Inputs	From other functions	From other processes
Outputs	To other functions	To other processes
Work	Task focused	Result focused
Teams	Departmental	Cross functional
Resources	Territorial	Shared
Ownership	Departmental manager	Shared
Procedures	Departmental based	Task based
Performance review	Departmental	Process

of other departments and yet, the functional structure has proved to be very successful primarily because it develops core competences and hence attracts individuals who want to have a career in a particular discipline. This is the strength of the functional structure but because work is always executed as a process it passes through a variety of functions before the desired results are achieved. This causes bottlenecks, conflicts and sub-optimization. A functional approach tends to create gaps between functions and does not optimize overall performance. One department will optimize its activities around its objectives at the expense of other departments. Some of the other differences are indicated in Table 1.7.

Functional outputs are indeed different from process outputs and obviously make an important contribution, but it is the outputs from business processes that are purchased by customers, not the functional outputs.

Managing processes

QS-9000 created a notion that quality management was about following procedures. This is regrettable because quality management has always focused on the result – the outputs of the organization and on whether they meet customer requirements. The endeavours of those involved in quality management have been focused on the engine that produces these results – trying to make it deliver results that met customer needs and expectations. This focus has become somewhat blurred by all the campaigns, initiative and slogans that have handicapped the field. They have caused us to take our eye off the ball. Often good intentioned initiatives have only focused on one aspect of

Table 1.8 *Contrasting procedures with processes*[18]

Procedures	Processes
Procedures are driven by completion of the task.	Processes are driven by achievement of a desired result.
Procedures are implemented.	Processes are operated.
Procedure steps are completed by different people in different departments with different objectives.	Process stages are completed by different people with the same objectives – departments do not matter.
Procedures are discontinuous.	Processes flow to conclusion.
Procedures focus on satisfying the rules.	Processes focus on satisfying the interested parties.
Procedures define the sequence of steps to execute a task.	Processes transform inputs into outputs through use of resources.
Procedures may be used to process information.	Information is processed by use of a procedure.
Procedures exist – they are static.	Processes behave – they are dynamic.
Procedures cause people to take actions and decisions.	Processes cause things to happen.

quality management. Many have only focused on the production operations and those that focused on management have not been associated with quality management. But that is changing – quality is now a strategic issue. We just have to change the way we perceive various aspects of quality management if these various tools, techniques and ideas are to be used successfully.

As indicated previously, the management system consists of a series of processes interconnected in a manner that enables the organization to achieve its objectives.

Until the spread of ISO 9000, organizations focused on fixing processes and problems with the output not on refining procedures. However, in view of the confusion that has arisen between processes and procedures a short diversion to explain the differences is appropriate. This is illustrated in Table 1.8.

A view that is emerging in the literature supporting ISO 9000:2000 is that the procedural approach is about how you do things and processes are about what you do. This is misleading as it places the person outside the process when in fact the person is part of the process. It also sends out a signal that processes are just a set of instructions rather than a dynamic mechanism for achieving results. This message could jeopardize the benefits that could be gained from using the process approach.

Finding a definition

There are different schools of thought on what constitutes a process.

<table>
<tr><td>

Procedures versus processes

The procedural approach is about doing a task, conforming to the rules, doing what we are told to do.

The process approach is about, understanding needs, doing whatever it takes to satisfy these needs, finding the best way of fulfilling these needs even if it means changing the way we do our job, checking whether the needs are being satisfied and in the best way and checking whether our understanding of these needs remains valid.

</td><td>

A process is defined in ISO 9000 as *a set of interrelated or interacting activities which transforms inputs into outputs* and goes on to state *that processes in an organization are generally planned and carried out under-controlled conditions to add value.* The inclusion of the word *generally* tends to suggest that organizations may have processes that are not planned, not carried our under controlled conditions and do not add value and indeed they do!

</td></tr>
</table>

Juran defines a process[19] as a systematic series of actions directed to the achievement of a goal. In Juran's model the inputs are the goals and required product features and the outputs are products possessing the features required to meet customer needs. The ISO 9000 definition does not refer to goals or objectives.

Hammer defines a process[20] as a collection of activities that takes one or more kinds of inputs and creates an output that is of value to the customer. Hammer places customer value as a criterion for a process unlike the ISO 9000 definition.

Davenport defines a "process[21] as a structured measured set of activities designed to produce a specified output for a particular customer or market".

The concept of adding value and the party receiving the added value is seen as important in these definitions. This distinguishes processes from procedures (see below).

It is easy to see how these definitions can be misinterpreted but doesn't explain why for many it results in flow charts they call processes. They may describe the process flow but they are not in themselves processes because they simply define transactions. A series of transactions can represent a chain from input to output but it does not cause things to happen. Add the resources, the behaviours, the constraints and make the necessary connections and you might have a process that will cause things to happen. Therefore any process description that does not connect the activities and resources with the objectives and results is invalid. In fact any attempt to justify the charted activities with causing the outputs becomes futile.

Business process re-engineering

Most organizations are structured into functions that are collections of specialists performing tasks. The functions are like silos into which work is passed and executed under the directive of a function manager before being passed into another silo where it waits its turn because the people in that silo have different priorities and were not lucky enough to receive the resources they requested. Each function competes for scarce resources and completes a part of what is needed to deliver product to customers. This approach to work came out of the industrial revolution influenced firstly by Adam Smith and later by Frederick Taylor, Henry Fayol and others. When Smith and Taylor made their observations and formulated their theories, workers were not as educated as they are today. Technology was not as available and machines not as portable. Transportation of goods and information in the 18th and 19th centuries was totally different from today. As a means to transform a domestic economy to an industrial economy the theory was right for the time. Mass production would not have been possible under the domestic systems used at that time.

Re-engineering is the fundamental rethinking and radical redesign of business processes to achieve dramatic improvements in critical contemporary measures of performance, such as costs, quality, service and speed.[22] Business process re-engineering is about turning the organization on its head. Abandoning the old traditional way of organizing work as a set of tasks to organizing it as a process. According to Hammer, re-engineering means scrapping the organization charts and starting again.

ISO/TS 16949 is not requiring re-engineering. It is requiring that work be managed as a process so that its performance is evaluated on resultant customer value. In identifying the business processes, the subprocesses, the tasks and the functions through which work passes opportunities for improvement may be detected. These can be prioritized for action as part of a continual improvement programme.

Process characteristics

We can conceive of three different types of processes:

* Processes that convert inputs into outputs without adding value (i.e. a process without a specified purpose, a procedure the purpose of which is to define a series of activities, the middleman in a supply chain who simply bills a handling charge but adds no value).
* Processes that convert inputs into outputs with perceived added value for the internal customer but no added value for external parties (i.e. the manager who sets up

processes to produce reports that protect the manager from blame but do not add value for customers).

- Processes that convert inputs into outputs of added value for the external interested party (i.e. a process that creates products or services possessing benefits that add value for customers, a process that provides employees with greater skills and adds value for employees, a process that reduces airborne emissions from the facility and adds value for the community).

Process purpose

From the definitions of a process it is clear that every process needs a purpose for it to add value. The purpose provides a reason for its existence. The purpose statement should be expressed in terms of what the process does and in doing so identify what is to be converted. The purpose of a sales process may be to covert prospects into orders for the organizations' products. Instead of calling the process a sales process you could call it the *prospect to order process*. Similarly the purpose of a design process may be to convert customer needs into product features that satisfy these needs.

Process objectives

Process objectives provide a means to measure the effectiveness with which the process fulfils its purpose. In this respect objectives could be though of as measures of the extent to which process purpose is fulfilled. The objectives of the sales process might be to maintain the rate at which prospects are converted into orders at or above 45%. Another objective might be to induce 50% of the market segment to be aware of organization's products and services. A human resource process objective might be to ensure 80% of staff receives the training that has been identified within the budget period. A technique often used to test the soundness of objectives is SMART (specific, measurable, achievable, realistic and timely). Further details of SMART are provided in Chapter 5 under the section Setting objectives.

Process inputs

The inputs of a process are considered to be those things that are transformed by the process into outputs. Under these conditions, the input has to change in some way. This implies that instructions, requirements, objectives or any documents are not *inputs* because they are not transformed. They are the same at the end of the process as they were when they entered it. It is not the requirement that is changed but a product generated from the requirement. However, if you enquire of a person as to what the process inputs are, they will no doubt tell you of all the things that are needed for the process to commence. Within the items mentioned would be the information that triggers the work – the instruction, the order, the telephone call, etc. In reality, therefore, anything received by a process is an input whether or not it is transformed.

Process outputs

The outputs of a process are considered to be the tangible or intangible results such as a product or advice. However, in addition to outputs, processes have outcomes. There is an effect that the process has on its surroundings. An outcome of a process may be a detrimental affect on the environment. Satisfaction of either customers or employees is an outcome not an output.

Process resources

The resources in a process are the supplies that can be drawn on when needed by the process. Resources are classified into human, physical and financial resources. The physical resources include materials, equipment, plans machinery but also include time. Human resources include managers and staff including employees, contractors, volunteers and partners. The financial resources include money, credit and sponsorship. Resources are used or consumed by the process. There is a view that resources to a process are used (not consumed) and are those things that don't change during the process. People and machinery are resources that are used (not consumed) because they are the same at the start of the process as they are at the end, i.e. they don't lose anything to the process. Whereas materials, components and money are either lost to the process, converted or transformed and are therefore process inputs. People would be inputs not resources if the process transforms them.

Process constraints

The constraints on a process are the things that limit its freedom. Actions should be performed within the boundaries of the law; regulations impose conditions on hygiene, emissions and the internal and external environment. They may constrain resources (including time), effects, methods, decisions and many other factors depending on the type of process, the risks and its significance with respect to the business and society. Some people call these things **controls** rather than constraints but include among them, the customer requirements that trigger the process and these could just as well be inputs. Customer requirements for the most part are objectives not constraints but they may include constraints over how those objectives are to be achieved. For instance they may impose sustainability requirements that constrain the options open to the designer (Figure 1.13).

Views differ and whilst a purist might argue that requirements are controls not inputs, and materials are inputs not resources, it matters not in the management of quality. All it might affect is the manner in which the process is described diagrammatically. The requirements would enter the process from above and not from the side if you drew the chart as a horizontal flow. The symbolic view of a process given in Figure 1.13 is only one view of a process. Demands in this model comprise requirements and constraints. The danger in drawing the process as a flow from input to output is that you might

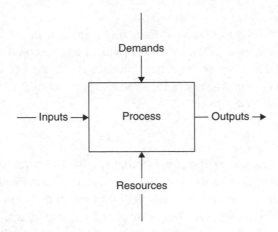

Figure 1.13 *Conceptual process*

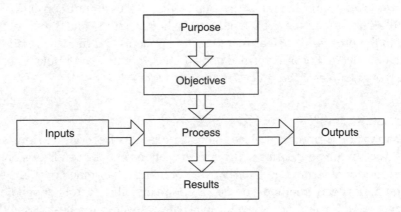

Figure 1.14 *A managed process (excluding resources and constraints)*

omit the provisions that need to be in place to manage the process. This is illustrated in Figure 1.14. The resources and demands are not shown for clarity.

Process results

The outputs are things that result from a process – of that there is no doubt – but the measurable results of the process are the outcomes:

- Is the process delivering outputs that meet the input requirements?
- Is the process operating efficiently?
- Is the process effective?

These questions are not answered by a single-process output but by monitoring and measurement taken of the outcomes – not the output.

Process classes
There are two classes of organizational processes – *macro-processes* and *micro-processes*. Macro-processes are multi-functional in nature consisting of numerous micro-processes. Macro-processes deliver business outputs and have been referred to as *Business Processes* for nearly a decade or more. For processes to be classed as business processes they need to be in a chain of processes having the same stakeholder at each end of the chain. The input is an input to the business and the output is an output from the business. This is so that the outputs can be measured in terms of the inputs. If the outputs were a translation of the inputs they could not be measured against the inputs.

There is a view that design is not a business process because the stakeholders are different at each end. On the input end could be marketing and the output end could be production. Under this logic, production would not be a business process because on the input could be sales and the output could be the customer. Therefore, the business process flow is: customer – sales – production – distribution – finance – customer.

The sales process takes the order from the customer and routes it to the production process. The production process supplies product to the distribution process and the distribution process delivers product to the customer, collects the cheque and routes it to the finance process where it is put into the bank and turned into cash. The business process is therefore 'order to cash' or Demand Fulfilment. With this convention, there would be only four business processes in most organizations. These are displayed diagrammatically in Figure 1.15 and the purpose of each process explained as follows with Table 1.9 showing the stakeholders:

* *Mission management process*: Determines the direction of the business, continually confirms that the business is proceeding in the right direction and makes course corrections to keep the business focused on its mission. The business processes are developed within mission management as the enabling mechanism by which the mission is accomplished.
* *Resource management process*: Specifies, acquires and maintains the resources required by the business to fulfil the mission and disposes of any resources that are no longer required.
* *Demand creation process*: Penetrates new markets and exploits existing markets with a promotional strategy that influences decision-makers and attracts potential customers to the organization. New product development would form part of this process if the business were market driven.

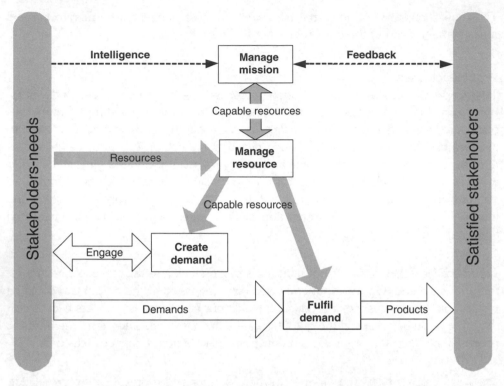

Figure 1.15 *Generic system model (the organization as a set of interconnected processes)*

Table 1.9 *Business process stakeholders*

Business process	Input stakeholder	Output stakeholder
Mission management	Shareholders, owners	Shareholders, owners
Demand creation	Customer	Customer
Demand fulfilment	Customer	Customer
Resource management	Resource user	Resource user

- *Demand fulfilment process*: Converts customer requirements into products and services that deliver customer satisfaction. New product development would form part of this process if the business were order driven.

Micro-processes deliver departmental outputs and are task oriented. In this book these are referred to as *Work Processes*. A management system is not just a collection of work processes, but also the interconnection of business processes. The relationship between these two types of processes is addressed in Table 1.10.[23]

Table 1.10 *Relationship of business process to work processes*

Scope	Business process	Work process
Relationship to organization hierarchy	Unrelated	Closely related
Ownership of process	No natural owner	Departmental head or supervisor
Level of attention	Executive level	Supervisory or operator level
Relationship to business goals	Directly related	Indirectly related and sometimes (incorrectly) unrelated
Responsibility	Multifunctional	Single function
Customers	Generally external or other business processes	Other departments or personnel in same department
Suppliers	Generally external or other business processes	Other departments or personnel in same department
Measures	Quality, cost delivery	Errors, quantities, response time
Units of measure	Customer satisfaction, shareholder value, cycle time	% defective, % sales cancelled, % throughput

Making the connections

It is important to visualize the complete picture when setting out to define and manage your processes. Processes exist in a context to deliver against an objective that also serves to deliver a strategic objective. As illustrated in Figure 1.12, there is a continuum from interested parties to mission through the system to results and back to the mission. The arrows form connections so that there is a clear line of sight from the results to the mission and this will only be accomplished if those carrying out strategic planning realize that processes cause results. Processes will not cause the right results unless the process objectives have been derived from the mission. The measures employed to indicate work process performance need to relate to the measures employed in the related business process so that when all the measures indicate that the system is performing as it should, the strategic objectives are being achieved.

Process effectiveness

A process should be effective but what determines its effectiveness? How would you know whether a process is effective? Effectiveness is about doing the right things – so what should a process do? Firstly and most obviously the process should deliver the required output – namely a decision, a document, a product or a service. But there is much more. It is not sufficient merely to deliver output. Output that is of poor quality is undesirable as is output that is late. But even when the output is of good quality and on time there are other factors to be considered. If in producing the output the laws of the land are breached, the process is clearly not effective. If in producing the output, the producers are exploited, are forced to work under appalling conditions or become de-motivated and only deliver the goods when stimulated by fear, the process is again not effective. So we could fix all these factors and deliver the required output on time and have satisfied employees.

Employees are but one of the stakeholders – customers are the most important; and although an output may be of good quality to its producers, it may not be a product that satisfies customers. The costs of operating the process may not yield a profit for the organization and its shareholders, and even if in compliance with current environmental laws, it may waste natural resources, dissatisfy the community and place unreasonable constraints on suppliers such that they decline to supply the process's material inputs. There is therefore only one measure of process effectiveness – that the process outcomes satisfy all interested parties.

Process management principles

As a result of the foregoing a set of seven principles has begun to emerge on which effective process management is based.[24] They all begin with the letter 'C' but that was not intentional until five of the seven turned out that way and then it seemed possible that '7Cs' were within reach.

Consistency of purpose

Processes will deliver the required outputs when there is consistency between the process purpose and the external stakeholders. When this principle is applied the process objectives, measures, targets, activities, resources and reviews will have been derived from the needs and expectations of the stakeholders.

Clarity of purpose

Clear measurable objectives with defined targets establish a clear focus for all actions and decisions, and enable the degree of achievement to be measured relative to stakeholder satisfaction. When this principle is applied people know what they are trying to do and how their performance will be measured.

Connectivity with objectives

The actions and decisions that are undertaken in any process will be those necessary to achieve the objectives and hence there will be demonstrable connectivity between the two. When this principle is applied the actions and decisions that people take will be those necessary to deliver the outputs needed to achieve the process objectives and no others.

Competence and capability
The quality of process outputs is directly proportional to the competence of the people, including their behaviour, and is also directly proportional to the capability of the equipment used by these people. When this principle is applied personnel will be assigned on the basis of their competence to deliver the required outputs and equipment will be selected on the basis of its capability to produce the required results.

Certainty of results
Desired results are more certain when they are measured frequently using soundly based methods and the results reviewed against the agreed targets. When this principle is applied people will know how the process is performing.

Conformity to best practice
Process performance reaches an optimum when actions and decisions conform to best practice. When this principle is applied work is performed in the manner intended and there is confidence that it is being performed in the most efficiency and effective way.

Clear line of sight
The process outputs are more likely to satisfy stakeholder expectations when periodic reviews verify whether there is a clear line of site between objectives, measures and targets, and the needs and expectations of stakeholders. When this principle is applied the process objectives, measures and targets will periodically change causing realignment of activities and resources thus ensuring continual improvement.

Developing an effective system

Every organization is different but has characteristics in common with others. Even in the same industry, the same market, producing the same type of products, each organization will be different in what it wants to do, how it goes about doing it and how it perceives its stakeholders and their needs. It is therefore not possible to design one system that will suit all organizations. Each has to be tailored to the particular characteristics of the organization and their stakeholders. However, just as human beings have the same organs and processes that differ in size and capability, organizations will have similar functions and processes that differ in size and capability.

In Figure 1.12 we showed that the mission is achieved by the organization that produces results that delight the stakeholders that place demands that shape the mission – and on its goes through a continuous cycle. In Figure 1.15 we showed how the organization (shown in Figure 1.12 in the cycle) could be viewed as a set of interconnected processes. These processes are present in all organizations, hence the label Generic

System Model. All organizations seek to *create a demand* – even non-profit organizations. All organizations seek to fulfil the demand, again even non-profit organizations and all organizations need resources to create and fulfil the demand and therefore have a resource management process. Finally all organizations have a purpose and a mission, even if it is not well defined, and seek to develop and improve their capability and their performance so as to achieve their mission; thus all organizations have a mission management process. This is of course at a very high level. The differences arise within the detail of each business process.

Whilst the purpose of these processes might be the same, the structure of them may well be different for each organization but a pattern of actions has emerged that can be used to flush out the information necessary to design these processes. Who or which function performs these activities is not important; in fact, letting the function get in the way, often changes the outcome such that instead of developing a process-based management system, you end up with a function-based management system that simply mirrors the organization structure.

Establishing the goals

Every organization has goals or what it wants to achieve, how it wants to be perceived and where it is going. These goals are often formed by looking both inwards and outwards, and are expressed relative to the needs and expectations of stakeholders or benevolent interested parties.

Purpose, mission and vision
The first step is to clarify the organization's purpose, mission and vision.

Purpose is clarified by top management confirming why the organization exists or for what purpose it has been established such as to exploit the gap in the market for personal communicators.

Mission is clarified by top management confirming the direction in which the organization is current proceeding such as to provide personal communicators that are high on reliability, security, safety and data accessibility.

Vision is clarified by top management confirming what they want the organization to become in the years ahead, what they want it to be known for or known as such as being a world class brand leader in personal communications.

Values
Values are confirmed by top management expressing what they believe are the fundamental principles that guide the organization in accomplishing its goals, what it stands

for such integrity, excellence, innovation, reliability, responsibility, fairness, etc. These values characterize the culture in the organization.

Stakeholder needs
As explained previously all organizations have stakeholders, those people or organizations on which the organization depends for its success. They include customers, shareholders, employees, suppliers and society as a number of discrete groups rather than individuals. In order to identify the stakeholder needs, you need to examine the stakeholders relative to the purpose, mission and vision, and the results will be a distinct set of needs and expectations:

- *Customers* might need on-time delivery, high reliability, low life cycle cost, disposable product, prompt after-sales response, etc.
- *Shareholders* might need financial return on investment and above average growth.
- *Employees* might need competitive pay and conditions, flexible working hours and crèche.
- *Suppliers* might need prompt payment of invoices, loyalty in exchange for flexibility.
- *Society* might need compliance with statutory laws and regulations, corporate responsibility, employment prospects for the local community.

Stakeholder satisfaction indicators
Each of the stakeholders will be looking for certain outcomes as evidence that their needs have been met. These become the organization's performance indicators and hence the corporate objectives that need to be achieved.

Develop the processes
Identifying business processes
In Figure 1.15, we showed the organization as a set of interconnected processes, four to be precise. It is these processes that deliver stakeholder satisfaction. The names can be different for specific organizations; it matters not what they are called but it is important what they deliver.

Deriving process control parameters
We need to control the processes so that they deliver the results required:

- *Repeatedly*, we get the same result every time we run the process with the same set-up conditions.
- *Consistently*, the results we get are those needed to meet the stakeholder needs.
- *Continually*, the process runs as planned without unexpected interruption.

In order to do this we need to define the process parameters that need to be controlled.

By examining the objectives we identified at the stage above and asking, *"What outputs would we look for as evidence that the objectives have been achieved?"* and we derive the process outputs.

If we examine the outputs and ask, *"Which of these processes delivers these outputs?"* We will derive the outputs for each process. These are the control subjects.

If we now recall the basic principles of quality control, we addressed previously by asking, *"How would we establish that these outputs are correct?"* We reveal the units of measure and by asking, *"What criteria will indicate whether our outputs are acceptable?"* We identify the performance standard which will indicate whether our performance is good or bad.

Taking an example, customers are likely to need products that perform like the specification that are delivered on time and represent value for money. This represents three outcomes. By asking the questions above we deduce that there are four outputs as shown in Figure 1.16 and that the Demand Fulfilment process can deliver these outputs. We also deduce that there are 10 different measures requiring seven sensors or measurement methods and nine different standards or targets to aim for. Alternative names for the parameters are give in parenthesis.

In reality there may be hundreds of things to measure but they all need to be linked to the stakeholder outcomes.

Identifying work processes
Once we know the process outputs and how success will be measured we can determine the main work processes. The work processes can be identified from determining the factors critical to success. If understanding customer requirements is key to success there will be a work process that focuses on understanding customer needs. If product innovation is key to success there will be a work process that focuses on product innovation, probably called product design. Using this method, outlines of each of the four business processes are given in Figures 1.17 and 1.18. Although every organization is different, the differences tend to be at the work process level rather than the business process level.

In the same way we derived the outputs, measures and targets for the business process we can take the business process objective and derive work process, outputs, measures and targets.

Determining work process activities
Once we have the defined work process parameters the next step is to determine the activities required to deliver the work process outputs. This can be accomplished by

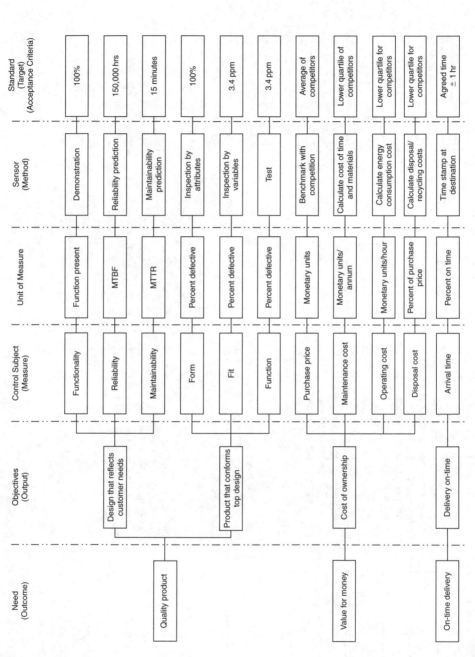

Figure 1.16 *Derivation of process parameters*

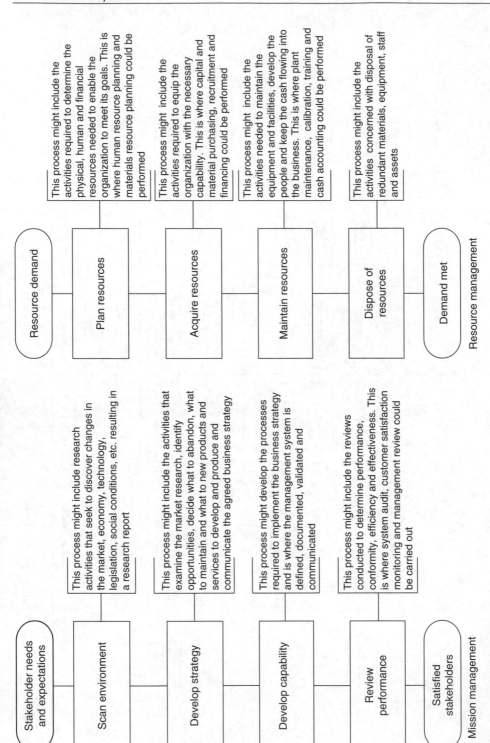

Figure 1.17 *Outline business processes 1*

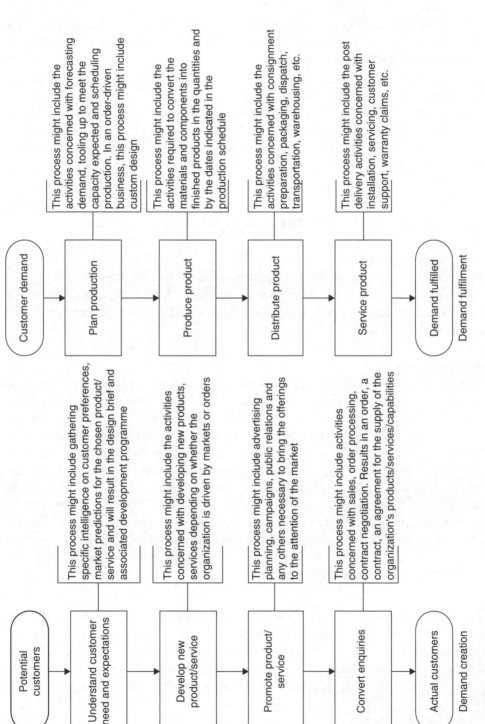

This process might include gathering specific intelligence on customer preferences, market predictions for the chosen product/service and will result in the design brief and associated development programme

This process might include the activities concerned with developing new products, services depending on whether the organization is driven by markets or orders

This process might include advertising planning, campaigns, public relations and any others necessary to bring the offerings to the attention of the market

This process might include activities concerned with sales, order processing, contract negotiation. Results in an order, a contract, an agreement for the supply of the organization's products/services/capabilities

This process might include the activities concerned with forecasting demand, tooling up to meet the capacity expected and scheduling production. In an order-driven business, this process might include custom design

This process might include the activities required to convert the materials and components into finished products in the quantities and by the dates indicated in the production schedule

This process might include the activities concerned with consignment preparation, packaging, dispatch, transportation, warehousing, etc.

This process might include the post delivery activities concerned with installation, servicing, customer support, warranty claims, etc.

Potential customers

Understand customer need and expectations

Develop new product/service

Promote product/service

Convert enquiries

Actual customers

Demand creation

Customer demand

Plan production

Produce product

Distribute product

Service product

Demand fulfilled

Demand fulfilment

Figure 1.18 *Outline business processes 2*

brainstorming, observation, past experience or theoretical analysis. The result will be a series of process flow charts with details of specific inputs and outputs. Examples of these are given in other parts of this book.

Determining the competence and capability
The next stage is to determine the competence required by those who are to carry out the actions and decisions on the flow charts. Competence is related to the work process outputs determined previously. Competence determination and measurement is addressed in Chapter 6.

Whilst many activities may be carried out using general office equipment and facilities, others may require specific capabilities and these need to be determined.

Resourcing processes
The numbers of people of differing competences will need to be determined and these people acquired in time to operate the work processes. Likewise the quantity of equipment and size of facilities needed should be determined and arrangements made to acquire them.

Installing processes
Process installation is concerned with bringing information, human resources and physical resources together in the right relationship so that all the components are put in place in readiness to commence operation. In many cases the process will be installed already because it existed before formalization. In some cases process installation will require a cultural change. There is little point in introducing change to people who are not prepared for it. Installing a dynamic process-based system into an environment in which people still believe in an element-based system or in which management still thinks in terms of functions is doomed to fail. Therefore, a precursor to *process installation* is the preparation of sound foundations. Everyone concerned needs to understand the purpose and objectives of what is about to happen; in other words, they all need to perceive the benefits and be committed to change and understand the concepts and principles involved.

The process of installing a new process or one that requires a change in practice is one that is concerned with the management of change. It has to be planned and resourced and account taken of attitudes, culture, barriers and any other resistance there may be.[25] You must remember that not all those who are to use the process may have participated in its development and may therefore be reluctant to change their practices.

Commissioning processes
Process commissioning is concerned with getting processes working following installation. The people will have been through reorientation and will have received all the necessary

process information. Any new resources will have been acquired and deployed and the old processes decommissioned. Installation and commissioning of new processes take place sequentially usually without a break so that current operations are not adversely affected.

Process integration
Process integration is concerned with changing behaviour so that people do the right things right without having to be told. The steps within a process become routine, habits are formed and beliefs strengthened. The way people act and react to certain stimuli becomes predictable and produces results that are required. Improvement does not come about by implementing requirements – it comes about by integrating principles into our behaviour.

System integration
System integration is concerned with making the interconnections between the processes, ensuring that all the linkages are in place and that the outputs from a process feed the interfacing processes at the right time in the right quality. The system will not be effective if the process linkages do not function properly.

Review performance
There are several reviews that need to take place some during system development and others afterwards on a planned frequency.

Reviews during development
Performance reviews are necessary to verify that development is proceeding to plan and the correct outputs have been generated for each of the stages above.

Risk assessments are necessary to verify that each of the process designs reflect a safe and cost-effective way of producing the required process outputs in a manner that satisfies the constraints with adequate failure prevention features inherent in the design.

Reviews after development
- Output reviews are necessary to establish whether the required outputs are being produced.
- Process reviews are necessary to establish whether the process outputs are being achieved in the most effective way.
- Effectiveness reviews are necessary to establish whether the outputs, measures and targets remain relevant to stakeholder needs.

Improve capability
Following each review, changes need to be made as necessary to bring about improvement by better control, better utilization of resources and better understanding of stakeholder needs.

Summary

In this chapter we have examined the basic concepts that underpin the body of knowledge of quality management relative to ISO/TS 16949. We have discovered that knowledge of ISO/TS 16949 alone is insufficient to be able to apply the concepts and principles of quality management that are expressed in ISO 9000. We demonstrated that the terms used in ISO 9000 are much more than words but labels for universal principles that help us discover the right things to do.

We have explored the meaning of quality and hopefully corrected some misconceptions along the way. We have also explored familiar terms like quality control and quality assurance and shown that these terms are not simply names for departments within an organization but much broader concepts that apply to the management of any activity.

We have exposed the gulf between the customer and management perceptions of quality and the importance of bringing these closer together and concluded that organizations have goals and these goals are achieved through a system of management processes. We have thus exposed the myth that organizations have multiple management systems and in reality have only one system – a system that enables the organization to satisfy all its stakeholders and accomplish its mission.

We have examined process management concepts, explored the terminology and introduced a set of principles on which the effective process management can be based. We have used these principles to show how an effective process-based management system can be developed and established. Finally we have found a level where the business cycle the system model and associated business processes are all generic paving the way for work processes to be developed to suit particular business needs within a sound framework that embraces all stakeholders.

With a clear understanding of these principles we should not only be able to determine the rationale for the ISO/TS 16949 requirements but determine the actions needed where the requirements are less detailed and in so doing use ISO/TS 16949 in a manner that will bring lasting benefits to the organization.

Basic concepts – Food for thought

1   Without a set of principles, achieving a common understanding in the field of quality management would be impossible.
2   In supplying products or services there are three fundamental parameters that determine their saleability – price, quality and delivery.
3   Organizations exist because of their ability to satisfy their customers and other interested parties.
4   It is quite possible for an organization to satisfy its customers and fail to satisfy the needs of the other interested parties.
5   The needs of all parties have to be satisfied in order for *quality* to be achieved.
6   Products or services that do not possess the right features and characteristics either by design or by construction are products of poor quality.
7   A gold-plated mousetrap that does not fail is not a success if no one needs a gold-plated mousetrap!
8   The more prescription we have the more we get immersed in the detail and lose sight of our objectives.
9   Customer focus means putting your energy into satisfying customers and understanding that profitability or avoidance of loss comes from satisfying customers.
10   People naturally concentrate on what they are measured – it is therefore vital that leaders measure the right things.
11   Processes are dynamic – they cause things to happen.
12   The behaviour of any part of a system has some effect on the behaviour of the system as a whole.
13   Everyone in the organization should be continually questioning its performance and seeking ways to reduce variation, improve their methods and seeking better ways of doing things.
14   The factual approach leads us to control activities based on fact rather than opinion or emotion.
15   Organizations depend on their suppliers as much they depend on their customers.
16   Control is sometimes perceived as undesirable as it removes freedom, but if everyone were free to do just as they liked there would be chaos.
17   Keeping the process under control is process control. Keeping the process within the limits of the customer specification is quality control.
18   Putting out fires is not improvement of the process; it only puts the process back to where it should have been in the first place.
19   Quality assurance activities do not control quality, they establish the extent to which quality will be is being or has been controlled.
20   Quality should be a strategic issue that involves the owners because it delivers fiscal performance.
21   We do not need quality objectives in addition to all the other objectives because all objectives are quality objectives.
22   All organizations possess a management system. Some are formal and some are informal.
23   The quality management system is the means by which the organization achieves its objectives and therefore no function or activity should exist outside the system.
24   If you think of the organization as a system it becomes the management system – the way things are done – the interconnection of processes that deliver business outcomes.
25   A series of transactions can represent a chain from input to output but it does not cause things to happen.
26   Only when you add the necessary resources, behaviours and constraints, and make the necessary connections will you have a process that will cause things to happen.
27   There is only one measure of process effectiveness – that the process outcomes satisfy all interested parties.

Chapter 2

Perceptions of ISO 9000 and its derivatives

*By three methods we may learn wisdom: first, by reflection, which is noblest; second,
by imitation, which is easiest; and third, by experience, which is the bitterest.*

Confucius

Introduction

Since the publication of the ISO 9000 family of standards in 1987 a new industry has
grown in its shadow. This industry is characterized by: Standards Bodies, Accreditation
Bodies, Certification Bodies, Consulting Practices, Training Providers, Software Providers
and a whole raft of publications, magazines, Web sites and schemes – all in the name
of quality! But has ISO 9000 and its derivates, such as ISO/TS 16949 fulfilled their
promise? There are those with vested interests that would argue that it has improved
the efficiency and effectiveness of organizations. Equally others would argue that it has
done tremendous damage to industry. One of the problems in assessing the validity of
the pros and cons of the debate is the very term ISO 9000 because it means different
things to different people.

Perceptions that have been confirmed time and again by consultants, other organizations and frequent audits from the certification bodies over the last 20 years makes
these perceptions extremely difficult to change. If ISO 9000 is perceived rightly or
wrongly, as a badge on the wall or a set of documents, then that is what it is. If this was
not the intent of ISO 9000 then clearly we have to do something about it. But why
should these perceptions be changed? After all, can over 500,000 organizations have
got it wrong? Some organizations in fact did use ISO 9000 wisely but they are likely to be
in the minority. Many organizations also chose not to pursue ISO 9000 certification and
focused on total quality management (TQM) but that too led to dissatisfaction with the
results. As an introduction to this handbook on ISO 9000:2000 it may be useful to take a
look at these perceptions – look at how we have come to think about ISO 9000: quality,

quality systems, certification and inspection. The perceptions are no different when we look at ISO/TS 16949 because it is an ISO 9000 derivative; it contains all the requirements of ISO 9001. A realization of these perceptions will hopefully enable us to approach the new standard with a different perspective or at least provide food for thought.

How we think about ISO 9000

To the advocate, ISO 9000 is a standard and all the negative comments have nothing to do with the standard but the way it has been interpreted by organizations, consultants and auditors. To the critics, ISO 9000 is what it is perceived to be and this tends to be the standard and its support infrastructure. This makes any discussion on the subject difficult and inevitably leads to disagreement.

Some people often think about ISO 9000 as a system. As a group of documents, ISO 9000 is a set of interrelated ideas, principles and rules, and could therefore be considered a system in the same way that we refer to the metric system or the imperial system of measurement. ISO 9000 is both an international standard and until December 2000 was a family of some 20 international standards. As a standard, ISO 9000 was divided into four parts with Part 1 providing guidelines on the selection and use of the other standards in the family. The family of standards included requirements for quality assurance and guidelines on quality management. Some might argue that none of these are in fact standards in the sense of being quantifiable. The critics argue that the standards are too open to interpretation to be standards – anything that produces such a wide variation is surely an incapable process with one of its primary causes being a series of objectives that are not measurable. ISO/TS 16949:1999 was not a family of standards (ignoring the fact that it was a Technical Specification). Only ISO 8402 of the ISO 9000 family was invoked but this has changed with the new version. However, if we take a broader view of standards, any set of rules, rituals, requirements, quantities, targets or behaviours that have been agreed by a group of people could be deemed to be a standard. Therefore, by this definition, ISO 9000 is a standard and yes ISO/TS 16949 is also a standard!

An auto industry view

The Society of Motor Manufactures and Traders (SMMT) conducted a study in 1995 of the relationship between ISO 9000 and product quality. An initial questionnaire was sent to 650 companies and 12 of these were selected for interview. The conclusion of the study was that "there was little evidence to suggest that ISO 9000 certification can give any indication of the actual quality performance of a company and also little evidence that it plays a significant part in determining the overall ability of a company to satisfy the needs of its customers". Although now nearly 10 years old and 4 years after the launch of a complete revision of the standard, there is little to suggest that anything has significantly changed but clearly another study is urgently needed.

ISO 9000 is also perceived as a label given to the family of standards and the associated certification scheme. However, certification was never a requirement of any of the standards in the ISO 9000 family – this came from customers. (This principle has been preserved in ISO/TS 16949 to some extent but not wholly. There is no requirement for ISO/TS 16949 certification in ISO/TS 16949:2002, that will come from specific vehicle manufacturers but there is a requirement for ISO 9001 certification of suppliers.) Such notions as "We are going for ISO 9000" imply ISO 9000 is a goal like a university degree, and like that there are those who pass, who are educated and those who merely pass the examination. You can purchase degrees from unaccredited universities just as you can purchase ISO 9000 certificates from unaccredited certification bodies. The acceptance criteria is the same, it is the means of measurement and therefore the legitimacy of the certificates that differ. Fortunately you cannot purchase ISO/TS 16949 certificates except from certification bodies accredited by the International Automotive Task Force (IATF).

As many organizations did not perceive they had a quality management system (QMS) before they embarked on the quest for ISO 9000 certification, the program, the system and the people were labelled 'ISO 9000' as a kind of shorthand. Before long, these labels became firmly attached and difficult to shed and consequently why people refer to ISO 9000 as a 'system'. The same practice is pervading ISO/TS 16949 implementation programs.

How we think about quality management systems

All organizations have a way of doing things. For some it rests in the mind of the leaders, for others it is translated onto paper and for most it is a mixture of both. Before ISO 9000 came along, organizations had found ways of doing things that worked for them. We seem to forget that before ISO 9000, we had built the pyramids, created the mass production of consumer goods, broken the sound barrier, put a man on the moon and brought him safely back to earth. It was organizational systems that made these achievements possible. Systems, with all their inadequacies and inefficiencies, enabled mankind to achieve objectives that until 1987 had completely revolutionized society. The next logical step was to improve these systems and make them more predictable, more efficient and more effective – optimizing performance across the whole organization – not focusing on particular parts at the expense of the others. What ISO 9000 did was to encourage the formalization of those parts of the system that served the achievement of product quality – often diverting resources away from the other parts of the system.

ISO 9000 did require organizations to establish a quality system as a means of ensuring product met specified requirements. What many organizations failed to appreciate was

Figure 2.1 *Bolt-on systems*

that they all have a management system; i.e. a way of doing things and because the language used in ISO 9000 was not consistent with the language of their business, many people did not see the connection between what they did already and what the standard required. People may think of the organization as a system, but what they do not do is to manage the organization as a system. They fail to make linkages between actions and effects and will change one function without considering the effects on another. (Neither ISO 9000 nor ISO/TS 16949 has brought about an improvement in this situation.)

New activities were therefore bolted onto the organization such as management review, internal audit, document control, records control, corrective and preventive actions without putting in place the necessary linkages to maintain system integrity. What emerged was an organization with warts as illustrated in Figure 2.1. This was typical of those organizations that merely pursued the 'badge on the wall'. Such was the hype, the pressure and the razzmatazz, that the part that was formalized using ISO 9000 became labelled as the ISO 9000 quality system. It isolated parts of the organization and made them less efficient. Other organizations recognized that quality was an important issue and formalized part of their informal management system. When ISO 14001 came along this resulted in the formalization of another part of their management system to create an environmental management system (EMS). In the UK at least, with the advent of BS 8800 on occupational health and safety management systems (SMS), a third part of the organization's management system was formalized. The effect of this piecemeal formalization is illustrated in Figure 2.2. This perception of ISO 9000, ISO 14000 and any other management system standard is also flawed – but it is understandable.

The 1994, edition of QS-9000 was characterized by its focus on procedures. In almost every element of ISO 9001 there was a requirement for the supplier to establish and maintain documented procedures to control some aspects of an organization's operations. So much did this requirement pervade the standard that it generated a belief that QS-9000 was simply a matter of documenting what you do and doing what you

Figure 2.2 *Separate systems*

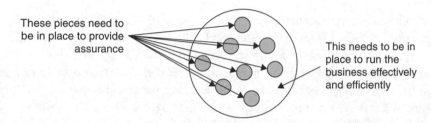

Figure 2.3 *Separating assurance activities from management activities*

document. This led to the perception that QS-9000 built a bureaucracy of procedures, records and forms with very little effect on quality.

The 1994 version of QS-9000 also created a perception that quality systems only exist to assure customers that product meets requirements. QS-9000 was often referred to as a quality assurance standard because customers used it for obtaining an assurance of the quality of products being supplied. This perception is illustrated in Figure 2.3, in which the organization is represented as a circle containing islands that serve the assurance of quality and with the remainder of the organization running the business. This is one reason why Toyota terminated its ISO 9000 certification program; it did not cover any important aspects of the business such as cost management.

Assurance equates with provision of objective evidence and this equates with the generation and maintenance of documentation, i.e. procedures and records. With the pressure from auditors to show evidence, organizations were persuaded to believe that if it was not documented it did not exist and this ultimately led to the belief that quality systems were a set of documents. These systems tended to be sets of documents that were structured around the elements of a standard. None of the standards required this, but this is how it was implemented by those who lacked understanding. However, ISO 9001 Clause 4.2.1 required suppliers to establish a quality system to ensure (not assure) that product met specified requirements. In other words, it required the system to cause conformity with requirements. A set of documents alone cannot cause product to conform to requirements. When people change the system they invariably mean that they update or revise the system documentation. When the system is audited invariably it is the documentation that is checked and compliance with documentation verified. There is often little consideration given to processes, resources, behaviors or results. As few people seem to have read ISO 8402, it is not surprising that the documents are perceived as a system. (*Note*: In talking with over 600 representatives of UK companies in 1999 and 2000 the author discovered that less than 10% had read ISO 8402.) But ISO 8402 defined a system rather differently. A quality system was defined as *the organization structure, procedures, processes and resources needed to implement quality management* – clearly not a set of documents. The 1994 version QS-9000 required a system to be established and documented. If the system was a set of documents, why then require it to be established as well as documented? (We have no evidence to show that the authors understood the difference so it is rather patronizing to speculate that they did!)

The persistence of the auditors to require documentation led to situations where documentation only existed in case something went wrong – in case someone was knocked down by a bus. While the unexpected can result in disaster for an organization it needs to be based on a risk assessment. There was often no assessment of the risks or the consequences. Unlike ISO/TS 16949, which requires contingency plans and failure mode and effects analysis (FMEA). This could have been avoided simply by asking the question: "So what?" So there are no written instructions for someone to take over the job but even if there were, would it guarantee there were no hiccups? Would it *ensure* product quality? Often the new person sees improvements that the previous person missed or deliberately chose not to make; often the written instructions are of no use without training; and often the written instructions are of no value whatsoever.

There has also been a perception in the service industries that ISO 9000 quality systems only deals with the procedural aspects of a service and not the professional aspects. For instance in a medical practice, the ISO 9000 quality system is often used only for processing patients and not for the medical treatment. In legal practices, the quality system again has been focused only on the administrative aspects and not on the legal issues. The argument for this is that there are professional bodies that deal with the professional side of the business. In other words, the quality system only addresses the non-technical issues, leaving the profession to address the technical issues. This is not *quality management*. The quality of the service depends on both the technical and non-technical aspects of the service. Patients who are given the wrong advice would remain dissatisfied even if their papers were in order or even if they were given courteous attention and advised promptly. To achieve quality one has to consider both the product and the service. A faulty product delivered on time, within budget and with a smile remains a faulty product!

How we think about certification

Pressure for certification

When an organization chooses not to pursue ISO 9000 certification or not to retain the ISO 9000 certificate, it should make no difference to the way the organization is managed. It is similar to the man who chooses not to take the course examination. He still has the knowledge he has acquired whether or not he takes the examination and gets a certificate. What he cannot do is to demonstrate to others that he has reached a certain level of education without having to prove it every time. People who know him do not care that he did not take the examination. It is only those who do not know him that he will have difficulty convincing.

Many organizations were driven to seek ISO 9000 certification or ISO/TS 16949 certification for that matter by pressure from customers rather than as an incentive to improve business performance and therefore sought the quickest route to certification. The critics called this coercion and like most command and control strategies, believed it resulted in managers cheating just to get the badge. What was out of character was that suppliers that were well known to customers were made to jump through this hoop in order to get a tick in a box in a list of approved suppliers. It became a 'necessary evil' to do business. Certainly when perceived as a means to get a badge, the standard was no more than a marketing tool. It could have been used as a framework for improvement but the way it was imposed on organizations generated fear brought about by ignorant customers who mistakenly believed that imposing ISO 9000 or ISO/TS 16949 would improve quality. To achieve anything in our society we inevitably have to impose rules and regulations – what the critics regard as *command and control* – but unfortunately, any progress we make masks the disadvantages of this strategy and because we only do what we are required to do, few people learn. When people make errors more rules are imposed until we are put in a straitjacket and productivity plummets. There is a need for regulations to keep sharks out of the bathing area, but if the regulations prevent bathing we defeat the objective, as did many of the customers that imposed ISO 9000.

ISO 9001 Certification is a requirement of ISO/TS 16949 but not ISO 9001, nor is it encouraged by the standard. It is however encouraged by governments and this is where the misunderstanding arises. Governments encouraged organizations to use ISO 9000 alongside product standards in their purchasing strategy so as to raise the standard of quality in national and international trade.[1] Certification became a requirement of customers – they mandated it through contracts. ISO 9000 was a convenient standard to use in order for customers to gain an assurance of quality. ISO 9000 was launched at a time when customers in the Western world took an adversarial approach to their suppliers. It came out of the defense industry where there was a long tradition of command and control. As a consequence, ISO 9000 followed the same pattern of imposing requirements to prevent failures that experience had shown led to poor product quality. ISO 9000 did not require purchasers to impose ISO 9000 on their suppliers. What it *did* require was for purchasers to determine the controls necessary to ensure whether purchased product met their requirements. But the easy way of meeting this requirement was to impose ISO 9000. (Unfortunately, this approach is being used in the automotive industry where second, third or fourth tier suppliers are being coerced into getting ISO/TS 16949 certification.) It saved the purchaser from having to assess for themselves the capability of suppliers. Unfortunately, the assessment process was ineffective because it led to suppliers getting the badge that were not capable of meeting their customer's requirements. ISO 9001 required suppliers to establish a quality system to ensure that product met specified requirements but it allowed organizations to specify their *own requirements*; provided they did what they said they did, they could receive the certificate. As there were no specific requirements in the standard that caused the auditors to verify that these requirements were those needed to meet the needs and expectations of customers, organizations could produce rubbish and still receive the badge. What was being checked was *consistency* – not *quality*.

Before ISO 9000, organizations were faced with meeting all manner of rules and regulations. Government inspectors and financial auditors frequently examined the books and practices for evidence of wrong doing but none of these resulted in organizations creating something that was not integrated within the routines they applied to manage the business. When ISO 9000 came along, many organizations embarked on a course of action that was perceived to have no value except to keep the badge – the ISO 9000 certificate. Activities were only documented and performed because the standard required it. Take away the certification and there was no longer a business need for many of these procedures and activities.

ISO 9000-1 in fact suggested that there were two approaches to using ISO 9000: 'management-motivated' and 'stakeholder-motivated'. It suggested that the supplier should consult ISO 9000-1 to understand the basic concepts but few organizations did this. It suggested that with the management-motivated approach organizations should firstly design their systems to ISO 9004-1 and then choose an appropriate assessment standard. It also suggested that with the stakeholder-motivated approach an organization should initially implement a quality system in response to the demands of customers and then select ISO 9001, ISO 9002 or ISO 9003 as appropriate for assessment. It suggested that having found significant improvements in product quality, costs and internal operating results from this approach, the organization would initiate a management-motivated approach based on ISO 9004. Those suppliers that actually obtained such benefits no doubt did initiate a management-motivated approach but many only focused on getting a certificate and therefore did not gain any benefits apart from the marketing advantage that ISO 9000 certification brought.

The approach to certification

Believing that ISO 9000 was only about 'documenting what you do', organizations set to work on responding to the requirements of the standard as a list of activities to be carried out. Again, this belief became so widespread that ISO co-ordinators or ISO 9000 project managers were appointed to establish and maintain the quality system. In some organizations, managers were assigned responsibility for meeting the requirements of a particular element of the standard even though there was not only no requirement to do so, but also no business benefit from doing so. Consultants were engaged to write the documents and apart from some new procedures governing internal audits, management review and document control, very little changed. There was a lot of money thrown at these projects in the quest to gain certification. However, none of the surveys conducted since 1987 have shown any significant improvement in an organization's overall performance – quite simply because nothing changed, not the processes, not the people nor the culture. The 'system' existed just to keep the badge on the wall. The ninth ISO survey[2] indicated that 9862 certificates had been withdrawn at the end of 1999 and of these 473 were for reasons of either insufficient return on investment or no business advantage. However some 7186 organizations discontinued certification for reasons unknown, indicating that certification was probably perceived as not adding value.

The approach to auditing

To make matters worse, the certification scheme established to assess the capability of organizations perpetuated this belief. These third-party auditors would reinforce the message by commencing their interviews with the question "Have you got a procedure for … ?" Audits would focus on seeking evidence that the organization was implementing its procedures – this technique was not limited to ISO 9001 assessments, it also pervaded QS-9000 assessments. Desperate to put the 'badge on the wall' organizations responded to the auditor's expectations and produced quality manuals that mirrored the structure of the standard – manuals containing nothing more than the requirements of Section 4 of ISO 9001 or ISO 9002, reworded as policy statements. (It is sadly no different with QS-9000 and ISO/TS 16949 certification. You only have to scour the Internet and alight on *ISO/TS 16949 Quality Manuals* following the clauses of the standard.) The auditor would therefore establish an organization's readiness for the audit by the closeness with which the quality manual addressed the requirements of the standard rather than by examining performance. A more sensible approach might have been to ask for the last 3 months, data for the key processes to establish if the processes were stable. Four years after the publication of ISO 9001:2000, whilst the habit of asking for a documented procedure is declining, there is little evidence that asking for performance data has yet become a habit.

Instead of using the whole family of standards as a framework, the standards became a stick with which to beat people. Managers would ask, where does it say that in the standard and if the auditor or consultant could not show them, the manager did nothing. The astute manager would ask, "Why would I want to do that?" and if the auditor or consultant could not give a sound business case for doing it, the manager did nothing.

Auditor training

Customers of auditor training courses behaved as though all they wanted was a training certificate. This led to lower standards. The auditors were poorly trained and the trainers became a victim of the system. Rules forced training bodies to cover certain topics in a certain time. Commercial pressure resulted in training bodies cutting costs to keep the courses running. Customers would not pay for more than they thought, they needed, but they did not know what they needed. Tell them what is required to convert a novice into a competent auditor and they wince! When there are providers only too willing to relieve them of their cash, customers opt for the cheaper solution. Had customers of training course been purchasing a product that failed to function there would have been an outcry, but the results of training were less likely to be measured. The training auditors received focused on auditing for conformity and led to auditors learning to catch people out. It did not lead to imparting the skills necessary for them to conduct audits that added value for organizations.

The effect of competition

Certification bodies were also in competition and this led to auditors spending less time conducting the audit that was *really* needed. They focused on the easy things to spot – not on whether the system was effective. Had the provision of certification services not been commercialized, there would not have been pressure to compromise quality. Organizations stayed with their certification body because they gave them an easy ride. What certification body would deliberately do things to lose customers? They will do everything they can to keep customers – even it means turning a blind eye. Certification bodies were also barred from making suggestions on improvement because it was considered to be consulting. They therefore stuck to familiar ground. The accreditation bodies were supposed to be supervising the certification bodies but they also needed revenue to be able to deploy assessors in sufficient numbers to maintain the integrity of the certification scheme. It had to be commercially viable at the outset otherwise the whole certification scheme would not have got off the ground because governments would not have been prepared to sponsor it. It is interesting that in the UK, there has been considerable protest against privatizing the National Air Traffic Service for fear that profits will compromise air space safety. There was no outcry against commercially operated quality system certification but equally unsafe products could emerge out of an ineffective quality system and enter the market. Thankfully the IATF has sought wise council and imposed an industry-led accreditation and witness audit scheme that might just make ISO/TS 16949 certification audits less prone to abuse.

Misplaced objectives

The certification scheme also added another dimension, that of scope. The scope of certification was determined by the organization so that only those parts of the quality system that were in the scope of certification were assessed. The quality system may have extended beyond the scope of certification and the scope of the standard but been far less than the scope of the business. This is illustrated in Figure 2.4.

Quality managers scurried around before and after the assessor and in doing so led everyone else to believe that all that were important to the assessor were documentation. This led others in the organization to focus on the things the auditor looked for, not on the things that mattered. They became so focused on satisfying the auditor that they lost sight of their objectives. They focused on surviving the audit and not on improving the performance. It has the same effect as the student who crams for an examination. The certificate may be won but an education is lost. What would the organization rather have: a certificate or an effective management system? Organizations had it in their power to terminate the contract with their certification body if they did not like the way they handled the assessment. They had it in their power to complain to the accreditation body if they were not satisfied with the certification body's service rendered but on both counts they failed to take any action. Certification bodies are suppliers, not regulators. What went wrong with ISO 9000 assessments is that the auditors lost sight of the objective to improve the quality of products and services. They failed to ask themselves whether the discrepancies they found had any bearing on the quality of the product. Many of the nonconformities were only classified as such because the organization had chosen to document what it did regardless of its impact

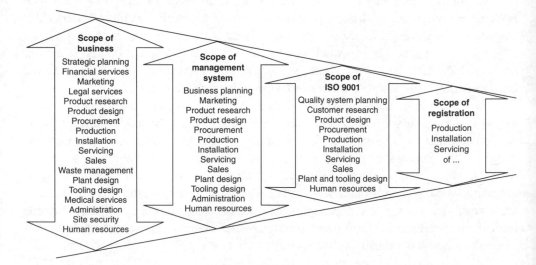

Figure 2.4 *The scoping effect*

on quality. Auditors often held the view that if an organization took the trouble to document *it*, *it* must be essential to product quality and therefore by not doing *it*, product quality must be affected!

It is to be hoped that ISO/TS 16949 qualified auditors are a cut above the rest, i.e. they certainly have to jump through more hoops but they often operate within an organization that provides ISO 9000 and ISO/TS 16949 certification services. Unless the management is separate, there is a risk that lax practices will pervade ISO/TS 16949 assessments until detected by the IATF witness audits.

> Is our goal to survive the audit or to improve our performance?

How ISO 9000 made us think about quality

ISO 9000 was conceived to bring about an improvement in product quality. It was believed that if organizations were able to demonstrate that they were operating a quality system that met international standards, customers would gain greater confidence in the quality of products they purchased. It was also believed that by operating in accordance with documented procedures, errors would be reduced and consistency of output ensured. If you find the best way of achieving a result, put in place measures to prevent variation, document it and train others to apply it; it follows that the results produced should be consistently good.

The requirements of the standard were perceived to be a list of things to do to achieve quality. The ISO co-ordinator would often draw up a plan based on the following logic:

- We have to identify resource requirements so I will write a procedure on identifying resource requirements.
- We have to produce quality plans so I will write a procedure on producing quality plans.
- We have to record contract reviews so I will write a procedure on recording contract reviews.
- We have to identify design changes so I will write a procedure on identifying design changes.

The requirements in the standard were often not expressed as results to be achieved. Requirements for a documented procedure to be established resulted in just that. Invariably the objectives of the procedure were to *define something* rather than to *achieve something*. This led to documentation without any clear purpose that related to the achievement of quality. Those producing the documentation were focusing on meeting the standard but not on achieving quality. Those producing the product were focusing on meeting the customer requirement but the two were often out of sync. As quality assurance became synonymous with procedures, so people perceived that they could achieve quality by following procedures. The dominance of procedures to the exclusion of performance is a misunderstanding of the implementers. The standard required a documented system *that ensured product met specified requirements* – a clear purpose. Once again the implementers lost sight of the objective. Or was it that they knew the objective but in order to meet it, the culture would have to change and if they could get the badge without doing so, why should they? Although the certification scheme is different for ISO/TS 16949, there is a risk that this approach will continue.

Issuing a procedure was considered to equate to the task completed. Unfortunately, for those on the receiving end, the procedures were filed and forgotten. When the auditor came around, the individual was found to be totally unaware of the 'procedure' and consequently found noncompliant with it. However, the auditor would discover that the individual was doing the right things so the corrective action was inevitably to change the procedure. The process of issuing procedures was not questioned, the individual concerned was blamed for not knowing the procedure and the whole episode failed to make any positive contribution to the achievement of quality. But it left the impression on the individual that quality was all about following the procedures. It also left the impression that quality was about consistency and providing you did what you said, you would do regardless of it being in the interests of satisfying customers, it was OK. One is left wondering whether anyone consulted the dictionary in which quality is defined as *a degree of excellence?*

Another problem was that those who were to implement requirements were often excluded from the process. Instead of enquiring as to the best way of meeting this requirement, those in charge of ISO 9000 implementation assumed that issuing procedures would in fact cause compliance with requirements. It requires a study of the way work gets done to appreciate how best to meet a requirement. Procedures were required to be documented and the range and detail was intended to be appropriate to the complexity of the work, the methods used, and the skills and training needed. The standard also only required work instructions where their absence would adversely affect quality. It is as though the people concerned did not read the requirement properly or had no curiosity to find out for themselves what ISO had to say about procedures – they were all too ready to be told what to do without questioning why they should be doing it.

More often than not, the topics covered by the standard were only a sample of all the things that need to be done to achieve the organization's objectives. The way the standard classified the topics was also often not appropriate to the way work was performed. As a consequence, procedures failed to be implemented because they mirrored the standard and not the work. ISO 9000 may have required documented procedures but it did not insist that they should be produced in separate documents, with titles or an identification convention that was traceable to the requirements. Unfortunately this insistence of documented procedures has not subsided entirely. There are still six mandatory documented procedures required by ISO 9001:2000 indicating a complete lack of imagination. They could have eliminated all documented procedures had they required a risk assessment be carried out. ISO/TS 16949:2002 has avoided adding to these by requiring several 'processes' but if people cannot see any difference they will simply produce more documented procedures.

Critics argue[3] that ISO 9000 did not enable organizations to reduce variation as a result of following the procedures. It is true that ISO 9000 did not explain the theory of variation – it could have done, but perhaps it was felt that this was better handled by the wealth of literature available at the time (ISO/TS 16949 certainly does not make the same mistake). However, ISO 9000 did require organizations to identify where the use of statistical techniques was necessary for establishing, controlling and verifying process capability but this was often misunderstood. Clause 4.14 of ISO 9001 required corrective action procedures – procedures to identify variation and eliminate the cause so this should have resulted in a reduction in variation. The procedures did not always focus on results – they tended to focus on transactions – sending information or product from A to B. The concept of corrective action was often misunderstood. It was believed to be about fixing the problem and preventive action was believed to be about preventing recurrence. Had users read ISO 8402 they should have been enlightened. Had they read Deming they would have been enlightened but in many cases the language of ISO 9000 was a deterrent to learning. Had the auditors understood variation, they too could have assisted in clarifying these issues but they too seemed ignorant, willing to regard Clause 4.20 as not applicable in many cases. But in the automotive industry, things were different. Statistical process control (SPC) techniques and process capability studies had been part of the quality programs for many years, although these techniques were often only applied to the production line.

Clause 4.6 of the undervalued and forgotten standard ISO 9000-1, dated 1994, starts with "The International Standards in the ISO 9000 family are founded on the understanding that all work is accomplished by a process". In Clause 4.7, it starts with "Every organization exists to accomplish value-adding work. The work is accomplished through a network of processes". In Clause 4.8, it starts with "It is conventional to speak of quality systems as consisting of a number of elements. The quality system is carried out by means of processes which exist both within and across functions". Alas, few people read ISO 9000-1 and as a result the baggage that had amassed was difficult to shed especially because there were few if any certification bodies suggesting that the guidance contained in ISO 9000-1 should be applied. Unfortunately, this message from ISO 9000-1 was not conveyed through the requirements of ISO 9001. ISO 9001 was not intended as a design tool. It was produced for contractual and assessment purposes but was used as a design tool instead of ISO 9000-1 and ISO 9004-1. This is also true of ISO/TS 16949:2002. It is an assessment tool and not a design tool but all too often, organizations look no further for inspiration.

How we think about reviews, inspections and audits

Audits of the quality system were supposed to determine its effectiveness but effectiveness seemed to be judged by the extent to which procedures were being followed. ISO 9001 Clause 4.1.3 did state that *the system should be reviewed for its continuing suitability and effectiveness in satisfying the requirements of the standard and the supplier's quality policy and objectives*. The words underlined were added in the 1994 revision. Clause 4.17 did require internal audits to *verify whether quality activities and related results comply with planned arrangements and to determine the effectiveness of the quality system*. Again the words underlined were added in the 1994 revision. But the original and modified wording seemed to have had no effect. Quality systems continued to be judged on product nonconformities, audit findings and customer complaints. Ten years later nothing has changed although there is a glimmer of light on the horizon that might suggest that some auditors are looking at performance.

The management review was supposed to question the validity of these procedures, the validity of the standards and the performance of the system. It was supposed to determine whether the system was effective, i.e. whether the system enabled people to do the right things right. But effectiveness was not interpreted as doing the right things; it was interpreted as conforming to the standard. It led to quality being thought of as conformity with procedures. The reviews and audits therefore focused on deficiencies against the requirements of the standard and deviations from procedure rather than the results the system was achieving. But as the system was not considered to be the way the organization achieved its results, it was not surprising that these totally inadequate management reviews continued in the name of keeping the badge on the wall. Audits did not establish that people were doing the right things. Had they done so the system would have been changed to one that *caused people to do the right things right without having to be told*.

It was often thought that the standard required review, approval, inspection and audit activities to be performed by personnel independent of the work. Critics argue that as a consequence both worker and inspector assumed the other would find the errors. ISO 9000 does not require independent inspection. There is no requirement that prohibits a worker from inspecting his or her own work or approving his or her own documents. It is the management that chooses a policy of not delegating authority for accepting results to those who produce them. There will be circumstances when independent inspection is necessary either as a blind check or when safety, cost, reputation or national security could be compromised by errors. What organizations could have done, and this would have met ISO 9000 requirements, is to let the worker decide on the need for independent inspection except in special cases. Alternatively, they could have carried out a risk assessment and imposed independent inspection where the risks warranted it. However, inspection is no substitute for getting it right first time and it is well known that you cannot inspect quality into an output if it was not there to start begin with.

Is ISO/TS 16949:2002 any different?

There are those who want to believe that the standard has not changed very much (if at all) and do not believe it has changed in its intent and as a consequence do not have to change their approach. The sad thing is that if the standard is perceived as not having significantly changed, it will continue to wreak havoc by being interpreted and used in the same inappropriate way that it has been for the last 17 years. But there is another way. By looking at ISO 9000 as a framework on which can be built a successful organization (rather than as a narrow set of minimum requirements) significant benefits can be gained. There are real benefits from managing organizations as a set of interconnected processes focused on achieving objectives that have been derived from an understanding of the needs of customers and other interested parties.

While the requirements of ISO/TS 16949 are expressed in a way that takes the reader through a cycle starting with the organization's purpose, leading onto quality policy and quality objectives, and ending with performance being reviewed against objectives, there remain many inconsistencies that could lead to confusion. Many of the linkages between purpose, policy, objectives, processes and results are inferred – they are not expressed unambiguously. It is only by searching for understanding that a clear logic emerges. The use of the word *quality* creates an anomaly and tends to represent the standard as simply a tool to meet customer quality requirements and no others. ISO/TS 16949 does venture into occupational safety and environment but primarily to safeguard the customer's interest. This is not to say that the standard is flawed. It is only saying that the concepts could be presented more clearly. Consequently, an approach has been taken in this book that requires the principles and requirements contained in the ISO/TS 16949 to be perceived as general business concepts and not simply limited to the achievement of *quality* in its narrowest sense. While the arguments for taking this approach are addressed in the book, the theme of the book is reflected in the following principles.

The *organization's purpose* is its reason for existing and can be expressed through vision and mission statements. The relationship between purpose, policy and objectives are blurred in the standard because from an assurance viewpoint *corporate values* seem to be irrelevant although with such revelations like Enron, Worldcom and Shell, it may not be long before values or corporate responsibility will be recognized by these standards.

The *quality policy* exists to shape behaviour and establishes the core values in an organization and therefore equates with the corporate policy – no benefits are gained from specifically expressing a quality policy and ignoring other policies because all policies influence the behaviours that are key to satisfying the needs of interested parties.

Quality is a strategic objective that is established to fulfil the needs and expectations of all interested parties and therefore equates with the corporate objectives – no benefits are gained from ranking quality equally with other objectives.

The *QMS* is *the* management system that enables the organization to fulfil its purpose and mission. Consequently, in this handbook the term *management system* is used throughout (except when referring to the requirements of the standard) rather than the terms *quality system*, *quality management system* or simply *QMS*. Organizations have only one system – no benefits are gained from formalizing part of a system that focuses on quality. By dropping the word quality from this term, it is hoped that the reader will begin to perceive a system that is significantly more beneficial than the quality system addressed by QS-9000 or ISO/TS 16949:1999.

The *adequacy, suitability and effectiveness* of the management system is judged by how well the system enables the organization to achieve its objectives, operate efficiently and fulfil its purpose and mission – no benefits are gained from simply focusing on one aspect of performance when it is a combination of factors that deliver organizational performance.

If you read the handbook from cover to cover you will discover that these principles are repeated regularly throughout in one form or another. Hopefully this is not too irritating but the handbook is intended as a reference. However, we rarely learn by a chance observation and it often requires frequent exposure to ideas presented in different forms and context before our beliefs or perceptions are changed.

Summary

In this chapter we have examined the various perceptions about ISO 9000 and its infrastructure and derivatives including ISO/TS 16949. These have arisen from personal observation, discussion with clients and colleagues and studying John Seddon's contribution in *The Case Against ISO 9000*. We have also compared and contrasted ISO 9001 with ISO/TS 16949.

Wherever appropriate the perceptions are challenged from a basis of what the standard actually requires. This is no excuse for the resultant confusion. The standard could have been better written but it is unfair to put all the blame on the standard. The standards bodies, certification bodies, accreditation bodies, training providers, consultants, software providers and many others have contributed to this confusion. Commercial interests have as usual compromised quality. We have followed like sheep, pursued goals without challenging whether they were the right goals but most of all we have forgotten why we were doing this. It was to improve quality, but clearly it has not.

ISO 9000 merely brings together concepts that have been applied in organizations for many years – not some unique concepts of management that only exist to put a 'badge on the wall' – but it appears that the use of international standards to consolidate and communicate these concepts has not been as effective as we believed it would be. The BNFL problems with fake quality control records, the Firestone problem with unqualified materials, the SA 80 rifle that jams in cold weather, laser guided bombs that miss the target and the recent spate of problems with the railways in the UK all send the signal that we have not solved the problem of effectively managing quality. This is despite ISO 9000 and the teachings of Juran, Deming, Feigenbaum, Ishikawa, Crosby and many others. Neither ISO 9000:2000 nor ISO/TS 16949:2002 is unlikely to change this situation because all these problems are caused by people who for one reason or another chose not to do the right things. All we can hope for is that ISO/TS 16949: 2002 will raise the bar enough to enable more organizations to satisfy more customers and do less harm to society.

At the end of 2003, the 13th ISO survey of ISO 9000 certificates indicated that the number had fallen from a peak of 561,747 in 2002 to 500,125 but as there were a number of organizations that had still to complete the transition and a number of certificates that had still to be issued, the true figures will be somewhat different. It would therefore be incorrect to conclude that ISO 9000 certification is declining, although it might well do so as organizations realize that certification is not essential for business in their sector of the market. As for the automotive sector, it is likely that the number of certifications will increase now that ISO 9001 certification is a requirement of ISO/TS 16949.

At the time of writing, Mercedes Benz is reported in the *Financial Times* as suffering quality problems. But how can this be, an organization that subscribes to ISO/TS 16949 having quality problems. Could this be more evidence that ISO 9000 and its derivatives is failing to change the culture. With all the money put into ISO/TS 16949 programs, one does not expect to see popular German and French Marques come bottom in a list of over 140 models in BBC Top Gear 2004 Survey, but I guess somebody has to!

Perceptions of ISO 9000 and its derivatives – Food for thought

1 Does ISO 9000 mean different things to different people?

2 If ISO 9000 is perceived rightly or wrongly, as a badge on the wall, a system, a label, a goal or a set of documents, is that what it is?

3 If any set of rules, rituals, requirements, quantities, targets or behaviours that have been agreed by a group of people could be deemed to be a standard, is ISO 9000 a standard?

4 Do managers think of the organization as a system? If so how come they do not manage the organization as a system?

5 Was ISO 9000:1994 or QS-9000 simply a matter of documenting what you do and doing what you document?

6 Do quality systems only exist to assure customers that product meets requirements?

7 Do you believe that if it is not documented it does not exist and that is why your quality system is a set of documents?

8 Do you believe that you can write instructions that do not rely on the user being trained?

9 Can a faulty product delivered on time, within budget and with a smile be anything other than a faulty product?

10 If your organization chooses not to pursue ISO 9000 certification or not to retain the certificate, will it make any difference to the way the organization is managed?

11 Did you cheat to get the ISO 9000 or QS-9000 certificate?

12 Did your application of ISO 9000/QS-9000 prevent you from producing nonconforming product or did it simply prevent you from producing product?

13 Is your organization one of those that coerced its suppliers into seeking ISO 9000 or ISO/TS 16949 certification because it was believed that the standard required it?

14 Did you establish a quality system to ensure that product met your customer's requirements or did you simply use it to ensure you met your own requirements?

15 If you were to take away the ISO 9000 or ISO/TS 16949 certification would there be a business need for all the procedures?

16 Did your third-party auditor establish your organization's readiness for the audit by the closeness with which the quality manual addressed the requirements of the standard?

17 Did you focus on the things the auditor looked for and not on the things that mattered?

18 Were your management more interested in surviving the audit than improving performance?

19 Were those producing the documentation focusing on meeting the standard or achieving quality?

20 Did your management believe the system was effective if it conformed to the standard?

21 Do you believe there are real benefits from managing organizations as a set of interconnected processes focused on achieving objectives that have been derived from an understanding of the needs of customers and other interested parties?

Chapter 3

Role, origins and application of ISO/TS 16949

One has to go rather slowly on fixing standards, for it is considerably easier to fix a wrong standard than a right one.

Henry Ford

The role of ISO/TS 16949

What is ISO/TS 16949?

ISO 9000/TS 16949 is a Technical Specification (hence the TS in the number) that, in conjunction ISO 9001:2000, defines the quality management system requirements for the design and development, production and, when relevant, installation and service of automotive-related products. It embodies ISO 9001, which is a generic standard for the assessment of quality management systems. It is not a product standard. ISO/TS 16949 does not contain requirements with which a product or a service can comply. There are no product acceptance criteria in ISO/TS 16949, so you cannot inspect a product against the standard.

What is the purpose of ISO/TS 16949?

The purpose of ISO/TS 16949 is to assist organizations supplying product or service into the automotive sector to operate systems that not only ensure whether these products and services meet customer requirements but also provide continual improvement, emphasize defect prevention, and reduce variation and waste in the supply chain.

The standard provides a vehicle for consolidating and communicating concepts in the field of quality management that have been approved by an international committee of representatives from the automotive industry as well as from national standards bodies.

It is not their purpose to fuel the certification, consulting, training and publishing industries. The primary users of the standards are intended to be organizations acting as either customers or suppliers.

Do we need to comply with other standards in the ISO 9000 family?

ISO/TS 16949:2002 embodies ISO 9001:2000; therefore, compliance with ISO/TS 16949:2002 implies compliance also with ISO 9001:2000. No separate ISO 9000 certification is necessary.

In addition to ISO/TS 16949:2002, users need to:

- interpret the terms used within the 'boxed text' in accordance with ISO 9000:2000;
- recognize that the principles on which the standard is based are defined in ISO 9004;
- consult ISO 9004:2000 when developing their systems beyond the requirements of ISO/TS 16949:2002;
- be guided by ISO 9004:2000 when undertaking continual improvement programs;
- take into account the guidance given in *International Automotive Task Force (IATF) Guidance to ISO/TS 16949:2002*;
- consult ISO 10012-1 and ISO 10012-2 relative to the control of measuring devices;
- invoke ISO/IEC 17025 in purchase orders for laboratory services used in connection with production standard parts;
- consult ISO 10019 when conducting internal audits.

What does ISO/TS 16949 apply to?

ISO/TS 16949 applies to sites of an organization where customer-specific parts for production or service are manufactured. The standard also applies to supporting sites, such as design centers, corporate headquarters and distribution centers; but these sites cannot obtain ISO/TS 16949 certification in their own right, they must be tied to a production site.

The term 'automotive' is to be understood as including cars, light, medium and heavy trucks, buses and motorcycles, and excludes industrial, agricultural and off-highway vehicles used in the mining, forestry and construction industries.

ISO/TS 16949 does not apply to organizations that make parts of its own design, i.e. proprietary parts. Hence if the customer orders parts to your organization's specification, the organization is not a candidate for ISO/TS 16949 certification but will require ISO 9001 certification.

The origins

Why is a sector standard needed?

The automotive industry has developed a number of standards since its inception at the end of the 19th century. There have been many product standards, but it was not until the emergence of QS-9000 that three giants in the automotive sector came together to create a set of common quality-system requirements. Now even more companies have come together to sign up to an international specification: ISO/TS 16949, but why? Why couldn't the automotive industry settle for ISO 9001:2000?

In principle the automotive industry could have settled for ISO 9001:2000. There are no additional principles embodied in ISO/TS 16949:2002. The difference is in the detail. Almost every additional requirement could be derived from the existing ISO 9001 requirement with a little imagination. But the industry cannot depend on people using their imagination. One of the key messages of ISO/TS 16949 is "the reduction of waste in the supply chain" thus echoing Toyota's strategy because whilst there is waste in the supply chain, costs will continue to rise unnecessarily.

As previously explained in Chapter 1, "there are those whose only need is a set of principles from which they are able to determine the right things to do. There are countless others who need a set of rules derived from principles that they can apply to what they do and indeed others who need a detailed prescription derived from the rules for a particular task".

A number of techniques have been developed within the automotive industry that they feel cannot be left to the imagination. Advanced Product Quality Planning (APQP), Failure Mode and Effects Analysis (FMEA), Production Part Approval Procedure (PPAP), Measurement Systems Analysis (MSA), Statistical Process Control (SPC) and error-proofing are among the primary techniques in this category. They are not unique to the automotive industry but it is that industry that has made most use of them and therefore if they are not prescribed, it sends out the signal that they are not important.

Another reason for the additional requirements is the way in which the automotive industry has developed. From the days of Henry Ford when everything was done in-house, even to the extent of mining the ore from which the steel was made, the industry has been transformed by a supply chain where the vehicle makers are no more than design and assembly plants. All the parts are supplied by the first, second or third tier suppliers in a chain that runs from body panels to fasteners, engine assemblies to circlips. Most parts are also designed for a specific vehicle. Some parts are proprietary, such as light bulbs, nuts and bolts, spark plugs and tires but even some tires are designed with specific vehicles in mind. The sheer turnover of vehicles every year is enormous. Customer expect new models every year or so; technology is forever pushing the frontiers and with it customer expectations forever increasing. Therefore, whether or not the ISO 9001 generic requirement could suffice, the industry needed a common vehicle to contain industry-specific requirements on which customer-specific requirements could hang.

The origins of ISO/TS 16949

> Many ISO 9000 registered organizations fail to satisfy their customers, but this is largely their own fault – they simply do not do what they say they will do.

For a document to become an international standard it must be acknowledged by many nations as defining good practice. This does not mean the standard defines all practices that one should adopt. Standards are 'minimums' not 'maximums'. Like hygiene standards, there is a minimum standard below which disease becomes virtually inevitable. Such standards do not and should not prohibit anyone in the group exceeding the standards. Within the business community, ISO 9000 represents the minimum system requirements for achieving quality in products and services. In other words, if you do all the things in ISO 9000 there is no reason why you should not *consistently* satisfy your customers.

Most ISO 9000 registered organizations claim to provide quality products and services, so why should there be so many dissatisfied customers when there are over 500,000 organizations[1] in the world certified to ISO 9001. One of the principle requirements in the standard is for the supplier to establish a quality management system as a means of ensuring that product or service meets the specified requirements. If an organization's products or services do not meet the specified requirements then, clearly the system has failed, but the failure is not a fault of the standard; it is a fault of the way the standard has been applied and interpreted both by the organizations themselves and the auditors who determine conformity. If the specified requirements are less than those of the customers, it is inevitable that products will bring dissatisfaction. This realization has, in the case of the automotive industry, led to two distinct needs:

* A need to harmonize fundamental supplier quality management system requirements and eliminate multiple interpretations.
* A need for a common certification scheme to ensure the integrity of the certification process worldwide.

Emergence of sector requirements

As a set of minimum standards, ISO 9000 addresses the business community. It was intended for purchasers as a means for them to obtain products and services of *consistent* quality from their suppliers. Taking the place of purchaser-specified general quality management requirements, ISO 9000 became the common requirement and hence eliminated the need for such requirements. As a consequence, it provides suppliers that meet its requirements with a demonstrable capability that others may not possess and hence, such capability becomes a persuasive marketing tool that will increase market share. ISO 9000 was also intended for application to all types of industry and therefore did not contain requirements for any specific industry sector, or type of products or services. Partially due to the scope of misinterpretation and the degree to which particular industries have common supplier requirements, certain industry sectors perceived the need for harmonizing such requirements in a form that added to those requirements in ISO 9000.

The drive for these additional requirements has come not from the suppliers but from users, such as the automotive, utilities, telecommunications, software and aerospace industries, that purchase millions of products and services used to produce the goods and services they provide to the consumer. Rather than invoke customer-specific conditions in each contract, the larger purchasers perceive real benefits from agreeing common quality management system requirements for their industry sector. Quite often a supplier will be supplying more than one customer in a particular sector and hence, costs increase for both the supplier and the customer if the supplier has to meet different requirements that serve the same objective. All customers desire products and services that *consistently* meet their requirements. While the physical and functional requirements for the product or service will differ, the requirements governing the manner in which their quality is to be achieved, controlled and assured need not differ. Differences in quality-system requirements may arise between industry sectors where the technology, complexity and risks are different.

There are those who see the emergence of sector standards as a retrograde step, having reached the stage where we have condensed all the world's national quality management system standards into a family of five standards. It may seem to be a retrograde step if these standards were regarded as the *crème de la crème* of standards. Unfortunately, ISO 9000 remains a 'minimum' and hence does not and was not intended to meet the needs of all the users. The alternative to suppressing sector standards at the international level, is to see them emerge at the national level or continue with the practice of purchasers invoking their own quality-system requirements within contracts, perpetuating fragmentation, duplication and driving up costs.

However, until ISO 9000 emerged in 1987, the automotive industry used a variety of customer-specific standards to govern a supplier's quality management practices.

The British contribution

Prior to the publication of ISO 9000, several nations had developed national quality-system standards, with many used only in the procurement of military equipment.

During the 1960s the larger corporations, such as British Aircraft Corporation, Central Electricity Generating Board, British Leyland and the General Post Office, developed a range of evaluation and vendor control tools to assure product quality. Most required second-party audit and surveillance programs of one sort or another based on Ministry of Defence (MoD) requirements. Then in 1968, MoD published an aviation publication (AvP) 92 that contained management requirements for the design and production of aircraft and guided-weapons systems but it was short lived as in 1973 the North Atlantic Treaty Organization (NATO) Allied Quality Assurance Publication (AQAP) standards were published. With all the fervor surrounding product quality in the aerospace sector, the Quality Panel of the UK Society of Motor Manufacturers (SMMT) set out to develop an equivalent standard for non-military applications. The result was BS 4891, which was published in 1972. In 1974, this was followed by BS 5179 with the title, "Operation and Evaluation of Quality Assurance Systems". However, BS 5179 was intended only as a guide and it was not until 1979 with the publication of BS 5750 that the major purchasers in the UK had a standard that could be invoked in contracts. A certification scheme was eventually established in 1983 following the UK Government's *White Paper* on standards and competitiveness.[2]

Why did QS-9000 embody ISO 9001?

No one told the task force they had to incorporate ISO 9001, but it is reported that faced with the decision to either adopt the standard and built on it or spend the rest of their lives telling people why they did not; they agreed it just made sense to adopt it.

In 1983, British Standards Institution (BSI) approached the International Organization for Standardization (ISO) in Geneva with a view to developing an international quality-system standard and eventually a committee was formed. Using BS 5750 as its basis the ISO 9000 series of standards was born.

Although the UK and in particular the UK automotive industry had been at the forefront of the development of non-military quality-system standards, harmonization within the automotive sector beyond BS 5750 was believed too difficult to achieve. Using BS 5750 only as a baseline, the UK motor manufactures continued to develop supplementary standards; many of which are still in use today. BS 5750 and its successor ISO 9000 was enforced by the UK automotive industry and no further harmonization took place.

The American contribution

In 1964, Ford began auditing its suppliers against their Q101 standard and General Motors (GM) began auditing their suppliers using Supplier Performance and Evaluation Report (SPEAR). With Chrysler following suit suppliers were being subject to multiple audits against varying standards. Then in the 1980s as Japan gained a foothold in Western markets with products of superior quality, the 'Big Three' (Ford, GM and

Chrysler) developed their supplier award programs (Ford: Q1 Award, GM: Targets for Excellence and Chrysler: Pentastar Award); in an attempt to win back customers. However, suppliers were still being subject to multiple audits and varying standards, and so they put pressure on the Automotive Industry Action Group (AIAG) to harmonize these different approaches as they were all trying to resolve the same problem, i.e. product quality.

In 1988, the Purchasing and Supply Vice Presidents of Chrysler, Ford and GM chartered a Task Force to standardize reference manuals, reporting formats and technical nomenclature resulting in five standardized reference manuals. In 1992 the Chrysler, Ford and GM Task Force set out to harmonize the fundamental supplier quality-system manuals and assessment tools, and produced QS-9000. This new standard embodied the requirements of ISO 9001:1994, added generic requirements, sector-specific requirements and customer-specific requirements. QS-9000 was first published in August 1994 and harmonized Chrysler's Supplier Quality Assurance Manual, Ford's 'Q101' and General Motor's 'Targets for Excellence' with some input from the truck manufacturers. It is pertinent that it was the Purchasing Vice Presidents of Chrysler, Ford and GM that set up the Task Force and that the initiative was driven by purchasing to improve the quality of supplies. Hence QS-9000 was not intended to apply to the design and assembly plants of Chrysler, Ford and GM.

In 1995, the first edition of QS-9000 was revised and by March 1998 the third edition was published.

The German contribution

In 1991, the Verband der Automobilindustrie e. V. (VDA) published VDA 6.1 'Quality System Audit', a questionnaire on quality-system evaluation based on DIN EN ISO 9004. VDA 6 is a series of guides covering the basics for quality audits, auditing and certification. They were therefore not intended as supplementary requirements to ISO 9000, but as guides for auditors performing audits of automotive suppliers. Their intention was to improve auditor competency in the industry by providing a uniform interpretation of ISO 9000 requirements and a common approach to automotive audits. Unlike QS-9000, VDA 6.1 does not incorporate the requirements of ISO 9001 Section 4.

The Italian contribution

In 1994, Associazione Nazionale Fra Industrie Automobilistiche (ANFIA) published Associazione Nazionale dei Valutatoridi Sistemi Qualita (AVSQ) '94 with the title "ANFIA Evaluation of Quality Systems – Guidelines for Use". This consisted of both a checklist and a user guide. The questions were derived from ISO 9001:1994 and for each question in the checklist there are guidelines on interpretation that are specific to the automotive industry. By 1995, AVSQ '94 was in the third edition in which VDA 6 second edition, EAQF '94 and ISO 9004-1:1994 have been used. Thus reciprocal recognition at the European level was achieved whereby certification to AVSQ '94 was recognized as equivalent to VDA 6.1 and EAQF '94 certification.

The French contribution

In 1990, PSA Peugeot – Citroen and Renault released a supplier quality assurance publication with the title Référentiel d'Evaluation d'Aptitude Qualité Fournisseurs (EAQF). The publication summaries the requirements of ISO 9001 Section 4 but as with VDA 6.1 and AVSQ, the requirements of ISO 9001 are not incorporated. The layout is very similar to AVSQ and there are many additional requirements to those in ISO 9001 and guidance on the application in the automotive industry. In 1994, the second edition was published with the integrated requirements of the German publication VDA 6.1, not previously contained in EAQF '90.

The motivation

In the last 10 years the motor industry has witnessed many mergers and joint ventures not just within national boundaries but also across nations. Table 3.1 illustrates 'who owns who' and shows that Britain is not alone among the countries that have sold its motor industry to foreign buyers. This does not mean that Britain and other countries do not have a motor industry – what it does signify is that the motor industry is now a global industry. The table also hides the fact that the European Union (EU) is the largest automotive production region (34%) in the world.

Buying the competition has been a way of entering foreign markets and is not a recent phenomenon. Ford chose another way by building manufacturing plants overseas, and designing and producing cars for the local market. In Europe that market has grown beyond a single country and although the cars may have different names they have the same body parts and engines. GM bought the British Company, Vauxhall in 1926 and the German Company Adam Opel in 1930, then in 1948 GM bought the Australian Company Holden. The 'Big Three' (Ford, GM and Chrysler) have been global players for many years. In Europe acquisitions have been rather slower. In 1969, VW bought Audi and then a long gap before acquiring Seat in 1986 and more recently Rolls Royce.

There are several joint ventures, such as the multipurpose vehicles (MPV) that Ford, Nissan and VW produce – exactly the same vehicle with slight modifications. Renault and Fiat also produce a common MPV and there are several partnerships, such as Rover engines being supplied to Proton, Ford and Honda. Lada gets its chassis from

Table 3.1 *Who owns who?*[3]

Manufacturer	Origin	Established	Owner or partner	Origin	Acquisition or merger
Peugeot	France	1904	PSA Peugeot Citroen	France	1999
Citroen	France	1919			
Nissan	Japan	1933	Renault	France	1999:36.6%
Rover Group	UK		BMW	Germany	1994
Skoda	Czechoslovakia	1936	VW	Germany	1990
Audi	Germany	1899	VW	Germany	1969
Bugatti	Italy	1881	VW	Germany	1998
Rolls Royce	UK	1904	VW	Germany	1998
Seat	Spain	1950	VW	Germany	1986:75% 1990:99.9%
Alfa Romeo	Italy	1906	Fiat	Italy	1986
Lancia	Italy	1906	Fiat	Italy	1979
Kia	Korea	1944	Hyundai	Korea	1998:51%
Lotus	UK	1948	Proton	Malaysia	1996
Simca	France	1930	Chrysler	USA	1967
Lamborghini	Italy	1921	Chrysler	USA	1987
Rootes Group	UK	1932	Chrysler	USA	1967
Daimler-Benz	Germany	1883	Chrysler	USA	1998:50%
Mazda	Japan	1920	Ford	USA	1979:25% 1996:33%
Volvo	Sweden	1927	Ford	USA	1999
Aston Martin	UK	1914	Ford	USA	1987:75% 1994:100%
Jaguar	UK	1936	Ford	USA	1989
Holden	Australia	1931	GM	USA	1931
Adam Opel AG	Germany	1898	GM	USA	1929
Isuzu	Japan	1937	GM	USA	1971
Saab	Sweden	1947	GM	USA	50%
Vauxhall	UK	1903	GM	USA	1925

Fiat and its engines from GM. Mitsubishi build the Carisma in Holland in the same plant that Volvo builds the S40. The Porche Boxster is assembled in Finland in the same company that assembles the Saab 9-3 cabriolet and so on. One cannot be sure who owns the company that makes your car, where the components come from and where it is assembled. What matters is that it meets your needs and expectations, and this can only be achieved if there are some common systems in use in each of the countries so that who owns and who and who builds, what become irrelevant to customer's confidence. It is reported that within 20 years there may only be six vehicle manufacturers left in the world.[4]

Harmonization

The automotive industry has comprised multinational corporations for many decades but there has been little a harmonization in quality-system requirements across all plants. QS-9000 harmonized these requirements not only in the USA but also in every country where GM, Ford and Chrysler had suppliers. With the emergence of VDA 6, AVSQ '94 and EAQF '94, GM, Ford and Chrysler suppliers were now being faced with up to four different quality-system standards; hence a Ford plant in Germany might have invoked VDA 6 on a supplier, whereas a Ford plant in the USA would invoke QS-9000 and a Ford plant in France would invoke EAQF '94. Consequently, the purchasing executives of the large European automakers approached GM, Ford and Chrysler with a view to harmonizing the USA, Italian, French and German automotive quality-system standards.

In 1996, the IATF was established comprising representatives of the vehicle manufactures and trade associations from the USA and Europe including:

- Fiat, ANFIA and IVECO from Italy.
- FIEV, PSA, Peugeot – Citroen and Renault from France.
- VDA, Adam Opel, Audi, BMW, Daimler-Benz, Ford Werke and VW from Germany.
- AIAG, Daimler-Chrysler, Ford and GM from the USA and
- SMMT from the UK.

These nations together with representatives from ISO/TC 176 developed a sector standard that became ISO/TS 16949. This specification incorporated Section 4 of ISO 9001:1994 and included requirements taken from QS-9000, VDA 6, AVSQ '94 and EAQF '94, and some new requirements all of which have been agreed by the international members. The evolution of ISO/TS 16949 is illustrated in Figure 3.1.

With the publication of ISO/TS 16949:1999 occurring during the period that ISO 9000 was undergoing a major revision, it was inevitable that by the end of the 3-year period it would be revised to embody ISO 9001:2000. Nearly all the automotive-specific requirements contained in the 1999 version have been carried over to the 2002 version with some minor revision to the text. The second edition has been produced through a partnership between the IATF and the Japanese Automobile Manufacturers Association (JAMA) but not Japanese vehicle manufacturers, which may join the IATF at a later date. QS-9000 VDA 6, AVSQ '94 and EAQF '94 are not being withdrawn.

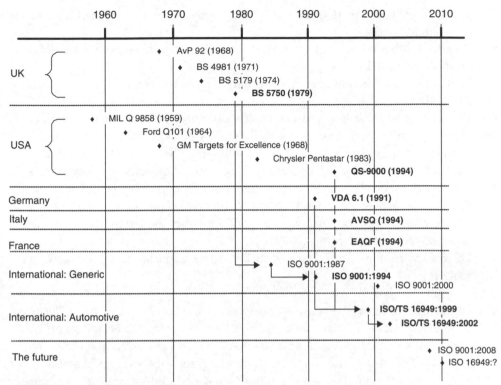

Figure 3.1 *Evolution of ISO/TS 16949*

Emergence of a common certification scheme

The ISO 9000 certification scheme

Although there are national variations, quality management system certification is performed by certification bodies (also known as 'registrars') that are accredited by one or more accreditation bodies that meet the requirements of ISO Guide 61:1996. The certification bodies are paid by the organization requesting certification and the accreditation body charges fees for the initial assessment of the certification body to ISO Guide 62:1996 in order to achieve accreditation, and for witness audits and other services to retain accreditation. There are many weaknesses with this arrangement and although there are national differences, in general:

- There is no law prohibiting an organization setting up as either an accreditation body or a certification body.
- Both accreditation and certification bodies are commercial organizations that operate without government funding and are therefore governed by supply and demand.

- There is no law requiring all certification bodies to be accredited by national accreditation bodies.
- Certification bodies are not compelled to deploy only those auditors qualified as meeting ISO 19011.
- There are gross differences in the interpretation of requirements between certification bodies.
- There are differences in interpretation of requirements between different offices of the same certification body especially those operating in different countries.
- Auditor registration bodies are not compelled to be accredited to EN 45013.
- Auditors are not compelled to register with an accredited auditor registration body.
- The supplier may choose the certification and the scope of registration.
- The supplier is deemed to be the client of the certification body and not the ISO.
- The customer does not have any power of veto over the issuing or withdrawal of either accreditation or certification.
- The customers have little influence in the training, qualification and selection of auditors.

The certification business has grown enormously in the last 20 years. As of December 2004, The International Accreditation Forum (IAF) recorded 42 accreditation bodies with 77 certification bodies registered with UKAS. The experience of the vehicle manufacturers with ISO 9000 certification led them to question the wisdom of so many certification bodies chasing the same business in a competitive market. The results seemed to indicate that cost reductions by the certification bodies led to a decline in the quality of auditing and that was the opposite of what the vehicle manufactures wanted. The vehicle manufacturers had *not* seen a significant rise in product quality as a result of ISO 9000, and they believed this was partially due to the quality of the accreditation and certification schemes being operated as well as inadequacies in the quality management system standard. When the four national automotive schemes were launched, great emphasis was placed on regulating more closely the accreditation and certification schemes, and from a customer perspective, the ISO 10011 scheme, had some particular problems:

- The competency of auditors in specific industry sectors is not verified – knowledge and experience is all that is necessary.
- The certification bodies adhere to EN 45012, which is a general standard that does not provide for specific industry sectors to tailor the requirements to their needs.
- The accreditation body rather than the industry determine which certification bodies are qualified to issue certificates.
- The certificates issued by the certification body are not subject to independent verification by the industry.
- Audits are carried out on behalf of the organization seeking certification not on behalf of the industry that created the requirements, and hence requirements are prone to an interpretation to suit clients and retain business.

- Although the schemes exist to satisfy the needs of industry, industry has no power to verify that standards are being maintained by the accreditation and certification bodies.

The IATF were intent on eliminating these weaknesses because neither ISO 9000 nor QS-9000 certified organizations were producing products that met the automotive customer's specified requirements for quality, price and delivery.

The ISO/TS 16949 certification scheme

The IATF have designed a certification scheme in which they are the regulator for the automotive-specific requirements, and hence have the ability to ensure greater uniformity in ISO/TS 16949 certification than has been hitherto the case with ISO 9000. The IATF regulates certification bodies and auditors performing ISO/TS 16949 certification audits. As a result 'opportunists' in the certification business, without the required automotive credentials will have nothing to offer except a certificate not recognized by subscribing IATF members. However, there is a penalty as it can also rule out credible certification bodies that do not have a sufficient number of clients in the automotive sector to qualify. The IATF scheme, will remove from the automotive sector third-party auditors who cannot demonstrate their competency to independent examiners. In effect, the scheme puts in place the conditions that require automotive auditors to possess a licence to practice and unlike ISO 9000 auditors, this licence has to be renewed every 3 years.

In brief, the auditor has to be qualified by IATF to perform the audits and to be eligible for qualification, the auditor has to be sponsored by an IATF approved certification body that is accredited for ISO 9001:2000 certification by a national accreditation body. In order to qualify, a certification body must conform to the Rules for Achieving IATF Recognition – a document specifying the conditions under which ISO/TS 16949 certificates are issued. The trade associations that are members of IATF perform an assurance function and have set up the International Automotive Oversight Body (IAOB) to administer accreditation and certification activities. This involves witness audits of certification bodies to verify that they are adhering to the IATF agreement. More information on certification can be found on the IAOB Web site and in the Rules for Achieving IATF Recognition.

Such measures will inevitably improve the quality of certification offered by certification bodies and will be good for the global automotive industry. As of December 2004 there are 53 certification bodies under contract to certify suppliers to ISO/TS 16949:2002. Witness audits have been underway since mid-2000 and have provided a wealth of information. One of the principal findings was that certification body auditors were failing to use the process approach and audit customer-specific requirements.

ISO/TS 16949 certification criteria

It is intended that organizations satisfy all the requirements of ISO/TS 16949, i.e. every sentence containing the word 'shall'. Of these there are 288 that break out into 407 separate requirements. In order to receive ISO/TS 16949:2002 certification the certification body needs to have gathered sufficient evidence to demonstrate compliance with all requirements relevant to the scope of registration. In the event that nonconformity is found it has to be eliminated before a certificate can be awarded. Nonconformities are classified in order to provide a basis for stopping or continuing with the assessment. The lead auditor is permitted to terminate the assessment on detection of a major nonconformity and therefore it is important to know what a major nonconformity is.

A major nonconformity is one or more of the following:

- The absence of or total breakdown of a system to meet an ISO/TS 16949:2002 requirement. A number of minor nonconformities against one requirement can represent a total breakdown of the system and thus be considered a major nonconformity.
- Any non-compliance that would result in the probable shipment of a nonconforming product. A condition that may result in the failure or materially reduce the usability of the products or services for their intended purpose.
- A non-compliance that judgment and experience indicate is likely either to result in the failure of the quality management system or to materially reduce its ability to assure controlled processes and products.

A minor nonconformity is a failure to comply with ISO/TS 16949:2002, which based on judgment and experience is not likely to result in the failure of the quality system or reduce its ability to assure controlled processes or products. It may be one of the following:

- A failure in some part of the organization's documented quality management system relative to ISO/TS 16949:2002.
- A single observed lapse in following one item of a company's quality management system.

The language in these definitions is not consistent with the language used in this book, therefore some interpretation is needed.

The absence of or total breakdown of a system to meet an ISO/TS 16949:2002 requirement implies that one might have not one system but 288 systems as there are 288 'shall' statements in the standard. It could also mean that failure to meet one requirement in one process is a total breakdown, but it is none of these.

The absence of a system means that there is neither any provision within the documented management system nor there is any practice being implemented that complies with the requirement.

A total breakdown of a system means that there is provision within the documented management system that complies with the requirement but no evidence was found of its implementation.

If an auditor finds one example in one process where one requirement ISO/TS 16949 is not being implemented but if he or she were to look elsewhere, evidence of conformity would be found; there is no justification for a major nonconformity unless, nonconforming product would be shipped. Emphasis is on the word 'would' because if there were stages downstream where the nonconformity would be detected it is simply not an issue. Apart from the first definition of a major nonconformity, the others quite sensibly regard all the nonconformities as having an impact on the product.

A failure of the quality management system arises where nonconforming product is shipped. It is not where a signature is missing from a document or some obscure record cannot be found. The focus should always be on results and capability rather than rules.

Regarding minor nonconformities, if you are doing what you say you will do, a minor nonconformity would be justified.

In all the cases you will need to submit root cause analysis with the solution for its elimination meaning that the analysis has to go beyond specific incidents.

ISO/TS 16949 and the EU

This is an extract from EU Directive 92/53/EEC of June 18, 1992 amending the Directive 70/156/EEC governing the type-approval of motor vehicles and their trailers. It must be assumed that Directive will be amended shortly to address ISO 9001:2000.

"The approval authority of a Member State must verify, before granting type-approval, the existence of satisfactory arrangements and procedures for ensuring effective control so that components, systems, separate technical units or vehicles when in production conform to the approved type. The approval authority must also accept the manufacturer's registration to harmonized standard, EN 29002 (whose scope covers the product(s) to be approved) or an equivalent accreditation standard as satisfying the requirements of Point 1.1. The manufacturer must provide details of the registration and undertake to inform the approval authority of any revisions to its validity or scope".

Applications of ISO/TS 16949 requirements

Using ISO/TS 16949

> What you are in effect doing is producing a description of the management system and it is only when the activities described are carried out as prescribed that the system begins to function.

Why 'application' and 'not implementation'? It is primarily because ISO/TS 16949 is an assessment standard. It is not a design standard. It is not intended that you design a quality management using ISO/TS 16949. The management system should be designed around the business and not the standard. As explained in Chapter 1, the management system is the way the organization functions. The terms 'organization' and 'management system' are synonymous. Both are created to fulfil goals and if the management system is created to fulfil different goals than the organization, there will be a continual conflict. This has been the problem with ISO 9000-based management systems, as was explained in Chapter 1. So, how then is ISO/TS 16949 applied? Quite simply, it is applied after you have designed your management system in order to verify that the system satisfies the relevant constraints.

It would be completely the wrong thing to do, to take each requirement of the standard and produce a written statement to match it. Just ask yourself, "What good would it do?" It is very unlikely that it would change anything although many systems have been documented in this way. Very few, if any organizations start to implement ISO/TS 16949 on a 'Greenfield site', – the organization has usually been in operation for some years and is producing product that some customers want; otherwise it would be a dormant business. Therefore, the best place to start is to define how the organization currently goes about its business but not by asking everyone to document what they do, which is purposeless. Do this in a structured manner. Follow the suggestion in Chapter 1 under the heading *Developing an effective system.*

Capture what you can, do not worry that you cannot answer some of the questions. The reviews will identify gaps and in filling them you will build a system using a process of progressive iteration.

Responding to ISO/TS 16949 requirements

The requirements of ISO/TS 16949 cannot be taken in isolation. There are specific cross-references as well as implied linkages through the wording as well as the duplication (see below). The document was designed as an assessment tool; therefore after the system description has passed through the first iteration you can ask, "Where in the system would you find activities being carried out in a way that satisfies this requirement in ISO/TS 16949?" Create a table in which you can insert the responses. If you find a related activity being carried out but not exactly like it is prescribed in the standard, ask yourself "Does the way we currently perform this activity produce an output that meets the customer requirement?" If it does, simply respond with a brief description of the way it is currently carried out. If the output produced does not meet the customer requirements, you have identified an opportunity for improvement that can be entered into an action plan. (Transition Support produces a tool for collecting responses to the requirements that has provision for generating an action plan and tracking actions.)[5]

Anomalies in ISO/TS 16949

ISO/TS 16949 contains a number of anomalies that users need to be aware of when attempting to respond to its requirements. Many of the automotive-specific requirements in ISO/TS 16949 could be derived from ISO 9001:2000; and therefore, when responding to both the ISO 9001 requirement and the TS requirement, you might think that two different responses are required when in fact the ISO 9001 response may well satisfy the TS requirement. In other cases, requirements appear to be duplicated often for clarity but it might give the impression that there are several requirements and not one requirement simply expressed differently. Some requirements only apply when specified in a contract; therefore if the customer has not written anything specifically in your contract on that topic, the requirement would not apply and so your response is simply '*not applicable*'.

Contract-specific requirements

The following requirements duplicate other requirements that would be stated in a contract:

Clause 7.1.2 requires approval of acceptance criteria wherever required.
Clause 7.3.6.2 requires a prototype program when required by the customer.
Clause 7.4.1.3 requires the organization to purchase products, materials and services from approved sources where specified in a contract.
Clause 8.5.2.4 requires correction action on rejected parts thus duplicating the general requirements for corrective action in Clause 8.5.2.

In all these cases the requirements add nothing to what would be stated in a contract unlike Clause 7.5.1.8, which says that when there is a servicing agreement, the organization shall verify its effectiveness.

Duplicated requirements

The 'Note' under Section 7.3 on design development applies the whole of Section 7.3 to manufacturing processes. However, it is odd that having inserted this note, the authors of ISO/TS 16949 proceeded to ignore it by adding such clauses as manufacturing process design inputs and manufacturing process design outputs. Could it be that they did not trust both suppliers and certification bodies to apply the requirements properly? This patchy treatment of Clause 7.3 gives the impression that perhaps the IATF expect process design inputs and outputs to be defined, but do not expect process design reviews as well as product design reviews or process design planning as well as product design planning. It might escape the eagle eye that while process verification and validation is not addressed in Clause 7.3, it finds a different place under Clause 8.2.3.1 where, if the 'Note' is to be believed, the requirements of Clauses 7.3.6 and 7.3.7 are duplicated, albeit in a different form but certainly having the same intent.

The authors of ISO 9001 did not place all the monitoring and measuring requirements in Section 8, and left a few in Section 7, such as design verification and validation. Was it an oversight or was it deliberate? There was a desire to place all the requirements where exceptions could be permitted in one section, thus conflicting with the process model they had so carefully constructed. Measurement is part of the realization process but the measurement requirements were placed in Section 8 because they cannot be excluded, but not all of them. The 'Note' making 7.3 apply to manufacturing processes meant that Section 1.2 had to modify the ISO 9001 exclusion so that none of Section 7.3 could be excluded from the manufacturing processes.

The requirement for control plans in Clause 7.5.1.1 is partially duplicated by the requirements for monitoring and measurement of product in Clause 8.2.4.

Order of requirements

The order in which the requirements of ISO TS 16949 are laid out in the standard should not be interpreted as being the order in which these requirement are to be fulfilled, or in fact indicative of being of any significance at all. For example, product and process approval is addressed in Clause 7.3.6.3. However, prior to process approval, process validation needs to be performed but requirements for this are not mentioned until one gets to Clause 7.5.2. Control plans, work instructions and other requirements are placed in Section 7.5 on production, when in reality the implementing actions are performed prior to production. Corrective action might be placed in Section 8 but you would not be able to control production or design, or purchasing processes unless corrective action is carried out within every process.

Exclusions

Although ISO 9001 does permit exclusions from Section 7 the only exclusion permitted from Section 7 in ISO/TS 16949 is design and development where the organization is not responsible for product design. Process design and development cannot be excluded.

Summary

In this chapter we have examined the history of quality-system requirements in the automotive industry that has led to the creation of ISO/TS 16949, its role, its purpose, scope and its application. We have positioned the *standard* as a vehicle for conveying requirements to suppliers and as an assessment tool rather than a design tool. We have done this deliberately in order to encourage organizations to develop their management systems around their business rather than the standard – simply because this is the way it was intended to be and because doing it the other way would not add any value. As portrayed in Chapter 1, basing your system on the standard simply produces a bolt-on system that any competent certification body would spot immediately as being totally ineffective. It is better to have a system that reflects how your organization functions and is non-compliant than one that is a mirror image of the standard and still non-compliant!

It is not ISO/TS 16949 that will cause improvement in business performance. ISO/TS 16949 is a standard and this type of standard cannot cause people to do the right things right first time. People produce the products and services supplied by organizations and standards merely serve to guide these people in choosing the right things to do. It is the quality of leadership in an organization that creates an environment in which people will do the right things right without having to be told.

**Customer-focused leadership
produces satisfied customers not standards.**

Role, origins and application – Food for thought

1 Do not simply go for the badge on the wall.

2 If you are going to use ISO/TS 16949, use it as an aid not a slave.

3 Go for certification only if it is essential for you to win business.

4 Do not label things or positions within your organization with ISO 9000 or ISO/TS 16949; it gives the impression that they exist only to serve the standard.

5 Do read and use all the referenced standards in ISO/TS 16949.

6 Do not get hung up on the words, seek the purpose of the requirement that transcends the words.

7 Do not assume the terms are used in the same way as you use them, apply the terms as they are defined in ISO 9000 and ISO/TS 16949.

8 Do not impose ISO 9001 on your suppliers unless you cannot verify the quality of the goods by yourself.

9 Do not rely on ISO 9001 certificates to give you an assurance of quality, check them out.

10 Do not assume that a non-accredited certification body is any worse than an accredited one; check them out.

11 Do not create separate systems, there is only one.

12 Do not assume that because a clause is in Section 7 you can exclude it, would it impact the organization if you did not do this?

13 Do not forget that contract review has gone, it is now product review and not dependent on you having a contract.

14 Do use the process approach to develop and audit your system.

15 Do not ask what benefit do we get from implementing ISO/TS 16949, ask what benefit do we get from managing our processes effectively.

16 If you choose to withdraw from ISO/TS 16949 certification, do not forget that you still need to manage your processes effectively.

Chapter 4

Quality management system

A system must have an aim. A system must create something of value, in other words results. Management of the system requires knowledge of the interrelationships between all the components within the systems and of the people that work in it.
W. Edwards Deming

Summary of requirements

Section 4 of ISO/TS 16949 contains the basic requirements for establishing a management system rather than any particular component of the system. In some instances they are duplicated in other clauses of the standard but this is not a bad thing because it emphasizes the principle actions necessary to develop, implement, maintain and improve such a system. Unlike previous versions, the focus has moved away from documentation towards processes and therefore these general requirements capture some of the key activities that are required to develop an effective system.

Although the clauses in Section 4 are not intended as a sequence there is a relationship that can be represented as a cycle, but first we have to lift some clauses from Section 5 to commence the cycle. The words in bold italics indicate the topics covered by the clauses within Sections 4 and 5 of the standard. The cycle commences with the **organization's purpose** (Clause 5.3 requires the quality policy has to be appropriate to the organization's purpose) through which are passed **customer requirements** (Clause 5.2 requires customer requirements to be determined) from which are developed **objectives** (Clause 5.4.1 requires objectives to be consistent with the quality policy). In planning to meet these objectives the **processes are identified** and their **sequence and interaction** determined. Once the relationship between processes is known, the **criteria and methods** for effective operation and control can be developed and **documented**. The processes are described in terms that enable their effective communication and a suitable way of doing this would be to compile the process descriptions into a *quality manual* that not only references the associated *procedures* and *records* but also shows how the processes interact. Before implementation the processes need to be *resourced* and the *information* necessary to operate and control them deployed and brought under *document control*. Once operational the processes need to be *monitored* to ensure they are functioning as planned. *Measurements* taken to verify that the processes are delivering the required output and actions taken to *achieve the planned results*. The data obtained from monitoring and measurement that is captured on *controlled records* needs to be *analysed* and opportunities for *continual improvement* identified and the agreed actions *implemented*. Here we have the elements of the process-development process that would normally be part of mission management but that process is largely addressed in the standard through Management Responsibility.

If every quality management system reflected the above linkages the organization's products and services would consistently satisfy customer requirements.

Establishing a quality management system (4.1)

The standard requires the organization *to establish a quality management system in accordance with the requirements of ISO/TS 16949.*

What does this mean?
To *establish* means to set up on a permanent basis, install, or create and therefore in establishing a management system, it has to be designed, constructed, resourced, installed and integrated into the organization signifying that a management system on paper is not a management system.

Establishing a system in accordance with the requirements of ISO/TS 16949 means that the characteristics of the system have to meet the requirements of ISO/TS 16949. However, the requirements of ISO/TS 16949 are not expressed as system requirements of the form "The system shall ..." but are expressed as organization requirements of the form "The organization shall ...". It would appear, therefore, that the system has to cause the organization to comply with the requirements and this will only happen if the system is the organization. Some organizations regard the management system as the way they do things but merely documenting what you do does not equate with establishing a system for the reasons given in Chapter 1 under the heading *Quality management systems.*

Why is this necessary?
This requirement responds to the System Approach Principle.

ISO/TS 16949 contains a series of requirements which if met will provide the management system with the capability of supplying products and services that satisfy the organization's customers. All organizations have a management system – a way of working but in some it is not formalized, in others it is partially formalized but not effective and in a few organizations the management system really enables its objectives to be achieved year after year. In such organizations, a management system has been established rather than evolved and if an organization desires year-after-year success, it needs a formal mechanism to accomplish this – it won't happen by chance or even by the brute force and determination of one man or woman – it requires an effective management system. This requires management to think of their organization as a system, as a set of interconnected processes that include tasks, resources and behaviors as explained in Chapter 1.

How is this implemented?

The terms 'establish', 'document', 'implement', 'maintain' and 'improve' are used in the standard as though this is a sequence of activities when in reality, in order to establish a system it has to be put in place and putting a system in place requires a number of interrelated activities that can be grouped into four stages of a System Development Project as outlined below and addressed in more detail in Chapter 1. These are consistent with the requirements of Clause 4.1 of ISO/TS 16949 and Clause 2.3 of ISO 9000 that address the Systems Approach.

Establish the goals

1 Clarify the organization's purpose, mission and vision (goals): this is what the organization has been formed to do and the direction in which it is proceeding in the short and long term. It is the 'Organization Purpose' that is referred to in Clause 5.3a) of the standard.
2 Confirm the values and principles that will guide the organization towards its goals: these are addressed by Clause 5.3 of the standard and in Chapter 5 under the heading *Quality policy*.
3 Identify stakeholder needs relative to the purpose, mission and vision: this is addressed by Clause 5.2 of the standard and in Chapter 5 under the heading *Customer focus*.
4 Identify stakeholder satisfaction indicators relative to these needs: these are addressed by Clause 5.4.1 of the standard and in Chapter 5 under the heading *Quality objectives*.

Develop the processes

5 Identify the processes that will achieve these objectives: this is addressed further on in this chapter under the heading Identifying processes.

6   Design processes to achieve these objectives: this is addressed further on in this chapter under a range of headings.
7   Install and resource the processes: this is addressed below.
8   Operate the processes as designed: this is addressed further on in this chapter under the heading *Implementing a quality management system and Managing processes*.

Review performance

9   Review the processes to establish whether the required objectives are being achieved, whether they are being achieved in the most effective way and whether they remain relevant to stakeholder needs; this is addressed further on in this chapter under the heading Maintaining a quality management system and in Chapter 5 under the heading Management review.

Improve capability

10 Change processes to bring about improvement by better control, better utilization of resources and better understanding of stakeholder needs. This is addressed further on in this chapter under the heading *Continual improvement in the quality management system and its processes*.

Identifying processes (4.1a)

The standard requires the organization *to identify the processes needed for the quality management system and their application throughout the organization.*

What does this mean?

Processes produce results of added value. Processes are not procedures (see Chapter 1 on *Basic concepts*). The results needed are those that serve the organization's objectives. Processes needed for the management system might be all the processes needed to achieve the organization's objectives and will therefore form a chain of processes from corporate goals to their accomplishment. However, as we discovered in Chapter 1, the business processes need to add value for the interested parties. The chain of processes is a *value chain* and therefore should extend from the needs of the interested parties to the satisfaction of these needs. This is illustrated by the process model of Figure 1.15.

The phrase "needed for the quality management system" in this requirement implies that there are processes that are needed for other purposes. It also implies that the quality management system is something that needs processes rather than something that comprises processes. The phrase is indeed ambiguous and needs rewording. A better way of expressing the intent would be to say "identify the processes needed to

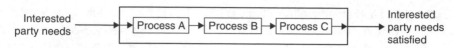

Figure 4.1 *End-to-end processes*

enable the organization to achieve its objectives and satisfy its customers". One might go further and suggest it should be 'all stakeholders', but ISO/TS 16949 limits requirements to customers as one of the stakeholders.

Traditionally work has been organized into functions of specialists, each performing tasks that serve functional objectives. This has the effect of sub-optimization: pursuit of local goals at the expense of organizational goals. By thinking of the organization as a collection of interconnected processes rather than a series of interconnected functions, each focused on the needs of interested parties, the chain of processes cuts across the functions. This is not to say that functions are unnecessary but we must recognize that work does not flow through functions, it flows across functions by means of a process. The processes needed for the management system are those processes with a purpose that is aligned to the organization's objectives (see Figure 4.1).

As illustrated in Table 1.7, the processes identified in the Generic System Model clearly show the same stakeholder at each end.

Why is this necessary?
This requirement responds to the Process Approach Principle.

The management system consists of a series of interconnected processes and therefore these processes need to be identified.

How is this implemented?
As the standard is not specific to the types of processes, we must assume it is all processes and this would therefore include both business processes and work processes. This distinction is important for a full appreciation of this requirement. The relationship between these two types of processes is addressed further in Chapter 1.

In many cases, organizations have focused on improving the work processes believing that as a result there would be an improvement in business outputs but often such efforts barely have any effect. It is not until you stand back that the system comes into view. A focus on work processes and not business processes is the primary reason why

ISO 9000, TQM and other quality initiatives fail. They resulted in sub-optimization, not optimization of the organizational performance. If the business objectives are functionally oriented, they tend to drive a function-oriented organization rather than a process-oriented organization. Establish process-oriented objectives, measures and targets focused on the needs and expectations of external stakeholders; the functions will come into line and you will be able to optimize organizational performance. When Texas Instruments re-engineered its processes in 1992, their process map showed only six processes.[1] Hammer remarks that hardly any company contains more than 10 or so principle processes.

There are several ways of identifying processes but the two that we will discuss here are the top-down approach and the bottom-up approach.

Top-down approach

From the organization's mission statement identify the stakeholders and their needs and expectations relative to this mission. Now determine the outputs needed to satisfy the stakeholders and group these outputs according to the following criteria:

1 Outputs that create demand for the organizations products and services.
2 Outputs that satisfy a demand for the organizations products and services.
3 Outputs that provide resources to the organization for creating or satisfying a demand.
4 Outputs that establish the goals and strategy for the organization, establish the enabling processes and review performance.

This will result in four business processes:

1 The group of outputs that create demand will be delivered by a Demand Creation Process. These are considered to be customer-oriented processes or COPs by the International Automotive Task Force (IATF).
2 The group of outputs that satisfy a demand will be delivered by a Demand Fulfillment Process. These are considered to be customer-oriented processes or COPs by the IATF.
3 The group of outputs that provide resources will be delivered by a Resource Management Process. These are considered to be support-oriented processes or SOPs by the IATF.
4 The group of outputs that establish goals, strategy, etc. will be delivered by a Mission Management Process. These are considered to be management-oriented processes or MOPs by the IATF.

Customer Oriented Processes have a customer at either end. A diagram called the octopus is used on IATF in Auditor Training to illustrate this concept (see Figure 4.2).

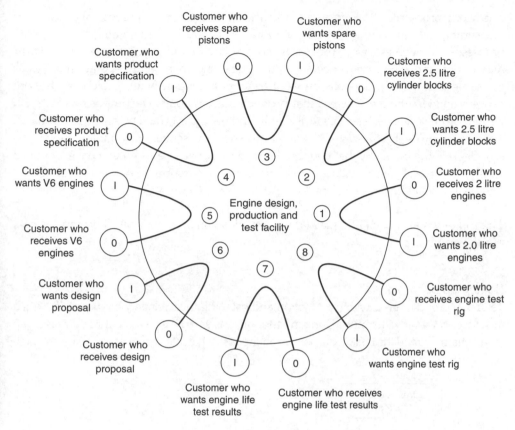

Figure 4.2 *The octopus Customer Oriented Processes*

In order to determine the key stages in each of these business processes you need to identify the factors on which delivery of the outputs depend – these are the critical success factors (CSF). The CSFs indicate the capabilities needed and consequently identify the processes required to delivery these capabilities. In order to fulfil demand, an organization might need a design capability, a production capability and a distribution capability. This results in a need for a design process, a production process and a distribution process. These are work processes in the Demand Fulfillment Process. These work processes depend on an available supply of capital, competent staff, equipment and materials, and well-equipped facilities. Delivery of this capability requires a capable resource management process that manages human, physical and financial resources. Another CSF would be the ability of the organization to identify customer needs and expectations in its chosen markets and consequently a need for an effective marketing process is identified. A further CSF would be the rate at which customer enquiries were converted into sales and consequently, a need for an effective sales process is identified. By identifying CSFs and the associated processes, a full list of work processes will emerge. One result of using this approach is given in Figures 1.17 and 1.18.

The work processes sit along a chain in the related business process and each has measurable outputs. However, not all work processes are triggered by other work processes. Sometimes a work process is triggered by a condition, a date, or a particular type of demand. The work processes in the chain shown in Figure 4.1 are not always the same processes, it depends on what triggers the input. Now, analyze each work process and identify the activities that produce these outputs. Decomposing the layers further, by analyzing each activity, will identify the tasks performed by specific individuals. The tasks may be illustrated in a flow chart to show the sequence and interaction. This is as deep in the hierarchy that you should need to go because when you have identified a task performed by an individual, the methods can be described in procedures or guides. The decomposition can therefore be illustrated as in Figure 4.3.

The top-down approach is explained in more detail in an e-book, *A Guide to Process Management*.[2]

Bottom-up approach

Take any group of tasks within a function and establish their relationships. Identify which task feeds another with inputs. Bring in other groups when it is realized that there is a gap in the chain. Extend the chain until you reach the tasks that interface with the external party

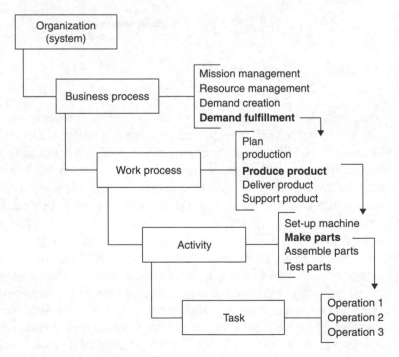

Figure 4.3 *System decomposition*

(customer, supplier, regulator, etc.). This approach can commence at the functional level and grow outwards until all groups are represented in the chain. In taking this approach, you may well find some tasks do not link with any other tasks or an external party and should be marked for action because clearly they appear to add no value. However, a task may not add value to one external party but may add value to another. If you focus only on customers, you will find a number of tasks that do not contribute to the achievement of customer needs but do contribute to satisfying the needs of other interested parties.

Now examine the chains and group those together that achieve a common objective or deliver a common output. You can call these groups 'processes'. The processes you identify may not be like those of any other organization – they don't have to be, as every organization is different, even those that appear to be in the same business. The processes you identify might include:

Work processes

Accounting	Internal auditing	New product planning	Promotion
Advertising	Inventory management	New product trials	Purchasing
Asset management	IT Infrastructure	Order processing	Recruitment
Business planning	maintenance	Organization	Resource planning
Calibration	IT Infrastructure	development	Service design
Design analysis	planning	Packing	Staff development
Despatch	Manufacturing	Plant maintenance	Strategic planning
Distribution	Manufacturing process	Process improvement	Tendering
Enquiry conversion	design	Product audit	Tooling design
Goods receiving	Market research	Product design	Training
Installation	Materials development	Production planning	Waste disposal

You could now ask, what do these processes have in common, and then put all those having the same things in common into a new set. There is a remote chance that you would discover the same four business processes as using the top-down method but it is, unlikely, simply because the focus is different.

The bottom-up approach involves everyone but has some disadvantages. As the teams involved are focused on tasks and are grouping tasks according to what they perceive are the objectives and outputs, the result might not align with the organizational goals; these groups may not even consider the organization goals and how the objectives they have identified relate to these goals. It is similar to opening a box of components and stringing them together in order to discover what can be made from them. It is not very effective if one's objective is to satisfy the external customer; therefore, the top-down approach has a better chance of linking the tasks with the processes that will deliver customer satisfaction.

Naming processes

"Identifying processes" means more than just naming them. A name is a label that triggers perceptions, some of which may differ from what we intended. It is therefore necessary to define the processes in terms that will form a clear understanding of the boundaries, purpose and outputs of each process. The above list identifies groups of processes and therefore there may be several processes that deliver marketing objectives, resource objectives and other objectives.

However, naming is important in order to avoid confusion between processes and functions. There are two conventions used for naming processes one is verb focused and the other is noun focused. The verb-focused construction of a process name commences with a verb, such as produce, define and acquire. The noun-focused construction may simply be one word, such as marketing, design and production. This distinction is illustrated in Figure 4.4. The noun-focused convention can give the impression that the chain of processes is a chain of functions having these titles, when in reality several functions will contribute to each process. In order to avoid the confusion, it may be more appropriate to use the verb-focused convention. In Figure 4.4, both product and service are treated differently because in practice the chain of end-to-end processes is different.

In every organization there are sets of activities but each set or sequence is not necessarily *a process*. If the result of a sequence of activities adds no value, continue the sequence until value is added for the benefit of customers – then you have defined a *business* or *work process*.

Another technique is IDEF.[3] (Integrated Definition) developed by the US Air Force. The objective of the model is to provide a means for completely and consistently modeling the functions (activities, actions, processes and operations) required by a system or enterprise,

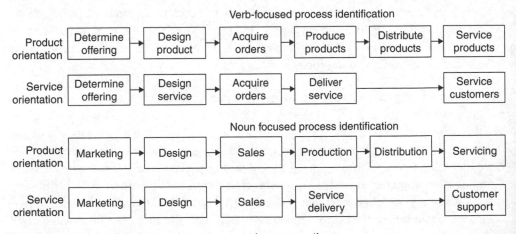

Figure 4.4 *Distinction between process naming conventions*

and the functional relationships and data (information or objects) that support the integration of those functions. IDEF is a systematic method of modeling that can reveal all there is to know about a function, activity, process or system. Considering its pedigree, it is more suited to very complex systems but can result in *paralysis from too much analysis!* Many management systems do not require such rigorous techniques. There can be a tendency to drill down through too many layers such that at the lowest level you are charting movements of a person performing an activity or identifying pens and pencils in a list of required resources. For describing the management system processes, it is rarely necessary to go beyond a task performed by a single individual. As a rough guide you can cease the decomposition when the charts stop being multifunctional.

Sequence and interaction of processes (4.1b)

The standard requires the organization to determine the sequence and interaction of the identified processes.

What does this mean?

Sequence refers to the order in which the processes are connected to achieve a given output. Interaction refers to the relationship between the processes and their dependencies and the source of inputs and destination of the outputs. The system model shown in Figure 2.15 illustrates the sequence and interaction of the business processes.

Why is this necessary?

This requirement responds to the System Approach Principle.

Objectives are achieved through processes, each delivering an output that serves as an input to other processes along a chain that ultimately results in the objective being achieved. It is therefore necessary to determine the sequence of processes. Some will work in parallel; others in a direct line but all feeding results that are needed to accomplish the objective. There will therefore be interactions between processes that need to be determined.

How is this implemented?

A practical way to show the sequence of processes is to produce a series of flow charts. However, charting every activity can make the charts appear very complex but by layering the charts in a hierarchy, the complexity is reduced into more digestible proportions.

Many processes will not only require inputs from other processes to start, but will require other inputs or conditions for tasks to be executed. For example, a verification process requires trained people, calibrated equipment and perhaps certain environmental conditions, and therefore relies on the Resource Management Process to deliver trained people, calibrated equipment and facilities in which the environment is maintained. The interface between the verification process and resource management process creates an interaction when the processes are active. The reliance on resource management to provide inputs creates a dependency. A verification process will also require documentation and therefore relies on the document control process feeding only up-to-date documents. The verification process has outputs and therefore relies on material handling processes and data storage processes to take away the outputs. Some of the interactions occur on demand and are therefore dynamic, others are passive and are often taken for granted but without which the process cannot deliver its required output. To test whether you have identified all the interactions, just ask yourself what would happen if a particular condition were not available – would the process still be able to deliver the required outputs?

Criteria and methods for effective operation and control (4.1c)

The standard requires the organization *to determine criteria and methods required to ensure the effective operation and control of the identified processes.*

What does this mean?
The criteria that ensure effective operation are the standard operating conditions, the requirements, targets or success criteria that need to be met for the process to fulfil its objectives.

The methods that ensure effective operation are those regular and systematic actions that deliver the required results. In some cases the results are dependent on the method used and in other cases, any method might achieve the desired results. Use of the word 'method' in this context is interesting. It implies something different than had the standard simply used word 'procedure'. Procedures may cover both criteria and methods but have often been limited to a description of methods. Methods are also ways of accomplishing a task that are not procedural. For example, information may be conveyed to staff in many ways – one such method might be an electronic display that indicates information on calls waiting, calls completed and call response time. The method of display is not a procedure although there may be an automated procedure for collecting and processing the data.

Why is this necessary?
This requirement responds to the Process Approach Principle.

A process that is operating effectively delivers the required outputs of the required quality, on time and economically, while meeting the policies and regulations that apply to the process. A process is not effective if it delivers the required quantity of outputs but the outputs do not possess the required characteristics, are delivered late, waste resources, breach policy and safety, environmental or other regulations. It is therefore necessary to determine the criteria for the acceptability of the process inputs and process outputs, and the criteria for acceptable operating conditions. Thus it is necessary to ascertain the characteristics and conditions that have to exist for the inputs, operations and outputs to be acceptable.

How is this implemented?

In order to determining the criteria for effective operation and control, you need to identify the factors that affect success. Just ask yourself the question: What are the factors that affect our ability to achieve the required objectives or deliver these outputs? In a metal machining process, material type and condition, skill, depth of cut, feed and speed affect success. In a design process input requirement adequacy, designer competency, resource availability and data access affect success. In an auditing process, objectives, method, timing, auditor competency, site access, data access and staff availability affect success. In a computer-activated printing process, the critical factor may be the compatibility of the input data with the printer software, the format of the floppy disk and the resolution of the image. There are starting conditions, running conditions and shutdown conditions for each process that need to be specified. If any one of these goes wrong, and whatever the sequence of activities, the desired result will not be achieved.

Determining the methods can mean, determining the series of actions to deliver the results or simply identifying a means to do something. For example, there are various methods of control:

- Supervisors control the performance of their work groups by being on the firing line to correct errors.
- Automatic machines control their output by in-built regulation.
- Manual machines control their output by people sensing performance and taking action on the spot to regulate performance.
- Managers control their performance by using information.

The method is described by the words following the word 'by' as in the above list. A method of preventing failure is by performing a FMEA. FMEA was developed by the aerospace industry, in the mid-1960s and first used by Ford Motor Company, in 1972. You don't have to detail how such an analysis is performed to have determined a method. However, in order to apply the method effectively, a procedure or guide may well be needed. The method is therefore the way the process is carried out which together with the criteria contributes to the description of the process.

Documenting a quality management system (4.1, 4.2.1)

The standard requires the organization *to document a quality management system in accordance with the requirements of ISO/TS 16949.*

What does this mean?
A document (according to ISO 9000 Clause 2.7.1) is *information and its supporting medium.* A page of printed information, a CD ROM or a computer file is a document, implying that recorded information is a document and verbal information is not a document. Clause 4.2 requires the management system documentation to include certain types of documents and therefore does not limit the management system documentation to the types of documents listed.

As a management system is the means to achieve the organization's objectives, and a system is a set of interrelated processes, it follows that what has to be documented are all the processes that enable the organization to achieve its objectives.

While there is a reduction in emphasis on documentation in ISO/TS 16949:2002 compared with the 1999 version, it does not imply that organizations will need less documentation to define their management system. What it means is that the organization is left to decide the documentation necessary for effective operation and control of its processes. If the absence of specific documentation does not adversely affect operation and control of processes, such documentation is unnecessary.

Why is this necessary?
This requirement responds to the System Approach Principle.

Before ISO 9000 came along, organizations prospered without masses of documentation and many still do. Those that have chosen not to pursue the ISO 9000 path often only generate and maintain documents that have a useful purpose and will not produce documents just for auditors unless there is a legal requirement. Most of the documentation that is required in ISO 9000 and ISO/TS 16949 came about from hindsight; the traditional unscientific way organizations learn and how management systems evolve.

ISO/TS 16949 contains a list of valid reasons for why documents are necessary and here is a list used in previous editions of this handbook:

* To communicate requirements, intentions, instructions, methods and results effectively.
* To convert solved problems into recorded knowledge so as to avoid having to solve them repeatedly.

- To provide freedom for management and staff to maximize their contribution to the business.
- To free the business from reliance on particular individuals for its effectiveness.
- To provide legitimacy and authority for the actions and decisions needed.
- To make responsibility clear and to create the conditions for self-control.
- To provide co-ordination for inter-departmental action.
- To provide consistency and predictability in carrying out repetitive tasks.
- To provide training and reference material for new and existing staff.
- To provide evidence to those concerned of your intentions and your actions.
- To provide a basis for studying existing work practices and identifying opportunities for improvement.
- To demonstrate after an incident the precautions which were taken or which should have been taken to prevent it or minimize its occurrence.

If only one of these reasons make sense in a particular situation, the information should be documented. In some organizations a view is taken that it is important to nurture freedom, creativity and initiative, and therefore there is a feeling that documenting procedures is counterproductive. Their view is that the documented procedures hold back improvement, forcing staff to follow routines without thinking and prevent innovation. While it is true that blindly enforcing procedures that reflect out-of-date practices coupled with bureaucratic change mechanisms is counter productive, it is equally short-sighted to ignore past experience, ignore decisions based on valid evidence and encourage staff to reinvent what were perfectly acceptable methods. Question by all means, encourage staff to challenge decisions of the past, but encourage them to put forward a case for change. That way it will cause them to study the old methods, select the good bits and modify the parts that are no longer appropriate. It is often said that there is nothing new under the sun – just new ways of packaging the same message. The *six sigma* initiative is a case in point (see Chapter 2). It is no different to the quality improvement programs of the 1980s or in fact the teachings of Shewart of the 1930s and Juran and Deming of the 1950s – it is just a different way of packaging the message.

How is this implemented?
Before deciding to document anything, there is something more important to consider. Documents, as stated previously are *information and its support medium*. The important matter is the transmission of information rather its documentation and therefore, the first thing you need to decide is the best method of transmitting the information. There are several choices:

- *Convey it through documents*: suitable for information that needs to be referred to when performing a task.
- *Convey it verbally*: suitable for instructions intended for immediate action.

- *Convey it so that it is observed visually*: suitable for warnings.
- *Convey it through education*: suitable for values, beliefs and principles.
- *Convey it through training*: suitable for methods and routines.
- *Convey it through example*: suitable for values, beliefs and methods.

There is no right or wrong answer; it depends on the simplicity or complexity of the information being transmitted and the degree of spontaneity required by the user. Written instructions on how to fire a gun are of no use in battle – you have to be trained. Values, such as honesty, integrity has to be internalized as a set of morals – no procedure would cause you to behave differently.

Clause 4.2.1 requires management system documentation to include five types of document:

(a) Quality policy and objectives
(b) Quality manual
(c) Documented procedures
(d) Documents needed to ensure the effective planning, operation and control of processes
(e) Records.

This list is somewhat inadequate for our purposes because it does not tell us what types of things we should document or provide criteria to enable us to decide what we need to document.

Obviously the size, type and complexity of the organization and the competency of the personnel will have an effect on the depth and breadth of the documentation but the subject matter other than that which is product, process or customer specific is not dependent on size, type and complexity of the organization, etc.

Control of anything follows a universal sequence of tasks illustrated by the generic control model in Figure 4.5.

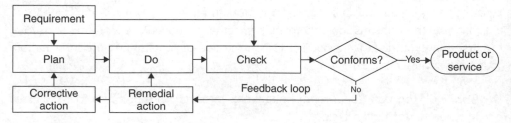

Figure 4.5 *Generic control model*

By asking some key questions derived from the model we can reveal the essential documents:

- What do we have to do? ... A statement of requirements, objectives and success criteria.
- How will we make it happen? ... A plan of action or work to be undertaken and preparation to be made.
- What provisions have we made to avoid failure? ... We have a risk assessment report showing the actions taken.
- What did we do? ... A record of the work carried out.
- How will we know it's right? ... A definition of the measurements that are to be performed.
- How can we prove its right? ... A record of the results of measurement.
- What gives us confidence in the result? ... The fact that we used a soundly based measurement technique or a calibrated device.
- What did we decide to do with the rejects? ... A plan of the remedial action to be taken.
- How do we know the reject was fixed? ... A record of the remedial action taken.
- How will we stop it happening again? ... A plan of the corrective action to be taken.
- How do we know it won't happen again? ... A record testifying to the effectiveness of the corrective action taken.

There are many subsidiary questions, each exploring the process in more detail. In the poem *"The Elephant Child"*, the English writer Rudyard Kipling wrote "I had six honest serving friends, they taught me all I knew, they are what, where, why, when, how and who." This I have called Kipling's Law because it is not just the words but also what can be obtained by using them. However, the six friends also need a trigger in order to focus on an object and this is where the process model in Figure 2.13 is useful. For each entry and exit apply Kipling's Law and establish Who does What, When, Where, Why and How such that you answer the questions in Table 4.1. The answers may not necessarily refer to whole documents. For example, in response to the question "where do the resources come from?" the answer might be found on a flow diagram in the form of an arrow linking two boxes.

The generic control model could be applied at four levels:

1 *At the system level*: so as to reveal the system requirements, system measurements and system changes and thus identify business requirements, business plans, business performance reviews and changes in the business strategy and structure. The system level is at the level of the whole organization.
2 *At the business process level*: so as to reveal the process requirements, process measurements and process changes and thus identify customer requirements, quality plans, compliance checks and customer feedback. The process level is at

Table 4.1 *Applying Kipling's law*

Inputs	Resources	Activities	Constraints	Outputs
What are the inputs?	What resources are required?	What activities are performed?	What are the constraints?	What are the outputs?
Where do they come from?	Where do they come from?	Where are they performed?	Where in this process are they applied?	Where do they go?
Who supplies them?	Who supplies them?	Who performs the activities?	Who imposes the constraints?	Who receives them?
How are they supplied?	How are they supplied?	How are the activities performed?	How are they addressed?	How are they supplied?
When are they supplied?	When are they supplied?	When are the activities performed?	When do they apply?	When are they supplied?
Why are they needed?	Why are they needed?	Why are the activities performed?	Why are the constraints necessary?	Why are they needed?

    the level of the functions that make up the business that may be discharged through one or more departments.

3   *At the activity level*: so as to reveal the activity requirements, activity measurements and activity changes thus identify the product requirements, product plans, inspections, test, reviews and modification actions. The activity level is at the level of a group or department that may assign people to perform the various activities that make up the activity.

4   *At the task level*: so as to reveal the task requirements, task measurements and task changes and thus identify work instructions, workplans, checks and rework actions. The task level is at the level of the individual.

Documented procedures (4.2.1c)

The standard requires the management system documentation *to include documented procedures required by ISO/TS 16949*.

What does this mean?
ISO 9000 defines a procedure as *a specified way to carry out an activity or a process*. This definition is ambiguous because an activity is on a different scale rather than a process. Process outputs are dependent on many factors of which activities are but one.

Several tasks accomplish an activity and several activities create a work process but there is more to a process than a series of activities as addressed in Chapter 1. This definition also results in a belief that procedures are documented processes but this, too, is inaccurate. Procedures tell us how to proceed; they are a sequence of steps to execute a routine activity and result in an activity or a task being performed regardless of the result.

There are very few procedures actually required by the standard but this does not imply you don't need to produce any others. The seven specific procedures required are a documented procedure for:

(a) document control,
(b) the control of records,
(c) identifying training needs,
(d) conducting audits,
(e) nonconformity control,
(f) corrective action,
(g) preventive action.

These areas all have something in common. They are what the authors of the early drafts of ISO 9000:2000 referred to as system procedures. They apply to the whole system and are not product, process or customer specific although it is not uncommon for customers to specify requirements that would impact these areas. Why procedures for these aspects are required and not for other aspects of the management system is unclear but it seems that the authors of ISO 9000, felt these were not processes or activities but tasks – a conclusion I find difficult to justify. They are certainly not business processes but could be work processes or activities within a work process. However, there is another message that this requirement conveys. It is that procedures are not required for each clause of the standard. Previously, countless organizations produced a manual of 20 procedures to match the 20 elements of ISO/TS 16949:1999 or the 23 elements of QS-9000. Some limited their procedures to the 26 procedures cited by the standard and others produced as many as were necessary to respond to the requirements.

Why is this necessary?
This requirement responds to the Process Approach Principle.

It is uncertain why the authors of ISO 9001:2000 deemed it necessary to require any specific procedures when there is a general requirement for the system to be documented. This should have been sufficient. Clearly procedures are required so that people can execute tasks with consistency, economy, repeatability and uniformity but there is no logical reason why procedures are required for only the above six subjects.

How is this implemented?

One solution is to produce the six procedures as required. In fact many ISO 9000:1994-based systems are likely to include such procedures so there is nothing new here. The question is, why would you want to do this when in all other aspects, you may have documented your processes?

Document control is an inherent part of every work process because the inputs pass through a number of stages, each adding value to result in the achievement of defined objectives. These are the acquisition, approval, publication, distribution, storage, maintenance, improvement and disposal tasks, but each of these may differ depending on where the information comes from and what it is used for.

Control of records is also an inherent part of every work process similar to document control. There are the preparation, storage, access, maintenance and disposal tasks. This is not one uninterrupted flow but a life cycle. There is not one task but several performed at different times for different reasons.

Auditing is certainly an activity with a defined objective. Without the provision of competent personnel and a suitable environment, audits will not achieve their objectives no matter how many times the procedure is implemented.

Nonconformity control like *records control* is an inherent part of certain work processes for the same reasons. The sequence of tasks is not in the form of an uninterrupted sequence. The sequence of stages may be identification, documentation, segregation, review, remedial action and disposal, but this is not a continuous sequence. There are breaks and different procedures may apply at each stage depending on what it is that is nonconforming.

The *identification of training needs* is certainly an important activity but why single this out for special attention? The main reason is that the text has been lifted straight from QS-9000 is because the authors of ISO/TS 16949 felt the provisions of ISO 9001:2000 on training were inadequate. If you have defined you human resource development process, you should have a stage where the competences of personnel are matched with those required for the job they do and training needs determined. It is just as important to have a procedure to determine and assess competence but ISO/ST 16949 leaves that to you to decide whether a procedure is needed for these activities.

There is little merit in having one corrective action procedure when the source of problems that require corrective action is so varied. One Corrective Action Form might be appropriate but its application will be so varied that it is questionable whether one size fits all. Presenting top management with a nonconformity report, because it has been detected that the organization charts are not promptly updated following a change, will not motivate them into action. Corrective action forms a part of every process rather than being a separate process but it could be viewed as a process in which the methods vary but the outcomes are the same, i.e. nonconformity is prevented from recurring. The designer, the producer, the supplier or a manager remembering they had a problem last time and doing it differently the next, i.e. they learn from their mistakes and prevents many problems from recurring not by following a procedure, but. No forms filled in, no procedures followed (just people using their initiative) this is why corrective action is part of every process operation.

Preventive action remains one of the most misunderstood requirements of ISO 9001 because it is mistaken for corrective action but more on this in Chapter 8. There is even less justification for one preventive action procedure because the source of potential problems is so varied. Preventive actions are taken in design, in planning, in training and in maintenance under the name of FMEA, Reliability Prediction, Quality Planning, Production Planning, Logistic Planning, Staff Development and Equipment Maintenance; preventive actions are built into these processes and similar to corrective action are part of every process design.

Documents that ensure effective planning, operation and control of processes (4.2.1d)

The standard requires management system documentation *to include documents required by the organization to ensure the effective planning, operation and control of its processes.*

What does this mean?
The documents required for effective planning, operation and control of the processes would include several different types of documents. Some will be product and process specific, and others will be common to all processes. Rather than stipulate the documents that are needed, other than what is required elsewhere in the standard, ISO/TS 16949 now provides for the organization to decide what it needs for *the effective operation and control of its processes.* This phrase is the key to determining the documents that are needed.

There are three types of controlled documents, as illustrated in Figure 4.6:

- Policies and practices, include process descriptions, control procedures, guides, operating procedures and internal standards.
- Documents derived from these policies and practices, such as drawings, specifications, plans, work instructions, technical procedures and reports.
- External documents referenced in either of the above.

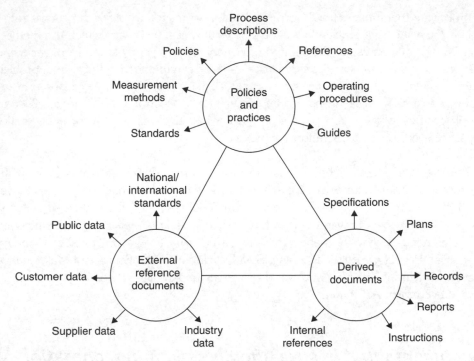

Figure 4.6 *Relationships between documents*

There will always be exceptions to this model but in general the majority of documents used in a management system can be classified in this way.

Derived documents are those that are derived by executing processes; e.g. audit reports result from using the audit process, drawings result from using the design process, procurement specifications result from using the procurement process. There are, however, two types of derived document: prescriptive and descriptive documents. *Prescriptive documents* are those that prescribe requirements, instructions, guidance, etc. and may be subject to change. They have issue status and approval status, and are implemented in doing work. *Descriptive documents* result from doing work and are not implemented. They may have issue and approval status. Specifications, plans, purchase orders, drawings are all prescriptive whereas audit reports, test reports, inspection records are all descriptive. This distinction is only necessary because the controls required will be different for each class of documents.

Why is this necessary?
This requirement responds to the Process Approach Principle.

The degree of documentation varies from a simple statement of fact to details of how a specific activity is to be carried out. To document *everything* you do would be impractical and of little value. Several good reasons for documenting information are listed under the heading *What should be documented*.

How is this implemented?

The identification of documentation needs was addressed under the heading *Documenting a quality management system*. In this heading, the specific types of documents are described in more detail.

Policies and practices

The relationship between these various types of policies and practices is illustrated in Figure 4.7.

Policies

Any statement made by management at any level that is designed to constrain the actions and decisions of those it affects is a *policy*. There does not have to be a separate document containing policies. Therefore policies can take many forms. The purpose and mission of the organization becomes a policy when expressed by the management, so do the principles or values guiding people's behavior; what is or is not permitted by personnel engaged by the organization. Policies are therefore essential in ensuring the effective planning of processes because they lay down the rules to be followed to ensure that actions and decisions taken in the design and operation of processes serve the business objectives and the needs and expectation of the stakeholders.

The policies can be integrated within any of the documents.

There are different types of policy that may impact the business processes:

- Government policy, which when translated into statutes applies to any commercial enterprise.
- Corporate policy, which applies to the business as a whole and may cover, e.g.:
  - *Environmental policy*: our intentions with respect to the conservation of the natural environment.
  - *Financial policy*: how the business is to be financed.
  - *Marketing policy*: into which markets the business is to supply its products.
  - *Investment policy*: how the organization will secure the future.
  - *Expansion policy*: the way in which the organization will grow, both nationally and internationally.
  - *Personnel policy*: how the organization will treat its employees and the labour unions.

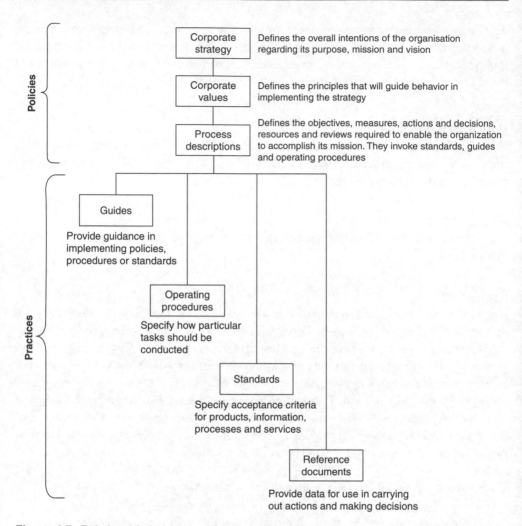

Figure 4.7 *Relationship between policies and practices*

- *Safety policy*: the organization's intentions with respect to hazards in the work place and to users of its products or services.
- *Social policy*: how the organization will interface with society.
- Operational policy, which applies to the operations of the business, such as design, procurement, manufacture, servicing and quality assurance. This may cover, e.g.:
  - *Pricing policy*: how the pricing of products is to be determined.
  - *Procurement policy*: how the organization will obtain the components and services needed.
  - *Product policy*: what range of products the business is to produce.
  - *Inventory policy*: how the organization will maintain economic order quantities to meet its production schedules.

- *Production policy*: how the organization will determine what it makes or buys and how the production resources are to be organized.
- *Servicing policy*: how the organization will service the products its customers have purchased.
- Department policy, which applies solely to one department, such as the particular rules a department manager may impose to allocate work, review output, monitor progress, etc.
- Industry policy, which applies to a particular industry, such as the codes of practice set by trade associations for a certain trade.

All policies set boundary conditions so that actions and decisions are channeled along a particular path to fulfil a purpose or achieve an objective. Many see policies as requirements to be met – they are requirements but only in so far as an enabling mechanism.

In organizations that have a strong value-based culture policies are often undocumented. Rules are appropriate to a command and control culture. In all cases you need to ask, what would be the effect on our performance as an organization if this were not documented? If the answer is nothing or a response such as, 'well somebody might do xyz' forget it! Is it likely? If you cannot predict with certainty that something will happen that should be prevented, leave people free to choose their own path unless it's a legal requirement.

However, even in a command and control culture one does not need to write *everything* down, as policies are needed only for important matters where the question of right or wrong does not depend on circumstances at the time, or when the relevance of circumstances only rarely come into the picture.

A common practice is to paraphrase the requirements of ISO/TS 16949 as operational policy statements. Whilst this approach does provide direct correlation with ISO/TS 16949 it renders the exercise futile because users can read the same things by referring to ISO/TS 16949. Operational policies should respond to the needs of the organization, not paraphrase the standard.

Process descriptions
Process descriptions are necessary in ensuring effective operation and control of processes because they contain or reference everything that needs to be known about a process.

Process descriptions may be maintained as discrete documents or combined into a manual. Process descriptions would include the following:

- Process objectives.
- Process owner.
- Process inputs in terms of the materials and information to be processed.

- Process outputs in terms of the products, services and information delivered.
- Set-up and shut-down conditions.
- Process flow charts indicating the sequence of actions and decisions identifying those responsible and the interacting processes, actions and references to supporting documentation.
- Resources: physical and human resources required to deliver process outputs.
- Dependencies: the known factors on which the quality of the process output depends (skills, competencies, behaviors and capabilities).
- Key performance indicators: the indicators by which the achievement of the process objectives will be measured.
- Performance measurement methods: measures to detect variation in product and process performance.
- Preventive measures: measures in place to prevent process and product error or failure.

Clearly this description goes well beyond the content of a procedure. It also goes well beyond flow charts. Flow charts depict the steps in a process but do not fully describe a process. We could have two processes each with the same sequence of steps and therefore the same flow chart. Process A performs well while Process B constantly underperforms, reject rates are high, morale is low, etc. The procedures, equipment and controls are identical. What causes such a difference in performance? In Process B the people have not been trained, the supervisors are in conflict and rule by fear and there is high absenteeism. There is poor leadership and because of the time spent on correcting mistakes, there is no time for maintaining equipment, cleanliness and documentation. In process A, the supervisors spend time building relationships before launching the process. They plan ahead and train their staff. They are constantly looking for opportunities for improvement and because the process runs smoothly, they have time for maintaining equipment, documentation and cleanliness – morale is high so there is no absenteeism. Therefore to describe a process in a way that will show how the process objectives are achieved, it is necessary to describe the features and characteristics of the process that cause success and this warrants more than can be depicted on a flow chart. It is not intended that behaviors should be documented but the activities that reflect appropriate behaviors, such as planning, preparing, checking, communicating, advising can be reflected in the process description.

The difference between a rule-based culture and a values-based culture provides another example of why flow charts alone do not describe processes. In a rule-based culture (or command and control culture), people will follow the rules or procedures regardless of the consequences. Conformity is paramount and produces consistency but not necessarily customer satisfaction. For example, a hotel has a procedure for maintaining the facilities and this requires annual maintenance of the air-conditioning system. Maintenance staff perform exactly as required by the procedure but take no account of the effect of their actions on customers using the conference facilities. The conference department performs exactly in accordance with the conference management procedure but takes no account of facility maintenance. The result is that a conference is held in the summer while the air-conditioning is out of action due to maintenance. This causes dissatisfaction to customers. Other examples result in the classic retort from service providers: "Its more than my job's worth to …". The staff only focus on doing what they are told and following the procedure. Their training concentrates on equipping them with the skills to perform the task. Initiative is not a requirement and can result in a reprimand from management when staff deviate from the rules. Staff are trained to perform the task and not trained to achieve objectives; otherwise they would weigh up the circumstances and apply their common sense.

Procedures
Procedures are necessary in ensuring the effective operation and control of processes because they layout the steps to be taken in setting up, operating and shutting down the process.

A procedure is a sequence of steps to execute a routine task. It prescribes how one should proceed in certain circumstances in order to produce a desired output. Sometimes the word can imply formality and a document of several pages but this is not necessarily so. A procedure can be five lines, where each line represents a step in the execution of a task (see also under the heading *Plans* below).

Procedures can only work, however, where judgment is no longer required or necessary. Once you need to make a judgment, you cannot prescribe what you might or might not do with the information in front of you. A form of judgment-based procedure is a decision tree that flows down a chain of questions to which either a yes or a no will route you down a different branch. The chart does not answer your questions but is a guide to decision-making.

Standards
Standards are essential in ensuring the effective operation and control of processes because they define the criteria required to judge the acceptability of the process capability and product quality.

Standards define the acceptance criteria for judging the quality of an activity, a document, a product or a service. There are national standards, international standards, standards for a particular industry and company standards. Standards may be in diagrammatic form or narrative form or a mixture of the two. Standards need to be referenced in control procedures or operating procedures. These standards are in fact *your* quality standards. They describe features and characteristics that *your* products and services must possess. Some may be type specific, others may apply to a range of products or types of products and some may apply to all products whatever their type. These standards are not the drawings and specifications that describe a particular product but are the standards that are invoked in such drawings and specifications, and are selected when designing the product.

In the process context, standards are essential for defining the acceptance criteria for a capable process.

Guides
Guides are necessary in ensuring the effective operation and control of processes because they provide information of use during preparation, operation, shutdown and troubleshooting.

Guides are aids to decision-making and to the conduct of activities. They are useful as a means of documenting your experience and should contain examples, illustrations, hints and tips to help staff perform their work as well as possible.

Derived documents
There are several types of derived documents:

Specifications
These are crucial in ensuring the effective planning and control of processes because they govern the characteristics of the inputs and outputs and are thus used in process and product design and measurement.

Plans
These are crucial in ensuring the effective operation and control of processes because they layout the work that is to be carried out to meet the specification. Unfortunately the word 'plan' can be used to describe any intent, will or future action so that specifications, procedures and process descriptions could be called plans when they are part of what you intend to do. A plan is therefore a statement of the provisions that have been made to achieve a certain objective. It describes the work to be done and the means by which the work will be done.

Reports
These are useful in ensuring the effective operation and control of processes because they contain information about the process or the product being processed. They may be used to guide decision-making both in the design and operation of processes, and in product realization.

Records
These are essential in ensuring the effective operation and control of processes because they capture factual performance from which decisions on process performance can be made. Records are defined in ISO 9000 as *documents stating the results achieved*

or providing evidence of activities performed. Records are therefore produced during an event or immediately afterwards. Records do not arise from contemplation. They contain facts, the raw data as obtained from observation or measurement and produced manually or automatically.

Instructions

These are crucial in ensuring the effective operation and control of processes because they cause processes to be initiated and define variables that are specific to the date and time, location, product or customer concerned. Work instructions define the work required in terms of who is to perform it, when it is to commence and when it is to be completed. They also include what standard the work has to meet and any other instructions that constrain the quality, quantity, delivery and cost of the work required. Work instructions are the product of implementing a control procedure, an operating procedure or a document standard (see further explanation below).

Internal references

These documents are useful in ensuring the effective operation and control of processes because they will contain data relevant to the equipment, people, facilities or other factors on which set-up or operation of the process depends. Reference documents differ from other types of documents in that they should be neither prescriptive nor instructional. They should not be descriptive like reports, proposals or records but should contain data that is useful in carrying out a task.

External reference documents

These documents are those not produced by the organization but used by the organization as a source of information, consequently the categories of external reference documents are identified by their source. There are several types including:

- National and international standards
- Public data
- Customer data
- Supplier data
- Industry data.

What should be documented? (4.2)

The standard requires *the extent of quality management system documentation to be dependent on the size and type of organization, complexity of the processes and the competency of personnel.*

Clause 4.9 of QS-9000 required procedures only where the absence of such procedures would adversely affect quality. This phrase, now omitted from the 2000 version, used to be taken out of context and used as a valid reason for not documenting certain aspects of the management system. The requirement only applied to procedures but could just as easily have applied to other types of documentation. The factors mentioned in the standard apply equally to documentation generated by a process and documentation supporting the process. For example, a person may need documented policies and practices to execute the processes reliably and also may produce documents that are required inputs for other processes. In both cases the factors of size, complexity, etc. apply to the extent of the supporting documentation as well as to the extent of the output documentation – therefore, documentation producers need to be aware of the documentation needs of the interconnecting processes.

Size of organization

If we think about it, what has size of the organization got to do with the amount of information you document? A large organization could be large because of the quantity of assets: 2000 offices with two people in each. Or it could be large because it employs 6000 people, 5500 of whom do the same job. Or we could find that of the 6000, there are 200 departments, each providing a different contribution and each staffed with people of different disciplines. Therefore, size in itself is not a factor and size without some units of measure is meaningless.

Type of organization

The type of organization will affect what you document and what documents you use but again not the amount of information you document. An organization that deals primarily with people may have little documentation. One that moves product may also have little documentation but one that processes information may have lots of documentation. A software house is different from a gas installation service, a bank is different from a textile manufacturer and therefore the content of the documentation will differ but they may use the same types of documents.

Complexity of processes

Complexity is a function of the number of processes and their interconnections in an organization. The more processes, the greater the number of documents. The more interconnections the greater the detail within documents. Complexity is also a function of the relationships. The more relationships, there are, the greater the complexity and channels of communication. Reducing the number of relationships can reduce complexity. Assigning work to fewer people reduces the number of transactions. Many documents exist simply to communicate information reliably and act as a point of reference should our memory fail us, which introduces another factor, that of the man's limited ability to handle unaided large amounts of data.

In the simplest of processes, all the influencing facts can be remembered accurately. As complexity increases, it becomes more difficult to remember all the facts and recall them accurately. A few extraordinary people have brilliant memories, some have learnt memory skills but the person of average ability cannot always remember a person's name or telephone number. The word 'password' is the most common password for Internet transaction because many of us would forget the password if it was something else. It would therefore be unreasonable to expect people to perform their work without the use of recorded information of some kind. What you should record and what you remember is often a matter of personal choice but in some cases you cannot rely on people remembering facts by chance. You therefore need to identify the dependencies in each process and perform a risk assessment to establish what must be documented.

Competence
Competence is the ability to demonstrate the skills, behaviors, attributes and qualifications to the level required for the job. Competency may depend on the availability of documentation. For example, a designer will refer to data sheets to assist in selecting components not because of a lack of competency but because of man's limited memory and a desire for accuracy. The designer can remember where to look for the relevant data sheets, but not the details. If the document containing the relevant data cannot be found, the designer is unable to do the job and therefore cannot demonstrate competence.

When personnel are new to a job, they need education and training. Documentation is needed to assist in this process for two reasons. Firstly to make the process repeatable and predictable and secondly to provide a memory bank that is more reliable than the human memory. As people learn the job they begin to rely less and less on documentation to the extent that in some cases, no supporting documentation may be used at all to produce the required output. This does not mean that once the people are competent you can throw away your documentation. It may not be used on a daily basis, but you will inevitably have new staff to train and improvements to make to your existing processes. You will then need the documentation as a source of information to do both.

Factors affecting the amount of information you document
There has to be a limit on what you document. At school we are taught reading, writing and arithmetic, so documents should not attempt to define these activities. But it depends on what you are trying to do. The documents in regular use need only detail what would not be covered by education and training. A balance should be attained between training and procedures. If you rely on training rather than employing documented procedures, you will need to show that you have control over the quality of training to a level that will ensure its effectiveness. We expect staff to know how to do the various tasks that comprise their trade or profession, how to write, how to design, how to type, answer the telephone, how to paint, lay bricks, etc. You may feel it necessary to provide handbooks with useful tips on how to do these tasks more economically and effectively, and you may also use such books to bridge gaps in education and training but these are not your procedures. If you need something to be done in a particular way because it is important to the outcome, the method will need to be documented so that others may learn the method.

QS-9000 required the quality manual to cover the requirements of the standard but this requirement has been removed. There is therefore no requirement to respond to the requirements of the standard in the order in which they are stated either in a quality manual or in documented procedures.

You can combine several procedures in one document, the size of which depends on the complexity of your business. The more complex the business the greater is the number of documents. The more variations in the ways that work is executed, the larger the description of management system will need to be. If you have a small business and only one way of carrying out work, your system description will tend to be small. Your management system may be described in one document of no more than 30 pages. On the other hand a larger business may require several volumes and dozens of documents of over 10 pages each to adequately describe the system.

Control procedures need to be user friendly and so should be limited in size. Remember you can use other documents, such as guides, standards and operating procedures to extend what you have written in the control procedures. The procedures should not, however, be so short as to be worthless as a means of controlling activities. They need to provide an adequate degree of direction so that the results of using them are predictable. If you neglect to adequately define what needs to be done and how to do it, don't be surprised that staff don't know what to do or constantly make mistakes. It is also important to resist the desire to produce manuals that are impressive rather than practical. Printing the documents on expensive paper with colored logo does not improve their effectiveness and if they are not written simply and understood by a person of average intelligence, they will not be used.

Reasons for not documenting information
There are also several reasons for not documenting information:

- If the course of action or sequence of steps cannot be predicted, a procedure or plan cannot be written for unforeseen events.
- If there is no effect on performance by allowing freedom of action or decision, there is no mandate to prescribe the methods to be employed.
- If it cannot be foreseen that any person might need to take action or make a decision using information from a process, there is no mandate to require the results to be recorded. (However you need to look beyond your own organization for such reasons if demonstrating due diligence in a product liability suit requires access to evidence).
- If the action or decision is intuitive or spontaneous, no manner of documentation will ensure a better performance.
- If the action or decision needs to be habitual, documentation will be beneficial only in enabling the individual reach a level of competence.

Implementing a quality management system (4.1)

The standard requires the organization *to implement a quality management system in accordance with the requirements of ISO/TS 16949.*

What does this mean?

The notion of implementing a management system seems to imply that the management system is a set of rules, a procedure or a plan. One implements procedures but the management system is far more than a collection of procedures. Also the standard requires a management system to be established and as stated previously, to establish a management system you need to design and construct it and integrate it into the organization. If you were to write a book and put it on a bookshelf, you would not refer to the book being established; created, designed or crafted perhaps but not established. Implementation therefore applies to the use and operation of the management system following its construction and integration and is therefore concerned with the routine operation of an already established, documented and resourced system. Effective implementation means adhering to the policies and practices, following what is stated, not changing your procedures after changing your practice.

Why is this necessary?

This requirement responds to the Leadership Principle.

It goes without saying that it is necessary to use the management system that has been established because the benefits will only arise from using the system.

How is this implemented?

There is no magic in meeting this requirement. You simply need to do what you said, you would do, you have to keep your promises, honor your commitments, adhere to the policies, meet the objectives, follow the procedures – in other words manage your processes effectively. Simply said but extremely difficult for organizations to do. Even if you documented what you do, your practices are constantly changing so little time would pass before the documents were out of date.

A common failing with the implementation of documented practices is that they are not sold to the workforce before they become mandatory. Also, after spending much effort in their development, documented practices are often issued without any thought given to training or to verifying that practices have in fact been changed. As a result, development is often discontinued after document release. It then comes as a shock to managers to find that all their hard work has been wasted. An effectively managed programme of introducing new or revised practices is a way of overcoming these shortfalls.

Managing processes (4.1)

The standard requires the organization *to manage the identified processes in accordance with the requirements of ISO/TS 16949.*

What does this mean?

The standard does not stipulate how processes should be managed in a coherent manner, it merely lays down a set of requirements which if met would ensure the organization's outputs would meet customer requirements. Managing the identified processes means managing those processes that have been identified as being necessary to achieve the organization's objectives. Managing these processes in accordance with the requirements of ISO/TS 16949 basically means that the way processes are managed should not conflict with the requirements. Strip away the jargon and managing processes means managing activities, resources and behaviors to achieve prescribed objectives.

The notes to Clause 4.1 of ISO/TS 16949 need some explanation. It is stated that the processes needed for the management system include *management activities, provision of resources, product realization and measurement.* This note could cause confusion because it suggests that these are the processes that are needed for the management system. It would be unwise to use this as the model and far better to identify the processes from observing how the business operates. The term *provision of resources* should be *Resource Management,* which is the collection of processes covering financial, human and physical resources. *Product realization* is also a collection of processes, such as design, production, service delivery, etc. that was previously referred to as Demand Creation. *Measurement* is not a single process but a work process, activity or even a task within each process. Grouping all the measurement processes together serves no useful purpose except it matches the standard – a purpose of little value in managing the organization.

Why is this necessary?

This requirement responds to the Process Approach Principle.

Desired results will not be achieved by chance – their achievement needs to be managed and as the processes are the means by which the results are achieved, this means managing the processes.

How is this implemented?

The first stage in managing a process is to establish what it is you are trying to achieve, what requirements you need to satisfy, what goals you are aiming at; then establish how you will measure your achievements, what success will look like. If the objective were customer satisfaction, what would the customer look for as evidence that the requirements have been met? Would it be product fulfiling the need, delivered on time and not early, at the agreed price with no hidden costs?

Managing processes is primarily about ensuring the following:

- Those involved in the process understand the objectives and how performance will be measured.
- Responsibility for actions and decision within the process is properly assigned and delegated.
- The processes providing the inputs are capable of meeting the demand required by the process.
- The required inputs are delivered when they need to be and are of the correct quality and quantity.
- The resources needed to perform the activities have been defined and conveyed to those who will deliver them.
- The forces that prevent, restrict limit or regulate some aspect of process performance are known and their effect minimized.
- The conditions affecting the behavior of personnel and equipment are under control.
- The activities deemed necessary to deliver the required outputs and achieve the process objectives are carried out in the prescribed manner.
- Sensors are installed to detect variance in performance.
- The measurement taken provide factual data on which to judge performance.
- Provisions have been made for communicating unusual changes in the inputs, operating conditions and process behavior and that these provisions are working effectively.
- Reviews are performed to verify outputs meet requirements.
- The causes of variation are determined and actions taken to restore the *status quo* and prevent recurrence of unacceptable variation.
- Reviews are performed to discover better ways of achieving the process objectives and actions are taken to improve the efficiency of operations.
- Objectives, targets and measures are reviewed and changed if necessary to enable the process to deliver outputs that continually serve the organization's objectives.

Ensuring information availability (4.1d)

The standard requires the organization *to ensure the availability of information necessary to support the operation and monitoring of the identified processes.*

What does this mean?
Information to support the operation of processes would include that related to:

- process inputs,
- planning activities,
- preparatory activities,
- result-producing activities,
- routing activities,
- process outputs.

Information to support the monitoring of processes would include that related to:

- past and current performance, throughput, response time, downtime, etc.;
- operating conditions;
- verification activities;
- diagnostic activities.

Such information would include plans, specifications, standards, records and any other information required to be used in operating and monitoring the process.

Why is this necessary?
This requirement responds to the Leadership Principle.

All processes require information whether they are automated or manually operated.

How is this implemented?
Ensuring the availability of information is part of process management and was also addressed above. One would expect personnel to have information available in order to plan their work, prepare facilities, or set up equipment, to perform operations, measure results, make decisions on the results, resolve problems, prepare the output for delivery and to route the output to its intended destination.

To ensure availability of information you need to provide access at point of use and this is addressed under the heading Control of documents.

Ensuring the availability of resources (4.1d)

The standard requires the organization *to ensure the availability of resources necessary to support the operation and monitoring of the identified processes.*

What does this mean?

The resources necessary to support the operation of processes would include:

- raw materials and consumables;
- personnel;
- utilities, such as heat, light, power and water;
- time;
- equipment, plant machinery, facilities and workspace;
- money to fund the needs of the process.

The resources necessary to support the monitoring of processes would include:

- instrumentation and equipment,
- verification and certification services to ensure measurement integrity,
- personnel to perform monitoring,
- computers and other tools to analyze results,
- utilities to energize the monitoring facilities,
- money to fund the needs of monitoring.

Why is this necessary?

This requirement responds to the Leadership Principle.

Without the necessary resources processes cannot function as intended. All processes consume resources. If there are insufficient resources to monitor the processes, it is hardly worthwhile operating them because you will not know how they are performing.

How is this implemented?

The process owner or manager is responsible for ensuring the availability of resources. This commences with identifying resource needs, securing an available and qualified supply, providing for their deployment into the process when required and monitoring their utilization. Further guidance is given in Chapter 6.

Measuring, monitoring and analyzing processes (4.1e)

The standard requires the organization *to measure, monitor and analyze the identified processes*.

What does this mean?

Measuring processes is rather different to measuring the output of processes – this is commonly referred to as inspection or product verification. Figure 4.8 illustrates this difference.

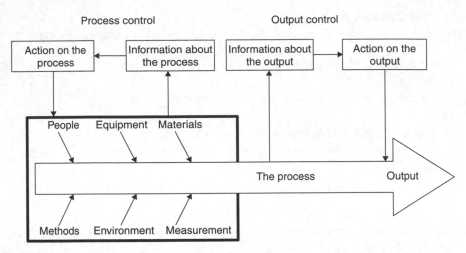

Figure 4.8 *Process control model*

Measuring is concerned with determination of the quantities of an entity, such as time, speed and capability indices, whereas monitoring is concerned with continual observation aside from periodic measurement.

Analyzing processes is concerned with understanding the nature and behavior of processes for the purpose of their design, development and improvement. Measuring and monitoring take place following installation of the process whereas process analysis can be used as a design tool.[4]

Why is this necessary?
This requirement responds to the Factual Approach Principle.

You can't manage a process unless equipped with facts about its performance. Observations from monitoring provide this useful information. You cannot claim success, failure or make improvements unless you know the current performance of your processes. It is therefore necessary to install sensors to gather this data. The facts may tell you where you are, but further analysis is needed to establish whether it is an isolated occurrence, an upward or a downward trend and whether improvement is feasible. Process analysis is performed to enable the decision makers to make decisions based on fact.

How is this implemented?
Process measurement
In order to measure the process you need:[5]

1   Process objectives (what the process is designed to achieve). For example, a sales process may be designed to convert customer enquiries into sales orders and

accurately convey customer requirements into the product or service generation process.

2   Indicators of performance (the units of measure). For example, a sales process measures may be:
   - The ratio of confirmed orders to enquiries.
   - The ratio of customer complaints relative or order accuracy to total orders completed.
   - The ratio of orders lost due to price relative to total order won. Similar indicators could be applied for quality and delivery.

   There may be other indicators related to the behavior of sales staff in dealing with customers and internal functions, such as whether the sales promise matched the true capability of the organization, etc.

3   Defined performance standard, i.e. the level above or below which performance is deemed to be substandard or inferior.

4   Sensors to detect variance before, during or after operations. There may be human or physical sensors, each of which has an element of measurement uncertainty.

5   Calibrated sensors, so that you can be assured the results are accurate and precise. There are two types of measurements to be made:
   - Measurements that tell us whether the process is operating as intended.
   - Measurements that tell us whether the process is effective.

   The former measurements are taken using the process indicators and the later measurements are taken using process analysis.

Process monitoring

For process monitoring to be effective, the staff involved need to understand the process objectives and how they are measured. They need to be vigilant to potential and actual variations from the norm. A typical type of process monitoring takes place in a process plant where there are lots of dials, gauges and data logging on a continuous basis. The watch engineers scan the log at the start of the watch for unusual occurrences that might account for variation in engine temperature. In monitoring a staff deployment process it may be noticed that staff are trained but there follows prolonged periods before the new skills are deployed. In the invoicing process it may be observed that a number of invoices go missing and have to be resent thus delaying receipt of revenue. In observing the design change process, it may be noticed that there is a burst in activity immediately prior to a holiday period without any additional resources being provided. Monitoring is looking for unusual occurrences or indicators of a potential change in performance.

Process analysis

Process analysis can be used to implement Clauses 4.1a–4.1c of ISO/TS 16949 as well as Clause 4.1e and 4.1f.

Analysis in process design   Process analysis is performed to design a process and understand the behavior of a process.[6] In this regard there are a number of activities that may be undertaken and there follows a sequence in which they could be implemented:

- Define the key performance indicators.
- Define the method of measurement.
- Establish current performance against the indicators.
- Produce a process flow chart.
- Perform a task analysis to determine who does what, when, where, how and why.
- Identify constraints on the process and test their validity.
- Perform a control analysis to determine/verify the controls to be/being applied.
- Deploy the system requirements (ISO/TS 16949 + any relevant regulations and statutes) to identify any gaps.
- Deploy known customer requirements (e.g. using Quality Function Deployment Chart (QFD)) to establish that the process will deliver the right output.
- Identify failure modes and effects to establish the issues that could jeopardize success.
- Install failure prevention features to reduce, contain or eliminate potential failure modes.
- Conduct relationship analysis to establish conflicts of responsibility and authority, and thus potential constraints.
- Perform productivity assessment to identify the number of transactions and their validity.
- Identify resources required to establish any deficiencies.
- Perform information needs analysis to identify/validate all the documentation needed.
- Perform a cultural analysis to establish the behavioral factors that will/are causing success or failure.

This analysis enables decisions to be made on the design or modification of processes and the conditions for their successful operation.

Analysis in process operation   In constructing the process, the identified measuring and monitoring stations should be installed. Process analysis is performed on the data generated by these sensors and includes several activities as follows:[7]

- Collect the data from monitoring activities.
- Sort, classify, summarize, calculate, correlate, present, chart and otherwise simplify the original data.
- Transmit the assimilated data to the decision-makers.
- Verify the validity of the variation.
- Evaluate the economical and statistical significance of the variation.
- Discover the root cause of the variation.
- Evaluate the alternative solutions that will restore the *status quo*.

This analysis enables decisions to be made on the continued operation of the processes and whether to modify the conditions under which they operate (see also the section Variation in Chapter 1).

Maintaining a quality management system (4.1)

The standard requires the organization *to maintain a quality management system in accordance with the requirements of ISO/TS 16949.*

What does this mean?
For many working to QS-9000 this was interpreted as maintaining documents, but as the management system is the means by which the organization's objectives are achieved, it clearly means much more than this. Maintenance is concerned with both retaining something in and restoring something to a state in which it can perform its required function. In the context of a management system this entails maintaining processes and their capability.

Why is this necessary?
This requirement responds to the Process Approach Principle.

Without maintenance, any system will deteriorate and management systems are no exception. A lack of attention to each of the factors mentioned above will certainly result in a loss of capability and therefore poor quality performance, financial performance and lost customers. Even to maintain performance a certain degree of improvement is necessary; in fact, even raising standards can be perceived as a means of maintaining performance in a dynamic environment in which you adapt or die.

How is this implemented?
In maintaining processes you need to keep:

* reducing variation;
* physical resources operational;
* human resources competent;
* financial resources available for replenishment of consumables, replace worn out or obsolete equipment;
* the process documentation up to date, as changes in the organization, technology, resources occur;
* space available to accommodate input and output;
* buildings, land and office areas clean and tidy – remove the waste;
* benchmarking processes against best in the field.

In maintaining capability you need to keep:

- replenishing human resources as staff retire, leave the business or are promoted;
- renewing technologies to retain market position and performance;
- surplus resources available for unforeseen circumstances;
- up to date with the latest industry practices;
- refreshing awareness of the vision, values and mission.

Another set of actions that can be used is the Japanese 5-S technique:[8]

1 *Seiri* (straighten up).
2 *Seiton* (put things in order).
3 *Seido* (clean up).
4 *Seiketsu* (personal cleanliness).
5 *Shitsuke* (discipline).

Continual improvement in the quality management system and its processes (4.1 and 4.1f)

The standard requires the organization *to continually improve the effectiveness of the quality management system in accordance with the requirements of ISO/TS 16949 and to implement action necessary to achieve planned results and continual improvement of the identified processes.*

What does this mean?
ISO 9000 defines continual improvement as a recurring activity to increase the ability to fulfil requirements. As the organization's objectives are its requirements, continually improving the effectiveness of the management system means continually increasing the ability of the organization to fulfil its objectives.

Why is this necessary?
This requirement responds to the Continual Improvement Principle.

If the management system is enabling the organization to accomplish its objectives when that is its purpose, why improve? The need for improvement arises out of a need to become more effective at what you do, more efficient in the utilization of resources so that the organization becomes best in its class. The purpose of measuring process performance is to establish whether or not the objectives are being achieved, and if not to take action on the difference. If the performance targets are being achieved, opportunities may well exist to raise standards, and increase efficiency and effectiveness.

How is this implemented?

Action necessary to achieve planned results is addressed under the heading *Corrective action* in Chapter 8.

If the performance of a process parameter is currently meeting the standard that has been established, there are several improvement actions you can take:

- *Raise the standard*: e.g. if the norm for the sales ratio of orders won to all orders bid is 60%, an improvement program could be developed for raising the standard to 75% or higher.
- *Increase efficiency*: e.g. if the time to process an order is within limits, identify and eliminate wasted resources.
- *Increase effectiveness*: e.g. if you bid against all customer requests, by only bidding for those you know, you can win, you improve your hit rate.

You can call all these actions improvement actions because they clearly improve performance. However, we need to distinguish between being better at what we do now and doing new things. Some may argue that improving efficiency is being better at what we do now, and so it is; but if in order to improve efficiency we have to be innovative we are truly reaching new standards. Forty years ago, supervisors in industry would cut an eraser in half in the name of efficiency rather than hand out two erasers. Clearly this was a lack of trust disguised as efficiency improvement and it had quite the opposite effect. In fact they were not only increasing waste but also creating a hostile environment.

Each of the improvement actions is dealt with later in the book and the subject of continual improvement addressed again under the heading *Quality management system planning* in Chapter 5.

There are several steps to undertaking continual improvement:[9]

1. Determine current performance.
2. Establish the need for change.
3. Obtain commitment and define the improvement objectives.
4. Organize diagnostic resources.
5. Carry out research and analysis to discover the cause of current performance.
6. Define and test solutions that will accomplish the improvement objectives.
7. Product improvement plans which specify how and by whom the changes will be implemented.
8. Identify and overcome any resistance to change.
9. Implement the change.
10. Put in place controls to hold new levels of performance and repeat step one.

Outsourcing (4.1 and 4.1.1)

The standard requires the organization *to ensure control of any outsourced processes that affects product conformity with requirements* and *to identify such control within the quality management system*. The automotive additions require the organization *to take responsibility for controlling outsourced processes that are carried out by customer-approved sources*.

What does this mean?

In purchasing products and services to the supplier's own specification, the organization is not outsourcing processes or subcontracting. It is simply buying products and services. An outsourced process is one that is managed by another organization on behalf of the parent organization. The most common outsourced processes are manufacturing processes, such as fabrication, assembly and finishing processes. But organizations also outsource information technology, human resources, cleaning, maintenance and accounting services. Often it is the activities that are outsourced not the process in its entirety. For a process to be outsourced, the supplier should be given an objective and given freedom to determine how that objective will be met. If the supplier merely performs activities dictated by the organization to the organization's specification it is not a process that has been outsourced.

Why is this necessary?

At one time, an organization would develop all the processes it required and keep them in-house because it was believed it had better control over them. As trade became more competitive, organizations found that their none-core processes were absorbing a heavy overhead and required significant investment, just to keep pace with advances in technology. They realized that if they were to make this investment, they would diminish the resources given to their core business and not make the advances they needed to either maintain or grow the business. A more cost-effective solution was to put the management of these non-core processes in the hands of organizations for which they were the core processes. However, the organization needs to have control over all the processes required for it to achieve its objectives otherwise it is not in a position to predict performance or provide confidence to its customers that it will satisfy customer requirements.

How is this implemented?

When managing an outsourced process, the organization is not simply placing orders for products and checking that the products received comply with the requirements but establishing process objectives and verifying that the supplier has developed a process that is capable of achieving those objectives. The processes should have a capability that enables the organization to avoid checking the outputs. All the rigor applied to the internal processes should be applied to the outsourced process. Data on the performance of these processes, their efficiency and effectiveness should be analyzed by the organization and measures put in place to cause improvement action should the outcomes of the process not satisfy the organization's objectives.

Preparing the quality manual (4.2.1b)

The standard requires *a quality manual to be established and maintained that includes the scope of the quality management system, the documented procedures or reference to them and a description of the sequence and interaction of processes included in the quality management system.*

What does this mean?
ISO 9000 defines a quality manual as *a document specifying the quality management system of an organization.* It is therefore not intended that the manual be a response to the requirements of ISO/TS 16949, but unfortunately many manuals do little more than this. As the top-level document describing the management system it is a system description describing how the organization is managed.

Countless quality manuals produced to satisfy QS-9000 were no more than 20 sections that paraphrased the requirements of the standard. Such documentation adds no value. They are of no use to managers, staff or auditors. Often thought to be useful to customers, customers would gain no more confidence from this than would be obtained from the registration certificate.

Why is this necessary?
This requirement responds to the System Approach Principle.

A description of the management system is necessary as a means of showing how all the processes are interconnected and how they collectively deliver the business outputs. It has several uses as:

- a means to communicate the vision, values, mission, policies and objectives of the organization;
- a means of showing how the system has been designed;
- a means of showing linkages between processes;
- a means of showing who does what;
- an aid to training new people;
- a tool in the analysis of potential improvements;
- a means of demonstrating compliance with external standards and regulations.

How is this implemented?

When formulating the policies, objectives and identifying the processes to achieve them, the manual provides a convenient vehicle for containing such information. If left as separate pieces of information, it may be more difficult to see the linkages.

The requirement provides the framework for the manual. Its content may therefore include the following:

- Introduction
  (a) Purpose (of the manual).
  (b) Scope (of the manual).
  (c) Applicability (of the manual).
  (d) Definitions (of terms used in the manual).
- Business overview
  (a) Nature of the business/organization; its scope of activity, its products and services.
  (b) The organization's interested parties (customers, employees, regulators, shareholders, suppliers, owners, etc.).
  (c) The context diagram showing the organization relative to its external environment.
  (d) Vision and values.
  (e) Mission.
- Organization
  (a) Function descriptions.
  (b) Organization chart.
  (c) Locations with scope of activity.
- Business processes (for (c)–(f) see Chapter 1 under the heading *Process classes*)
  (a) The system model showing the key business processes and how they are interconnected.
  (b) System performance indicators and method of measurement.
  (c) Mission management process description.
  (d) Resource management process description.
  (e) Demand creation process description.
  (f) Demand fulfilment processes description.
- Function matrix (relationship of functions to processes)
- Location matrix (relationship of locations to processes)
- Requirement deployment matrices
  (a) ISO/TS 16949 compliance matrix.
  (b) ISO 14001 compliance matrix.
  (c) Regulation compliance matrices (environment, health, safety, information security, etc.).
- Approvals (list of current product, process and system approvals)

The process descriptions can be contained in separate documents and should cover the topics identified previously (see the section Documents that ensure effective planning, operation and control of processes above).

As the manual contains a description of the management system a more apt title would be a Management System Manual (MSM) or maybe a title reflecting its purpose might be Management System Description (MSD) or Business Management System Description (BMSD).

In addition a much smaller document could be produced that does respond to the requirements of ISO/TS 16949, ISO 14001 and the regulations of regulatory authorities. Each document would be an exposition produced purely to map your management system onto these external requirements to demonstrate how your system meets these requirements. When a new requirement comes along, you can produce a new exposition rather than attempt to change your system to suit all parties. A model of such relationships is illustrated in Figure 4.9. The process descriptions that emerge from the Management System Manual describe the core business processes and are addressed in this chapter under the heading *Documents that ensure effective planning, operation and control of processes*.

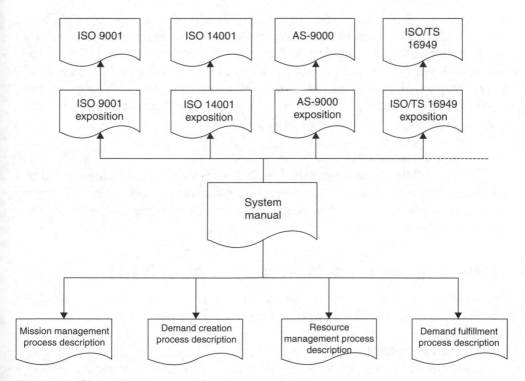

Figure 4.9 *System relationship with external standards*

If you choose to document your management system and make it accessible through a web browser, such as Internal Explorer or Netscape, the 'quality manual' will not be a definable entity unless for the purposes of the external audit you refer to the complete Management System Description as the Quality Manual.

Scope of the quality management system (4.2.2)

The standard requires the quality manual *to include the scope of the quality manage-ment system including details of justification for any exclusion.*

What does this mean?
The system may not cover all activities of the organization and therefore, those that are addressed by the quality management system or excluded from it need to be identified. The standard may also address activities that may not be relevant or applicable to an organization. The permissible exclusions are explained in Section 1.2 of ISO/TS 16949 and addressed in Chapter 3.

Why is this necessary?
This requirement responds to the System Approach Principle.

It is sensible to describe the scope of the management system so as to ensure effective communication. The scope of the management system is one area that generates a lot of misunderstanding particularly when dealing with auditors, consultants and cus-tomers. When you claim you have a management system that meets ISO/TS 16949 it could imply that you design, develop, install and service the products you supply, when in fact you may only be a distributor. Why you need to justify specific exclusions is uncertain because it is more practical to justify inclusions.

How is this implemented?
As ISO/TS 16949 is a sector-specific standard, you may have a quality manual that describes only those aspects of your management system that are used in the supply of automotive products. The quality manual might therefore define a scope that is less than that of the organization.

Referencing procedures in the quality manual (4.2.2)

The standard requires the quality manual *to include the documented procedures established for the quality management system or reference to them.*

What does this mean?
As the standard now only requires seven documented procedures it is unclear whether it is these procedures that should be included or referenced or all procedures. A practical

interpretation is to include or reference the Process Descriptions which themselves reference all the other documents used to manage processes.

Why is this necessary?

This requirement responds to the Process Approach Principle.

By including or referencing the documentation that describes the management system, you are providing a road map that will help people navigate through the system. Take the road map away and people won't know which documents to use. All documentation in the management system should be related and serve a defined purpose. By expressing the documentation in a hierarchy you provide a baseline and thus a means of configuration control. This will prevent new documents being created or existing documents withdrawn without reference to the baseline.

How is this implemented?

The retention of a requirement for the manual to include or refer to documented procedures indicates that management system documentation is still perceived by the standard makers as primarily comprising procedures. It is as though every requirement constitutes an activity that requires a documented procedure rather than it forming a part of a process. However, the requirement for the interaction of processes to be described in the manual provides a means for correcting this inconsistency. If the manual is structured as suggested previously, each business process will be described and in doing so the relevant procedures for performing tasks within the process can be listed or included. One way is to refer to procedures when detailing some aspect of a process. Another way is to list the procedures in an annex to each process description. A third way is to include an appendix in the manual that lists all the procedures. Alternatively a reference can be made to a database or number of databases that contain the procedures. With electronic documentation it is often not practical to duplicate lists of documents that would appear in a directory of a computer. Duplication creates a need for the synchronization of two or more lists to be maintained, thus causing additional effort and the possibility of error. Adding a new document to a file structure on a computer is the same as adding a new document to a list. A problem with file structures is that the configuration is changed when a document is added or deleted and therefore the status at any time in the past cannot be established unless a record is kept. Auditors and investigators certainly need to be able to establish the status of the system documentation at intervals, so that they can determine the documentation that was current when a particular event took place. Ideally, the database containing the documents needs the capability to reconstruct the file structure that existed at a particular date in the past.

Describing the processes and their interaction (4.2.2)

The standard requires the quality manual *to include a description of the interaction between the processes of the quality management system.*

What does this mean?
Each of these processes within the management system interacts with the others to produce the required outputs. A description of this interaction means that the linkages between processes, the source and destination of these linkages and what passes along these channels should be described. In describing an air-conditioning system, e.g., the system drawing would show all the components and how they linked together to form the system. Any component not linked into the system would have no function in the system and therefore would not be essential to its performance. Similarly with a management system, any process not connected to the system cannot perform a useful purpose within the system and can therefore be ignored.

Why is this necessary?
This requirement responds to the System Approach Principle.

The management system comprises the processes required to achieve the organization's objectives and therefore they are linked together. It is necessary to describe these linkages so that it can be demonstrated that a coherent system exists and how it operates.

How is this implemented?
The sequence and interaction of processes was addressed previously. In this case the interaction is required to be described and one way of doing this is through process flow charts arranged in a hierarchy. You could include all the process descriptions in the manual or merely include the business process flow charts and reference the process descriptions. The advantage of including all the process descriptions is that the manual becomes much more of a useful document and avoids duplicating information already contained in the process descriptions.

Figure 1.15 illustrates a high level description of the organization when viewed as a system of interconnected processes. Expanding each of the four processes in this model would create a series of diagrams that adequately responds to this requirement and avoids following the undesirable practice of showing clauses of the standard in boxes with the boxes linked together to form some type of interrelationship. The clauses of the standard do not reflect the organization's processes, they simply comprise groups of requirements.

Control of documents (4.2.3)

Documents required for the management system (4.2.3)

The standard requires documents required by the quality management system to be controlled.

What does this mean?

Documents required by the management system are those documents used by or generated by a process that forms part of the management system. The documentation used by a process could include policy documents, process descriptions, procedures, work instructions, contractual documents and standards. Those generated by a process could include product specifications, subcontracts, plans, orders, instructions, reports, and records. However, records are subject to different controls from the other types of documents because they are time-related and once produced they must not be changed, unless they contain errors (see under the heading *Control of records*).

The term document should be taken to include data or any information that is recorded and stored either on paper or magnetic media in a database or on a disk. It may be both an audio and visual record although the controls that will be applied will vary depending on the media. The requirement is not limited to documented procedures. These are only one type of document that needs to be controlled. There is often confusion between quality system documents and quality documents, and also between technical documents and quality documents. Such fine distinctions are unnecessary. Whether the document has the word *quality* in its title is irrelevant. The only question of interest is "Is the document used or generated by this process?" If the answer is "yes" the document should be controlled in some respects. Notes that you make when performing a task and then discard are not documents, generated by a process – they are merely are an aide-mémoire. Someone else may not make any notes at all or make different notes.

Controlling documents means regulating the development, approval, issue, change, distribution, maintenance, use, storage, security, obsolescence or disposal of documents. A controlled document is a document for which one or more of these attributes are controlled. Therefore if document security is controlled but not change, the subject document can still be classed as being controlled.

Why is this necessary?

This requirement responds to the Process Approach Principle.

It is necessary to regulate the documentation to ensure that:

- documents fulfil a useful purpose in the organization;
- resources are not wasted in the distribution of nonessential information;
- only valid information is used in the organization's processes;
- people have access to appropriate information for them to perform their work;
- information is kept up to date;
- information is in a form that can be used by all relevant people;
- classified information is restricted to only those with a need to know;
- information is important to the investigation of problems, improvement opportunities or potential litigation is retained.

How is this implemented?

For those using document control software with the documents located on a secure server, many of the recommendations in this part of the book may seem unnecessary. However, even proprietary software claiming to meet the requirements of ISO/TS 16949 for document control may not contain all the features you need. There are also many organizations that still use paper for good reasons. Paper does not crash without warning. Paper can be read more easily at a meeting or on a train. Comments can be added more easily to paper.

In the world of documents there are two categories: those that are controlled and those that are not controlled. A controlled document is one where requirements have been specified for its development, approval, issue, change, distribution, maintenance, use, storage, security and obsolescence or disposal. You do not need to exercise control over each of these elements for a document to be designated as a controlled document. Controlling documents may be limited to controlling their revision. On the other hand, you cannot control the revision of national standards but you can control their use, storage, obsolescence, etc. Even memoranda can become *controlled documents*, if you impose a security classification on them.

The standard acknowledges that records are indeed documents but require different controls to those that apply to other documents.

In order to control documents a document control process needs to be established that provides an adequate degree of control over all types of documents generated and used in the management system.

The process stages are common to all documents but the mechanisms for controlling different types of documents may differ. There are many software packages available that can be used to develop documents and control their issue, access, storage, maintenance and disposal. Unfortunately few can handle all types. You may not wish to trust all your documentation to one package. The software has automated many of the procedural issues such that, it is no longer the "Achilles Heel" of a management system.

Document control procedures (4.2.3)

The standard requires that *a documented procedure be established to define the controls needed*.

What does this mean?

This requirement means that the methods for performing the various activities required to control different types of documents should be defined and documented.

Although the standard implies that a single procedure is required, if you choose to produce several different procedures for handling the different types of documents, it is doubtful that any auditor would deem this noncompliant. Where this might be questionable is in cases where there is no logical reason for such differences and where merging the procedures and settling on a best practice would improve efficiency and effectiveness.

Why is this necessary?

This requirement responds to the Process Approach Principle.

Documents are recorded information and the purpose of the document control process is to firstly ensure that the appropriate information is available wherever needed and secondly to prevent the inadvertent use of invalid information. At each stage of the process are activities to be performed that may require documented procedures in order to ensure consistency and predictability. Procedures may not be necessary for each stage in the process.

How is this implemented?

Document development and maintenance

Every process is likely to require the use of documents or generate documents and it is in the process descriptions that you define the documents that need to be controlled. Any document not referred to in your process descriptions is therefore, by definition, not essential to the achievement of quality and not required to be under control. It is not necessary to identify uncontrolled documents in such cases. If you had no way of tracing documents to a governing process, a means of separating controlled from uncontrolled may well be necessary.

The procedures that require the use or preparation of documents should also specify or invoke the procedures for their control. If the controls are unique to the document, they should be specified in the procedure that requires the document. You can produce one or more common procedures that deal with the controls that apply to all documents. The stages in the process may differ depending on the type of document and organizations involved in its preparation, approval, publication and use. One procedure may cater for all the processes but several may be needed.

The aspects you should cover in your document control procedures, (some of which are addressed further in this chapter) are as follows:

- planning new documents, funding, prior authorization, establishing need, etc.;
- preparation of documents, who prepares them, how they are drafted, conventions for text, diagrams, forms, etc.;

- standards for the format and content of documents, forms and diagrams;
- document identification conventions;
- issue notation, draft issues and post approval issues;
- dating conventions, date of issue and date of approval or date of distribution;
- document review, who reviews them and what evidence is retained;
- document approval, who approves them and how approval is denoted;
- document proving prior to use;
- printing and publication, who does it and who checks it;
- distribution of documents, who decides, who does it and who checks it;
- use of documents, limitations, unauthorized copying and marking;
- revision of issued documents, requests for revision, who approves the request and who implements the change;
- denoting changes, revision marks, reissues, sidelining and underlining;
- amending copies of issued documents, amendment instructions, and amendment status;
- indexing documents, listing documents by issue status;
- document maintenance, keeping them current and periodic review;
- document accessibility inside and outside normal working hours;
- document filing, masters, copies, drafts and custom binders;
- document storage, libraries and archive, who controls location and loan arrangements;
- document retention and obsolescence.

With electronically stored documentation, the document database may provide many of the above features and may not need to be separately prescribed in your procedures. Only the tasks carried out by personnel need to be defined in your procedures. A help file associated with a document database is as much a documented procedure as a conventional paper-based procedure.

Document security

While many of the controls are associated with developing new documents and managing change, they address controls for causing the right things to happen. Controls are also needed to prevent, unauthorized changes, copying and disposal as well as computer viruses, fire and theft.

One solution used by many organizations is to publish their documents in a portable document format (PDF) as this provides built-in security measures. Users access the documents through a web browser or directly to a server with controlled access. The users can't change the document but may be permitted to print it and naturally printed versions would be uncontrolled. Some organizations state this on the document but it is really unnecessary. Such a practice suggests the system has been documented for the auditors rather than the employees. However, you obviously need to make sure everyone understands what the management system is and how to access and use the associated documents.

If original documents are available for users, inadvertent change can be a real problem. A document that has been approved might easily be changed simply because the 'current date' has been used in the approval date field. Every time a user accesses the document, the 'approval date' changes.

Whatever the controls, they need rigorous testing to ensure that the documents are secure from unauthorized change.

Document approval (4.2.3a)

The standard requires that *documents be approved for adequacy prior to issue.*

What does this mean?
Approval prior to issue means that designated authorities have agreed the document before being made available for use. Whilst the term *adequacy* is a little vague it should be taken as meaning that the document is judged as fit for the intended purpose. In a paper-based system, this means approval before the document is distributed. With an electronic system, it means that the documents should be approved before they are published or made available to the user community.

Why is this necessary?
This requirement responds to the Factual Approach Principle.

By subjecting documentation to an approval process prior to its use you can ensure that the documents in use have been judged by authorized personnel and found fit for purpose. Such a practice will also ensure that no unapproved documents are in circulation, thereby preventing errors from the use of invalid documents.

How is this implemented?
Adequacy of documents
The document control process needs to define the process by which documents are approved. In some cases it may not be necessary for anyone other than the approval authority to examine the documents. In others it may be necessary to set up a panel of reviewers to solicit their comments before approval is given. It all depends on whether the approval authority has all the information needed to make the decision and is therefore 'competent'. One might think that the CEO could approve any document in the organization but just because a person is the most senior executive it does not mean he or she is competent to perform any role in the organization.

Users should be the prime participants in the approval process so that the resultant documents reflect their needs and are fit for the intended purpose. If the objective is stated in the document, does it fulfil that objective? If it is stated that the document applies to certain equipment, area or activity, does it cover that equipment, area or activity to the depth expected of such a document? One of the difficulties in soliciting comments to documents is that you will gather comment on what you have written but not on what you have omitted. A useful method is to ensure that the procedures requiring the document specify the acceptance criteria so that the reviewers and approvers can check the document against an agreed standard.

To demonstrate documents have been deemed as adequate prior to issue, you will need to show that the document has been processed through the prescribed document approval process. Where there is a review panel, a simple method is to employ a standard comment sheet on which reviewers can indicate their comments or signify that they have no comment. During the drafting process you may undertake several revisions. You may feel it necessary to retain these in case of dispute later, but you are not required to do so. You also need to show that the current issue has been reviewed so your comment sheets need to indicate document issue status.

Approval authorities

The standard no longer contains a specific requirement for documents to be approved by authorized personnel. The person approving a document derives his or her authority from the process. The process descriptions or procedures should identify who the approval authorities are, by their role or function, preferably not their job title and certainly not by their name because both can change. The procedure need only state that the document be approved, e.g., by the Chief Designer prior to issue. Another method is to assign each document to an owner. The owner is a person who takes responsibility for its contents and to whom all change requests need to be submitted. A separate list of document owners can be maintained and the procedure need only state that the Owner approves the document. It is not necessary for all approval authorities to be different from the author. You only need separate approval authorities where there is added value by having an extra pair of eyes. Admiral Rickover US Navy formulated some "Basic rules for doing your job", in 1948, in which he wrote: "An essential element of carrying out my work is the need to have it checked by an independent source. Even the most dedicated individual makes mistakes."

Denoting approval

The standard doesn't require that documents visibly display approval but it is necessary to be able to show that the designated authorities have in fact approved documents in use. Electronic systems of control differ significantly in this area. With paper-based systems, approval can be denoted directly on the document, on a change or issue record, in a register or on a separate approval record. The presence of a colored header or the stamp of the issuing authority can substitute for actual signatures on documents. Providing signatures and front sheets often adds an extra sheet but no added value. The objective is to employ a reliable means of indicating to users that the document is approved. Some organizations maintain a list of authorized signatories where there are large numbers of people whose signatures and names are unknown to users. If you are dealing with a small group of people who are accessible and whose signatures are known, a list of authorized signatures is probably unnecessary. All you need is a means of checking that the person who signed the document was authorized to do so. If below the signature you indicate the position of the person and require his or her name to be printed alongside his or her signature, you have exercised due diligence.

With electronic systems, indication of approval is accomplished by electronic signature captured by the software as a function of the security provisions. These can be set up to permit only certain personnel to enter data in the approval field of the document. The software is often not as flexible as paper-based systems and therefore provisions need to be made for dealing with situations where the designated approval authority is unavailable. If you let competency determine authority rather than position, other personnel will be able to approve documents because their electronic signature will provide traceability.

With most electronic file formats you can access the document properties from the toolbar. Document properties can tell you when the document was created, modified, accessed and printed. But it can also tell you who the author was and who approved it, providing the author entered this information before publication. Open the 'properties' of any document and you may get some surprises. The information is often inserted automatically and is not erased when a file is moved from one computer to another. If you use an old document as the basis for creating a new document you will carry over all the old document's properties so don't be surprised if you get a call one day questioning why you are using a competitors information.

Issuing documents

The term *issue* in the context of documents means that copies of the document are distributed. You will of course wish to *issue* draft documents for comment but obviously they cannot be reviewed and approved beforehand. The sole purpose of issuing draft documents is to solicit comments. The ISO/TS 16949 requirement should have been that the documents are reviewed and approved prior to *use*. Some organizations insist that even drafts are approved for issue. Others go further and insist that copies cannot be taken from unapproved documents. This is nonsense and not what is intended by the standard. Your draft documents need to look different from the approved versions either by using letter issue notation (a common convention) or by printing the final versions on colored or watermark paper. If the approved document would carry signatures, the absence of any signature indicates that the document is not approved. With electronic systems, the draft documents should be held on a different server or in a different directory and provisions made to prohibit draft documents being published into the user domain.

Approving external documents

The requirements for document approval do not distinguish between internal and external documents. However, there is clearly a need to review and approve external documents prior to their internal release in order to establish their impact on the organization, the product, the process or the management system. The external document control procedure should make provision for new documents and amendments to such documents to be reviewed and approved for use prior to their issue into the organization. This aspect is also addressed in this chapter under the heading *Reviewing engineering specifications*.

Approving data

ISO 9000 defines a document *as information and its support medium*. This means that databases containing contacts, problems, sales, complaints, inventory, etc., are documents and yet we don't call them 'document bases'. We prefer the term database. The term data is not defined in ISO 9000 but is commonly understood to be information organized in a form suitable for manual or computer analysis. When data is recorded it becomes information and should therefore be controlled. All data should be examined before use otherwise you may inadvertently introduce errors into your work. The standard does not require common controls for all information so you are at liberty to pitch the degree of control appropriate to the consequences of failure.

Regarding approval of data, you will need to define which data needs approval before issue as some data may well be used as an input to a document which itself is subject to approval. It all depends on how we interpret 'approved prior to issue'. This should be taken to mean 'issue to someone else'. Therefore if you use data that you have generated yourself it does not need approval prior to use. If you issue data to someone else, it should be approved before distributing in a network database. If your job is to run a computer program in order to test a product, you might use the data resulting from the test run to adjust the computer or the program. You should be authorized to conduct the test and therefore your approval of the data is not required because the data has not in fact been issued to anyone else. The danger hiding in this requirement is that an eagle-eyed auditor may spot data being used without any evidence that it has been approved. As a precaution, ensure you have identified in your procedures those types of data that require formal control and that you know the origin of the data you are using.

Document review (4.2.3b)

The standard requires that *documents be reviewed*.

What does this mean?

A review is another look at something. Therefore document review is a task that is carried out at any time following the issue of a document.

Why is this necessary?
This requirement responds to the Continual Improvement Principle.

Reviews may be necessary when:

- taking remedial action (i.e. correcting an error);
- taking corrective action (i.e. preventing an error recurring);
- taking preventive action (i.e. preventing the occurrence of an error);
- taking maintenance action (i.e. keeping information current);
- validating a document for use (i.e. when selecting documents for use in connection with a project, product, contract or other application);
- taking improvement action (i.e. making beneficial change to the information).

How is this implemented?
Reviews may be random or periodic. Random reviews are reactive and arise from an error or a change that is either planned or unplanned. Periodic reviews are proactive and could be scheduled once in a year to review the policies, processes, products, procedures, specification, etc., for continued suitability. In this way obsolete documents are culled from the system. However, if the system is being properly maintained there should be no outdated information available in the user domain. Whenever a new process or a modified process in installed the redundant elements including documentation and equipment should be disposed of.

Revision of documents (4.2.3b)

The standard requires that *documents be updated as necessary and re-approved following their review.*

What does this mean?
Following a document review, action may or may not be necessary. If the document is found satisfactory, it will remain in use until the next review. If the document is found unsatisfactory there are two outcomes:

- The document is no longer necessary and should be withdrawn from use – this is addressed by the requirement dealing with obsolescence.
- The document is necessary but requires a change – this is addressed by this requirement.

The standard implies that updating should follow a review. The term *update* also implies that documents are reviewed only to establish whether they are current when in fact document reviews may be performed for many different reasons. A more appropriate term to *update* would be *revise*. Previously the standard addressed only the review and approval of changes and did not explicitly require a revision process. However, a revision process is executed before a document is subject to re-approval.

Why is this necessary?
This requirement responds to the Continual Improvement Principle.

It is inevitable that during use a need will arise for changing documents and therefore provision needs to be made to control not only the original generation of documents but also their revisions.

How is this implemented?
The change process
The document change process consists of a number of key stages some of which are not addressed in ISO/TS 16949:

- identification of need (addressed by document review);
- request for change (not addressed in the standard);
- permission to change (not addressed in the standard);
- revision of document (addressed by document updates);
- recording the change (addressed by identifying the change);
- review of the change (addressed under quality planning);
- approval of the change (addressed by document re-approval);
- issue of change instructions (not addressed in the standard);
- issue of revised document (addressed by document availability).

As stated previously, to control documents it is necessary to control their development, approval, issue, change, distribution, maintenance, use, storage, security, obsolescence or disposal and we will now address those aspects not specifically covered by the standard.

What is a change?
In controlling changes it is necessary to define what constitutes a change to a document. Should you allow any markings on documents, you should specify those that have to be supported by change notes and those that do not. Markings that add comment or correct typographical errors are not changes but annotations. Alterations that modify instructions are changes and need prior approval. The approval may be in the form of a change note that details the changes that have been approved.

Request for change
Anyone can review a document but approved documents should only be changed/revised/amended under controlled conditions. The document review will conclude that either a change is necessary or unnecessary. If a change is necessary, a request for

change should be made to the issuing authorities. Even when the person proposing the change is the same as would approve the change, other parties may be affected and should therefore be permitted to comment. The most common method is to employ Document Change Requests. By using a formal change request it allows anyone to request a change to the appropriate authorities.

Change requests need to specify:

- the document title, issue and date;
- the originator of the change request (who is proposing the change, his or her location or department);
- the reason for change (why the change is necessary);
- what needs to be changed (which paragraph, section, etc. is affected and what text should be deleted);
- the changes in text required where known (the text which is to be inserted or deleted).

By maintaining a register of such requests, you can keep track of who has proposed what, when and what progress is being made on its approval. You may of course use a memo or phone call to request a change but this form of request becomes more difficult to track and prove you have control. You will need to inform staff where to send their requests.

Permission to change

On receipt of the request you need to provide for its review by the change authority. The change request may be explicit in what should be changed or simply report a problem that a change to the document would resolve. Someone needs to be nominated to draft the new material and present it for review but before that, the approval authorities need to determine whether they wish the document to be changed at all. There is merit in reviewing requests for change before processing in order to avoid abortive effort. You may also receive several requests for change that conflict and before processing you will need to decide which change should proceed. While a proposed change may be valid, the effort involved may warrant postponement of the change until several proposals have been received – it rather depends on the urgency (see below).

As with the review and approval of data, you need to be careful about how you control changes to data. Data that have not been issued to anyone does not require approval if changed. Only the data that have been issued to someone other than its producer need to be brought under change control. If you are using data provided by someone else, in principle you can't change it without that person's permission. However, there will be many circumstances where formal change control of data is unnecessary and many where it is vital as with scientific experiments, research, product testing, etc. One way of avoiding seeking approval to change data is to give the changed data a new identity thereby creating new data from old data. It is perfectly legitimate for internal data (but not copyrighted data) because you have not changed the original data provided that others can still access it. If you use a common database for any activities you will need to control changes to the input data.

Making the change

The technology available for producing and controlling documents has changed dramatically over the last 50 years. There are four levels of technology in use:

- Documents produced, stored and distributed manually (handwritten or typed on paper).
- Documents produced and stored electronically but distributed manually (printed on paper).
- Documents produced, stored, distributed locally and controlled electronically (Intranet).
- Documents produced, stored, distributed worldwide and controlled electronically (Internet).

Each technology requires its own controls such that the controls applied to one type of technology would be totally inappropriate for another technology. Although we live in an age of Information Technology, all four types operate concurrently. The pen and paper are not obsolete and have their place alongside more sophisticated technologies. Maintenance personnel require documentation that may only be available in paper form although many might be equipped with laptop computers with a radio link to a central database. Document controls therefore need to be appropriate to the technology used to produce, store, distribute and control the documents.

The 1987 version of the ISO 9001 required that *documents be re-issued after a practical, number of changes have been made* but this provision was removed in the 1994 version. The requirement stems from the days before word-processing when changes were promulgated by amendment leaflet or change notes and one had to stick additional paragraphs over ones which were crossed out. In such circumstances there were only so many changes of this nature that you could make before the document became unusable and consequently a potential source of error. If you still operate in this fashion, the number of changes may well be a limiting factor but if you use word processors, other factors ought to be taken into account. However, there are practical reasons with documents distributed in paper medium (whether or not they are electronically produced documents), why it may not be prudent to reissue a document after each change. There are several types of changes you may need to consider:

- changes affecting a whole range of documents;
- changes affecting many pages of a single document;
- changes affecting a few pages of a single document.

For the change that affects a whole range of documents you will either need to reissue the complete set of documents or employ a Global Change Notice (GCN). When the cost and time required to process a change that affects many documents is prohibitive, something like a Global Change Notice is a useful tool to have in your management system. With a GCN you can instruct document holders to make changes to certain documents in their possession without having to identify every document. For example, if a component specification changes, a GCN can authorize the new information to be added to any documents that specify that particular component without having to process hundreds of documents. When the document is subsequently revised for other reasons, the GCN can be embodied so that over a period of time all documents will eventually be brought up to date. You will need a means of alerting staff to the issue of a GCN but if you control your distribution lists this should not present a problem. With electronic systems, a macro can be run on the database to update all references to a particular aspect thus updating automatically all the affected documents. Where is mechanism gets complicated is in cases where there are different forms of data capture and storage. For example, the computer aided design (CAD) data will probably not be generated using the same software tools as the management procedures. Advertising literature may be generated using drawing packages or DTP software and not word processing software. Flow charts may not be generated using word processing software. The technology is not yet available to search and replace information held in different forms on multiple platforms.

Where a change affects many pages, the document should be reissued. Even if the substantive change is minor, the knock-on effect in double-sided documents with diagrams, etc. can result in a change to every page. With modern word processing techniques, even adding a full stop can throw out several pages.

Where a change affects only a few pages, you can issue the changed pages with an amendment instruction informing the recipient which pages to change. Alternatively you can use the Document Change Notice (DCN) to add new pages and amend text at the same time.

If only a few words or figures are affected, the DCN is by far the least expensive and the quickest method.

As an alternative to actually issuing changes, you may wish to process the change requests to the master and hold reissue of the document until a suitable number of changes, or a significant proportion of the document has been changed. It is not the number of changes that is significant because a single change could have far greater effect than 20 minor changes. With small documents, say 3–6 pages, it is often easier to reissue the whole document for each change.

Identifying the change (4.2.3c)

The standard requires that *changes to documents be identified.*

What does this mean?
The requirement means that it should be possible to establish what has been changed in a document following its revision.

Why is this necessary?
This requirement responds to the Factual Approach Principle.

There are several benefits in identifying changes:

- Approval authorities are able to identify what has changed and so speed up the approval process.
- Users are able to identify what has changed and so speed up the implementation process.
- Auditors are able to identify what has changed and so focus on the new provisions more easily.
- Change initiators are able to identify what has changed and so verify whether their proposed changes were implemented as intended.

How is this implemented?
There are several ways in which you can identify changes to documents:

- by sidelining, underlining, emboldening or similar technique,
- by a change record within the document (front or back) denoting the nature of change,
- by a separate change note that details what has changed and why,
- by appending the change details to the initiating change request.

If you operate a computerized documentation system, your problems can be eased by the versatility of the computer. Using a database you can provide users with all kinds of information regarding the nature of the change, but be careful. The more you provide the greater the chance of error and the harder and more costly it is to maintain. Staff should be told the reason for change and you should employ some means of ensuring that where changes to documents require a change in practice, adequate instruction is provided. A system that promulgates change without concern for the consequences is out of control. The changes are not complete until everyone whose work is affected by them both understands them and are equipped to implement them when necessary. Although not addressed under document control, the requirement for the integrity of the management system to be maintained during change in Clause 5.4.2 implies that changes to documents should be reviewed before approval to ensure that the compatibility between documents is maintained. When evaluating the change you should assess the impact of the requested change on other areas and initiate the corresponding changes in the other documents.

Identifying the current revision of documents (4.2.3c)

The standard requires the *current revision status of documents to be identified.*

What does this mean?

When a document is revised its status changes to signify that it is no longer identical to the original version. This status may be indicated by date, by letter or by number or may be a combination of issue and revision. Every change to a document should revise the revision index. Issue 1 may denote the original version. On changing the document an incremental change to the revision index is made so that the new version is now Issue 2 or Issue 1.1 depending on the convention adopted.

Why is this necessary?

This requirement responds to the Factual Approach Principle.

It is necessary to denote the revision status of documents so that firstly, planners can indicate the version that is to be used and secondly, that users are able to clearly establish which version they are using or which version they require and so as to avoid inadvertent use of incorrect versions.

How is this implemented?

There are two aspects to this requirement. One is the identity denoted on the document itself and the other is the identity of documents referred to in other documents.

Revision conventions

Changes may be *major* causing the document to be reissued or re-released, or they may be *minor* causing only the affected pages to be revised. You will need to decide on the revision conventions to use. Software documents often use a different convention to other documents such as Release 1.1, or Version 2.3. Non-software documents use conventions such as Issue 1, Issue 2 Revision 3 and Issue 4 Amendment 2. A convention often used with draft documents is the letter revision status whereby the first draft is Draft A, second draft is Draft B and so on. When the document is approved, the status changes to Issue 1. During revision of an approved document, drafts may be denoted as Issue 1A, 1B, etc. and when approved the status changes to Issue 2. Whatever the convention adopted, it is safer to be consistent so as to prevent mistakes and ambiguities.

Revision letters or numbers indicate maturity but not age. Dates can also be used as an indication of revision status but dates do not indicate whether the document is new or old and how many changes there have been. In some cases this is not important, but in others there are advantages in providing both date and revision status therefore denoting date and revision status is often the simplest solution.

Document referencing

Staff should have a means of being able to determine the correct revision status of the documents they should use. You can do this through the work instructions, specification or planning documents, or by controlling the distribution, if the practice is to work to the latest issue. However, both these means have weaknesses. Documents can get lost, errors can creep into specifications and the cost of changing documents sometimes prohibits keeping them up-to-date. The issuing authority for each range of documents should maintain a register of documents showing the progression of changes that have been made since the initial issue. With configuration documents (documents which prescribe the features and characteristics of products and services) the relationship between documents of various issue states may be important. For example, a Design Specification at Issue 4 may equate with a Test Specification at Issue 3 but not with the Test Specification at Issue 2. This record is sometimes referred to as a Master Record Index or MRI but there is a distinct difference between a list of documents denoting issue state and a list of documents denoting issue compatibility state. The former is a Document Record Index and the later a Configuration Record Index. If there is no relationship between the document issues care should be taken not to imply a relationship by the title of the index.

The index may be issued to designated personnel or so as to preclude the use of obsolete indices, it may be prudent not to keep hard copies. With organizations that operate on several sites using common documentation it may well be sensible to issue the index so that users have a means of determining the current version of documents.

It is not necessary to maintain one index. You can have as many as you like. In fact if you have several ranges of documents it may be prudent to create and index for each range.

Regarding electronically controlled documents, arranging them so that only the current versions are accessible is one solution. In such cases and for certain type of documents, document numbers, issues and dates may be of not concern to the user. If you have configured the security provisions so that only current documents can be accessed, providing issue status, approval status, dates, etc., adds no value for the user, but is necessary for those maintaining the database. It may be necessary to provide access to previous versions of documents. Personnel in a product-support function may need to use documentation for various models of a product as they devise repair schemes and perform maintenance. Often documentation for products no longer in production carries a different identity but common components may still be utilized in current models.

Re-approving documents after change (4.2.3b)

The standard requires that documents *be re-approved after revision*.

What does this mean?
Following a change the revised document needs to be subject to approval as verification of its fitness for purpose.

Why is this necessary?
This requirement responds to the Factual Approach Principle.

As the original document was subject to approval prior to issue it follows that any changes should also be subject to approval prior to issue of the revised version. The approval does not have to be by the same people or functions that approved the original although this may be the case in many situations. The criteria are not whether the people or functions are the same, but whether the approvers are authorized. Organizations change and therefore people and functions may take on different responsibilities.

How is this implemented?
Depending on the nature of the change, it may be necessary to provide the approval authorities with factual information on which a decision can be made. The change request and the change record should provide this information. The change request provides the reason for change and the change note provides details of what has changed.

The change should be processed in the same way as the original document and submitted to the appropriate authorities for approval. If approval is denoted on the front sheet of your documents, you will need to reissue the front sheet with every change. This is another good reason to use separate approval sheets. They save time and paper. With electronically controlled documents, archived versions provide a record of approvals providing they are protected from revision.

Ensuring the availability of controlled documents (4.2.3d)

The standard requires that *relevant versions of applicable documents are available at points of use.*

What does this mean?
The relevant version of a document is the version that should be used for a task. It may not be the latest version because you may have reason to use a different version of a document, such as when building or repairing different versions of the same product. Applicable documents are those that are needed to carry out work. *Availability at points of use* means the users have access to the documents they need at the location where the work is to be performed. It does not mean that users should possess copies of the documents they need, in fact this is undesirable because the copies may become outdated and not withdrawn from use.

Why is this necessary?

This requirement responds to the Process Approach Principle.

This requirement exists to ensure that access to documents is afforded when required. Information essential for the performance of work needs to be accessible to those performing it otherwise they may resort to other means of obtaining what they need that may result in errors, inefficiencies and hazards.

How is this implemented?

Availability

In order to make sure that documents are available you should not keep them under lock and key (or password protected) except for those where restricted access is necessary for security purposes. You need to establish who wants which documents and when they need them. The work instructions should specify the documents that *are* required for the task so that those documents not specified are not essential. It should not be left to the individual to determine which documents are essential and which are not. If there is a need for access out of normal working hours, access has to be provided. The more copies there are the greater the chance of documents not being maintained, so minimize the number of copies. A common practice is to issue documents to managers only and not the users. This is particularly true of management system documents. One finds that only the managers hold copies of the Quality Manual. In some firms all the managers reside in the same building, even along the same corridor and it is in such circumstances that one invariably finds that these copies have not been maintained. It is therefore impractical to have all the copies of the Quality Manual in one place. Distribute the documents by location, not by named individuals. Distribute to libraries, or document control centers so that access is provided to everyone and so that someone has responsibility for keeping them up-to-date. If using an *Intranet*, the problems of distribution are less difficult but there will always be some groups of people who need access to hard copy.[10]

The document availability requirement applies to both internal and external documents alike. Customer documents, such as contracts, drawings, specifications and standards need to be available to those who need them to execute their responsibilities. Often these documents are only held in paper form and therefore distribution lists will be needed to control their location. If documents in the public domain are required, they only need be available when required for use and need not be available from the moment they are specified in a specification or procedure. You should only have to produce such documents when they are needed for the work being undertaken at the time of the audit. However, you would need to demonstrate that you could obtain timely access when needed. If you provide a lending service to users of copyrighted documents, you would need a register indicating to whom they were loaned so that you can retrieve them when needed by others.

A document that is not ready for use or is not used often may be archived. But it needs to be accessible otherwise when it is called for it won't be there. It is therefore necessary to ensure that storage areas, or storage mediums provide secure storage from which documents can be retrieved when needed. Storing documents off-site under the management of another organization may give rise to problems if they cannot be contacted when you need the documents. Archiving documents on magnetic tape can also present problems when the tape cannot be found or read by the new technology that has been installed! Electronic storage presents very different problems to conventional storage and gives rise to the retention of 'insurance copies' in paper, if the retrieval mechanism fails.

Relevant versions of internal documents

A question often asked by Assessors is "How do you know you have the correct issue of that document?" The question should not arise with an electronically controlled documentation system that prohibits access to archived versions. If your system is not that sophisticated, one way of ensuring the latest issue is to control the distribution of documents so that each time a document changes, the amendments are issued to the same staff who received the original versions. If you identify authentic copies issued by the issuing authority in some way, by colored header, red stamp or other means, it will be immediately apparent which copies are authentic and under control, and which are uncontrolled. Another way is to stamp uncontrolled documents with an 'Uncontrolled Document' stamp. All paper documents should carry some identification as to the issuing authority so that you can check the current issue if you are in doubt. The onus should always rest with the user. It is his or her responsibility to check that he or she has the correct issue of a document before work commences. One way of signifying authenticity is to give documents copy numbers in red ink as a practical way of retaining control over their distribution. If documents are filed in binders by part or volume, the binder can be given a copy number, but you will need a cross-reference list of who holds which copy.

Where different versions of the same document are needed, you will need a means of indicating which issue of which document is to be used. One method is to specify the pertinent issues of documents in the specifications, drawings, work instructions or planning documents. This should be avoided if at all possible because it can cause maintenance problems when documents change. It is sometimes better to declare that staff should use the latest version unless otherwise stated and provide staff with a means of determining what the latest version is.

Relevant versions of external documents

In some cases the issues of public and customer specific documents are stated in the contract and therefore it is important to ensure that you possess the correct version before you commence work. Where the customer specifies the issue status of public domain documents that apply you need a means of preventing their withdrawal from use in the event that they are revised during the term of the contract.

Where the issue status of public domain documents is not specified in a contract you may either have a free choice as to the issue you use or, as more likely, you may need to use the latest issue in force. Where this is the case you will need a means of being informed when such documents are revised to ensure you can obtain the latest version. The ISO 9000 series for instance is reviewed every 5 years so could well be revised at 5-year intervals. With national and international legislation the situation is rather different because this can change at *any* time. You need some means of alerting yourself to changes that affect you and there are several methods from which to choose:

- subscribing to the issuing agency of a standards updating service;
- subscribing to a general publication which provides news of changes in standards and legislation;
- subscribing to a trade association which provides bulletins to its members on changes in the law and relevant standards;
- subscribing to the publications of the appropriate standards body or agency;
- subscribing to a society or professional institution that updates its members with news of changes in laws and standards;
- joining a business club which keeps its members informed of such matters;
- as a registered company you will receive all kinds of complementary information from government agencies advising you of changes in legislation.

As an ISO/TS 16949 registered company you will receive bulletins from your certification body on matters affecting registration and you can subscribe to ISO Management Systems[11] to obtain worldwide news of events and changes in the management systems arena.

The method you choose will depend on the number and diversity of external documents you need to maintain and the frequency of usage.

Issuing change instructions

If you require an urgent change to a document, a legitimate means of issuing change instructions is to generate a Document Change Note. The Change Note should detail the changes to be made and be authorized by the appropriate authorities. On receipt of the Change Note the recipients make the changes in manuscript or by page replacement, and annotate the changes with the serial number of the Change Note. This practice primarily applies to paper systems but with electronically controlled documents, changes can be made to documents in a database without any one knowing and therefore it is necessary to provide an alert so that users are informed when a change has been made that may affect them. If the information is of a type that users invariably access rather than rely on memory, change instructions are may be unnecessary.

Ensuring documents are legible and identifiable (4.2.3e)

The standard requires documents *to remain legible and readily identifiable*.

What does this mean?

Legibility refers to the ease with which the information in a document can be read or viewed. A document is readily identifiable if it carries some indication that will quickly distinguish it from similar documents. Any document that requires a reader to browse through it looking for clues is clearly not readily identifiable.

Why is this necessary?

This requirement responds to the Process Approach Principle.

The means of transmission and use of documents may cause degradation such that they fail to convey the information originally intended. Confusion with document identity could result in a document being misplaced, destroyed or otherwise being unobtainable. It can also result in incorrect documents being located and used.

How is this implemented?

Legibility

This requirement is so obvious, it hardly needs to be specified. As a general rule, any document that is printed or photocopied should be checked for legibility before distribution. Legibility is not often a problem with electronically controlled documents. However, there are cases where diagrams cannot be magnified on screen so it would be prudent to verify the capability of the technology before releasing documents. Not every user will have perfect eyesight! Documents transmitted by fax present legibility problems due to the quality of transmission and the medium on which the information is printed. Heat-sensitive paper is being replaced with plain paper but many organizations still use the old technology. You simply have to decide your approach. For any communication required for reference, it would be prudent to use photocopy or scan the fax electronically and dispose of the original. Documents used in a workshop environment may require protection from oil and grease. Signatures are not always legible, so it is prudent to have a policy of printing the name under the signature. Documents subject to frequent photocopying can degrade and result in illegible areas.

Identification

It is unusual to find documents in use that carry no identification at all. Three primary means are used for document identification – classification, titles and identification numbers. Classification divides documents into groups based on their purpose – policies, procedures, records, plans, etc., are classes of documents. Titles are acceptable providing there are no two documents with the same title in the same class. If you have hundreds of documents it may prove difficult to sustain uniqueness. Identification can be made unique in one organization but outside it may not be unique. However, the title as well as the number is usually sufficient. Electronically controlled documents do not require a visible identity other than the title in its classification. Classifying documents with codes enables their sorting by class.[12]

Control of external documents (4.2.3f)

The standard requires *documents of external origin to be identified and their distribution controlled*.

What does this mean?

An external document is one produced externally to the organization's management system. There are two types of external documents, those in the public domain and those produced by specific customers and/or suppliers. Controlling distribution means designating those who need external documents and ensuring that any change of ownership is known about and approved.

Why is this necessary?

External documents are as much part of the management system as any other document, and hence require control although the control that can be exercised over external documents is somewhat limited. You cannot for instance control the revision, approval or identification of external documents therefore all the requirements concerning document changes will not apply. You can, however, control the use and amendment of external documents by specifying which versions of external documents are to be used and you can remove invalid or obsolete external documents from use or identify them in a way that users recognize as invalid or obsolete. You can control the amendment of external documents by controlled distribution of amendment instructions sent to you by the issuing agency.

How is this implemented?

External documents are likely to carry their own identification that is unique to the issuing authority. If they do not carry reference number, the issuing authority is normally indicated which serves to distinguish them from internal documents. Where no identification is present other than a title, the document may be invalid. This sometimes happens with external data and forms. If the source cannot be confirmed and the information is essential, it would be sensible to incorporate the information into an appropriate internal document.

In order to control the distribution of external documents you need to designate the custodian in the appropriate process descriptions or procedures and introduce a mechanism for being notified of any change in ownership. If the external documents are classified, prior approval should be granted before ownership changes. This is particularly important with military contracts because all such documents have to be accounted for. Unlike the internal documents, many external documents may only be available in paper form so that registers will be needed to keep track of them. If electronic versions are provided, you will need to make them 'read only' and put in place safeguards against inadvertent deletion from the server.

Preventing unintended use of obsolete documents (4.2.3g)

The standard requires *the unintended use of obsolete documents to be prevented and a suitable identification to be applied to obsolete documents retained for any purpose.*

What does this mean?
In simple terms an obsolete document is one that is no longer required for operational purposes. If an obsolete document has not been removed from the workplace there remains a possibility that it could be used. A suitable identification is one that readily distinguishes a current version of a document from an obsolete version.

Regrettably the standard no longer refers to invalid documents as well as obsolete documents. Invalid documents may not be obsolete and may take several forms. They may be:

* documents of the wrong issue status for a particular task;
* draft documents which have not been destroyed;
* documents which have not been maintained up to date with amendments;
* documents which have been altered or changed without authorization;
* copies of documents which have not been authenticated;
* unauthorized documents or documents not traceable through the management system;
* illegal documents.

Why is this necessary?
This requirement responds to the Process Approach Principle.

A means of distinguishing current documents from obsolete documents is needed to prevent their unintended use. Use of obsolete information may lead to errors, failures and hazards.

There are several reasons why you may wish to retain documents that are replaced by later versions:

- As a record of events. (What did we do last time?)
- For verifying that the correct changes were made. (What did it say in the last version or the first version?)
- For justifying rejection of a discarded solution. (We've had this one before – what did we decide on that occasion?)
- For investigating problems that did not occur previously. (What was it that we did differently?)
- For preparing a defense in a product liability case. (What controls were in place when we made this product?)
- For learning from previous designs. (How did we solve that problem?)
- For restoring/refurbishing a product. (They don't make them like they used to – do they?)
- For reference to explanations, descriptions, translations, etc., in order to preserve consistency in subsequent revisions or new documents. (What wording did we use last time we encountered this topic or term?)

How is this implemented?

With an electronic documentation system, access to obsolete documents can be barred to all except the chosen few. All it needs is for operational versions to be held in an operational directory and archived versions to be transferred into an archive directory automatically when a new version is released. On being transferred the revision status should be changed automatically to 'obsolete' indicating that later versions have been released.

With paper documents, it is more difficult. There are two options. You either remove the obsolete documents or mark them in some way that they are readily recognizable as obsolete.

It is unnecessary to remove invalid or obsolete documents if you provide staff with the means of determining the pertinent issues of documents to use. There are often valid reasons for retaining obsolete documents. One may need to remove copies of previous versions of a document but retain the master for reference purposes. You cannot demonstrate to an assessor that you corrected a deficiency if you don't retain the version that contained the deficiency as well as the subsequent version. If you do not have a means of readily distinguishing the correct version of a document, amendment instructions should require that the version being replaced is destroyed or returned to the document controller. If you allow uncontrolled copies to be taken, you will need to provide a means of distinguishing controlled and uncontrolled documents.

One way of identifying obsolete documents is to write SUPERSEDED or OBSOLETE on the front cover, but doing this requires that the custodian be informed. When a version of a document is replaced with a new version, the withdrawal of the obsolete version can be accomplished in the amendment instructions that accompany the revision. When documents become obsolete by total replacement, their withdrawal can also be accomplished with the amendment instruction. However, where a document becomes obsolete and is not replaced, there needs to be a Document Withdrawal Notice which informs the custodian of the action to be taken and the reason for withdrawal.

There is no simple way of identifying invalid documents because the reasons that they are invalid will vary. By printing authentic documents on colored paper or providing paper with a special header one can inject a degree of control. Placing the approval signatures on the front sheet will immediately identify an unapproved document. However, the onus must rest with the user who if properly trained and motivated will instinctively refrain from using invalid documents.

With electronically controlled documents the invalid documents can be held in a database with access limited to those managing the documents. In such cases, an approved document becomes invalid when its status is denoted as draft or withdrawn.

Reviewing engineering specifications (4.2.3.1)

The standard requires the organization *to have a process to assure the timely review, distribution and implementation of all customer engineering standards/specifications.*

What does this mean?
Customer engineering standards and specifications are external documents therefore your 'procedure' for controlling external documents should also cover these documents. *Timely review* means days not weeks or months; therefore immediately a new customer document is received, it should be routed to a person authorized through procedures to carry out a review. As to this being a process, it is perhaps part of a configuration management process established to ensure that the required standard, the design standard and build standard are one and the same.

Why is this necessary?
Should the organization receive engineering specifications and standards following acceptance of a contract and proceed to implement these requirements without due regard to program and technical impact, the consequences might be quite catastrophic.

How is this implemented?
The requirement could be placed under 'Review of requirements related to the product' since any documents issued by customers form part of the contract and should go through *contract review* before acceptance and implementation.

These documents should always be routed to the customer representative (see Clause 5.5.2.1) so that the same routines are initiated each time. The review should establish the applicability of the document and its impact on the contract, and any design and production documentation that has already been produced. Any changed documents should be treated as an amendment to the contract and processed accordingly. Design work, FMEAs, MSA, APQP and other work may have to be redone and this will naturally impact time schedules and costs.

As with all controlled documents, a distribution list for customer documents should be maintained so that copies can be withdrawn, replaced or amended when required.

Maintaining modification records (4.2.3.1)

The standard requires the organization *to maintain a record of the date on which each change is implemented in production*.

What does this mean?

When a change is implemented in production the revision level of the parts affected changes so that if it is June 1 today we should be building Rev B not Rev A.

The date on the PPAP fixes the required build configuration so if the design changes the PPAP has to be re-approved and this might have a knock-on effect on control plans, FMEAs, MSA, etc. After a PPAP has been changed, parts produced after the date of the PPAP should be to the revision level stated on the PPAP.

Why is this necessary?

It is important to know when changes have been made in order to control the build configuration of the product. These records are often date driven so that on a certain date we can determine what the configuration was required to be and what it was. Dates are useful too when investigating problems. The record might state Rev B and be dated May 15 but we know it can't be Rev B because Rev B was not released until June 1. We know the design was changed on June 1 but the last Process FMEA review was December 15 so it can't have been performed against the lasted design.

How is this implemented?

There are two types of records that need to be maintained. One deals with changes to documents and the other deals with changes to products resulting from changes to documents. The master document register or list should list all controlled internal and external documents including customer documents in terms of their title, date and revision status. The product modification record should define the design standard of the products to be built. This would show the batches produced and the revision state of the specifications to which each batch of product has been produced. It would also show the date and batch when changes were embodied. In addition to this record you will need inspection records that denote the configuration of the build so that the two records would match.

Control of records (4.2.1e and 4.2.4)

Establishing and maintaining records (4.2.4)

The standard requires records *to be established and maintained to provide evidence of conformity to requirements and the effective operation of the quality management system.*

What does this mean?

A record is defined in ISO 9000 as a document stating results achieved or providing evidence of activities performed.

Although a record is a document, the document control requirement of Clause 4.2.3 do not apply to *records* primarily because records are not issued, neither do they exhibit revision status simply because they are results that are factual when recorded. If the facts change, a new record is usually created rather than the previous record revised. Even where a record is revised and new facts added, the old facts remain identified as to their date. The only reason for revising facts contained in a record without changing the identity of the record or the date when they were collected is where the facts were incorrectly recorded. This subtle difference demands different treatment for documents that are classed as records to those that are classed as informative. As with other types of documents, records result from processes and may be used as inputs to other processes.

Records required by the management system means records used or generated by the management system. There is therefore no requirement to produce records solely to satisfy an auditor. The records required are those for the effective operation of the organization's processes (see Clause 4.1d of ISO/TS 16949). If a record has no useful purpose within the management system there is no requirement that it be established or maintained.

This does not mean that a record is required to prove conformity with every single requirement for every product and service. There are those records required by the standard (see below) that are required for every product and service where applicable; and in all other cases, audit records showing compliance on a sample of operations would be sufficient.

Compliance with many requirements can be demonstrated by observation, analysis, interview or examination of documentation. If a result or an activity is not required to be recorded in order to manage the organization effectively and satisfy the interested parties, it is not necessary to maintain records of it.

Why is this necessary?
This requirement responds to the Factual Approach Principle.

The reason for establishing records is to provide information necessary for managing processes, meeting objectives and demonstrating compliance with requirements – both customer requirements and legal requirements.

The reason for maintaining records is to secure their access, integrity, durability and disposal. If information of this nature cannot be located, or is illegible, or its integrity is suspect, or is obsolete, the decisions that require this information cannot be made on the basis of fact and will therefore be unsound or not as robust as intended. Lost, altered or illegible records can result in an inability to prove or disprove liability.

How is this implemented?
Establishing records
Records that are governed by the requirements of Clause 4.2.4 should be classified as *controlled records* so as to distinguish them from other records. If information needs to be recorded in order to manage the organization effectively, these records should be identified in the governing procedures as being required. This will then avoid arguments on what is or is not a controlled record, because once you have chosen to identify a record as a controlled record you have invoked all the requirements that are addressed in Clause 4.2.4. The references to Clause 4.2.4 give some guidance to the types of records required but should not be interpreted as definitive. There are 24 references to records as indicated below:

1 Engineering change records (Clause 4.2.3.1).
2 Management review records (Clause 5.6.3).
3 Records of education, experience, training and qualifications (Clause 6.2.2).
4 Records needed to provide evidence that realization processes and resultant product meet requirement (Clause 7.1).
5 Customer requirement review records (Clause 7.2.2).
6 Design and development inputs (Clause 7.3.2); this is new and does not fit the criteria for a record (see Chapter 7).
7 Design and development review records (Clause 7.3.4).
8 Design verification records (Clause 7.3.5).
9 Design validation records (Clause 7.3.6); this is new.

10 Design and development change review records (Clause 7.3.7); this is new.
11 Supplier evaluation records (Clause 7.4.1).
12 Validation arrangements for processes to include requirements for records (Clause 7.5.2).
13 Product identification records (Clause 7.5.3).
14 Records of unsuitable customer property (Clause 7.5.4).
15 Calibration records (Clause 7.6).
16 Internal audit results are to be recorded (Clause 8.2.2).
17 Results of process capability studies (Clause 8.2.3.1).
18 Significant process event and change records (Clause 8.2.3.1).
19 Product verification records (Clause 8.2.4).
20 Nonconformity records (Clause 8.3).
21 Customer waiver records (Clause 8.3.4).
22 Results of corrective actions taken are to be recorded (Clause 8.5.2).
23 Rejected product analysis records (Clause 8.5.2.4).
24 Results of preventive actions taken are to be recorded (Clause 8.5.3).

Such an analysis is only useful to illustrate that not all records required for effective quality management are identified in ISO/TS 16949. It is unnecessary to produce a record testifying that each requirement of ISO/TS 16949 has been met. For example, there is no obvious benefit from *maintaining* records that a document is legible (Clause 5.5.6e), that a customer enquired about some information that was later provided (Clause 7.2.3) or that a particular filing system was improved (Clause 8.1). At the time, the facts may be recorded temporarily but disposed of as the event passes into history. Such information may not be needed for future use. However, there are obvious benefits from requiring records to be established for:

* customer complaints,
* warranty claims,
* failure analysis reports,
* process capability studies,
* service reports,
* concessions,
* change requests,
* subcontractor assessments,
* performance analysis,
* deviations and waivers,
* contract change records,
* quality cost data,
* external quality audit records.

Regarding the effectiveness of the management system, the very existence of a document is not evidence of effectiveness but it could be regarded as a record. To be a *record*, the document would need to contain results of an examination into the effect-iveness of the system.

One can demonstrate the effective operation of the management system in several ways:

* by examination of customer feedback;
* by examination of system, process and product audit results;
* by examination of the management review records;
* by examination of quality cost data;
* by examination of results against the organization's objectives.

Showing records that *every* requirement of the standard has been met will not how-ever, demonstrate that the system is effective. You may have met the requirement but not carried out the right tasks or made the right decisions. The effectiveness of the management system should be judged by how well it fulfils its purpose. Although there is no specific requirement for you to do this, you can deduce this meaning from the requirement in Clause 5.6.1 for the system to be reviewed for its continuing suitability. Some Assessors may quote this requirement when finding that you have not recorded a particular activity that is addressed in the standard. They are not only mistaken but also attempting to impose an unnecessary burden on companies that will be perceived as bureaucratic nonsense. One can demonstrate the effectiveness of the system simply by producing and examining one or more of the above records.

The subcontractor records that are delivered to you should form part of your records. However, the controls you can exercise over your subcontractor's records are some-what limited. You have a right to the records you have paid for but no more unless you invoke the requirements of this clause of the standard in your subcontract. Your rights will probably only extend to your subcontractor's records being made available for your inspection on their premises therefore, you will not be able to take away copies. It is also likely that any subcontractor records you do receive are *copies* and not originals. Before placing the contract you will need to assess what records you will require to be delivered and what records the contractor should produce and retain.

Maintaining records

The standard requires records to be maintained.

There are three types of maintenance regarding records:

* keeping records up to date;
* keeping the information in the records up to date;
* keeping the records in good condition.

Some records are designed to collect data because they pass through the process and need to be promptly updated with current information.

The filing provisions should enable your records to be readily retrievable, however, you need to maintain your files if the stored information is to be of any use. In practice, records will collect at the place they are created and unless promptly removed to secure files may be mislaid, lost or inadvertently destroyed. Once complete, records should not be changed. If they are subsequently found to be inaccurate, new records should be created. Alterations to records should be prohibited because they bring into doubt the validity of any certification or authentication because no one will know whether the alteration was made before or after the records were authenticated. In the event that alterations are unavoidable due to time or economic reasons, errors should be struck through in order that the original wording can still be read, and the new data added and endorsed by the certifying authority.

Records held electronically present a different problem and why a requirement for the protection of records is introduced (see later).

Establishing a records procedure (4.2.4)

The standard requires records *to remain legible, readily identifiable and retrievable and that a procedure defines the controls needed for the identification, storage, protection, retrieval, retention time and disposition of records.*

What does this mean?
Records have a life cycle. They are generated during which time they acquire an identity and are then assigned for storage for a prescribed period. During use and storage they need to be protected from inadvertent or malicious destruction and as they may be required to support current activities or investigations, they need to be brought out of storage quickly. When their usefulness has lapsed, a decision is made as to whether to retain them further or to destroy them.

Readily retrievable means that records can be obtained on demand within a reasonable period (hours not days or weeks). Readily identifiable means that the identity can be discerned at a glance.

Although the requirement implies a single procedure, several may be necessary because there are several unconnected tasks to perform. A procedure cannot in fact ensure a result. It may prescribe a course of action which if followed may lead to the correct result, but it is the process that ensures the result not the procedure.

Customers may specify retention times for certain records as might regulations applicable to the industry, process or region in which the organization operates. It may be a criminal offence to dispose of certain records before the limit specified in law.

Why is this necessary?

This requirement responds to the Process Approach Principle.

The requirement for a procedure is not as important as the topics the procedure is required to address. Records are important in the management of processes and building customer confidence, therefore it is important that they be maintained, retained and accessible.

How is this implemented?

Records procedures

You may only need one procedure which covers all the requirements but this is not always practical. The provisions you make for specific records should be included in the documentation for controlling the activity being recorded. For example, provisions for inspection records should be included in the inspection procedures; provisions for design review records should be included in the design review procedure. Within such procedures you should provide the forms, (or content requirement for the records), the identification, collection/submission provisions, the indexing and filing provisions. It may be more practical to cover the storage, disposal and retention provisions in separate procedures because they may not be type-dependent. Where each department retains their own records, these provisions may vary and therefore warrant separate procedures.

Legibility of records

Unlike prescriptive documents, records may contain handwritten elements and therefore it is important that the handwriting is legible. If this becomes a problem, you either improve discipline or consider electronic data capture. Records also become soiled in a workshop environment so may need to be protected to remain legible. With electronically captured data, legibility is often not a problem. However, photographs and other scanned images may not transfer as well as the original and lose detail so care has to be taken in selecting appropriate equipment for this task.

Identification of records

Whatever the records, they should carry some identification in order that you can determine what they are, what kind of information they record and what they relate to. A simple way of doing this is to give each record a reference number and a name or a title in a prominent location on the record.

Records can take various forms: reports containing narrative, computer data, and forms containing data in boxes, graphs, tables, lists and many others. Where forms are used to collect data, they should carry a form number and name as their identification. When completed they should carry a serial number to give each a separate identity. Records should also be traceable to the product or service they represent and this can be achieved either within the reference number or separately, provided that the chance of mistaken identity is eliminated. The standard does not require records to be identifiable to the product involved but unless you do make such provision you will not be able to access the pertinent records or demonstrate conformance to specified requirements.

Retrieving records

You need to ensure that the records are accessible to those who will need to use them. This applies not only to current records but also to those in the archive and any 'insurance copies' you may have stored away. A balance has to be attained between security of the records and their accessibility. You may need to consider those who work outside normal working hours and those rare occasions when the troubleshooters are working late, perhaps away from base with their only contact via a computer link. In providing for record retrieval you need to consider two aspects. You need to enable authorized retrieval of records and prohibit unauthorized retrieval. If records are held in a locked room or filing cabinet, you need to nominate certain persons as key holders and ensure that these people can be contacted in an emergency. Your procedures should define how you provide and prohibit access to the records. With electronically held records, password protection will accomplish this objective provided that you control the enabling and disabling of passwords in the records database. For this reason it is advisable to install a personnel termination/movement process that ensures passwords are disabled or keys returned on departure of staff from their current post.

Remember these records are not personal property or the property of a particular department. They belong to the organization and are a record of the organization's performance. Such records should not be stored in personal files. The filing system you create should therefore be integrated with the organization's main filing system and the file location should either be specified in the procedure that defines the record or in a general filing procedure.

If you operate a computerized record system, filing will be somewhat different although the principles are the same as for paper records. Computerized records need to be located in named directories for ease of retrieval and the locations identified in the procedures.

Storage of records

Records soon grow into a mass of paper and occupy valuable floor space. To overcome this problem you may choose to microfilm the records but keep them in the same location or archive them in some remote location. In both cases you need to control the process and the conditions of storage. With paper archives you will need to maintain records of what is where and if the archive is under the control of another group inside or outside the organization, you will need adequate controls to prevent loss of identity and inadvertent destruction.

A booking in/out system should be used for completed records when they are in storage in order to prevent unauthorized removal.

You will need a means of ensuring that you have all the records that have been produced and that none are missing or if they are, you know the reason. One solution is to index records, and create and maintain registers listing the records in numerical order as reference or serial numbers are allocated. The records could be filed in sequence so that you can easily detect if any are missing, or you can file the records elsewhere providing your registers or your procedures identify the location.

Records should also be stored in a logical order (filed numerically or by date) to aid retrieval. With electronically held records their storage should be secure from inadvertent deletion. If archived on CD ROM, floppy disk or tape, protection methods should be employed (see below).

Protection of records
The protection of records applies when records are in use and in storage, and covers such conditions as destruction, deletion, corruption, change, loss and deterioration arising from willful or inadvertent action.

On the subject of loss, you will need to consider loss by fire, theft and unauthorized removal. If using computers you will also need to consider loss through computer viruses and unauthorized access, deletion or the corruption of files.

It is always risky to keep only one copy of a document. If computer-generated, you can easily take another copy provided you always save it, but if manually generated, its loss can be very costly. It is therefore prudent to produce additional copies of critical records as an insurance against inadvertent loss. These 'insurance copies' should be stored in a remote location under the control of the same authority that controls the original records. Insurance copies of computer disks should also be kept in case of problems with the hard disk or the file server. Data back-up at defined periods should be conducted and the backed-up data stored securely at a different location than the original data.

Records, especially those used in workshop environments can become soiled and therefore, the provisions should be made to protect them against attack by lubricants, dust, oil and other materials which may render them unusable. Plastic wallets can provide adequate protection whilst records remain in use.

Retention of records (4.2.4 and 4.2.4.1)

ISO 9001 requires record retention time to be defined in a documented procedure and the automotive additions require records control to satisfy regulatory and customer requirements.

It is important that records are not destroyed before their useful life is over. There are several factors to consider when determining the retention time for records:

- *The duration of the contract*: some records are only of value whilst the contract is in force.
- *The life of the product*: access to the records will probably not be needed for some considerable time, possibly long after the contract has closed. In some cases the organization is required to keep records for up to 20 years and for product liability purposes, in the worst-case situation (taking account of appeals) you could be asked to produce records up to 17 years after you made the product.
- *The period between management system assessments*: assessors may wish to see evidence that corrective actions from the last assessment were taken. If the period of assessment is 3 years and you dispose of the evidence after 2 years, you will have some difficulty in convincing the assessor that you corrected the deficiency.

You will also need to take account of the subcontractor records and ensure adequate retention times are invoked in the contract.

Where the retention time is actually specified can present a problem. If you specify it in a general procedure you are likely to want to prescribe a single figure, say 5 years for all records. However, this may cause storage problems – it may be more appropriate therefore to specify the retention times in the procedures that describe the records. In this way you can be selective.

You will also need a means of determining when the retention time has expired so that if necessary you can dispose of the records. The retention time doesn't mean that you must dispose of them when the time expires, only that you must retain the records for at least that stated period. Not only will the records need to be dated but the files that contain the records need to be dated and if stored in an archive, the shelves or drawers also dated. It is for this reason that all documents should carry a date of origin and this requirement needs to be specified in the procedures that describe the records. If you can rely on the selection process a simple method is to store the records in bins or computer disks that carry the date of disposal.

While the ISO/TS 16949 requirement applies only to records, you may also need to retain tools, jigs, fixtures and test software; in fact, anything that is needed to repair or reproduce equipment in order to honor your long-term commitments.

Should the customer specify a retention period greater than what you prescribe in your procedures, special provisions will need to be made and this is a potential area of risk. Customers may choose not to specify a particular time and require you to seek approval before destruction. Any contract that requires you to do something different creates a problem in conveying the requirements to those who are to implement them. The simple solution is to persuade your customer to accept your policy. You may not want to change your procedures for one contract. If you can't change the contract, the only alternative is to issue special instructions. You may be better off storing the records in a special contract store away from the normal store or alternatively attach special labels to the files to alert the people looking after the archives.

Disposition of records

Disposition in this context means the disposal of records once their useful life has ended. The requirement should not be confused with that on the retention of records. Retention times are one thing and disposal procedures quite another.

As stated previously, records are the property of the organization and not personal property so their destruction should be controlled. Controls should ensure that records are not destroyed without prior authorization and, depending on the medium on which data are recorded and the security classification of the data, you may also need to specify the method of disposal. The management would not be pleased to read details in the national press of the organization's performance, collected from a waste disposal site by a zealous newspaper reporter – a problem often reported as encountered by government departments!

Validation of records

The standard does not specifically require records to be authenticated, certified or validated other than product verification records in Clause 8.2.4. A set of results without being endorsed with the signature of the person who captured them or other authentication lacks credibility. Facts that have been obtained by whatever means should be certified for four reasons:

1   They provide a means of tracing the result to the originator in the event of problems.
2   They indicate that the provider believes them to be correct.
3   They enable you to verify whether the originator was appropriately qualified.
4   They give the results credibility.

If the records are generated by computer and retained in computerized form, a means needs to be provided for the results to be authenticated. This can be accomplished through appropriate process controls by installing provisions for automated data recording or preventing unauthorized access.

Summary

In this chapter we have examined the requirements contained in Section 4 of ISO/TS 16949. Although Clause 4.1 is quite short, the six requirements addressing processes encapsulate the essence of the standard. We have explored the ways in which a management system may be defined and documented, and discovered a range of methods that can be used. Perhaps one of the most significant differences between previous versions of ISO/TS 16949:2002 is that organizations are now free to choose the documents they need to manage the business. This simple change should signal an end to mountains of paperwork created simply to satisfy auditors. It should result in a change in perception. The system should no longer be perceived as a set of documents but as a means to achieve the organization's objectives. Effort should be seen to be directed towards improving performance rather than towards improving documents.

Quality management system requirements checklist

These are the topics that the requirements of ISO/TS 16949 address consecutively.

General requirements
1   Establishing and implementing a documented quality management system
2   Implementing a documented quality management system
3   Maintaining a documented quality management system
4   Continually improving the effectiveness of the quality management system
5   Identifying processes
6   Determining process sequence and interaction
7   Determining criteria and methods for operation and control of processes
8   Availability of resources necessary to support the operation and control of processes
9   Availability of information necessary to support the operation and control of processes
10   Monitoring, measurement and analysis of processes
11   Restoring the *status quo*
12   Continual process improvement
13   Process management
14   Outsourcing processes
15   Responsibility for outsourced processes

Documentation requirements
General
16   Documenting the quality policy and objectives
17   Documenting the quality manual
18   Documenting the quality management system procedures
19   Documenting the information needed for the effective operation and control of processes
20   Documenting records

Quality manual
21   Establishing and maintaining a quality manual

Control of documents
22   Controlling documents required by the quality management system
23   Establishing document control procedures
24   Approval of documents
25   Review and revision of documents
26   Re-approval of documents after revision
27   Identifying revision status
28   Availability of documents
29   Identification and legibility of documents
30   Control of documents of external origin
31   Control of obsolete documents
32   Identifying obsolete documents retained for use

Engineering specifications
33   Review, distribution and implementation of customer engineering information
34   Change implementation records

Control of records
35   Establishing and maintaining records of conformity
36   Establishing and maintaining records of effectiveness
37   Ensuring legibility of records
38   Ensuring identification of records
39   Retrieval of records
40   Establishing a records control procedure

Records retention
41   Defining record retention periods

Quality management system – Food for thought

1 Is your system thought of as a set of documents or a set of interconnected processes that deliver the organizations objectives?

2 Is your system integrated into the organization so that people do the right things right without having to be told?

3 Is your system a collection of interconnected processes rather than a series of interconnected functions?

4 Does every process in the chain of processes from requirements to their satisfaction add value?

5 Are your business objectives functionally oriented driving a functional-oriented organization or are they process-oriented?

6 If you have simply changed the names of your procedures to reflect processes where have you defined the process objectives, the resources and the behaviors required to cause these objective to be achieved and the measures required to determine process adequacy, efficiency and effectiveness?

7 Are your processes described simply as a series of transactions or are there provisions in the process to manage its performance?

8 Do the outputs from one process connect with other processes?

9 Do all the inputs to a process have their origin in other processes or external organizations?

10 Do you have measures that enable you to determine how well each process is performing and how these measures are known to those controlling the process?

11 Is there any activity, task or processes that exists only to meet the requirements of ISO/TS 16949?

12 Are you sure that all the documentation in place is needed for the effective operation and control of your processes?

13 There are six ways of conveying information. Before you document, have you eliminated the other five ways as being unsuitable for the particular situation?

14 If a person moves onto another job, how much of what is removed from the process is essential for the process to maintain its capability?

15 How much of what affects your ability to achieve results is dependent on staff following documented procedures?

16 How much of what affects your ability to achieve results is dependent on the physical and human environment in which your staff work?

17 How much of what affects your ability to achieve results is dependent on your staff's ability to do the right things right?

18 How much of what affects your ability to achieve results is dependent on your staff's motivation to do the right things right?

19 Do you question the effectiveness of established policies and procedures first when the customer complains or do look for someone to blame?

20 What causes the actions and decisions for which there are no documented policies or procedures?

21 Are you managing a set of functions or a series of processes and do you know the difference?

22 Do you know what each process aims to achieve?

23 Do you know how each process causes the observed results?

24 Do you know whether the process is producing outcomes that satisfy the process objectives?

25 Do you know how to change process performance to bring it into line with the objectives?

Chapter 5

Management responsibility

Attention to quality can become the organization's mind-set only if all of its managers – indeed all of its people – live it.

Tom Peters

Summary of requirements

While the implementation of *all* requirements in ISO/TS 16949 is strictly management's responsibility, those in Section 5 of the standard are indeed the responsibility of top management. All clauses in this section commence with the phrase "Top management shall …". The first four clauses clearly apply to the strategic-planning processes of the organization rather than to specific products. However, it is the board of directors that should take note of these requirements when establishing their vision, values, mission and objectives. These requirements are amongst the most important in the standard. There is a clear linkage between customers needs, policy, objectives and processes. One leads to the other in a continuous cycle as addressed previously in Chapter 1. Although the clauses in Section 5 are not intended as a sequence, each represents a part of a process that establishes direction and keeps the organization on course. If we link the requirements together in a cycle (indicating the headings from ISO/TS 16949 in bold italics type) the cycle commences with a **Vision** – a statement of what we want to be or do – and then a **Focus on customers** for it is the customer that will decide whether or not the organization survives. It is only when you know what your market is, who your customers will be and where they will be that you can define the **Purpose** or **Mission** of the organization. From the purpose or mission you can devise a **Vision** (where you want to get to, i.e. what you want to become) and from the mission come the **Policies or Values** that will guide you on your journey. These policies help frame the **Objectives**, the milestones en route towards your destination. The policies won't work unless there is **Commitment** so that everyone pulls in the same direction. **Plans** have to be made to achieve the objectives, and these plans need to identify and layout the

Processes that will be employed to deliver the results – for all, work is a process and without work nothing will be achieved. The plans also need to identify the **Responsibilities and Authority** of those who will be engaged in the endeavour. As a consequence it is essential that effective channels of **Internal Communication** be established to ensure that everyone understands what they are required to achieve and how they are performing. No journey should be undertaken without a means of knowing where you are, how far you have to go, what obstacles are likely to lie in the path ahead or what forces will influence your success. It is therefore necessary to collate the facts on current performance and predictions of what lies ahead so that a **Management Review** can take place to determine what action is required to keep the organization on course or whether any changes are necessary to the course or the capability of the organization for it to fulfil its purpose and mission, and so we come a full circle. What the requirements of Section 5 therefore address is the mission management process with the exception of process development, which happens to be addressed in Section 4 of the standard.

Management commitment (5.1)

Commitment to the QMS (5.1)

The standard requires that top management *provide evidence of its commitment to the development and implementation of the quality management system and continually improving its effectiveness.*

What does this mean?
Top management
Throughout this section of the standard the term *Top management* is used. Previously QS-9000 used the term Management with executive responsibility. Although it is customary to only grant executive responsibility to department-level managers and above, the requirement did not compel the top managers and directors to take any of the actions implied by the standard because executive responsibility could in fact be held by personnel at any level in an organization's hierarchy. This change is therefore significant because it does bring the actions and decisions of top management into the QMS and makes them full partners in the development, operation and improvement of the system. Top management is defined in ISO 9000 as *the person or group of people who direct and control an organization* and therefore sit at the top of the tree.

Commitment
A commitment is an obligation that a person (or a company) takes on in order to do something. It is very easily tested by examination of the results. It is also easy for you to make promises to resolve immediate problems hoping that the problem may go away

in due course. This is dishonest and although you may mean well, the problem will return to haunt you if you can't deliver on your promise.

A commitment exists if a person agrees to do something and informs others of their intentions. A commitment that is not communicated is merely a personal commitment with no obligation except to one's own conscience.

Commitment therefore means:

- Doing what you need to do to meet the organization's objectives.
- Doing what you say you will do.
- Not accepting work below standard.
- Not shipping product below standard.
- Not walking by problems, not overlooking mistakes.
- Improving processes.
- Honouring plans, procedures, policies and promises.
- Listening to the workforce.
- Listening to the interested parties.

Commitment to a management system
Although the standard does indicate how commitment to a management system should be demonstrated it presupposes that top management understands what it is committing itself to. There is a presumption is that a management that is committed to the development, implementation and continual improvement of a management system will be committed to quality because it believes that the management system is the means by which quality will be achieved. This is by no means obvious because it depends on top management's perception of the management system. (This argument was also addressed with respect to Clause 0.1 of the standard that we discussed in Chapter 1 under the heading *Level of attention*.) To some the management system may be a set of documents, instructions for performing administrative tasks, a collection of procedures, etc. Many managers have given a commitment to quality without knowing what impact it would have on their business. Many have given a commitment without being aware that their own behaviour may need to change let alone the behaviour of their managers and staff. The phrase "continually improving its effectiveness" clearly implies that top management has to ensure the management system continually fulfils its purpose and hence need to understand what its purpose is. A study of Clauses 4.1 and 5.1 together with the definitions in ISO 9000 should leave one in no doubt that the purpose of the management system is to enable the organization to satisfy its customers. Expressing a commitment to the development and improvement of the management system means a belief that the management system is the means by which quality is achieved, and that resources will be provided for its development, implementation, maintenance and improvement.

As the revised requirement now moves the focus from quality to the means for its achievement it has the tendency for creating even greater misunderstanding than previously. The perceptions of quality may well be customer focused but the managers may perceive the management system as only applying to the quality department and consequently may believe that a commitment to the development of a management system implies that they have to commit to maintaining a quality department. This is not what is intended. A quality department is the result of structuring a particular organization to meet its business objectives. There are several other solutions that would not result in the formation of a department dedicated to quality. There is certainly no requirement in ISO/TS 16949 for a department that is dedicated to quality.

Why is this necessary?

This requirement responds to the Leadership Principle.

Management must be seen to do what it says it will do so as to demonstrate its understanding. There is no doubt that actions speak louder than words for it is only when the words are tested that it is revealed whether the writers are serious. It is not about whether management can be trusted but whether they understand the implications of what they have committed themselves to. The wording of this requirement is no accident. The requirements of QS-9000 were clearly not interpreted as intended, and it has therefore been necessary to strengthen the message that the management system is supposed to make the right things happen and not simply be regarded as a set of procedures.

How is this implemented?

Defining top management

In order to clarify who in the organization is a member of top management it will be advantageous to specify this in the Quality Manual. It is then necessary to ensure that the positions of the personnel performing the specified actions are from top management.

A search of Section 5 of the standard will reveal that Top management are required to:

*   provide evidence of commitment to the development and improvement of the management system;
*   communicate the importance of meeting customer requirements;
*   establish the quality policy and ensure it meets prescribed criteria;
*   ensure that customer needs and expectations are determined, converted into requirements and fulfilled to achieve customer satisfaction;
*   establish quality objectives and ensure they are established at each relevant function and level;
*   include the quality objectives in the business plan;
*   designate personnel to ensure that customer requirements are addressed;

- ensure that resources needed to achieve the objectives are identified, planned and available;
- appoint members of management who shall have responsibility for ensuring the management system is established, etc.;
- review the management system to ensure its continuing suitability and effectiveness.

Merging requirements in Sections 5.1 to 5.6 of the standard produced this list. It is interesting that in some of the above references top management is required to *ensure* something rather than *do* something. To ensure means to make certain and top management can't make certain that something will happen unless it is in control. However, it does mean that it can delegate to others the writing of the policy, the objectives and provision of resources.

In some organizations, there are two roles, one of Management Representative and another of Quality Manager with the former only being in top management. It is clear that in such cases, the Quality Manager should not be the person carrying out these actions but the person who is a member of top management.

Commitment

Once communicated, a commitment can be tested by establishing:

- if resources have been budgeted for discharging the commitment;
- that resources are allocated when needed;
- that performance of the tasks to which the person has given his or her commitment are progressed, monitored and controlled;
- that deviations from commitment are not easily granted.

In managing a management system, such tests will need to be periodically carried out even though it will be tedious to both the person doing the test and the person being subjected to it. It is less tedious if such tests are a feature of the programme that the management has agreed to, thereby making it impersonal and by mutual consent.

The management has to be committed to quality, in other words it must not knowingly ship defective product, give inferior service or in any other way dissatisfy its stakeholders. It must do what it says it will do and what it says it will do must meet the needs and expectations of the stakeholders. A manager who signs off waivers without customer agreement is not committed to quality whatever the reasons. It is not always easy, however, for managers to honour all their commitments when the customer is screaming down the phone for supplies that have been ordered or employees are calling for promised pay rises. Unlike QS-9000, the standard now requires proof that managers are committed to quality through their actions and decisions. When they start spending time and money on quality, diverting people to resolve problems, motivating their staff to achieve performance standards, listening to their staff and to customers, there is *commitment*. It will also be evident from customer feedback, internal and external audits and sustained business growth. Increased profits do not necessarily show that the company is committed to quality. Profits can rise for many reasons not necessarily because of an improvement in quality. Managers should not just look at profit results to measure the success of the improvement programme. Profits may go down initially as investment is made in management system development. If managers abandon the programme because of short-term results, it shows not only a lack of commitment but also a lack of understanding. Every parent knows that a child's education does not bear fruit until he or she is an adult. It is therefore much better to tailor the programme to available resources than abandon it completely.

Commitment to a management system

Although the standard defines how top management should demonstrate its commitment there are some obvious omissions. Commitment to the development and improvement of a management system could be demonstrated by evidence that top management:

* understands the role of the management system in relation to achieving business objectives and customer satisfaction;
* is steering management system development and improvement effort, and monitoring its performance;
* is promoting core values, adherence to policy, best practice and continual improvement;
* undertakes the top management actions in a timely manner;
* does not defer decisions that are impeding progress in management system development and improvement;
* is implementing the processes that have been developed to achieve business objectives;
* is actively stimulating system improvements.

Communicating the importance of requirements (5.1a)

The standard requires that top management *communicate to the organization the importance of meeting customer as well as statutory and regulatory requirements.*

What does this mean?

Communication

This requirement is the primary application of the *customer focus principle*. Effective communication consists of four steps: attention, understanding, acceptance and action. It is not just the sending of messages from one source to another. It is therefore not enough for top management to publish a quality policy that declares the importance of meeting customer, regulatory and legal requirements – this is only the first step and depending on how this is done, it may not even accomplish the first step.

Statutory and regulatory requirements

Although the ISO 9000 family addresses all interested parties, ISO/TS 16949 focuses on customers and therefore the statutory and regulatory requirements referred to in this requirement are those pertaining directly or indirectly to the product or service supplied. However, there are few regulations that in some way would not impact the customer if not complied with. Customers not only purchase products and services but also often desire a sustaining relationship so that they can rely on consistent quality, delivery and cost. If the operations are suspended through non-compliance with environmental, health and safety legislation, existing customers will be affected and potential customers lost. The absence of key personnel due to occupational health reasons may well impact deliveries and relationships. Some customers require their suppliers to operate ethically and conserve the environment. If the financial laws are breached causing the organization to cut costs to pay the fines, credit worthiness may also be affected. Such breaches also distract management from focusing on customers; therefore it is not only the product-related regulations that can affect customer satisfaction.

There is also a secondary meaning. Some customers may require the organization to provide products and services that do not comply with statutory and regulatory requirements.

Why is this necessary?

This requirement responds to the Customer Focus Principle.

Top management need to attract the attention of the workforce and this requires much more then pinning-up framed quality policy statements all over the place. In organizations in which the producers are remote from the customer, it is necessary to convey customer expectations down the line so that decisions are made primarily on the basis of satisfying these expectations and not on the secondary internal requirements. It is also necessary for management to signal its intentions with respect to customer requirements so that personnel know of the factors affecting the work priorities.

How is this implemented?

Before embarking on a communication programme top management should examine its culture relative to profit. If asked why the organization exists, many would say it is to make a profit, clearly indicating they are a profit-focused organization (not a customer-focused organization). Profits are important but are the *result* of what we do – not the *reason* for doing it. Profits are needed to pay off loans, for investment in new technology, new plant and new skills, and to award higher salaries but they come from satisfying customers, and operating efficiently and effectively. Focusing on profit creates a situation where people constantly observe the bottom line rather than the means to achieve it so that most decisions are based on cost rather than quality. Managers in such organizations rule by fear and treat a lack of capability as an excuse rather than a valid reason why targets are not met. It is right to question the value of any activity but with the focus on customer satisfaction not profit. It is vital therefore that top management consider the culture they have created before communicating the importance of customer requirements. Where profit is the first priority, stressing the importance of customer requirements will create a conflicting message. Management needs to make employees understand that the priorities have changed so that the aim is to satisfy customers profitably.

If the customer's requirements cannot be satisfied profitably, business should not be transacted unless there are opportunities for recovering the loss in another transaction. It is also important that before accepting a commitment to supply, a clear understanding of customer needs and expectations as well as regulatory and legal requirements is established – hence sales personnel need to understand that they should not make promises the organization cannot honour.

Meeting this requirement may therefore require a realignment of priorities and careful consideration as to the way management will attract the attention, understanding, acceptance and correct action of the workforce.

Establishing quality policy (5.1b)

The standard requires that top management *establish the quality policy*.

What does this mean?
ISO 9000 defines a quality policy as the overall intentions and direction of an organization related to quality as formally expressed by top management. It also suggests that the policy be consistent with the overall policy of the organization and provide a framework for setting quality objectives. Furthermore ISO 9000 advises that the eight quality management principles be used as a basis for forming the quality policy. The quality policy can therefore be considered as the values, beliefs and rules that guide actions, decisions and behaviours. A value may be 'integrity' and expressed as: *we will be open and honest in our dealings with those inside and outside the organization*. A rule may be 'confidentiality' and expressed as *Company information shall not be shared with those outside the organization*. Both these are also beliefs because it might be believed that deceiving people only leads to failure in the long run. It might also be believed that disclosing confidential information fuels the competition and will drive the organization out of business. Both values guide actions, decisions and behaviours, and hence may be termed *policies*. They are not objectives because they are not achieved – they are demonstrated by the manner in which actions and decisions are taken and, the way your organization behaves towards others. The relationship between policy and objectives is important as ISO 9001 implies one is derived from the other but in reality the relationship is quite different as illustrated in Figure 5.1.

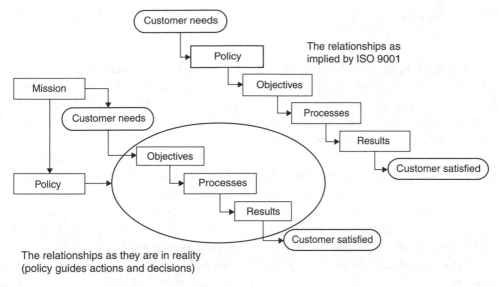

Figure 5.1 *Policy–objectives–process relationship*

The detail of quality policy will be addressed later. What is important in this requirement is an understanding of why a quality policy is needed, what is required to *establish* a quality policy and where it fits in relation to other policies.

Corporate terminology

There are so many similar terms that are used by upper management that it is not surprising that their use is inconsistent. We use the term *goal* when we mean purpose or perhaps we really did mean goal or should we use the term *mission* or *objective*, or perhaps a better term would be *target* … and so on. The problem is that we often don't know the intension of the user. Did he or she carefully select the term to impart a specific meaning or would the alternatives have been equally appropriate?

Purpose	Why we exist, why we do what we do.
Mission	Where we are going: the journey.
Goals	Our intended destination.
Vision	What we want to become.
Values	What beliefs will guide our behaviour.
Strategy	How we are going to get there.
Policy	Rules that guide our actions and decisions: the signposts en route.
Principles	Fundamental truth.
Objectives	What we want to achieve: the results.
Measures	What will indicate achievement.
Targets	What we aim at to achieve objectives.

Why is this necessary?

This requirement responds to the Leadership Principle.

Defining the purpose or mission of the business is one thing but without some guiding principles, the fulfilment of this mission may not happen unless effort is guided in a common direction. If every manager chooses his or her direction, and policies, the full potential of the organization would not be realized. A shared vision is required that incorporates shared values and shared policies.

The purpose of corporate policies is to influence the short- and long-term actions and decisions and to influence the direction in which the mission will be fulfilled. If there were policies related to the organization's customers, they could be fulfilled at the expense of employees, shareholders and society. If there were policies related to profit, without other policies being defined, profit is positioned as a boundary condition to all actions and decisions. Clearly this may not direct the organization towards its mission.

How is this implemented?

To establish means to put in place permanently. A quality policy that is posted in the entrance hall – published yes, but not established. For a policy to become established, it has to reflect the vision of the organization and underpin every conscious thought and action. This will only arise if everyone believes in the policy. For this to happen, managers need to become the role model so that by their actions and decisions they exemplify the policy. Belief in the policy is unlikely if the quality policy is merely perceived as something written only to satisfy ISO/TS 16949.

With QS-9000, the focus was on product quality and consequently quality was associated with order-driven processes that delivered goods to customer requirements. The new definition of quality in ISO 9000 clearly goes beyond relations with the purchaser and embraces all interested parties. The eight quality management principles also go far beyond the mechanical processes for achieving product quality and embrace the softer factors that influence the behaviour of people in an organization. Clearly, a quality policy that addresses all these issues comes close to reflecting all the policies of the organization and hence can be termed the corporate policy. While organizations may have a safety policy, an environmental policy, a personnel policy, a servicing policy, etc., these are really topics within the corporate policy. Quality is not just another topic but a term that embraces all the topics. It is hard to think of any policy that could not be classed as a quality policy when quality is defined as *the degree to which a set of inherent characteristics fulfils requirements*. This definition does not limit *quality* to the fulfilment of customer requirements, but extends it to the fulfilment of any requirements including employees, suppliers, in fact any interested party.

The phrase "a set of inherent characteristics" might be perceived as being limited to products and services, but an organization, a person, a process, a decision, a document or even an environment has a set of inherent characteristics.

Safety policies are quality policies because they respond to employees and customers as interested parties. Environmental policies are quality policies because they respond to society as an interested party. There is therefore no advantage in issuing a separate quality policy – it is more effective if the organization formulates its corporate policies and within them addresses the topics covered by the eight quality management principles.

Establishing quality objectives (5.1c)

The standard requires that top management *ensure that quality objectives are established*.

What does this mean?
ISO 9000 defines quality objectives as *results sought or aimed for related to quality*. It also suggests that these objectives be based on the quality policy and be specified at different levels in the organization, being quantified at the operational level. As with quality policy the details will be addressed later and here we will focus on what it means to establish quality objectives, and how they relate to other objectives.

As the quality policy equates to the corporate policy, it follows that quality objectives equate to corporate objectives. All of the organization's objectives should in some way serve to fulfil the requirements of customers and other interested parties. It is also interesting to note that in ISO 9000, the term *requirement* is defined as *a need or expectation that is stated, customarily implied or obligatory*. While an investor may not specify a requirement for growth in share value, it would certainly be an expectation. While an employee does not express requirements for salary increases when profits rise, it would certainly be an expectation and while society has no way other than to protest or invoke the law to impose its desires on an organization, it certainly has the power to make organization's comply and even change the law in extreme cases. So quality objectives *do* equate to corporate objectives.

Why is this necessary?
This requirement responds to the Leadership Principle.

Management needs to ensure that the objectives are established as a basis for action. All work serves an objective and it is the objective that stimulates action. The reason for top management setting the objectives is to ensure that everyone channels their energies in a positive direction that serves the organizations purpose and mission.

How is this implemented?

For an objective to be established it has to be communicated, translated into action and become the focus of all achievement. Objectives are not wish lists. The starting point is the purpose and mission statement, and first among the factors identified as affecting the ability of the organization to accomplish its mission; is *customer satisfaction*. If you do not satisfy your customers you will have no business. Second will be satisfying all other stakeholders. If you do not satisfy the needs and expectations of society, employees, suppliers and particularly shareholders they will withdraw their stake in the business and you won't have any resource with which to operate the business. It is in these areas the organization needs to excel and therefore they become the focus for action and consequently the setting of objectives. Although ISO 9000 suggests that the quality objectives should be based on the quality policy, it is more likely to be current performance, competition and opportunities arising from new technology that drive the objectives. The setting of objectives is addressed further under the heading Quality objectives (5.4.1).

Conducting management reviews (5.1d)

The standard requires that top management *conduct management reviews.*

What does this mean?

The term *review* is defined in ISO 9000 as *an activity undertaken to ensure the suitability, adequacy, effectiveness and efficiency of the subject matter to achieve established objectives.* The addition of the term *management* means that the management review can be perceived as a review *of* management rather than a review *by* management, although both meanings are conveyed in the standard. The rationale for this is that the examples given in ISO/TS 16949 such as design review and nonconformity review clearly indicate it is design and nonconformity that is being reviewed. If the system was to be reviewed then the action should be called a *system review.* It is no doubt unintentional in the standard but, if the management system is perceived as the way in which the organization's objectives are achieved, a review of management is in fact a review of the way achievement of objectives is being managed because the organization exists to achieve objectives and so both meanings are correct.

Why is this necessary?

This requirement responds to the Factual Approach Principle.

Top management has set the policies and agreed the objectives, and means for their achievement, i.e. the management system. It follows therefore that periodic checks are needed to establish that the management system continues to fulfil its purpose. Failure to do so will inevitably result in deterioration of standards and performance as inherent weaknesses in the system build up to eventually cause catastrophic failure, i.e. customer satisfaction will decline and orders, markets and business will be lost. It may be argued that this won't happen because people won't let it happen – they will take action. If these conditions persist, what will emerge is not a managed system but an unmanaged system, that is unpredictable, unreliable with erratic performance. A return to the days before the management system (as defined in this book) was established.

How is this implemented?

Top management will not regard the management review as important unless they believe it is essential to running the business. The way to do this is to treat it as a business performance review. This is simpler than it may appear. If the quality policy is now accepted as corporate policy and the quality objectives are accepted as corporate objectives, any review of the management system becomes a performance review and no different to any other executive meeting. The problem with the former management reviews was that they allowed discussion on the means for achieving objectives to take place in other management meetings leaving the management review to a review of errors, mistakes and documentation that no one was interested in anyway. The management system *is* the means for achieving objectives therefore it makes sense to review the *means* when reviewing the *ends* so that actions are linked to results and commitment secured for all related changes in one transaction.

The requirement emphasizes that top management conduct the review (not the quality manager, not the operational manager but *top management*), those who direct and control the organization at the highest level. In many ISO 9000 and ISO/TS 16949 registered organizations, the management review is a chore, an event held once each year, on a Friday afternoon before a national holiday – perhaps a cynical view but nonetheless often true. The reason the event has such a low priority is that management have not understood what the review is all about. Tell them it's about reviewing nonconformities, customer complaints and internal audit records, and you will be lucky if anyone turns up. The quality manager produces all the statistics so the other managers are free of any burden. By careful tactics, these managers may come away with no actions, having delegated any in their quarter to the quality manager.

In order to provide evidence of its commitment to conducting management reviews, management would need to demonstrate that it planned for the reviews, prepared input material in the form of performance results, metrics and explanations, decided what to do about the results and accepted action to bring about improvement. It would also show that it placed the burden on all managers and did not single out the quality manager as the person responsible and accountable for the management review.

Ensuring availability of resources (5.1e)

The standard requires that top management *ensure the availability of necessary resources*.

What does this mean?

A resource is something that can be called on when needed and therefore includes time, personnel, skill, machines, materials, money, plant, facilities, space, information, knowledge, etc. *To ensure* means to make certain. For top management *to ensure the availability of necessary resources*, it has to know:

- what objectives it is committed to achieve,
- what resources are required to achieve these objectives,
- when the resources will be needed,
- of the availability of such resources,
- the cost of acquiring these resources,
- that all of the above is feasible.

The resources referred to here are not only those needed to document the management system, audit and update it and verify product conformity as was implied in QS-9000. They are the resources needed to run the business.

Why is this necessary?

This requirement responds to the Leadership Principle.

No plans will be achieved unless the resources to make them happen are provided. It is therefore incumbent on top management to not only require a management system to be developed, implemented and maintained but also provide the means by which this is accomplished. In many organizations, the quest for ISO 9000 or ISO/TS 16949 certification has resulted in an existing manager being assigned additional responsibilities for establishing the management system without being provided with additional resources. Even if this person did have surplus time available, it is not a job for one person. Every manager should be involved and they too need additional resources. In the long term, the total resources required to maintain the organization will be less with an effective management system than without but to start with, additional time and skills are required and need to be made available.

How is this implemented?

One of the problems with implementing the QS-9000 was that it did not require management to link resource provision with objectives. It is not uncommon for management to budget for certain resources and when the time comes to acquire them, the priorities have changed and the ambitious plans are abandoned. Top management will need to be more careful when agreeing to objectives and plans for their accomplishment. It will need to have confidence that, excluding events beyond its control, funds will be available to acquire the resources committed in the improvement plans. This does not mean that management will be forced to fund plans when it clearly has no funding available but such circumstances need to be monitored. Any lack of funding should be reviewed to establish whether it was poor estimating, forces outside the organization's control or a genuine lack of commitment. Opportunities often change and it would be foolish to miss a profitable business opportunity while pursuing an improvement programme that could be rescheduled. The risks need to be assessed and the objectives adjusted. It could also be that such business opportunities remove the need for the improvement programme because it removes the process or product line that is in need of improvement. However, managers at all levels need to be careful about approving plans which are over ambitious, impractical or not feasible with the anticipated resources that will be available.

Process efficiency (5.1.1)

The standard requires top management to *review processes to assure their efficiency and effectiveness.*

What does this mean?

This requirement means that top management determine the extent to which the processes reflect best practice and the extent to which the process objectives align with stakeholder needs and expectations. It is not a review to establish that the processes are achieving their objectives, i.e. the purpose of the measurements undertaken in Clause 8.2.3.

Why is this necessary?

There are three measures of performance. Are we doing it OK, can we do it better and are we doing the right things. This requirement addresses the second and third question because the first question is answered by measurements performed to satisfy Clause 8.2.3.

How is this implemented?

Process efficiency is a function of conformity and consumption. Firstly, we need to know if the processes are running as planned. The people are doing what we stated they should do: following the policy, the procedures and the instructions as documented. We determine this by audit (see Chapter 8). The results of the audit will tell us the extent to which policies and procedures are being followed, not necessarily whether the results are acceptable. There are several outcomes for which data needs to be presented to top management:

- We are doing it OK, and are following the documented procedures. If this is the outcome then can we do it by consuming fewer resources (time, money, materials, people, etc.).
- We are doing it OK, but are not following the documented procedures. If this were the case the documented procedures are ineffective so need to be re-designed.
- We are not doing it OK, and are not following the documented procedures. If this were the case, would it make any difference if we did follow the procedures? If not the process needs to be re-designed.

If we are doing it OK, and are following the documented procedures, the question we must ask ourselves is, are we using the right success measures, are we aiming at the right targets, do we have the right objectives? The answers to these come out of market research, benchmarking or simply customer focus meetings.

Customer focus (5.2)

Determining customer requirements (5.2)

The standard requires top management *to ensure that customer requirements are determined.*

What does this mean?
A customer is defined in ISO 9000 as an organization or person that receives a product. This could be a consumer, client, end user, retailer, beneficiary or purchaser. The meaning has therefore changed considerably from the 1987 version of ISO 9000 that was primarily focused on purchasers.

In the Committee Drafts of ISO 9001:2000, this requirement was expressed differently. It required that customer needs and expectation be determined – which to many is very different from determining customer requirements. The former implies that the organization should be proactive and seek to establish customer needs and expectations before commencing the design of products and services, and offering them for sale. The latter implies that the organization should react to the receipt of an order by determining what the customer wants. However, the original meaning has not been lost because ISO 9000 defines a requirement as *a need or expectation that is stated, customarily implied or obligatory.* We can therefore use the term *requirement* or the terms *needs, expectations and obligations* as synonymous. As there is no indication in the statement that such requirements are limited to those in an order or contract, it can be interpreted as requiring both proactive and reactive action.

The requirement is specific in that it requires top management *to ensure* customer requirements are determined – meaning that it has to ensure an effective process is in place for determining customer requirements.

Why is this necessary?
This requirement responds to the Customer Focus Principle.

All organizations have customers. Organizations exist to create and retain satisfied customers; those that do not do so, fail to survive. Not-for-profit organizations have customers even though they may not purchase anything, they give and they take, and if the organization fails to fulfil their needs, it ceases to exist. Governments are a prime example.

If they fail to satisfy the voters they fail to be re-elected. It is therefore essential for the survival of an organization that it determines customer requirements.

How is this implemented?

The requirement is implemented through the 'Demand Creation Process'. This requirement extends the management system beyond the processes required to satisfy current customers and clearly brings the marketing process as well as the sales process into the management system.

The marketing process is primarily concerned with finding out what customers want and attracting them to the organization so that these wants are satisfied. In this process it is important to keep the organization's purpose and mission in focus because all too easily the organization may become entangled in pursuing opportunities that others may be far better equipped to satisfy – "stick to the knitting" as Tom Peters would say.[1] There are millions of opportunities out there. The key is to discover those that your organization can exploit better than any other and generate a profit by doing so.

The sales process is primarily concerned with making contact with customers for existing products and services, and converting enquiries into firm orders. In this process, it is important that the customer requirements are determined so as to match the benefits of existing products and services with the needs and expectations of customers. The tender/contract review process is therefore important in ensuring needs are understood before a commitment to supply is accepted.

Understanding customer needs

In order to determine customer needs and expectations you need to answer some important questions:

* What is our business?
* What will our business be?
* What should our business be?

Answers to questions above are obtained by answering the following questions:

* Who are our customers?
* Where are our customers?
* What do customers buy?
* What is value to the customer?
* Which of the customers wants are not adequately satisfied?

The change in direction of the American people that quickly came about with the mass production of the automobile-made Amtrak think long and hard about what business they were in. If they stayed in the railroad business for much longer they would cease to exist. They therefore came to the conclusion that they were not in the railroad business but in the transportation business and as a result they adapted and prospered.

The answers to the above questions will enable marketing objectives to be established for:

- existing products and services in present markets;
- abandonment of obsolete products, services and markets;
- new products and services for existing markets;
- new markets;
- service standards and service performance;
- product standards and product performance.

The results of market research will be a mix of things. It will identify:

- new potential customers for existing products and services;
- new potential markets;
- opportunities for which no technology exists;
- opportunities for which no product or service solution exists;
- enhancements to existing products and services.

The organization needs to decide which of these to pursue and this requires a marketing process that involves all the interested parties. Marketing processes that only involve the marketers will not exploit the organization's full potential. The contributions from design, production, service delivery, legal and regulation experts are vital to formulating a robust set of customer requirements from which to develop new markets, new products and new services. The research may identify a need for improvement in specific products or a range of products, but the breakthroughs will come from studying customer behaviour. For example, research into telecommunications brought about the mobile phone and technology has reduced it in size and weight so that the phone now fits into a shirt pocket. Further research on mobile phones has identified enhancements such as access to e-mail and the Internet through the mobile phone but whether these are essential improvements is debatable. The fear of radiation and driving laws means that a breakthrough will arise by eliminating manual interaction so that the communicator is worn like a hat, glove or a pair of spectacles, being voice activated and providing total hands-free operation.

Determining customer needs and expectations should not be limited to your present customer-base. Customers may want your products but may be unable to obtain them. If your products and services are limited to the home market either due to import regulations or distribution policies you could satisfy a new sector of the market with your existing products and extend their life. Austin Morris did this in the 1960s with their Morris Cowley and Oxford by exporting the technology to India. Forty years later,

Rover is exporting its technology to China and perhaps finally closing the door on the British Motor Industry.

From analysing the results of the research a design brief or requirement can be developed that translates customer needs and expectations into performance, physical and functional characteristics for a product or service. This forms the basis of the input into the product and service design processes. Often this requirement is no more than a couple of lines on a memorandum. The process of converting needs into requirements can therefore be quite protracted and iterative. With some projects, establishing requirements is a distinct phase that is put out for tender and where the winner may not necessarily win a subsequent contract to develop the product.

If you misunderstand customer needs and expectations you will produce an inadequate set of requirements, often not knowing they are inadequate until you launch the product into the market. It is therefore the most important stage in the product realization process where ideas and beliefs are tested and retested to ensure they really do reflect customer needs and expectations.

Gathering the data

Decisions affecting the future direction of the organization and its products and services are made from information gleaned through market research. Should this information be grossly inaccurate, over optimistic or pessimistic the result may well be the loss of many customers to the competition. It is therefore vital that objective data is used to make these decisions. The data can be primary data (data collected for the first time during a market research study) or secondary data (previously collected data). However, you need to be cautious with secondary data because it could be obsolete or have been collected on a different basis than needed for the present study.

The marketing information primarily identifies either *problems* or *opportunities*. *Problems* will relate to your existing products and services, and should indicate why there has been a decline in sales or an increase in returns. In order to solve these problems a search for possible causes should be conducted and one valid method for doing this is to use the Fishbone Diagram or Cause and Effect Diagram. *Opportunities* will relate to future products and services, and should indicate unsatisfied wants. There are three ways of collecting such data: by observation, survey and experiment.

Teruyuki Minoura, who in 2003 was Managing Director of Global Purchasing for Toyota Motor Corporation warned that suppliers needed to shift their focus to the car user instead of the carmaker. "You are going to have to start analysing the needs and wants of the end user. You're going to be finding out what end users want and working to develop suitable components. Then you're going to be offering what you've developed to carmakers like us, who are going to incorporate these components into our designs."

Observation studies are conducted by actually viewing the overt actions of the respondent. In the automotive industry this can either be carried out in the field or in the factories where subcontractors can observe their customer using their materials or components.

Using surveys is the most widely used method for obtaining primary data. Asking questions that reveal their priorities, their preferences, their desires, their unsatisfied wants, etc., will provide the necessary information. Information on the profile of the ultimate customers with respect to location, occupation, life style, spending power, leisure pursuits, etc., will enable the size of market to be established. Asking questions about their supplier preferences and establishing what these suppliers provide that you don't provide is also necessary. Knowing what the customer will pay more for is also necessary, because many may expect features that were once options, to be provided as standard.

A method used to test the potential of new products is the *controlled experiment*, using prototypes, alpha models, etc., distributed to a sample of known users. Over a limited period these users try out the product and compile a report that is returned to the company for analysis.

A source of secondary data can be trade press reports and independent reviews. Reading the comments about other products can give you some insight into the needs and expectations of potential customers.

Meeting requirements (5.2)

The standard requires *customer requirements to be met with the aim of enhancing customer satisfaction*.

What does this mean?

This requirement changes the focus from *doing what you say you do* to *doing what you need to do to satisfy your customers*. It also means that if your interpretation of customer requirements is incorrect in some way, you have an obligation to go beyond the requirements and aim for customer satisfaction. It does not mean however, that you must satisfy customers regardless of their demands. Some customers are unreasonable and expect something for nothing so it is your choice not to supply them if that is the case.

Why is this important?

This requirement responds to the Customer Focus Principle.

Organizations only stay in business by satisfying their customers. ISO/TS 16949 focuses on customers but organizations can fail to survive if they do not also satisfy other interested parties such as their employees, regulators and society. People will only continue to work for organizations that treat them fairly. Regulators can close down businesses if they breach the rules and if all else fails, society can influence government and so change the law to stop organizations behaving in a way that harms the population not only of the host country but even the earth itself.

How is this implemented?
The whole standard addresses the elements of a management system that aims to achieve customer satisfaction and therefore by constructing and operating a system that meets the intent of ISO/TS 16949, this goal will be achieved.

Quality policy (5.3)

Ensuring policy is appropriate (5.3a)

The standard requires the quality policy *to be appropriate to the purpose of the organization.*

What does this mean?
The purpose of an organization is quite simply the reason for its existence and as Peter Drucker so eloquently put it, "there is only one valid definition of business purpose: to create a customer".[2] In ensuring that the quality policy is appropriate to the purpose of the organization, it must be appropriate to the customers the organization desires to create. It is therefore necessary to establish who the customers are, where the customers are, what they buy or wish to receive and what these customers regard as value.

Why is this necessary?
This requirement responds to the Leadership Principle.

As stated above, the quality policy is the corporate policy and such policies exist to channel actions and decisions along a path that will fulfil the organization's purpose and mission. A goal of the organization may be the attainment of ISO/TS 16949 certification and thus a quality policy of meeting the requirements of ISO/TS 16949 would be consistent with such a goal, but goals are not the same as purpose as indicated in the box to the right. Clearly no organization would have ISO/TS 16949 certification as its purpose because certification is not a reason for existence, i.e. maybe a constraint but not a purpose.

Policies expressed as short catchy phrases such as "to be the best" really do not channel actions and decisions. They become the focus of ridicule when the organization's fortunes change. There has to be a clear link from mission to policy.

How is this implemented?
Policies are not expressed as vague statements or emphatic statements using the words *may*, *should* or *shall*, but clear intentions by use of the words *"we will"* – thus expressing

a commitment or by the words "*we are, we do, we don't, we have*" expressing shared beliefs. Very short statements tend to become slogans which people chant but rarely understand the impact on what they do. Their virtue is that they rarely become outdated. Long statements confuse people because they contain too much for them to remember. Their virtue is that they not only define what the company stands for but how it will keep its promises.

Examples of corporate policies

On customers
We will listen to our customers, understand and balance their needs and expectations with those of our suppliers, employees, investors and society, and endeavour to give full satisfaction to all parties.

On leadership
We will establish and communicate our vision for the organization and through our leadership exemplify core values to guide the behaviour of all to achieve our vision.

On people
We will involve our people in the organization's development, utilize their knowledge and experience, recognize their contribution and provide an environment in which they are motivated to realize their full potential.

On processes and systems
We will take a process approach towards the management of work and manage our processes as a single system of interconnected processes that delivers all the organization's objectives.

On continual improvement
We will provide an environment in which every person is motivated to continually improve the efficiency and effectiveness of our products, processes and our management system.

On decisions
We will base our decisions on the logical and intuitive analysis of data collected where possible from accurate measurements of product, process and system characteristics.

On supplier relationships
We will develop alliances with our suppliers and work with them to jointly improve performance.

On profits
We will satisfy our stakeholders in a manner that will yield a surplus that will be used to develop our capabilities and our employees, reward our investors and contribute to improvement in our society.

On the environment, health and safety
We will operate in a manner that safeguards the environment, and the health and safety of those who work with us.

In the ISO 9000 definition of quality policy it is suggested that the eight quality management principles be used as a basis for establishing the policy. One of these principles is the *Customer Focus principle*. By including in the quality policy the intention to identify and satisfy the needs and expectations of customers and other interested parties and the associated strategy by which this will be achieved, this requirement would be fulfilled. The inclusion of the strategy is important because the policy should guide action and decision. Omitting the strategy may not ensure uniformity of approach and direction.

Expressing a commitment (5.3b)

The standard requires that the quality policy *include a commitment to comply with requirements and continually improve the effectiveness of the quality management system.*

What does this mean?

A commitment to comply with requirements means that the organization should undertake to meet the requirements of all interested parties. This means meeting the requirements of customer, suppliers, employees, investors, owners and society. Customer requirements are those either specified or implied by customers or determined by the organization and these are dealt with in more detail under Clauses 5.2 and 7.2.1. The requirements of employees are those covered by legislation such as access, space, environmental conditions, equal opportunities and maternity leave but also the legislation appropriate to minority groups such as the disabled and any agreements made with unions or other representative bodies. Investors have rights also and these will be addressed in the investment agreements. The requirements of society are those obligations resulting from laws, statutes, regulations, etc.

An organization accepts such obligations when it is incorporated as a legal entity, when it accepts orders from customers, when it recruits employees, when it chooses to trade in regulated markets and when it chooses to use or process materials that impact the environment.

The effectiveness of the management system is judged by the extent to which it fulfils its purpose. Therefore improving effectiveness means improving the capability of the management system. Changes to the management system that improve its capability, i.e. its ability to deliver outputs that satisfy all the interested parties, are a certain types of change and not all management system changes will accomplish this. This requirement therefore requires top management to pursue changes that bring about an improvement in performance.

Why is this necessary?

This requirement responds to the Leadership Principle.

Policies guide action and decision. It is therefore necessary for top management to impress on their workforce that they have entered into certain obligations that commit everyone in the enterprise. Such commitments need to be communicated through policy statements in order to ensure that when taking actions and making decisions, staff give top priority to meeting the requirements of the interested parties. This is not easy. There will be many difficult decisions where the short-term interests of the organization may need to be subordinated to the needs of customers. Internal pressures may tempt people to cut corners, break the rules and protect their own interests. Committing the organization to meet requirements may be an easy decision to take, but difficult to honour.

Over 500,000 organizations worldwide have obtained ISO 9001 certification but many (perhaps the majority) have not improved their performance as a result. They remain mediocre and not top performers primarily because the management system is not effective. If it were effective the organization would meet its objectives year on year and grow the number of satisfied customers. Resources, technology, market conditions, economical conditions, and customer needs and expectations continually change; thus impacting the organization's capability. The effectiveness of the management system in meeting this challenge therefore needs to be continually improved.

How is this implemented?

A policy containing a commitment to meeting the needs and expectations of all interested parties would meet the first part of this requirement. In making a commitment to meet all these requirements the organization is placed under an obligation to:

- identify the relevant requirements;
- design and install processes that will ensure the requirements are met;
- verify compliance with the identified requirements;
- demonstrate to relevant authorities that the requirements have been met.

The second part can be dealt with by including a policy that commits the organization to improve the effectiveness of the system by which the organization's objectives are achieved. Policies are more easily understood when expressed in terms that are understood by the employees. Some organizations use the terms *internal* and *external* customers but even this can be ambiguous because not everyone will think of themselves as internal customers. The term *interested parties* is 'ISO speak' and may not be readily understood. Spell it out if necessary; in fact it is highly desirable where relevant to state exactly what you mean rather than use the specific words from the standard.

Providing a framework for quality objectives (5.3c)

The standard requires the quality policy *to provide a framework for establishing and reviewing quality objectives*.

What does this mean?

The quality policy represents a set of guiding principles and therefore when setting as well as reviewing quality objectives, these principles should be employed to ensure the objectives are appropriate to the purpose of the organization. It does not mean that the words used in the quality policy should somehow be translated into objectives (see below).

Why is this necessary?

This requirement responds to the Leadership Principle.

The quality policy statements arising from QS-9000 were often stand-alone statements with little or no relationship to the operations of the business. The following are some typical quality policy statements:

- We will perform exactly like the requirements or cause the requirements to be officially changed.
- We will satisfy our customers requirements on time, every time and within budget.
- Our aim is to give customer satisfaction in everything we do.
- We shall not knowingly ship defective product.

If we take just one of the above policy statements *"Our aim is to give customer satisfaction in everything we do"* on its own it is a motherhood statement. Nice, looks good in the lobby, visitors are impressed – but the bottom line is that actual performance does not meet the expectations set. The reason is that no one thought out the process for accomplishing this; the links between the policy, the objectives and the processes for realizing it were not put in place.

Without being linked to the business processes, these policies remain dreams. There has to be a means to make these policies a reality and it is by setting objectives that are derived from the policy that this is accomplished. For the first time in these standards, a link has been made between policy and objectives so that policies are not merely motherhood statements but intentions for action. By deriving objectives from the policy you initiate a process for bringing about compliance with policy.

How is this implemented?

ISO 9000 recommends that the eight principles of quality management are used as the basis for establishing the quality policy and therefore these can provide the basis for setting objectives. However, it is not common practice for top management to derive its objectives from policies. Objectives are normally derived from needs, not guiding principles. The relationship between objectives and the management system was explained in Chapter 4 and here it was shown that objectives are derived from the mission statement resulting in objectives for marketing, innovation, productivity, human organization, financial resources, social responsibility and profit. If the objectives are based on the policy and the policy is based on the eight quality management principles, there

Table 5.1 *Fabricating quality objectives from a quality policy*

Quality policy	Quality objectives
We are committed to providing products that are delivered on *time*	98% on-time delivery as measured by the customer
and meet *customer requirements*	99.9% of monthly output to be defect free as measured by customer returns
while yielding a *profit* and increasing *sales*.	5% profit on annual sales 25% increase in annual sales volume
We accomplish this through *product and process innovation,*	20% of our product range will contain new products and 50 improvement teams will be set up to seek process improvement
cost reduction activities,	15% reduction in cost of poor quality as a percentage of sales
compliance with ISO/TS 16949 requirements.	Certified to ISO/TS 16949 by July 2005

is a mismatch because there are no principles covering profit (see also under the heading *Expressing quality objectives*).

Purely as an exercise and not as a real policy we could fabricate a policy that does provide a framework for setting objectives but it is not recommended.

Let us assume the policy is as follows:

> We are committed to providing products that are delivered on time and meet customer requirements while yielding a profit and increasing sales. We accomplish this through product and process innovation, cost reduction activities and compliance with ISO/TS 16949 requirements.

If we use the policy as a framework it might look like that in Table 5.1.

This seems to be derived from what top management wants. It passes the scrutiny of the external auditors simply because it meets the requirement. One cannot deny that, but does it meet the intent? Isn't it the intent of the organization to satisfy its stakeholders and therefore wouldn't one expect the objectives to be derived from the needs and expectations of these stakeholders rather than a motherhood statement that could apply to any organization?

Ensuring policy is communicated and understood (5.3d)

The standard requires that the quality policy *is communicated and understood within the organization*.

What does this mean?

For a policy to be communicated it has to be brought to the attention of personnel. Personnel have to be made aware of how the policy relates to what they do so that they understand what it means before action is taken. Without action there is no demonstration that communication has been effective. If you are already doing it, publishing the policy merely confirms that this is your policy. If the organization does not exhibit the right characteristics, there will need to be a change in culture to make the policy a reality.

Why is this necessary?

This requirement responds to the Leadership Principle.

As has been stated previously, a policy in a nice frame positioned in the lobby of an organization may impress the visitors but unless it is understood and adhered to, it will have no effect on the performance of the organization. This requirement is also duplicated by the requirements in Clause 6.2.2.3 for personnel to be informed about the consequences to the customer of nonconformity to quality requirements.

How is this implemented?

It is difficult to imagine how a policy could be understood if it wasn't communicated but it signifies that the understanding has to come about by top management communicating the policy rather than the policy being deployed via the grape vine.

Whilst it is important that management shows commitment towards quality, policy statements can be one of two things: *worthless* or *obvious*. They are worthless if they do not reflect what the organization already believes and is currently implementing. They are obvious if they do reflect the current beliefs and practices of the organization. It is therefore foolish to declare in your policy what you would like the organization to become.

This is perhaps the most difficult requirement to achieve. Any amount of documentation, presentations by management, and staff briefings will not necessarily ensure that the policy is understood. Communication of policy is about gaining understanding but you should not be fooled into believing that messages delivered by management are effective communication. Effective communication consists of four steps: attention, understanding, acceptance and action. It is not just the sending of messages from one source to another. So how do you *ensure* that the policy is understood?

Within your management system you should prescribe the method you will employ to ensure that all the policies are understood at all appropriate levels in the organization. There will be levels in the organization where a clear understanding of the corporate policy is necessary for the making of sound decisions. At other levels, staff may work to instructions, having little discretion in what they can and cannot do. At these levels relevant aspects of the policy may be translated by the local manager into words that the staff understand. These can be conveyed through local procedures or notices.

One method to ensure understanding is for top management to do the following:

- Debate the policy together and thrash out all the issues. Don't announce anything until there is a uniform understanding among the members of the management team. Get the managers to face the question, "Do we intend to adhere to this policy?" and remove any doubt before going ahead.
- Ensure the policy is presented in a user-friendly way.
- Announce to the workforce that you now have a policy that affects everyone from the top down.
- Publish the policy to the employees (including other managers).
- Display the policy in key places to attract peoples' attention.
- Arrange and implement training/instruction for those affected.
- Test understanding at every opportunity, e.g. at meetings, when issuing instructions/procedures, when delays occur, when failures arise and when costs escalate.
- Audit the decisions taken that affect quality and go back to those who made them if they do not comply with the stated policy.
- Take action every time there is misunderstanding. Don't let it go unattended and don't admonish those who may have misunderstood the policy. It may not be their fault!
- Every time there is a change in policy, go through the same process. Never announce a change and walk away from it. The change may never be implemented!
- Give time for the understanding to be absorbed. Use case studies and current problems to get the message across.

The audit programme is another method of testing understanding and is a way of verifying whether the chosen method of ensuring understanding is being effective.

In determining whether the policy is understood, auditors should not simply ask, "What is the quality policy?" All this will prove is whether the auditee remembers it! The standard does not require that everyone knows the policy, only that it be communicated and understood. To test understanding therefore, you need to ask, e.g.:

- How does the quality policy affect what you do?
- What happens if you can't accomplish all the tasks in the allotted time?
- What would you do if you discovered a nonconformity immediately prior to delivery?
- How would you treat a customer who continually complains about your products and services?
- What action would you take if someone asked you to undertake a task for which you were not trained?
- What are your objectives and how do they relate to the quality policy?
- What action would you take if you noticed that someone was consuming food and drink in a prohibited area?
- What action would you take if you noticed that a product for which you were not responsible was in danger of being damaged?

Ensuring that the policy is reviewed (5.3e)

The standard requires the quality policy *to be reviewed for continuing suitability.*

What does this mean?

This requirement means that the policy should be examined in light of planned changes in the organization to establish whether it will remain suitable for guiding the organization towards its mission, i.e. will this policy guide us to where we now want to go?

Why is this necessary?

This requirement responds to the Leadership Principle.

Nothing remains static for very long. As the organization grows and seeks new opportunities, its size and characteristics will need to change as it responds to the markets and economic climate in which it currently operates. A policy established under different circumstances may therefore not be appropriate for what the organization needs to become to meet these challenges – it may not be suitable for guiding the future organization towards its mission. The policy is required to be appropriate to the organization's purpose and while the purpose may not change, the environment in which the organization operates does change. These changes will impact the corporate policy. The policy will need to be reviewed in light of changes in the economic, social and technological environment for its suitability to enable the organization to fulfil its purpose.

How is this implemented?

The quality policy should be reviewed whenever a change in the market, the economic climate, statutory and legal requirements, technology or a major change in the organizational structure is contemplated. This review may be conducted through *management review* but it rather depends on the type of review that is (see also under the heading *Management review*).

The review may conclude that no change is needed to the actual words but the way they are being conveyed might need to change. If the environment or the organization has changed, the policy might be acceptable but needs to be interpreted differently, conveyed to different people using different examples than were used previously.

Changes in policy have wide impact and therefore should not be taken lightly. They should be reviewed by top management with the full participation of the management team and therefore should be debated at the Corporate Planning Review or Business Review meeting. We are not talking about tinkering with the wording but a real change in direction. Changes in technology might mean that the workforce ceases to be predominantly on site as it becomes more effective to promote home or remote working. This change will impact the policy regarding leadership and people. Changes in the economic climate might mean that the workforce ceases to consist primarily of employees as it becomes more effective to outsource work to subcontractors and consultants. This change will impact the policy regarding leadership, suppliers and people.

Once the decision is made to change the policy it has to be communicated and the process for educating the workforce initiated.

Quality objectives (5.4.1)

Establishing objectives (5.4.1)

The standard requires that top management *ensure that quality objectives, including those needed to meet requirements for product, are established at relevant functions and levels within the organization.*

What does this mean?
An objective is a result that is aimed for and is expressed as a result that is to be achieved. Objectives are therefore not policies. The requirement should also not be interpreted as applicable only to organizational functions and levels. Objectives are required at levels within the organization not levels within the organization *structure.* This is clarified by the requirement for objectives to include those needed to meet requirements for product. There are therefore five levels at which control and improvement objectives need to be established:

1   Corporate level where the objectives are for the whole enterprise to enable it to fulfil its vision.
2   Process level where the objectives are for specific processes to enable them to fulfil corporate goals.
3   Product or service level where the objectives are for specific products or services, or ranges of products or services to enable them to fulfil *or create* customer needs and expectations.
4   Departmental or function level where the objectives are for an organizational component to enable it to fulfil corporate goals.
5   Personal level where the objectives are for the development of individual competency.

Why is this necessary?
This requirement responds to the Leadership Principle.

The requirement for defining objectives is one of the most important requirements. Without quality objectives there can be no improvement and no means of measuring how well you are doing. Without objectives any level of performance will do. If you don't know where you are going, any destination will do! Objectives are therefore necessary as a basis for measuring performance, to give people something to aim for, to maintain the *status quo* in order to prevent decline and to advance beyond the *status quo* for the enterprise to grow.

How is this implemented?

Objectives for control and improvement

A management system is not a static system but a dynamic one and if properly designed and implemented can drive the organization forward towards world-class quality. All managerial activity is concerned either with maintaining performance or with making change. Change can retard or advance performance. That which advances performance is beneficial. In this regard, there are two classes of quality objectives, those serving the control of quality (maintaining performance) and those serving the improvement of quality (making beneficial change).

The objectives for quality control should relate to the standards you wish to maintain or to prevent from deteriorating. To maintain your performance and your position in the market you will have to continually seek improvement. Remaining static at whatever level is not an option if your organization is to survive. Although you will be striving for improvement it is important to avoid slipping backwards with every step forwards. The effort needed to prevent regression may indeed require innovative solutions. While to the people working on such problems, it may appear that the purpose is to change the *status quo*, the result of their effort will be to maintain their present position not raise it to higher levels of performance. Control and improvement can therefore be perceived as one and the same thing depending on the standards being aimed for and the difficulties in meeting them.

The statements of objectives may be embodied within business plans, product development plans, improvement plans, process descriptions and even procedures.

Process for establishing objectives

Achievable objectives do not necessarily arise from a single thought even when the policies provide a framework. There is a process for establishing objectives. At the strategic level, the subjects that are the focus for setting objectives are the factors that affect the organization's ability to accomplish its mission – the critical success factors, such as marketing, innovation, human resources, physical and financial resources, productivity and profit. There may be other factors such as the support of the community, of unions, of the media as certain businesses depend on continued support from society (see Table 5.2). Customer needs, regulations, competition and other external influences shape these objectives and cause them to change frequently. The measures arise from an analysis of current performance, the competition and there will emerge the need for either improvement or control. The steps in the objective setting process are as follows:[3]

- Identifying the need.
- Drafting preliminary objectives.
- Proving the need to the appropriate level of management in terms of:
  - whether the climate for change is favourable,
  - the urgency of the improvement or controls,
  - the size of the losses or potential losses,
  - the priorities.
- Identifying or setting up the forum where the question of change or control is discussed.
- Conducting a feasibility study to establish whether the objective can be achieved with the resources that can be applied.
- Defining achievable objectives for control and improvement.
- Communicating the objectives.

The standard does not require that objectives be achieved but it does require that their achievement be planned and resourced. It is therefore prudent to avoid publishing objectives for meeting an unproven need and which has not been rigorously reviewed and assessed for their feasibility. It is wasteful to plan for meeting objectives that are unachievable and it diverts resources away from more legitimate uses.

Objectives are not established until they are understood and therefore communication of objectives must be part of this process. Communication is incomplete unless the receiver understands the message but a simple yes or no is not an adequate means of measuring understanding. Measuring employee understanding of appropriate quality objectives is a subjective process. Through the data analysis carried out to meet the requirements of Clause 8.4 you will have produced metrics that indicate whether your quality objectives are being achieved. If they are being achieved you could either assume your employees understand the quality objectives or you could conclude that it doesn't matter. Results alone are insufficient evidence. The results may have been achieved by pure chance and in 6 months time your performance may have declined significantly. The only way to test understanding is to check the decisions people make. This can be done with a questionnaire but is more effective if one checks decisions made in the workplace. Is their judgement in line with your objectives or do you have to repeatedly adjust their behaviour?

For each objective you should have a plan that defines the processes involved in its achievement. Assess these processes and determine where critical decisions are made and who is assigned to make them. Audit the decisions and ascertain whether they were contrary to the objectives. A simple example is where you have an objective of decreasing dependence on inspection. By examining corrective actions taken to prevent recurrence of nonconformities you can detect whether a person decided to increase the level of inspection in order to catch the nonconformities or considered alternatives. Any person found increasing the amount of inspection has clearly not understood the objective.

Corporate goals

At the corporate level the objectives are concerned with business performance and will often be expressed as business goals or a mission statement addressing markets, the environment and society.

Goals reflect the intended destination of the organization. They could be such destinations as:

* to be a world-class supplier of car windscreens;
* to capture 50% of the market in high-temperature lubricants;
* to be first to market with innovative solutions in automobile safety;
* to be an organization that people like doing business with.

These destinations capture the imagination but without planning they are mere pipe dreams. You need ask, "How will we know when we have got there?" If you can't define what success looks like, you have the wrong objectives.

These are objectives for improvement – for changing the *status quo*. They also focus on intentions that are optional. For instance, meeting customer needs and expectations is not an option and therefore not a goal. If you make it a goal you would send out the wrong signal. It gives the impression that you do not currently intend to meet customer needs and expectations but intend to do so at some point in the future. This is an intention, not a destination and therefore a policy.

In order to establish your corporate goals you need to:

* analyse competitor products where available,
* benchmark inside and outside the industry.

There are many books[4] and organizations you can turn to for advice on benchmarking. With benchmarking you analyse your current position, find an organization that is performing measurably better and learn from them what they are doing that gives them the competitive edge. You then change your processes as a result of what you learn and then implement the changes.

Corporate objectives for control might include those for maintaining: market share, market penetration, the values system, supplier relationships and ISO/TS 16949 registration.

Process objectives

There are two types of processes: business processes and work processes. Business processes deliver business outputs and work processes deliver outputs required by business processes. At the process level the objectives are concerned with process performance – addressing process capability, efficiency and effectiveness, use of resources, and controllability. As a result objectives for control may focus on reducing errors and reducing waste, increasing controllability but may require innovative solution to achieve such objectives. Objectives for improvement might include increasing throughput, turnaround times, response times, resource utilization, environmental impact, process capability and use of new technologies, etc.

Product objectives

At the product level, objectives are concerned with product or service performance addressing customer needs and competition. Again these can be objectives for control or improvement. Objectives for control might include removing nonconformities in existing products (improving control) whereas objectives for improvement might include the development of new products with features that more effectively satisfy customer needs (improving performance), use of new technologies, and innovations. A product or service that meets its specification is only of good quality if it satisfies customer needs and requirements. Eliminating all errors is not enough to survive – you need the right products and services to put on the market. With QS-9000 quality management was perceived as applying after the organization had decided where it is going, what customers needed and what products and services will be supplied. However, with ISO/TS 16949 this perception needs to change because the purpose of standard is to enable organizations to satisfy their customers not just at the point of sale but in the product and service features they offer. This means finding out what customers want, what they need and what benefits they expect to gain from owning the organization's products or using their services. Objectives for satisfying the identified needs and expectations of customers with new product features and new service features are thus quality objectives.

Departmental objectives

At the departmental-level objectives are concerned with organizational performance – addressing the capability, efficiency and effectiveness of the organization, its responsiveness to change, the environment in which people work, etc. Control objectives might be to maintain expenditure within the budget, to keep staff levels below a certain level, to maintain moral, motivation or simply to maintain control of the department's operations. Objectives for improvement might be to improve efficiency by doing more with less resources, improving internal communication, interdepartmental relationships, information systems, etc.

Personal objectives

At the personal-level objectives will be concerned with worker performance addressing the skills, knowledge, ability, competency, motivation and development of people. Objectives for control might include maintaining timekeeping, work output and objectivity. Objectives for improvement might include improvement in work quality, housekeeping, interpersonal relationships, decision-making, computer skills, etc.

Expressing quality objectives (5.4.1 and 5.4.1.1)

The standard requires quality objectives *to be measurable and consistent with the quality policy and these objectives and measurements to be included in the business plan and used to deploy the quality policy.*

What does this mean?

There should be a tangible result from meeting an objective and a defined time period should be specified. The objective should therefore be expressed in the form: *what is to be achieved,* and there should be success measures that indicate when the objective has or has not been achieved – the passing of a date, a level of performance, or the absence of a problem, a condition, a situation.

It implies that each statement within the quality policy should have an associated quality objective and each lower-level quality objective should be traceable to higher-level quality objectives that have a clear relationship with a statement within the quality policy. Through this relationship, the objectives deploy the quality policy. In this respect the quality policy would appear to be the mission statement of the organization rather than some vague expression of goodness.

It implies that these quality objectives are not minor, departmental targets but strategic goals that are included in a business plan.

Why is this necessary?

This requirement responds to the Leadership Principle.

Where objectives are not measurable there is often some difficulty in establishing whether they have been achieved. Achievement becomes a matter of opinion and therefore variable. Measures provide consistency and predictability, and produce facts on which decisions can be made.

Objectives need to be consistent with the quality policy so that there is no conflict. For example, if the policy is "We will listen to our customers, understand and balance their needs and expectations with those of our suppliers, employees, investors and society, and endeavour to give full satisfaction to all parties" an objective which penalizes suppliers for poor performance would be inconsistent with the policy.

How is this implemented?
Setting objectives
A technique has evolved to test the robustness of objectives and is identified by the letters SMART meaning that objectives should be Specific, Measurable, Achievable, Realistic and Timely.

S Specific	Objectives should be **specific** actions completed in executing a strategy. They should be derived from the mission and relevant to the process or task to which they are being applied. They should be specified to a level of detail that those involved in their implementation fully understand what is required for their completion – not vague or ambiguous and defining precisely what is required.	
M Measurable	Objectives should be **measurable** actions that have a specific end condition. Objectives should be expressed in terms that can be measured using available technology. When setting objectives you need know how achievement will be indicated, the conditions or performance levels that will indicate success.	
A Achievable	Objectives should be **achievable** with resources that can be made available – they should be achievable by average people applying average effort.	
R Realistic	Objectives should be **realistic** in the context of the current climate and the current and projected workload. Account needs to be taken of the demands from elsewhere that could jeopardize achievement of the objective.	
T Timely	Objectives should be **time-phased** actions that have a specific start and completion date. Time-phased objectives facilitate periodic review of progress and tracking of revisions. The specific date or time does not need to be expressed in the objective unless it is relevant; in other cases the timing for all objectives might be constrained by their inclusion in the 2005 Business Plan, implying all the objectives will be achieved in 2005. The Business Plan for 2005–2008 implies all objectives will be achieved by 2008.	

Although the SMART technique for objective setting is used widely, there is some variance in the words used. If you search the Internet on the key words 'SMART objectives' you will discover several variations on this theme. The S of SMART has been used to denote 'Small' meaning not too big to be unachievable – one small step at a time. The A of SMART has been used to denote 'Attainable, Accountable and Action oriented' and the R of SMART has been used to denote 'Resource-consuming action and Relevant'. These differences could arise out of different uses of the technique.

Consistency with policy

The subject matter of quality objectives arises from an analysis of the factors that affect the organization's ability to accomplish its mission as stated previously and these arise from an analysis of stakeholder needs and expectations. The quality policy may influence the wording of these objectives to some extent but it is doubtful that you would want to derive specific objectives from the policy itself. Let us say you have a policy that addresses customer focus. Your objectives would include marketing objectives that were customer focused thereby linking the policy with the objectives. You may have a human resource objective for improving employee motivation. However, in this instance the process designed to achieve this objective would need to demonstrate adherence to a policy for the involvement of people. Here the process and not the objective links with the policy.

Policies condition behaviour so that objectives are achieved so they may each cover different topics. Drucker identified eight categories of corporate objectives[5] and these have been matched with the eight quality management principles in Table 5.2. The match is not perfect because there is no match for profit requirement, however, if you were to establish your objectives based on Drucker's categories, you would be able to show that your objectives were consistent with a quality policy that was based on the eight quality management principles.

Measures

It is important to determine the measures that will be used to verify achievement of the objectives. If you have an objective for being World Class, what measures will you use that indicate when you are World Class? You may have an objective for improved delivery performance. What measures will you use that indicates delivery performance has improved? You may choose to use percent delivered on time. You will also need to set a target relative to current performance. Let us say that currently you achieve 74% on-time delivery so you propose a target of 85%. However, targets are not simply figures better than you currently achieve. The target has to be feasible and therefore it is necessary to take the steps in the process described previously for setting objectives.

In the last 30 years or so there has emerged an approach to management that focuses on objectives. Management by objectives (MBO) or management by results (MBR) has dominated boardrooms and management reports. In theory MBO or MBR is a sensible way to manage an organization but in practice this has led to internal competition, sub-optimization and punitive

> **Objectives**
>
> Atsushi Niimi, CEO of Toyota Motor Manufacturing North America is reported to have said in October 2003 "We have some concerns about sustaining high quality because North American parts suppliers average 500 defects per million parts versus 15 per million in Japan. But if it works, and Lexus's made in USA are equivalent to those from Japan, Toyota will have exported a major upgrade of its already respected production system."

Table 5.2 *Matching objectives with the quality management principles*

Objective category	Subject	Customer focus	Leadership	Involvement of	Process approach	Systems approach	Continual	Factual approach	Supplier relationships
Marketing	Existing products in current markets	•							
	Abandoning products	•							
	New products in current markets	•							
	New markets	•							
	New products in new markets	•							
	Standards and performance	•					•		
Innovation	Reaching market goals in near future						•		
	Taking advantage of technological advances in the distant future						•		
Human resources	Supply of managers and their development			•					
	Supply of staff and their development			•					
	Relationships with representative bodies			•					
	Relationship with suppliers								•
	Employee attitudes and competencies				•				
Physical resources	Supply of raw materials and components					•			
	Supply of capital equipment					•			
	Supply of buildings and facilities					•			
Financial resources	Investment and attracting capital		•						
	Obtaining financial resources					•			
Productivity	Utilization of knowledge				•				
	Utilization of physical resources					•			

Table 5.2 *Continued*

Objective category	Subject	Quality management principles							
		Customer focus	Leadership	Involvement of	Process approach	Systems approach	Continual	Factual approach	Supplier relationships
	Utilization of time				•				
	Utilization of financial resources								
	Making workers productive		•						
	Utilizing experience and ability			•					
	Utilizing reputation								
Social responsibility	Disadvantaged people			•					
	Protection of the environment	•							
	Education of potential employees Contribution to professions			•					
	Health and safety of employees at work			•					
	Minimizing impact on society, economy, community and individual	•							
Profit requirement	Producing the minimum profit needed to accomplish the other objectives								

measures being exacted on staff that fail to perform. Deming's 14 points[6] included the elimination of MBO for the simple reason that management derives the goals from invalid data. They observe that a goal was achieved once and therefore assume it can be achieved every time. If they understood the process they would realize that the highs and lows are a characteristic of natural variation. They observe what the competition achieves and raise the target for the organization without any analysis of capability or any plan for its achievement. Management sets goals and targets for results that are beyond the capability of staff to control. Targets for the number of invoices processed, the number of orders won, the hours taken to fix a problem. Such targets not only ignore the natural variation in the system but also are set without any knowledge about the processes that deliver the results. If a process is unstable, no amount of goal setting will change its performance. If you have a stable process, there is no point in setting a goal beyond the capability of the process; you will get what the process delivers.

Business plan

There are likely to be different business plans for different time periods: a 1-, 3- and 5-year business plan. Each will take a view of the external environment as it is assumed to be over these periods. Within the plan the objectives may be expressed under specific headings so that there are financial objectives, product development objectives, safety objective, etc. You can take the view that all objectives are quality objectives and therefore no special heading are needed or there is a need for indicating which objectives are quality objectives. If you take the latter course, you need to define what message is being conveyed so as to avoid the implication that the objectives under the heading 'Quality objectives' are not perceived as only applying to the Quality Department unless of course that is what you intend to convey.

Whilst the standard also requires the measures to be included in the business plan, it would create some inconsistency if the other objectives were expressed without the corresponding measures. So it might be an opportunity to define the measures for all objectives thus revealing the criteria for determining when each objective has been achieved.

Quality management system planning (5.4.2)

Planning to meet quality objectives (5.4.2a)

The standard requires top management *to ensure that the planning of the quality management system is performed to meet the quality objectives and the requirements in Clause 4.1.*

What does this mean?

Planning is performed to achieve objectives and for no other purpose and therefore the requirement clearly indicates that the purpose of the management system is to enable the organization to meet its quality objectives. This is reinforced by the definition of quality planning in ISO 9000 which states that *it is part of quality management focused on setting objectives and specifying necessary operational processes and related resources to fulfil the quality objectives.*

We have deduced that quality objectives are corporate objectives and therefore quality planning will be synonymous with *strategic planning*. (Remember that Clause 0.1 advises us that the adoption of a quality management system was a strategic decision!)

The additional requirement for management system planning to meet the requirements of Clause 4.1 means that in planning the processes of the management system, you need to put in place provisions to measure, monitor and analyse processes, determine their sequence and interaction, and determine criteria and methods to ensure effective operation and control. In addition you will need to provide resources and information necessary to support the operation and monitoring of these processes. The focus has clearly moved away from procedures as a means to establish a management system.

Why is this necessary?

This requirement responds to the Process Approach Principle.

The link between planning of the management system and quality objectives was not clearly expressed in QS-9000 and consequently there was often a disconnect between the management system and quality objectives and the quality policy. This resulted in systems of documentation that served no useful purpose apart from appeasing the auditors. Now there is a clear linkage from policy to objectives and from objectives to processes that are established to meet such objectives. This means that the management system should now be more results oriented with the objectives employed as measures of performance.

How is this implemented?

The process of defining objectives was outlined above indicating that planning proceeds only after the feasibility of achieving an objective has been established. One plans only to achieve an objective and remembering that planning consumes resources, an effective management system would need to ensure that dreams, wish lists and ambitions do not become the subject of any formal planning.

As objectives are required to be defined at relevant functions and levels, it follows that planning is also required at relevant functions and levels thereby requiring planning at corporate, divisional and department levels, product, process and system level.

The planning referred to in this clause is focused on that needed to meet the organization's objectives. It is not focused on that needed to meet specific contracts or orders or for specific products and services – this type of planning is addressed in Chapter 7, under the heading *Product realization*. The organizational and resource planning needed for developing a new range of products or services would be considered to be part of corporate planning.

Objectives are achieved through processes and therefore in planning to meet an objective, the planner should identify the process or processes involved. At a high level this may be no more than an outline strategy for achieving the objectives, minimizing risks and measuring success. Responsibilities may identify no more than a function or department although in some organizations ownership is deemed paramount and individual managers are named. At the lower level, the plan may extend to detail activities with a bar chart showing timescales and responsibilities.

Processes for planning

In Taylor's Scientific Management, there was a planning function that did all the planning. In the complex organizations created since Taylor's time, it has become too unwieldy for one function to do all the planning. Planning has been deployed to the function that will achieve the objective. At the corporate level there will be a corporate-planning process that runs on a cycle of 1, 3 and 5 years. Every function will be involved in providing their inputs. As each cycle ends a new one begins. The planning process includes the objective setting process and it is quite common for the ideas to come from below, float to the top where selections are made and passed down again for feasibility studies which go the Board for approval and return to the source for detailed budgets and justification. Depending on the resources involved and the urgency the process may take months or even years to gain approval for the plans. The sanction to spend is often based on approval levels requiring budgets and detail plans before approval can be given. Even after such a lengthy process, there is likely to be another process for acquiring the resources that also requires approval, indicating that approved plans do not necessarily signify that permission to spend has been granted. This is often because of timing. When the time comes to acquire the resources, the priorities may have changed and plans once approved may be put on ice. It is interesting that ISO/TS 16949 does not require plans to be implemented. What it does requires is for the *system* to be implemented and within this system may be provisions to abort plans when circumstances dictate that necessity for the survival of the organization.

It is therefore necessary to define the planning processes so that there is a clear linkage between objectives and plans to meet them.

Corporate planning

Corporate plans should contain provisions made to accomplish corporate objectives. It is quite common to produce separate business plans of the following types:

- annual business plan,
- 3-year business plan,
- 5-year plus business plan.

Such plans may exist for each profit centre and consolidate the plans of all functions within that profit centre. These plans typically contain the budgets and other provisions such as head count and inventory required for meeting the declared objectives. Corporate planning is not usually referred to as corporate quality planning, although in the larger enterprises, corporate quality planning may be one part of the corporate plan. In such cases, the corporate quality plan may address objectives related to improvement by better control leaving the objectives related to improvement by innovation to be defined in product development or process development plans. The labels are not important. The scope of the set of plans should address all the defined objectives regardless of what they are called.

Department planning

Corporate objectives need to be deployed to each relevant department. Some objectives may be achieved wholly within the confines of one department whereas other objectives may have one department as the primary responsibility with other departments providing a contribution. In some cases, objectives will cascade to all departments and subdivisions within each department. Departmental budgets form part of this planning and contribute to corporate planning. Departmental plans should define the provisions made for achieving departmental objectives and this may typically include the acquisition and development of physical and human resources, re-organization of staff, development of new practices, application of new technologies.

Product and process development planning

Although each department will include in its budgets provisions associated with product and process development, for large developments, it is often necessary to co-ordinate these budgets on a corporate level to ensure nothing is overlooked. Such developments apply to innovations targeted at improving the organization's capability and not those arising from specific contracts. Product development plans will typically define provisions for research, design, development, production and launch of new products and services. At the corporate level these plans will be of a strategic nature with resources budgets, risk analysis, assumptions, dependencies, major work packages and timescales. Process development plans will typically define provisions for research, design, development, acquisition, installation and commissioning of new processes. At the corporate level these plans will again be of a strategic nature with resources budgets, risk analysis, assumptions, dependencies, major work packages and timescales. Detail plans should be created from these corporate plans to address the operational aspects from research through to in-service. These plans are addressed under *Product realization*, Chapter 7.

Product and service objectives will clearly involve the design, development, production/ delivery processes but other objectives may require re-organization, new facilities, new technology, etc. The processes for achieving such objectives should form part of the management system even though there may be no clauses in ISO/TS 16949 that directly refer to such processes.

Personal planning

The organization's objectives should be deployed down to individuals where relevant and translated into the knowledge, skills and competencies required. This often takes place during annual appraisals but it is important that the timing of such appraisals matches the corporate planning cycle so that any personnel development serves the corporate objectives. Personnel planning may be carried out at corporate, departmental and individual level. The organization may have a need for improving its capability in a new area and may therefore require all its people trained in particular subjects and skills. The quest for ISO/TS 16949 certification may well create such a demand. Equally at departmental level, new technology may be planned for installation and all staff will require instruction and training. Both these types of change may occur outside the staff appraisal cycle and require additional personnel development planning.

Planning for change (5.4.2b)

The standard requires the integrity of the quality management system *to be maintained when changes to the quality management system are planned and implemented.*

What does this mean?
This requirement refers to *change* in general not simply changes to the management system documentation. As the management system is the means by which the organization's objectives are achieved (not just a set of documents), it follows that any change in the enabling mechanism should be planned and performed without adversely affecting its capability. Changes needed to accomplish these objectives should be managed and the processes required to execute the changes should be part of the management system.

It means that the linkages and the compatibility between interfaces should be maintained during a change. By being placed under *planning*, there is recognition that the plans made to meet the defined objectives may well involve changing the organization, the technology, the plant, machinery, the processes, the competency levels of staff and perhaps the culture.

Why is this necessary?
This requirement responds to the Process Approach Principle.

If changes in the management system are permitted to take place without consideration of their impact on other elements of the management system, there is likely to be a deterioration in performance. In the past, it may have been common for changes to be made and some months later the organization charts and procedures to be updated indicating that these documents are perceived as historical records – certainly not documents used in executing the change. The updating activities were often a response to the results of the change process creating in people's minds that it was a housekeeping or administrative chore. To meet this requirement, change management processes need to be designed and put in place. The integrity of the management system will be maintained only if these processes are made part of the management system so that in planning the changes, due consideration is given to the impact of the change on the organization, its resources, processes and products, and any documentation resulting from or associated with these processes.

How is this implemented?

To control any change there are some basic steps to follow and these are outlined in Chapter 1 under the heading *Quality improvement*. To maintain the integrity of the management system you need to do several things:

- Use the change processes defined in the management system documentation to plan and execute the change. If they don't exist, the management system does not reflect how the organization operates. These processes should be part of the business management subsystem.
- Determine the impact of the change on the existing system and identify what else need to change to maintain system effectiveness.
- Plan and execute the change concurrently with associated changes to documentation.
- Don't remove the old processes until the new processes have been proven effective.
- Measure performance before, during and after the change.
- Don't revert to routine management until the changes have been integrated into the culture, i.e. people perform the new tasks without having to be told.

The management system should not be perceived as a set of discrete processes – all should be connected; therefore change in one process is likely to have an effect on the others. For example, if new technology is to be introduced, it may not only affect the process in which it is to be used but also the staff development process, the equipment maintenance process and the design process. It will also affect the marketing process because the new technology will improve the organization's capability and thus enabling it to create new markets, attract different customers, etc. On a more mundane level, if a new form is introduced, it is not only the process in which the form is used that may be affected but also the interfacing processes that receive the form when complete. It is the information management processes that make the form available and secure its contents.

Responsibility and authority (5.5.1)

Defining responsibilities and authority (5.5.1)

The standard requires that *the responsibilities and authority be defined*.

What does this mean?

The term responsibility is commonly used informally to imply an obligation a person has to others. However, the term authority has increasingly become associated with power and public bodies but in principle one cannot have responsibility without authority and vice versa. Problems arise when these two are not matched, where one is greater or less than the other.

Responsibility is, in simple terms, an area in which one is entitled to act on one's own accord. It is the obligation of staff to their managers for performing the duties of their jobs. It is thus the obligation of a person to achieve the desired conditions for which they are accountable to their managers. If you cause something to happen, you must be responsible for the result just as you would if you cause an accident; so to determine a person's responsibility ask, "What can you cause to happen?"

Authority is, in simple terms, the right to take actions and make decisions. In the management context it constitutes a form of influence and a right to take action, to direct and co-ordinate the actions of others and to use discretion in the position occupied by an individual, rather than in the individuals themselves. The *delegation of authority* permits decisions to be made more rapidly by those who are in more direct contact with the problem.

Why is this necessary?
This requirement responds to the Leadership Principle.

It is necessary for management to define who should do what in order that the designated work is assigned to specific individuals for it to be carried out. In the absence of the delegation of authority and assignment of responsibilities, individuals assume duties that may duplicate those duties assumed by others. Thus jobs that are necessary but unattractive will be left undone. It also encourages decisions to be made only by top management that can result in an increased management workload but also engender a feeling of mistrust by the workforce.

How is this implemented?
A person's job can be divided into two components: actions and decisions. Responsibilities and authority should therefore be described in terms of the actions assigned to an individual to perform and discretion delegated to an individual, i.e. the decisions they are permitted to take together with the freedom they are permitted to exercise. Each job should therefore have core responsibilities that provide a degree of predictability and innovative responsibilities that in turn provide the individual with scope for development.

In defining responsibilities and authority there are some simple rules that you should follow:

- Through the process of delegation, authority is passed downward within the organization and divided among subordinate personnel whereas responsibility passes upwards.

- A manager may assign responsibilities to a subordinate and delegate authority, however, he or she remains responsible for the subordinate's use of that authority.
- When managers delegate responsibility for something, they remain responsible for it. When managers delegate authority they lose the right to make the decisions they have delegated but remain responsible and accountable for the way such authority is used. Accountability is one's control over the authority one has delegated to one's staff.
- It is also considered unreasonable to hold a person responsible for events caused by factors that they are powerless to control.
- Before a person can be in a state of control they must be provided with three things:
  - Knowledge of what they are supposed to do, i.e. the requirements of the job, the objectives they are required to achieve.
  - Knowledge of what they are doing, provided either from their own senses or from an instrument or another person authorized to provide such data.
  - Means of regulating what they are doing in the event of failing to meet the prescribed objectives. These means must always include the authority to regulate and the ability to regulate both by varying the person's own conduct and varying the process under the person's authority. It is in this area that freedom of action and decision should be provided.
- The person given responsibility for achieving certain results must have the right (i.e. the authority) to decide how those results will be achieved, otherwise, the responsibility for the results rests with those who stipulate the course of action.
- Individuals can rightfully exercise only that authority which is delegated to them and that authority should be equal to that persons' responsibility (not more or less than it). If people have authority for action without responsibility, it enables them to walk by problems without doing anything about them. Authority is not power itself. It is quite possible to have one without the other! A person can exert influence without the right to exert it.

Communicating responsibilities and authority (5.5.1)

The standard requires that the responsibilities and authority *be communicated within the organization*.

What does this mean?
Communication of responsibility and authority means that those concerned need to be informed and to understand their obligations so that there is no doubt about what they will be held accountable for.

Why is this necessary?
This requirement responds to the Leadership Principle.

There are several reasons why is it necessary to communicate this information:

- to convey consistency and avoid conflict;
- to show which functions make which contributions and thus serve to motivate staff;
- to establish channels of communication so that work proceeds smoothly without unplanned interruption;
- to indicate from whom staff will receive their instructions, to whom they are accountable and to whom they should go to seek information to resolve difficulties.

How is this implemented?

There are several ways in which responsibilities and authority can be communicated:

- in an organization structure diagram or *organigram*,
- in function descriptions,
- in job descriptions,
- in terms of reference,
- in procedures,
- in flow charts.

The standard does not stipulate which method should be used. In very small companies a lack of such documents defining responsibility and authority may not prove detrimental to quality provided people are made aware of their responsibilities and adequately trained. However, if you are going to rely on training, then there has to be some written material that is used so that training is carried out in a consistent manner.

Organigrams are a useful way of showing interrelationships but imprecise as a means of defining responsibility and authority. They do illustrate the lines of authority and accountability but only in the chain of command. Although it can define the area in which one has authority to act, it does not preclude others having responsibilities within the same area; e.g. the title Design Manager–Computer Products implies that the person could be responsible for all aspects of computer product design when in fact they may not have any software, mechanical engineering or reliability engineering responsibilities. Titles have to be kept brief because they are labels for communication purposes and are not usually intended for precision on the subject of responsibilities and authority. One disadvantage of organigrams is that they do not necessarily show the true relationships between people within the company. There are also no customers or suppliers on the charts thereby omitting external relationships. Horizontal relationships can be difficult to depict with clarity in a diagram. They should therefore not be used as a substitute for policy.

Function descriptions are useful in describing the role and purpose of a function, its objective and its primary responsibilities and authority. Function in this context refers to business functions rather than product functions and is a collection of activities that make a common and unique contribution to the purpose and mission of a business. Function is determined by the contribution made rather than the skill that the contributors possess. The marketing function in a business generates revenue and the people contributing to marketing may possess many different skills, e.g. planning, organizing, selling, negotiating, data analysis, etc. It is quite common to group work by its contribution to the business and to refer to these groupings as functions so that there is a marketing function, a design function, a production function, etc. However, it should not be assumed that all those who contribute to a function reside in one department. The marketing department may contain many staff with many skills, but often the design staff contributes to marketing. Likewise, the design function may have the major contribution from the design department but may also have contributors from research, test laboratory, trials and customer support. Therefore the organization chart may in fact not define functions at all but a collection of departments that provide a mixture of contributions. In a simple structure the functions will be clear but in a complex organization, there could be many departments concerned with the marketing function, the design function, the production function, etc. For example, the Reliability Engineering Department may be located in the Quality Department for reasons of independence but contributes to the design function. A Function/Department/Group description is needed to define the role the function executes in each process to which it contributes. These become useful in staff induction as a means of making new staff aware of who is who and who does what without getting into too much detail. They are also useful to analysts and auditors because they enable a picture of who does what to be quickly assimilated.

Job descriptions or job profiles are useful in describing what a person is responsible for, however, it rather depends on the reason for having them as to whether they will be of any use in managing quality. Those produced for job evaluation, recruitment, salary grading, etc. may be of use in the management system if they specify the objectives people are responsible for achieving and the decisions they are authorized to take.

Terms of reference are not job descriptions but descriptions of the boundary conditions. They act as statements that can be referred to in deciding the direction in which one should be going and the constraints on how to get there. They are more like rules than a job description and more suited to a committee than an individual. They rarely cover responsibilities and authority except by default.

Procedures are a common way of defining peoples' responsibilities and authority because it is at the level of procedures that one can be specific as to what someone is required to do and what results they are responsible for. Procedures specify individual actions and decisions. By assigning actions or decisions to a particular person or role a person carries out you have assigned to them a responsibility or given them certain authority. They do present problems, however. It may be difficult for a person to see clearly what his or her job is by scanning the various procedures because procedures often describe tasks rather than objectives. When writing procedures never use names of individuals because they will inevitably change. The solution is to use position or role titles and have a description for a particular position or role that covers all the responsibilities assigned through the procedures. Individuals only need to know what positions they occupy or roles they perform. Their responsibilities and authority are clarified by the procedures and the position or role descriptions.[7]

Flow charts are becoming a more popular choice for defining responsibility and authority. Assigning job titles to actions and decisions in a flow chart makes it clear that anyone with the indicated title carries responsibility for performing action or decision described within the shape. Colour coding could be used to make global changes less tedious when job title change. It does have the advantage of being concise but this could be a disadvantage as it does not allow qualification or detail to be added.

In organizations that undertake projects rather than operate continuous processes or production lines, there is a need to define and document project-related responsibilities and authority. These appointments are often temporary, being only for the duration of the project. Staff are assigned from the line departments to fulfil a role for a limited period. To meet the requirement for defined responsibility, authority and interrelationships for project organizations you will need Project Organization charts and Project Job descriptions for each role, such as Project Manager, Project Design Engineer, Project Systems Engineer, Project Quality Engineer, etc.

As project structures are temporary, there need to be systems in place that control the interfaces between the line functions and project team. Such a system would include:

- policies that govern the allocation of work to the divisions;
- policies that govern the allocation of work to staff in these divisions;
- job descriptions for each role stating responsibilities, authority and accountability;
- procedures that identified the roles responsible for each task and for ensuring that information is conveyed to and from these staff at the appropriate time;
- procedures that consolidate information from several disciplines for transmission to the customer when required;
- monitoring procedures to track progress and performance;
- procedures that ensure the participation of all parties in decisions affecting the product, its development and production;
- procedures for setting priorities and securing commitment;
- procedures that include the management of subcontractor programmes during development and deal with the transmission of information to and from the subcontractors, what is to be transmitted, by whom, in what form and with whose approval.

Some organizations have assigned responsibility for each element of the standard to a person, but such managers are not thinking clearly. Now the elements have been reduced from 20 to 5 with ISO/TS 16949, such allocations will need to be reviewed. There are 51 clauses and many are interrelated. Few can be taken in isolation therefore such a practice is questionable. When auditors ask, "Who is responsible for purchasing?" ask them to specify the particular activity they are interested in. Remember you have a system which delegates authority to those qualified to do the job.

Responsibility for quality (5.5.1.1)

Alerting authorities to nonconformities (5.5.1.1)

The standard requires managers with responsibility for corrective action *to be promptly informed of nonconforming products or processes.*

What does this mean?

All managers have responsibility for corrective actions, i.e. action intended to prevent the recurrence of nonconformities but each manager will only carry responsibility for the outputs of the specific processes under his or her control. Therefore the requirement implies that should an operator observe that a process is producing nonconforming product or the process is not meeting the defined capability targets, the process owner should be informed.

Why is this necessary?

It is no longer sufficient to correct nonconforming product, fix the problems, reset the equipment, etc. What has to happen is that the root cause needs to be determined and action taken to prevent a recurrence. If capable processes begin producing nonconforming product, it might signal a change in state that needs to be corrected.

How is this implemented?

There is variation in all processes, even capable ones; but the variation is usually within acceptable limits. When a process produces nonconforming product it may require some adjustment but the numbers nonconforming may be within the limits so no action is needed. Informing the process owner of nonconformities does not imply that corrective action will be taken but it gives the manager the opportunity to take appropriate action.

Process instruction sheets should include the name of the person responsible for the process, setting it up and monitoring its performance. The limits should be specified on the charts used for monitoring performance and instructions provided as to what action is to be taken when limits are exceeded.

Authority to stop production (5.5.1.1)

The standard requires personnel responsible for product quality *to have the authority to stop production to correct quality problems.*

What does this mean?
Personnel with responsibility for product quality in this context means those personnel who are able by their actions and decisions to affect the characteristics of the product and to cause those characteristics to meet the specified requirements. So this could include designers who specify the product, operators who make the product, inspectors who verify the product but more often it is just the operators.

> **Stopping production**
>
> Gary Convis President of Toyota's Georgetown factory says, "Toyota doesn't run 100% of the schedule time". "I've never really had criticism over lost production and putting a priority on safety and quality over hitting production targets", he says. "This is because Toyota learned that solving quality problems at source saves time and money downstream."

It is not presumed that such personnel would act irresponsibly. In production systems that employ preventive error-proofing, (autonomation) machines may be equipped with devices to stop production in the event of abnormal conditions arising such as a jammed mechanism, defect part or lubrication problem. This allows operators to control multiple machines. Where machines are not equipped with such devices or alarms only are employed, operators need to be empowered to perform the same function and stop the machine when something goes wrong.

Why is this necessary?
This requirement should not be necessary because anyone assigned a task should have the authority to start and stop it if they judge the outcome will not be correct. However, if the operator is unable through illness, accident or some other cause to stop production when the process becomes unstable, other people also need powers to stop production. Toyota seem to understand that stopping the machine when there is trouble forces awareness on everyone[8] and when properly understood saves time and money.[9]

How is this implemented?
People cannot be responsible for anything over which they have no control. Operators can only take responsibility for what they do or cause to happen. It is imperative that you avoid the situation whereby management has told someone he or she is responsible for quality without delegating authority to make the decisions necessary for carrying out the necessary actions. Therefore having assigned a responsibility for certain results, management should also delegate authority to personnel to control the processes which

produce the results for which they are responsible; thereby authorizing them to stop production if needed. It would be necessary to provide these personnel with adequately defined criteria and training so that they will be able to judge when abnormal conditions arise. If they have job descriptions then this would be the vehicle for defining this authority, but it could as easily be defined in a set of values that apply to everyone in the organization. But you have to be careful not to imply that anyone can stop production if they think there is something wrong as the walk by the machinery. Clearly having the authority to stop production on one production line does not necessarily mean you know what you are doing when it comes to another production line. What might look like a quality problem might be quite normal for that machine or product.

Staffing production shifts (5.5.1.1)

The standard requires production operations across all shifts *to be staffed with personnel in charge of or delegated responsibility for ensuring product quality.*

What does this mean?

It is difficult to visualize a situation in which shift operators would not have responsibility for ensuring product quality, but there might be cases where the responsibility of a machine minder on an automated production line is limited to filling the hopper, oiling the joints, stacking the outputs, calling someone when the alarm sounds or shutting down the line.

Why is this necessary?

Shift personnel might not have the competence to set up the line, adjust the machinery or measure the output.

How is this implemented?

You cannot assume that if the process is stable at the end of the day shift it will remain so throughout the night shift. Tools may wear out or break, the process may go out of control, materials may need to be replenished, etc. All of these require decisions. The reason for this requirement is so that there are staff on each shift who are authorized to:

- make process acceptance decisions;
- make machine set-up decisions;
- make product acceptance decisions;
- stop production in the event of an out-of-control situation developing;
- change the sampling criteria in the event of an out-of-control situation developing.

Management representative (5.5.2)

Appointing management representatives (5.5.2)

The standard requires the top management *to appoint a member of management who has certain defined responsibility and authority for ensuring the quality management system is established, implemented and maintained.*

What does this mean?

This means that one manager in particular is delegated the authority and responsibility to plan, organize and control the management system. Clearly this is not a one-man job and cannot be performed in isolation because the management system comprises all the processes required to create and retain satisfied customers. It does not mean that this manager must also manage each of the processes but should act as a co-ordinator, a facilitator and a change agent, and induce change through others who in all probability are responsible to other managers.

There is a note in Clause 5.5.3 of ISO/TS 16949, which states: *The responsibility of a management representative may also include liaison with external parties on matters relating to the supplier's quality management system.*

Logically a representative carries the wishes of the people they represent to a place where decisions are taken that affect them: Members of Parliament, Union Representatives, Committee Members, etc. The 'note' would appear to address the need for representation outside the business. Inside the business, the person represents management to the workforce but not in the same sense. The person carries the wishes of management (i.e. the policies) to the workforce so that the workforce makes decisions that take into account the wishes of management. However, the requirement matches more closely the role of a director rather than an external representative because this person is not only representing management but also directing resources in a way that will enable the company to achieve its objectives. What the organization needs is not so much a representative but a director who can represent management when necessary and influence other managers to implement and maintain the system. Such a person is unlikely to be under the direct authority of anyone other than the CEO.

In the standard the term 'management representative' appears only in the title of the requirement. The emphasis has been put on management appointing a member of its own management, indicating that the person should have a managerial appointment in the organization. This implies that a contractor or external consultant cannot fill the role. It also implies that the person should already hold a managerial position and be on the pay roll. However, it is doubtful that the intention is to exclude a person from being promoted into a managerial position as a result of a person being available for the appointment or in fact preclude the authority of the management representative being delegated to a contractor, provided that responsibility for the tasks is retained within the company.

Why is this necessary?

This requirement responds to the Systems Approach Principle.

As everyone in some way contributes to the quality of the products and services provided, everyone shares the responsibility for the quality of these products and services. The achievement of quality, however, is everyone's job but only in so far as each person is responsible for the quality of what they do. You cannot hold each person accountable for matters over which they have no control. It is a trait of human nature that there has to be a leader for an organization to meet an objective. It does not do it by itself or by collective responsibility – someone has to lead.

In principle the Management Representative or Quality Director role is similar to the roles of the Financial Director, Security Director, Safety Director, etc. It is a role that exists to set standards and monitor performance thus giving an assurance to management, customers and regulators that specified objectives are being achieved. The role takes the title from the subject that is vital to the survival of the company. If security is vital then, a director is given the task of establishing security policy and putting in a system that will ensure security is not compromised. The security staff do not implement the system – the duty of all other managers. The same is true for finance, personnel, quality and any other critical success factor. If quality is vital to survival then it makes sense to appoint someone to direct the programme that will ensure quality is not compromised. As with finance, security and personnel these directors do not implement the policies, they regulate compliance. The other functional managers are appointed to deal with other factors critical to the company's survival and each is bound by the others' policies. This way of delegating authority works because it establishes a champion for each key factor who can devote resources to achieving specified objectives. Each manager is responsible for some aspect of security, finance, quality, personnel, etc. Their responsibilities extend to implementing policy and achieving objectives. This means that the Production Director, for example, is responsible for implementing the quality policy and achieving quality objectives within a system that is under the control of the Quality Director. Likewise the Production Director is responsible for implementing the design solution that is under the control of the Design Director.

If you were to make every manager responsible for setting policy, setting up systems and ensuring compliance then you would have as many management systems as there were managers. This is not an effective way to run a business. In such a structure, you would not have one company but as many companies as there were managers. If each manager is to serve common objectives, then we have to divide the objectives between them and permit one manager to impose requirements on other managers. This is what is known as functional authority.

How is this implemented?

There are two schools of thought: one is that the management representative is a figure-head rather than a practitioner and has a role established solely to meet ISO 9000. It is doubtful that any organizations not registered to ISO 9000 will have made such an appointment. Those organizations not registered would not perceive there was a system to be managed. The CEO would either take on the role or would appoint one of the executive directors as the management representative in addition to his or her regular job – the role being to ascertain that a QMS is being established, implemented and maintained. Such a person may not necessarily employ the resources to do this. These resources would be dispersed throughout the organization. While the system is being developed, a project manager is assigned to co-ordinate resources and direct the project towards its completion. After the system is fully operational, a management system manager takes over to maintain and improve the system who, with a small staff, manages the audit and improvement programmes.

The other school of thought views the management representative as a practitioner and not a figurehead. Here you would appoint a senior manager as a quality director and assign him or her the role of management representative. This director takes on the role of project manager during the development phase and management system manager during the maintenance and improvement phase. He or she acts as the management representative with the customer and registrar, and in effect is the eyes of the customer inside the organization. Depending on the size and complexity of the organization, there may be one person doing all of these jobs. In some cases a fairly large team of engineers, auditors, analysts, statisticians, etc. may be appropriate. Before ISO 9000, organizations appointed quality managers not management representatives. The difference is that being a quality manager was a job whereas being a management representative is a role.

To give this appointment due recognition, an appointment at executive level would be appropriate. The title chosen should reflect the position and as stated previously need not be a full time job. Often companies appoint a member of the executive to take on the role in addition to other responsibilities. It could be the Marketing, Sales, Engineering, Production or any other position. The notion that there has to be independence is one that is now dated and a reflection of an age when delivery was more important than quality. A person with responsibility for delivery of product or service also carries a responsibility for the quality of his or her actions and decisions. A person who therefore subordinates quality to delivery is unfit to hold the position and should be enlightened or replaced.

If you have one management system, the role of management representative and job of quality director become difficult to separate, and can cause a conflict of interest unless the management representative is the CEO. In large organizations with multiple sites, each with separate ISO 9000 registrations, a more appropriate solution is to have a management representative for each site and one quality director for the whole organization.

As with all assignments of responsibility one has to:

- define the actions and decisions for which the person is to be responsible ensuring no conflict with others;
- define the competency needed;
- select a person with the necessary abilities;
- ensure that you give the person the necessary authority to control the results for which they are responsible;
- provide and environment in which the person is motivated to achieve the results for which he or she is responsible;
- evaluate and develop his or her competency to perform the role effectively.

Responsibility for establishing and maintaining processes (5.5.2)

The standard requires the management representative *to ensure that processes needed for the quality management system are established, implemented and maintained.*

What does this mean?
The management system consists of interconnected processes each of which needs to be established, implemented and maintained. This requirement means that top management being responsible for the system, delegates authority to one manager (the Quality Director/Manager/Management Representative) to orchestrate the design, development, construction, maintenance and improvement of these processes.

Why is this necessary?
This requirement responds to the Leadership Principle.

If the CEO assigned responsibility for this task to each functional manager, it is likely that a fragmented system would emerge rather than a coherent one. Someone has to lead the effort required, to direct resources and priorities and judge the resultant effectiveness.

How is this implemented?
Primarily, the designated person is the system designer for the management system appointed by top management. This person may not design all the processes and produce the documentation but may operate as a system designer. He or she lays down the requirements needed to implement the corporate quality policy and verifies that they are being achieved. As system designer, the person would also define the requirements for processes so as to ensure consistency and lead a team of process owners who

develop, implement and validate business processes. Previously the emphasis was on the management representative establishing the system that had the implication that the task was concerned with documenting procedures. This requirement clearly changes the focus to establishing processes and with it brings in a responsibility for process management. In this regard the person needs the authority to:

- manage the design, development, construction and evaluation of the processes of the management system including the necessary resources;
- determine whether the processes meet the requirements of the standard, are suitable for meeting the business needs, are being properly implemented and cause non-compliances to be corrected;
- manage the change processes for dealing with changes to the processes of the system.

Responsibility for reporting on QMS performance (5.5.2)

The standard requires the management representative *to report to top management on the performance of the quality management system and the need for improvement.*

What does this mean?

This requirement means that the Quality Director collects and analyses factual data across all the organization's operations to determine whether the quality objectives are being achieved and if not, to identify opportunities for improvement.

Why is this necessary?

This requirement responds to the Factual Approach Principle.

Each manager cannot measure the performance of the company relative to quality although individually they carry responsibility for the utilization of resources within their own area. The performance of the company can only be measured by someone who has the ability and authority to collect and analyse the data across all company operations. All managers may contribute data, but this needs to be consolidated in order to assess performance against corporate objectives just as a Finance Director consolidates financial data.

How is this implemented?

To report on management system performance and identify opportunities for improvement in the management system the Quality Director needs the right to:

- determine the effectiveness of the management system;
- report on the quality performance of the organization;
- identify opportunities for improvement in the management system;
- cause beneficial changes in quality performance.

By installing data collection and transmission nodes in each process, relevant data can be routed to the Quality Director for analysis, interpretation, synthesis and assessment. It can then be transformed into a language suitable for management action and presented at the management review. However, this requirement imposes no reporting time period, therefore performance should also be reported when considered necessary or on request of top management.

Responsibility for promoting awareness of customer requirements (5.5.2)

The standard requires the management representative *to ensure the awareness of customer requirements throughout the organization.*

What does this mean?

This means that the management representative encourages and supports initiatives by others to make staff at all levels aware of customer requirements. This is not necessarily the detail requirements as would be contained in specifications, but their general needs and expectations, what is important to them, what the product being supplied will be used for and how important the customer is to sustaining the business.

Why is this necessary?

This requirement responds to the Customer Focus Principle.

Unless staff are aware of customer requirements it is unlikely they will be achieved. Customer satisfaction is the aim of the management system and hence it is important that all staff at all levels do not lose sight of this. Clearly all managers are responsible for promoting awareness of customer requirements but this does not mean it will happen as internal pressures can cause distractions. Constant reminders are necessary when making decisions in which customer satisfaction may be directly or indirectly affected. Often staff at the coalface are remote from the end user and unaware of the function or role of their output relative to the final product or service. Heightened awareness of customer requirements and the role they play in achieving them can inject a sense of pride in what they do and lead to better performance.

How is this implemented?

There are general awareness measures that can be taken and awareness for specific customers. General awareness can be accomplished through:

- the quality policy and objectives;
- induction and training sessions;
- instructions conveyed with product and process documentation;

- bulletins, notice boards and staff briefings;
- brochures of the end product in which the organization's product features;
- videos of customer products and services featuring the organization's products.

It is also a responsibility of designers to convey (through the product specifications) critical features and special customer characteristics. Also production planners or service delivery planners should denote special requirements in planning documents so that staff are alerted to requirements that are critical to customers (see also Chapter 6 under the heading *Training, awareness and competence*).

Responsibility for external liaison
Although a note in Clause 5.5.3 of the standard, it is necessary to have someone who can liaise with customers or other stakeholders on quality issues, who can co-ordinate visits by stakeholders or their representatives and who can keep abreast of the state of the art in quality management through conferences, publications and exhibitions. However, it does have to be the same person dealing with all external liaison.

Customer representative (5.5.2.1)

The standard requires top management *to designate personnel with responsibility and authority to ensure that customer requirements are addressed.*

What does this mean?
This means that within your organization there is someone who is appointed by top management to pay particular attention to the requirements of a specific customer: someone who understands their needs and expectations, someone who gets inside the mind of the customer and who can translate their requirements into a language you can easily understand. It is sometimes a sales person but often an engineer or project manager. It is not necessarily a Director as the interface will be highly technical and specific, sometimes even tedious.

Why is this necessary?
Whatever your business you cannot operate as though you are a field of corn, letting the wind blow you in different directions. Each customer may have slightly different requirements with many often having no impact on product quality but on the presentation of information. If you characterize products and processes too closely to specific customer requirements, you run the risk of introducing inefficiencies and reducing productivity. You can however, maintain productivity and respond to your customer's varying demands through an interface function. Appointing a person as your customer liaison representative provides an opportunity to develop someone in your organization who knows as much about what the customers need and why it is needed than the

customers themselves. This person is then able to translate specific customer requirements into your language and back again. This may not be translating English into Polish but translating the customer jargon into terms you understand. The customer calls it a nonconformity report but you call it a problem, the customer calls it a complaint but you call it a concern. So rather than change all your processes to suit all your customers, translate customer requirement onto your own paperwork and use this throughout the process.

How is this implemented?

The person appointed not only needs the competence to liaise with customers on technical issues but the authority to make things happen so it is not a job for a lowly paid individual nor is it a job for a corporate seagull. The person has to gain the respect of the customer and the workforce, and have the ability to build and rebuild relationships: a diplomat, perhaps linguist, competent engineer and realist.

The appointed customer representative will need to spend sometime with the customer to learn their ways, understand their language/jargon, needs and expectations. Hence if native tongue of your staff is English and you do business with Swedish, Italian and French companies you may need people who can speak these languages who are familiar with the appropriate subject vocabulary. Beware however, that in appointing such a person you choose wisely. It also has to be someone you can trust to represent your interests. You will need a means of calibrating this person so that he or she does not get carried away with enthusiasm and start to impose requirements that are no more than personal likes and dislikes. Where a customer wants something that others have not yet demanded, consider the overall benefits and if it does provide added value change your processes. If not, find a compromise that is mutually beneficial.

It is important that this appointed individual controls communication channels at the project level. However, other individuals will become familiar with the customer's own representatives thus opening up several potential channels of communication so one needs to ensure no conflicting signals are transmitted. E-mail is such an easy way of communicating but also such an easy way to overlook people, send unchecked messages, and make unfounded claims. E-mail has got to a stage that it can sometimes look like it is taking on the mantle of a chat room. So be careful and lay down the protocols to maintain control.

Internal communication (5.5.3)

Establishing communication processes (5.5.3)

The standard requires *appropriate communication processes to be established within the organization.*

What does this mean?

Communication processes are those processes that convey information and impart understanding upwards, downwards and laterally within the organization. They include the people transmitting information, the information itself, the receivers of the information and the environment in which it is received. They also include all auditory and visual communication, the media that convey the information and the infrastructure for enabling the communication to take place. The medium of communication such as telephone, e-mail and video is the means of conveying information and forms part of the process of communication. Examples of such processes are briefings, announcements, meetings, conferences, presentations, internal publishing and distribution process (paper and Intranet), internal mailing process, (paper and electronic) and display boards.

Why is this necessary?

This requirement responds to the Leadership Principle.

The operation of a management system is dependent on effective transmission and reception of information and it is the communication processes that are the enablers. Information needs to be communicated to people for them to perform their role as well as possible. These processes need to be effective otherwise:

- the wrong information will be transmitted;
- the right information will fail to be transmitted;
- the right information will go to the wrong people;
- the right information will reach the right people before they have been prepared for it;
- the right information will reach the right people too late to be effective;
- the communication will not be understood;
- the communication will cause an undesirable result.

How is this implemented?

As the requirement focuses on processes rather than the subject of communication it follows that whatever the information that needs to be communicated, the communicator needs to select an appropriate communication process. There needs therefore to be some standard processes in place that can be used for communicating the majority of information. There needs to be a communication policy that facilitates downwards communication and encourages upwards and lateral communication.

A simple solution would be to identify the various types of information that need to be conveyed and the appropriate process to be used. In devising such a list you need to consider the audience and their location along with the urgency, sensitivity, impact and permanency of the message:

- Audience influences the language, style and approach to be used (Who are they?).
- Location influences the method (Where are they?).

- Urgency influences the method and timing when the information should be transmitted (When it is needed?).
- Sensitivity influences the distribution of the information (Who needs to know?).
- Impact influences the method of transmission and the competency of the sender (How should they be told and who should tell them?).
- Permanency influences the medium used (Is it for the moment or the long term?).

Communication processes should be established for communicating:

- *The vision, mission and values of the organization*: While displaying this information acts as a reminder, this does not communicate. You need to establish a process for gaining understanding, getting commitment and forming the culture.
- *Operating policies*: These are often conveyed through manuals and procedures but a communication process is needed to ensure they are understood at all levels.
- *The corporate objectives*: A process is needed for conveying these down the levels in the hierarchy with translation possibly at each level as they are divided into departmental, group, section and personal objectives.
- *Plans for entering new markets, for new products and processes and for improvement*: A process is needed for communicating plans following their approval so that all engaged in the project have a clear understanding of the strategy to be adopted.
- *Customer requirements, regulations and statutory requirements*: A process is needed for ensuring that these requirements reach the point at which they are implemented and are understood by those who will implement them.
- *Product and process objectives*: These are often conveyed through plans but a communication process is needed to ensure the plans are understood by those who are to use them or come into contact with them.
- *Product and process information*: A process is needed to ensure that all product and process information gets to those who need it, when they need it and in a form that they can understand.
- *Problems*: A process is needed for communicating problems from their source of detection to those who are authorized to take action.
- *Progress*: Managers need to know how far we are progressing through the plan and therefore a communication process is needed to ensure the relevant managers receive the appropriate information.
- *Change*: All change processes should incorporate communication processes in order to gain commitment to the change.
- *Results and measurements*: A process is needed for communicating financial results, quality and delivery performance, accomplishments, good news, bad news and customer feedback.

Clearly not everything can be communicated to all levels because some information will be sensitive, confidential or simply not relevant to everyone. Managers therefore need to exercise a 'need to know' policy that provides information necessary for people to do their job as well as creating an environment in which people are motivated. Other than national and commercial security, too much secrecy is often counterproductive and creates an atmosphere of distrust and suspicion that affects worker performance.

In communication processes there needs to be a feedback loop to provide a means for conveying questions and queries as well as acceptance. These feedback loops should be short – to the next level only. An Executive who demands to be kept informed of progress will soon stop reading the reports and if the process continues without change, the reports will just pile up in his or her office.[10] This is not an uncommon phenomenon. A manager may demand reports following a crisis but fail to halt further submissions when the problem has been resolved. The opposite is also not uncommon where a local problem is not communicated outside the area and action subsequently taken which adversely affects the integrity of the management system.

Communicating the effectiveness of the QMS (5.5.3)

The standard requires communication *to take place regarding the effectiveness of the quality management system*.

What does this mean?

This means that there should be communication from top management and from those on the scene of action (two way communication) as to whether the management system is enabling achievement of the organization's objectives. Information from above should initiate improvement action. Information from below should prompt investigation and analysis in order to identify improvement actions.

Why is this necessary?

This requirement responds to the Leadership Principle.

It is important that staff are kept informed of how effective the management system is to encourage continuation of the *status quo* or to encourage change. Also staff should be encouraged to report system effectiveness or ineffectiveness whenever it is encountered.

How is this implemented?

After each management review, the results can be communicated to staff but care should be taken with the format of the message. Charts against each major objective showing how performance has changed are the most effective. New improvement initiatives should also be communicated indicating the project name, the project leader or champion, the project objectives and timescales. However, the application of this requirement should not be restricted to annual communication briefings. Update the charts monthly and display on notice boards or on the Intranet. Provide means for staff to alert management of ineffectiveness in the management system by opening channels through to the Quality

Director. It is not uncommon for a particular practice to be changed locally or ignored altogether and subsequently discovered on periodic audit. There should be free communication so no one takes such action without consultation and agreement.

Management review (5.6)

Purpose of review (5.6)

The standard requires top management *to review the quality management system to ensure its continuing suitability, adequacy and effectiveness.*

What does this mean?

A review is another view of something. Although termed a management review the requirement is strictly referring to a review of the management system (see also under the heading *Management commitment*).

The three terms: adequacy, suitability and effectiveness are not included as three alternatives but as three different concepts. However, their meanings vary as illustrated in Table 5.3.

The adequacy of the management system is judged by its ability to deliver product or service that satisfies requirements, standards and regulations. It does what it was designed to do. In some cases this condition is referred to as effectiveness.

The suitability of the management system is judged by its ability to enable the organization to sustain current performance. If the management system is inefficient, the organization may not be able to continue to feed a resource hungry system. In such cases we would be justified in claiming that the management system is not suitable for its purpose even though customers may be satisfied by the outputs, other interested parties will soon express dissatisfaction. A better term would be efficiency.

The effectiveness of the management system is judged by how well it enables the organization to fulfil the needs of society. The system may deliver satisfied customers and

Table 5.3 *Elements of management system performance*

Concept	ISO term	Other terms
Output meets requirements	Adequacy	Effectiveness
Results achieved in best way	Suitability	Efficiency
System fulfils needs	Effectiveness	Adaptability

minimize use of resources but if it is not responding to the changing needs of society, of customers, of regulators and of other interested parties, it is not an effective system. In some cases this concept is referred to as adaptability. However, effectiveness is about doing the right things, not doing things right. Doing things right is about satisfying the customer. Doing the right things is choosing the right objectives. If the corporate objectives change or the environment in which the organization operates changes, will the system enable the organization to achieve these new objectives or operate successfully in the new environment? If the purpose of the system is merely to ensure customers are supplied with products and services which meet their requirements, then its effectiveness is judged by how well it does this and if not how much it costs to do it. If the purpose of the management system is to enable the organization to fulfil its purpose, its effectiveness is judged by how well it does this. The measures of effectiveness are therefore different.

Why is this necessary?

This requirement responds to the Systems Approach Principle.

There is a need for top management, as the sponsors of the system, to look again at the data the system generates and determine whether the system they installed is actually doing the job they wanted it to do. It is necessary that top management rather than the management representative review the system because the system serves the organization's objectives not just those of the management representative. Financial performance is reviewed regularly and a statement of accounts produced every year. There are significant benefits to be gained if quality performance is treated in the same way because it is quality performance that actually causes the financial performance. Underperformance in any area will be reflected in the financial results.

Treating the review as a chore, something that we do because we want to keep our ISO/TS 16949 certificate, will send out the wrong signals. It will indicate that members of top management are not serious about quality or about the system they commissioned to achieve it – it will also indicate they don't understand, and if they don't understand they clearly cannot be committed.

How is this implemented?

In any organization, management will conduct reviews of performance so as to establish how well the organization is doing in meeting the defined objectives. As the objectives vary it is often more practical to plan reviews relative to the performance characteristic being measured. As a result, organizations may convene strategic reviews, divisional reviews, departmental reviews, product reviews, process reviews, project reviews, etc. They all serve the same purpose that of establishing:

- whether performance is in line with objectives;
- whether there are better ways of achieving these objectives;
- whether the objectives are relevant to the needs of the stakeholders.

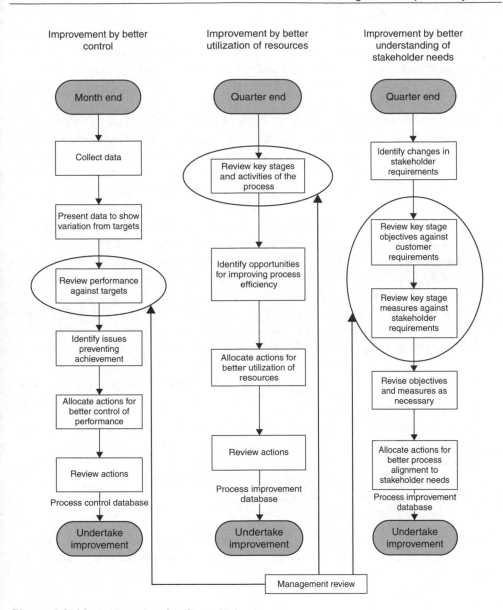

Figure 5.2 *Management review in context*

The review(s) are part of the improvement process and where it (they) fit in this process is illustrated in Figure 5.2.

Such reviews should be part of the management system and will examine the capability of the system to deliver against objectives and fulfil the mission. The management review referred to in ISO/TS 16949 is a review of the complete management system by top management and in a hierarchy of reviews is the top level review which should capture the results of all the lower-level reviews. The standard does not require only one review. In some organizations, it would not be practical to cover the complete system in one review. It is often necessary to consolidate results from lower levels and feed into intermediate reviews so that departmental reviews feed results into divisional reviews that feed results into corporate reviews. The fact that the lower-level reviews are not performed by top management is immaterial providing the results of these reviews are submitted to top management as part of the system review. It is also not necessary to separate reviews on the basis of ends and means. A review of financial performance is often separated from technical performance and both of these separate from management system reviews. This situation arises in cases where the management system is perceived as procedures and practices. The management system is the means the organization employs to achieve the ends. A review of results without review of the capability to achieve them (the means) would therefore be ineffective. For these reasons the management review as referred to in ISO/TS 16949 could well be the Strategic Review or Business Performance Review with separate committees or focus groups targeting specific aspects. In organizations that separate their performance reviews from their management system reviews, one has to question whether they are gaining any business benefit or in fact whether they have really understood the purpose of the management system.

In determining the effectiveness of the management system you should continually ask:

- Does the system fulfil its purpose?
- To what extent are our customers satisfied with our products and services?
- Are the corporate objectives being achieved as intended?
- Do measurements of process performance indicate the processes are effective?
- Do the results of the audits indicate that the system is effective?
- Are procedures being used properly?
- Are policies being adhered to?

If the answer is "Yes" your system is operating effectively. If your answer is "No" to any of these questions, your management system has not been effectively designed or is not being effectively implemented.

The management review is not a meeting. Management review is an activity aimed at assessing information on the performance of the management system. When you have a real understanding of the intentions of the review you will realize that its objectives cannot be accomplished entirely by a meeting. The review should be in three stages. Stage one is collecting and analysing the data, stage two is reviewing the data and stage three is a meeting to discuss the results and decide on a course of action. One of the reasons that the management review may not work is when it is considered something separate to management's job, separate to running the business, a chore to be carried out just to satisfy the standard. This is partially due to perceptions about quality. If managers perceive quality to be about the big issues like new product or service development, major investment programmes for improving plant, for installing computerization, etc., then the management review will take on a new meaning. If on the other hand it looks only at audit results it will not attract a great deal of attention unless of course the audits also include the big issues.

Planning the review (5.6.1)

The standard requires *management reviews at planned intervals.*

What does this mean?

Planned intervals means that the time between the management reviews should be determined in advance, e.g. annual, quarterly or monthly reviews. The plan can be changed to reflect circumstances but should always be looking forward.

Why is this necessary?

Previously the standard required reviews at defined intervals that allowed for reviews to be scheduled after each review on an *ad hoc* basis. The change indicates that more forethought is needed so that performance is measured on a regular basis thus enabling comparisons to be made.

How is this implemented?

This requirement responds to the Process Approach Principle.

A simple bar chart or table indicating the timing of management reviews over a given period will meet this requirement. The frequency of management reviews should be matched to the evidence that demonstrates the effectiveness of the system. Initially the reviews should be frequent say monthly, until it is established that the system is effective. Thereafter the frequency of reviews can be modified. If performance has already reached a satisfactory level and no deterioration appears within the next 3 months, extend the period between reviews to 6 months. If no deterioration appears in 6 months extend the period to 12 months. It is unwise to go beyond 12 months without a review because something is bound to change that will affect the system. Shortly after reorganization, (the launch of a new product or service, breaking into a new market, securing new customers, etc.) a review should be held to establish if performance has changed. After new technology is planned, a review should be held before and afterwards to measure the effects of the change.

Scope of review (5.6.1 and 5.6.1.1)

The standard requires the review *to include all requirements of the Quality management system and include assessing opportunities for improvement and the need for changes to the quality management system including quality policy and quality objectives.*

What does this mean?

The review is of current performance and hence there will be some parameters where objectives or targets have not been accomplished thus providing opportunities for improvement and some areas where the *status quo* is not good enough to grow the organization or meet new challenges.

Why is this necessary?

This requirement responds to the Continuous Improvement Principle.

Top management should never be complacent about the organization's performance. Even maintaining the *status quo* requires improvement, just to maintain market position, keep customers and retain capability. If the management review restricts its agenda to examining audit results, customer complaints and nonconformities month after month without a commitment to improvement, the results will not get any better – they will more than likely get worse.

There will be reports about new marketing opportunities, reports about new legislation, new standards, the competition and benchmarking studies. All these may provide opportunities for improvement. In this context improvement means improvement by better control (doing this better) as well as improvement by innovation (doing new things). These changes may affect the quality policy (see also under the heading *Ensuring that the policy is reviewed*) and will certainly affect the objectives. Objectives may need to change if they proved to be too ambitious or not far reaching enough to beat the competition.

How is this implemented?

The implication of this requirement is that performance data on the implementation of quality policy and the achievement of quality objectives should be collected and reviewed in order to identify the need to change the system, the quality policy and quality objectives.

As the management system is the means by which the organization achieves its objectives, it follows that the management review should evaluate the need for changes in the objectives and the processes designed to achieve them. It is therefore insufficient to limit the review to documentation, as was often the case when implementing QS-9000. In fact as no function, process or resource in the organization would exist outside the management system, the scope of the review is only limited by the boundary of the organization and the market and environment in which it operates.

The approach you take should be described in your quality manual – but take care! What you should describe is the process by which you determine the suitability, adequacy and effectiveness of the management system, and in doing so describe all the performance reviews conducted by management and show how they serve this objective. Describing a single management review without reference to all the other ways in which performance is reviewed sends out the signal that there isn't an effective process in place.

Records of management reviews (5.6.1)

The standard requires records from management reviews *to be maintained*.

What does this mean?
A record from the review means the outcome of the review but the outcome won't be understood unless it is placed in the right context. The records therefore also need to include the criteria for the review and who made what decisions.

Why is this necessary?
This requirement responds to the Factual Approach Principle.

Recorded results of management reviews are necessary for several reasons to:

- convey the actions from the review to those who are to take them;
- convey the decisions and conclusions as a means of employee motivation;
- enable comparisons to be made at later reviews when determining progress;
- define the basis on which the decisions have been made;
- demonstrate system performance to interested parties.

How is this implemented?
The records from management reviews need to contain:

- Date of review (location might be necessary if the review is carried out at a meeting).
- Contributors to the review (the process owners, functional managers, management representative, auditors, etc.).
- Criteria against which the management system is being judged for effectiveness (the organization's objectives).
- Criteria against which the management system is to be judged for continued suitability (future changes in the organization, legislation, standards, customer requirements, markets).
- The evidence submitted, testifying the current performance of the management system (charts, tables and other data against objectives).

- Identification of strengths, weaknesses, opportunities and threats (SWOT, the analysis of: What are we good at? What we are not good at? What can we change? What can't we change? What must we change?).
- Conclusions (Is the management system effective or not and if not in what way?).
- Actions and decisions (What will stay the same and what will change?).
- Responsibilities and timescales for the actions (Who will do it and by when will it be completed?).

Review inputs (5.6.1.1, 5.6.2 and 5.6.2.1)

The standard requires inputs to management review *to include information about various aspects of the system*. The automotive additions in Clause 5.6.2.1 require *inputs to include an analysis of actual and potential field failures and their impact on quality, safety or the environment* and Clause 7.3.4.1 requires *management review inputs to include design and development measurements*. In addition, Clause 5.6.1.1 requires *the reviews to include all requirements of the QMS and its performance trends as an essential part of the continual improvement process*.

What does this mean?

This means that data from audits, product measurements, process measurements, customers, end users, suppliers, regulators, etc. has to be analysed relative to defined objectives to establish current performance (How are we doing?) and identify improvement opportunities (Can we do better?). Data on planned changes in the organization, resources, the infrastructure, legislation and standards have to be examined for their impact.

Why is this necessary?

This requirement responds to the Factual Approach Principle.

It is necessary to gather sufficient data relative to the objectives being measured to provide a sound basis for the review. Any review of the management system needs to be based on fact. The factors identified in this clause cover most of the parameters influencing the effectiveness of the management system.

How is this implemented?

The key questions to be answered are, "Is the system effective?" and "Is it suitable for continued operations without change?" At every meeting of the review team these questions should be answered and the response recorded. In order to answer these questions certain inputs are required. The standard identifies several inputs to the review, which are addressed below, but these should not be seen as limiting. The input data should be that which is needed to make a decision on the effectiveness of the system. Rather than use a generic list as is presented in the standard, a list tailored to business needs should be developed and continually reviewed and revised as necessary.

Monitoring quality objectives (5.6.1.1)

Although omitted from the list of inputs in Clause 5.6.2 of ISO 9001, ISO/TS 16949:2002 addresses the omission in Clause 5.6.1.1.

System performance data should be used to establish whether the defined quality objectives are being met. It may also be used to establish whether there is conflict between the stated quality policy, the quality objectives and the organizational purpose, and the expectations and needs of the interested parties. There may be a small number of factors on which the performance of the organization depends and these above all others should be monitored. For example, in an electronic component manufacturer, cleanliness is paramount: the smallest of particles can cause a deterioration in performance. In a measuring laboratory accuracy and precision is paramount. In a car seating company, the appearance of the leather is critical to end-user perceptions, i.e. the slightest scuff is easily detected. Analysis of the data that the system generates should reveal whether the targets are being achieved. The requirement for performance trends implies that the evidence submitted for review should cover several time periods so that performance trends can be observed over the time period relevant to the objectives.

Progress on development projects should be submitted showing planned and achieved milestones with explanations where milestones have not been met.

It is also important to establish whether the system provides useful data with which to manage the business. This can be done by providing evidence showing how business decisions have been made. Those made without using available data from the management system show either that poor data is being produced or management is unaware of its value. One of eight quality management principles is the *factual approach to decision-making* and therefore implies decisions should be made using data generated from the management system.

Improvement opportunities may cover:

* the identification of major projects to improve overall performance;
* the setting of new objectives and targets;
* the revision of the quality policy;
* the adequacy of the linkages between processes.

Cost of poor quality
Costs of poor quality are failure costs and comprise the costs incurred in recovering from failure such as rework, repair, modification, warranty claims as well as the loss incurred

as a result of the failure, such as the cost of scrapped material, the labour and energy wasted in producing the scrap.

A more comprehensive treatment is given in ISO/TR 10014:1998 *Guidelines for managing the economics of quality*.

Audit results

Audit results should be used to establish whether the system is being used properly and whether the commitments declared in the quality policy are being honoured. You can determine this by providing the results of all quality audits of the system, of processes and of products. An analysis of managerial decisions should reveal whether there is constancy of purpose or lip service being given to the policy. Audit results should also be used to establish whether the audit programme is being effective and you can determine this by providing the evidence of previous audit results and problems reported by other means. Current performance from audit results should compare the results with the quality objectives you have defined for the system as a whole and for the audit programme in particular.

Improvement opportunities relative to audit results may cover:

- the scope and depth of the audit programme;
- the suitability of the audit approach to detecting problems worthy of management attention;
- the competency of the auditors to add value and discover opportunities that enhance the organization's capability;
- the relevance of audit results to the organization's objectives.

Customer feedback

Customer feedback should be used to establish whether customer needs are being satisfied. You can determine this by providing the evidence of customer complaints, market share statistics, competitor statistics, warranty claims, customer satisfaction surveys, etc. Current performance from customer feedback should compare the results with the quality objectives you have set for customer needs and expectations.

Improvement opportunities relative to customer feedback may cover:

- the extent to which products and services satisfy customer needs and expectations;
- programmes to eliminate the root cause of an increasing trend in customer rejects or returns;
- the adequacy of the means used to assess customer satisfaction and collect data;
- the need to develop new or enhanced products or services;
- the need to explore new markets or obtain more accurate data of current markets.

Process performance

Process performance data should be used to establish whether process objectives are being achieved. Current performance of processes should compare process data with the quality objectives you have set for the processes. Improvement opportunities relative to process performance may cover:

- the efficiency of processes relative to the utilization of resources (physical, financial and human resources and the manner in which they are structured);
- the effectiveness of processes relative to the utilization of knowledge, experience in achieving process objectives;
- the need to change process design, methods and techniques including process measurement;
- the need to reduce variation;
- the need to meet or exceed new legal and regulatory requirements that apply to the process.

Product performance (5.6.2.1)

Product performance data should be used to establish whether products fulfil their intended purpose in both design and build quality. Current performance of products should compare product data with product specifications and product specifications with design intent (what the product was intended to accomplish). The data may be obtained from actual and potential field failures and studies undertaken to assess the impact on safety and the environment. Examples might include engine seizure, tyre blow out, exhaust manifold blow out, premature airbag inflation and fuel tank explosion in collision.

Improvement opportunities relative to product performance may cover:

- the need to change product design, technology and materials;
- the need to change product literature to match actual performance (reset expectations);
- the adequacy of the means used to measure product performance and collect data;
- the conditions of use and application.

Corrective actions

Corrective action data should be used to establish whether the recurrence of problems is being prevented. Current performance on the status of corrective actions should compare the results with the quality objectives you have set for dealing with corrective actions such as closure time and degree of recurrence. Improvement opportunities relative to corrective actions may cover:

- the adequacy of problem analysis and resolution techniques;
- the need for new training programmes;
- the capability of the system to maintain performance in line with objectives (its sensitivity to change).

Preventive actions

Current performance on the status of preventive actions should compare the results with the quality objectives you have set for dealing with preventive actions such as closure time and degree of occurrence. Remember that preventive actions are supposed to prevent the occurrence of problems therefore a measure of status is the extent to which problems occur. However, this is an area that often causes confusion. Preventive action is often not an action identified as preventive but an action identified under guise of planning, training, research and analysis. Why else would you plan but to achieve objectives and hence to prevent failure? Why else do you perform a FMEA, but to prevent failure? Therefore don't just look for actions with the label *preventive*.

Improvement opportunities relative to preventive actions may cover:

* the adequacy of techniques to identify potential problems;
* the need for new tools and techniques, training programmes, etc.;
* the re-organization of departments, resources, etc.

Actions from management reviews

Current performance on follow-up actions from earlier management reviews should address not only whether they are open or closed but how effective they have been and how long they remain outstanding as a measure of planning effectiveness.

Improvement opportunities relative to actions from management reviews may cover:

* the prioritization of actions;
* the re-classification of problems relative to current business needs;
* the need to re-design the management review process.

Changes affecting the management system

It is difficult to foresee any change inside the organization that would not affect the management system in some way or other. However, the management system should be designed to cope with a degree of change without top management intervention. The change management processes to bring in new products, new processes, new people, new resources and new organization structures should be part of the management system. The system should be designed to handle such changes as a routine. If the management system is perceived as a set of documents, there are many changes that might not affect the management system but as stated previously, the management system is much more than this. Also changes in products, processes, organization structures, etc. will all affect the management system documentation but there should be processes in place to manage these changes under controlled conditions. In an environment in which perceptions of the management system have not been harmonized, it is likely that some change mechanisms will be outside the documented management system and in such circumstances, these changes need to be brought to the management review.

Review outputs (5.6.3)

The standard requires the outputs from the management review *to include decisions and actions related to the improvement of the effectiveness of the quality management system and its processes, improvement of product and actions related to resource needs.*

What does this mean?

Improving the effectiveness of the management system is not about tinkering with documentation but enhancing the capability of the system so that it enables the organization to fulfil its objectives more effectively. The management system comprises processes therefore the effectiveness of these too must be improved. Improvement of product related to customer requirements means not only improving the degree of conformity of existing product but enhancing product features so that they meet changing customer needs and expectations.

Why is this necessary?

This requirement responds to the Continual Improvement Principle.

The outcomes of the management review should cause beneficial change in performance. The performance of products is directly related to the effectiveness of the process that product them. The performance of these processes is directly related to the effectiveness of the system that connects them. Without resources no improvement would be possible.

How is this implemented?

The implication of this requirement is that the review should result in decisions being made to improve products, processes and the system in terms of the actions required.

Actions related to improvement of the system and its processes should improve the capability of the system to achieve the organization's objectives. Such actions should therefore focus on making beneficial changes in methods, techniques, relationships, measurements, behaviours, capacity, competency, etc. A quick fix to overcome a problem is neither a system change nor a system improvement because it only acts on a particular problem. If the fix not only acts on the present problem but also will prevent its recurrence, it can be claimed to be a system improvement. This may result in changes to documentation but this should not be the sole purpose behind the change – it is performance that should be improved.

Actions related to improvement of products should improve:

- *The quality of design*: The extent to which the design reflects a product that satisfies customer needs.

- *The quality of conformity*: The extent to which the product conforms to the design.
- *The quality of use*: The extent to which the user is able to secure continuity of use and low cost of ownership from the product.

Such actions may result in providing different product features or better-designed product features as well as improved reliability, maintainability, durability and performance. Product improvements may also arise from better packaging, better user instructions, clearer labelling, warning notices, handling provisions, etc.

Actions related to resource needs are associated with the resource-planning process that should be part of the management system. If this process were operating effectively, no work would commence without adequate resources being available. If the resources cannot be provided, the work should not proceed. It is always a balance between time, effort and materials. If the effort cannot be provided the time has to expand accordingly.

Summary

In this chapter we have examined the requirements contained in Section 5 of ISO/ TS 16949. All these requirements apply to top management and their implementation reflects the leadership style of the organization. We have shown that an understanding of customer needs and expectations is vital for organizations to prosper and essential for developing an effective management system. We have also discovered that success is not simply about getting results. The manner in which these results are obtained is important if we are to satisfy all the benevolent interested parties. We have learnt that this is where policies are needed to guide our actions and decisions and our behaviour in accomplishing the purpose and mission of the organization. We have seen that objectives are important in focusing the efforts of people so that they all pull in the same direction. But we have also seen that setting objectives and targets without a plan for meeting them is futile just as setting an objective or a target beyond the capability of the system will create frustration and low moral. Finally, we have realized the importance of evaluating system effectiveness and recognized that we must change the approach taken towards management review. We now know there is only one system and therefore our management review is a review of the way we manage the organization.

Management responsibility requirement checklist

These are the topics that the requirements of ISO/TS 16949 address consecutively. Topics 1–41 appeared in Chapter 4.

Management commitment

42 Evidence of management commitment to developing a QMS

43 Evidence of management commitment to continually improving the effectiveness of the QMS

44 Evidence of communicating the importance of meeting customer and regulatory requirements

45 Evidence of management establishing the quality policy

46 Evidence of management establishing quality objectives

47 Evidence of management commitment by conducting management reviews

48 Evidence of management commitment by ensuring the availability of necessary resources

Process efficiency

49 Monitoring product realization processes

Customer focus

50 Determining customer requirements

51 Meeting customer requirements

Quality policy

52 Purpose of organization

53 Commitment to comply with requirements

54 Commitment to continual improvement

55 Framework for quality objectives

56 Communication of quality policy

57 Review of quality policy

Planning
Quality objectives

58 Establishing quality objectives

59 Measurement of quality objectives

60 Consistency of quality objectives

61 Objectives for meeting product requirements

62 Inclusion of quality objectives in business plan

Quality management system planning

63 Planning of quality management system in line with process management principles

64 Planning of quality management system to meet quality objectives

65 Maintaining integrity of quality management system

Responsibility, authority and communication
Responsibility and authority

66 Defining and communicating responsibility and authority

Responsibility for quality
67   Notifying management of nonconformities
68   Authority to stop production
69   Staffing of shifts

Management representative
70   Appointment of management representative
71   Responsibility and authority of management representative

Customer representative
72   Designation of customer representatives

Internal communication
73   Establishing communication processes
74   Communicating the effectiveness of the QMS

Management review
General
75   Top management review of quality management system
76   Assessing opportunities for improvement
77   Records of management review

Quality management system performance
78   Scope of management review
79   Monitoring quality objectives
80   Results of management review

Review input
81   Performance information for input to management review
82   Changes affecting the QMS
83   Analysis of field failures

Review output
84   Decisions and action arising from management review

Management responsibility – Food for thought

1 Does your QMS make the right things happen or is it just a set of procedures?
2 Does your management perceive the QMS as the means by which the organization's objective are achieved?
3 Does your organization exist to make profit or to create and retain satisfied customers?
4 Does your quality policy affect how people behave in your organization or is it simply a slogan?
5 Do all your quality objectives relate to the organization's purpose and mission or are they focused only on what the quality department will achieve?
6 Does your management review examine the way the organization is managed or does it simply focus on conformity issues?
7 Do you struggle to obtain the necessary resources to do your job, or have you designed your job so that you get all the resources necessary for you to achieve your objectives?
8 Do you wait to receive a customer enquiry before identifying customer needs and expectations or have you researched the market in which you operate so that your offerings respond to customer needs?
9 What made you think that by simply publishing your quality policy, anything would change?
10 If you didn't know your current performance, how did you manage to set meaningful objectives?
11 When you set your objectives, was there any discussion on how they might be achieved and did it result in changing the way you do things?
12 If you don't think you need to change, how come you didn't meet these objectives last year?
13 How will you know when you have achieved your objectives?
14 Are you sure that none of the managers' objectives relate to extracting more performance from unstable processes?
15 Are you sure that none of the managers are tasked with meeting objectives for which no plans have been agreed for their achievement?
16 How many plans are the managers working on that have an objective that is not derived from the business plan?
17 Are you sure that managers are not pursing objectives that will cause conflict with those of others managers?
18 Are you sure you have not imposed a target on a member of staff for performance improvement when it is the system that requires improvement?
19 Do you always consider the impact of change on other processes before you proceed?
20 Does your staff know of the results for which they are accountable and are their job descriptions limited to such responsibilities?
21 If your management representative is unable to influence the other managers to implement, maintain and improve the management system, are you sure you have appointed the right person?
22 How often do you check that messages conveyed from management are actually understood by those they are intended to affect?
23 How often do you check that messages conveyed by workers are actually given due consideration by management and interpreted as the consequence of their own actions and decisions?
24 If the person with the most interest in the effectiveness of the management system is not your CEO, is there not something wrong with the way the management system is perceived in the organization?

Chapter 6

Resource management

Only the orchestra playing a joint score makes music
Peter F Drucker

Summary of requirements

Section 6 of ISO/TS 16949 draws together all the resources-related requirements that were somewhat scattered in the 1999 version. Resource management is a key business process in all organizations. In practice, resource management is a collection of related processes that are often departmentally oriented:

- financial resources are controlled by the Finance Department;
- purchased materials, equipment and supplies are controlled by the Purchasing Department;
- measuring equipment maintenance is controlled by the Calibration Department;
- plant maintenance is controlled by the Maintenance Department;
- staff development is controlled by the Human Resources or Personnel Department;
- building maintenance is controlled by the Facilities Management Department.

These departments control the resources in as much that they might plan, acquire, maintain and dispose of them but do not manage them totally because they are not the sole users or customers of the resource. They therefore only perform a few of the tasks necessary to manage resources. Collectively they control the human, physical and financial resources of the organization.

The resource management process has a number of distinct stages as shown in Figure 6.1.

Whatever the resource, firstly it has to be planned, then acquired, deployed, maintained and eventually disposed of. The detail of each process will differ depending on the type of resource being managed. Human resources are not 'disposed of' but their

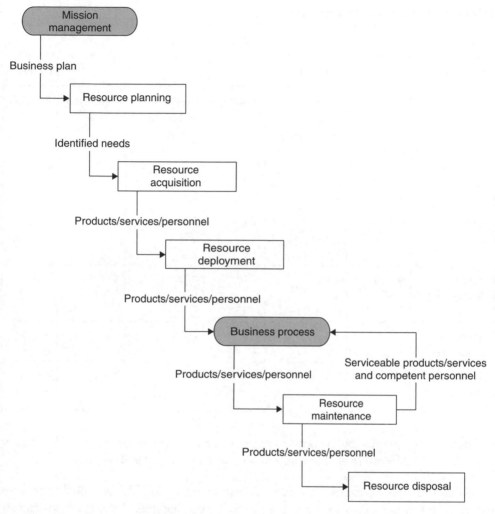

Figure 6.1 *Resource management process stages*

employment or contract terminated although the standard does not address disposal of any resources. Resource disposal impacts the environment and other interested parties, and if an automotive company discharges waste into the ground water, it could lead to prosecutions that displease the customer.

Only measuring devices need calibration but other devices need verification. The standard does not address financial resources specifically but clearly they are required to implement and maintain the management system, and hence run the organization. Purchasing is not addressed under resource management but under product realization. However, the location of clauses should not be a barrier to the imagination because their

Table 6.1 *Clause alignment with process model*

Resource management processes	Types of resource					
	Human	Physical				Finance
	Personnel	Plant, equipment and materials	Buildings	Utilities and Support Services	Measuring devices	Finance
Resource planning	6.1 6.2.2a 7.1	6.1 6.3 7.1	6.1 6.3 7.1	6.1 6.3 7.1	7.1 7.6	6.1 7.1
Resource acquisition	6.1 7.4	6.1 6.3 7.4	6.1 6.3	6.1 6.3 7.4	7.4	6.1
Resource deployment	4.1d 6.1 6.2.1	4.1d 6.1 6.3	4.1d 6.1 6.3	4.1d 6.1 6.3	4.1d 7.5.1	4.1d 6.1
Resource maintenance	6.2.2b	6.3	6.3	6.3	7.6	NA
Resource disposal	NA	NA	NA	NA	NA	NA

location is not governed by the process approach but by user expectations. The clauses of ISO/TS 16949 are related to these processes as indicated in Table 6.1.

Note that there are no clauses that address resource disposal. This is probably because the standard only focuses on intended product, whereas ISO 14001 would address resource disposal and unintended product.

Determining resources (6.1)

The standard requires the organization *to determine the resources needed to implement and maintain the quality management system, continually improve its effectiveness, and enhance customer satisfaction by meeting customer requirements.*

What does this mean?
Determination of resources means identifying resource needs. The resources needed to implement the management system include all human, physical and financial resources

needed by the organization for it to function. The resources needed to maintain the management system are those needed to maintain a level of performance. The resources needed to continually improve the effectiveness of the management system are those needed to implement change in the organization's processes. Those resources needed to enhance customer satisfaction are no more than those needed to achieve the organization's objectives because of the linkage between customer requirements, policy, objectives and processes. What is missing from the requirement is the determination of resources to establish the management system that includes the human, physical and financial resources to design and construct the processes needed to enable the organization to achieve its objectives. As Clause 4.1 requires the organization to establish, implement, maintain and continually improve the management system this could be more of an oversight than intentional.

If you put a boundary around the management system and perceive it only as a part of the overall management system, there will be those resources which serve the management system and those which serve other purposes, but such boundaries are not useful because they divert attention from the basic goal of satisfying the interested parties. If you are faced with making a decision as to what to include and exclude, the questions you need to answer are: "Why would we want to exclude a particular resource from the management system? What business benefit is derived from doing so?" Hopefully, you will conclude that there are no benefits from their exclusion and many benefits from their inclusion. Whether or not you exclude a part of the management system from ISO/TS 16949 certification is an entirely separate matter as was dealt with in Chapter 1.

Why is this necessary?
This requirement responds to the Process Approach Principle.

Without adequate resources, the management system will not function. As addressed previously, the management system is more than a set of documents. It is a dynamic system and requires human, physical and financial resources for it to deliver the required results. Starve it of resources and the planned results will not be achieved.

How is this implemented?
Resource management is a common feature of all organizations and while it may be known by different titles, the determination and control of resources to meet customer needs is a fundamental requirement and is fundamental to the achievement of all other requirements.

There are two types of resource requirements: those needed to set up and develop the organization, and those needed to execute particular contracts or sales. The former is addressed in Clause 6.1 and the latter in Clause 7.1.

The way many companies identify resource requirements is to solicit resource budgets from each department covering a 1–5-year period. However, before the managers can prepare budgets they need to know what requirements they will have to meet. They will need access to the corporate plans, sales forecasts, new regulations and statutes, new product development plans, marketing plans, production plans, etc. as well as the policies, objectives, process descriptions and procedures.

In specifying resource needs for meeting product requirements there are three factors that need to be defined – the quantity, delivery and quality expressed by questions, such as: How many do you want? When do you want them? To what specification or standard do they need to be? These factors will affect cost directly and if not determined when establishing the budgets, you could have difficulty later when seeking approval to purchase. In specifying resource needs for meeting organizational requirements there are three factors that need to be defined. These are the objectives for maintaining the *status quo*, for improving efficiency and for improving effectiveness.

Although resources are normally configured around the organization's departments and not its processes, departmental resource budgets can divert attention away from process objectives. A more effective way to determine resources is by process and not by department. In this way the resources become focused on process objectives and overcome conflicts that can arise due to internal politics and the power structure. It then becomes less of a problem convincing top management of the need when it can be clearly demonstrated that the requested resource serves the organization's objectives.

A practical way of ensuring that you have adequate resources is to assign cost codes to each category of work and divide them into two categories: *maintenance* and *improvement*. Include all costs associated with maintaining the *status quo* under maintenance and all costs associated with change under improvement. You can then focus on reducing maintenance costs for the same level of sales without jeopardizing improvement. It is often difficult to obtain additional resources after the budget has been approved but provided they can be justified against the organization's objectives, it should not be a problem. Either the management is serious about achieving the objective or it isn't. It is lunacy to set goals then object to providing the resources to achieve the goals. Arguments are perhaps more about estimating accuracy than need.

Determining resources is not simply about quantities: the number of people, equipment, machines, etc. It is also about capability and competence. It is of no benefit to possess the right number of people if they are not competent to deliver the outputs needed – no benefit to own the right number of machines is they can't produce product to the required accuracy.

There are four types of change that affect resources.

1 The unplanned loss of capability (staff leave or die, equipment or software obsolescence, major breakdown, fuel shortage and man-made or natural disaster).
2 An increase or reduction in turnover (doing more or less of what we do already).
3 A change in the organization's objectives (aiming at new targets, new products and new processes).
4 A change in the external standards, regulations, statutes, markets and customer expectations (we have to do this to survive).

Providing resources (6.1 and 7.5.5.1)

The standard requires the organization *to provide the resources needed to implement and maintain the quality management system, and continually improve its effectiveness and enhance customer satisfaction by meeting requirements.* The standard also requires *an inventory management system, the optimization of stock turns and obsolete product to be controlled as nonconforming product.*

What does this mean?
Providing resources means acquiring and deploying the resources that have been identified as being needed. The acquisition process, should deliver the resource in the right quantity and quality when they are needed. The deployment process should transport and prepare the resource for use. Therefore, if there is an identified need for human resources, they have been provided only when they are in a position ready to assume their duties, i.e. deemed competent or ready to take up a position under supervision. Likewise with equipment, it has been provided only when it is installed, commissioned and ready for use. The maintenance process should maintain stock levels, equipment, people, facilities, etc. so that there is no shortage of supply of capable and suitable physical and human resources.

The mention of obsolete product in this context reminds us that providing resources that are needed implies the removal of resources that are no longer needed which includes removal of obsolete product as well as surplus staff, equipment and space, etc.

Why is this necessary?
This requirement responds to the Process Approach Principle.

For plans to be implemented, resources have to be provided but the process of identifying resources can be relatively non-committal. It establishes a need, whereas the provision of resources requires action on their acquisition which depending on prevailing circumstances may not lead to all needs being provided for when originally required.

How is this implemented?

As indicated above, providing resources requires the implementation of the acquisition processes. The physical resources will be initiated through the purchasing process, the human resources through the recruitment process and the financial resources through the funding process. The purchasing process is dealt with further in Chapter 7, under the heading *Purchasing*. An example of a recruitment process is illustrated in Figure 6.2.

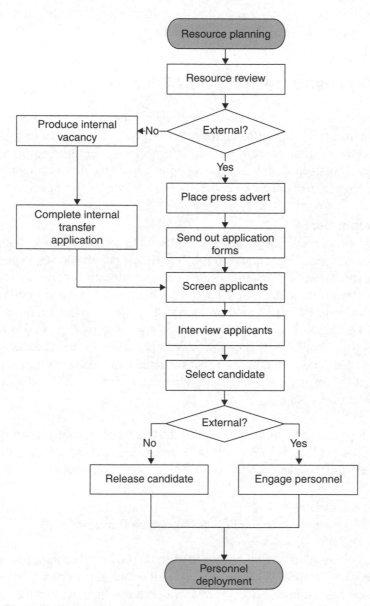

Figure 6.2 *Personnel acquisition process flow*

The acquisition of financial resources is beyond the scope of this book because such methods are so varied and specialized. For example, funds can be acquired by cutting costs, eliminating waste, downsizing, selling surplus equipment, stocks and shares, or seeking a bank loan.

Inventory

Inventory is not addressed in ISO 9001 but clearly an adequate supply of materials and components is necessary for the organization to produce the required products and deliver the required services. This omission has been corrected in ISO/TS 16949:2002 with the requirements inserted under Production and Service Provision rather than Resource management but will be dealt with here for consistency.

To enable you to achieve delivery requirements you may need adequate stocks of parts and materials to make the ordered products in the quantities required. In typical commercial situations, predicting the demand for your products is not easy – organizations tend to carry more inventory than needed to cope with unexpected demand. The possibility of an unexpected increase in demand leads to larger inventories as an out-of-stock situation may result in lost customer orders. Most companies have to rely on forecasts and estimates. Some customers may protect you to some extent from fluctuations in demand by giving you advanced notification of their production and service requirements in order that your production schedule can be 'order driven'. In the event that an increase in demand is necessary you should be given adequate warning in order that you can increase your inventory in advance of the need. If adequate warning cannot be given, you need suitable clauses in your contract to protect you against any unexpected fluctuations in demand that may cause you to fail to meet the delivery requirements.

Inventory management is concerned with maintaining economic order quantities in order that you order neither too much stock nor too little to meet your commitments. The stock level is dependent on what it costs both in capital and in space needed to maintain such levels. Even if you employ a 'ship-to-line' principle, you still need to determine the economic order quantities. Some items have a higher value than others thereby requiring a higher degree of control. Use of the Pareto Principle will probably reveal that 20% of inventory requires a higher degree of control to enable you to control 80% of the inventory costs.

Whether or not 100% on-time delivery is a requirement of your customers, you won't retain customers for long if you continually fail to meet their delivery requirements regardless of the quality of the products you supply. It is only in a niche market that you can retain customers with a long waiting list for your products. In competitive markets you need to exceed delivery expectations as well as product quality expectations to retain your market position.

In addition to establishing an inventory management system, the standard requires inventory turns (or stock turns) to be optimized. Inventory turns is a measure of how frequently the stock, raw material, work-in-progress and finished goods are turned over in relation to the sales revenue of a product. It is the ratio of sales turnover of product to value of stock. Or

$$\frac{\text{sales turnover of product}}{\text{value of stock}}$$

A high figure indicates better control over the manufacturing process. A low figure indicates excessive stock to accommodate process variations. Clearly to calculate the inventory turns you need to know both the sales over a given period and the value of the stock over the same period. Most organizations perform stock taking at least annually and therefore would be able to compute the inventory turns annually. However, this may not be good enough for some automotive customers. With resource planning tools, such as MRP and MRPII, inventory is computerized and its value known at any given time. Optimizing inventory turns requires accurate forecasting of needs and a commitment to reduce process variation. Stable processes operating within required limits yield optimum inventory turns thereby signifying a strong link between SPC, continual improvement and inventory control.

The standard also requires stock rotation; meaning that parts and materials are used on a first-in-first-out (FIFO) basis. The picking system will therefore need to be date sensitive to operate FIFO. Not a problem if inventory is computerized but can be difficult to maintain with any accuracy if manual because when the heat is on the paperwork gets ignored.

Competence of personnel (6.2.1)

The standard requires *personnel performing work affecting product quality to be competent on the basis of appropriate education, training, skills and experience with particular attention to the satisfaction of customer requirements.*

What does this mean?
If a person has the appropriate education, training and skills to perform a job the person can be considered qualified. If a person demonstrates the ability to achieve the desired results the person can be considered competent. Qualified and competent are therefore not the same. Qualified personnel may not be able to deliver the desired results. While they may have the knowledge and skill, they may exhibit inappropriate behaviours. (The expert who knows everything but whose interpersonal skills cause friction with staff to such an extent that it adversely affects productivity!)

A competent person may not be appropriately educated, trained or possess what are perceived to be the required skills. (The pragmatist who gets the job done but can't explain how he does it.) This requirement therefore presents a dichotomy because you can't be competent on the basis of appropriate education, training and skills unless you can also demonstrate you can achieve the required results. The standard fails to specify whether competence in this context is about what individuals *know* or what individuals *can do*. Taking a pragmatic approach, it would not serve the intent of the standard to simply focus on what people know therefore competence in the context of ISO/TS 16949 must be about what people *can do*.

Competence is the ability to demonstrate *use* of knowledge, skills and behaviours to achieve the results required for the job. It is the ability to perform the whole work role not just specific tasks, the ability to meet standards that apply in the particular job not in a classroom or examination but an ability to perform in the real working environment with all the associated variations, pressures, relationships and conflicts. Competence is not a probability of success in the execution of one's job; it is a real and demonstrated capability. ISO 9000 defines competence as the demonstrated ability to apply knowledge and skills. The subtle difference in this definition is that the frame of reference is not present by there being no reference to the results required for the job. Behaviours have been omitted but we must assume this is an oversight and not intentional.

Competence is concerned with outcomes rather than attributed abilities. Competence is more than a list of attributes. A person may claim to have certain ability but proof of competence is only demonstrated if the desired outcomes are achieved. The opposite of competence is incompetence, but this is perceived as a pejorative term so we tend to use the phrase "not yet competent" to describe those who have not reached the required standard of competence. Competence does not mean excellence. Competence is meeting the established performance or behavioural standards. Excellence is exceeding the established standards. Competence is a quality of individuals, groups and organizations. Corporate competence empowers an organization to build a competitive advantage and is the result of having capable processes that deliver corporate goals. Core competences are the things the organization is good at and these are dependent on employing competent human resources.

The requirement makes a distinction between those personnel whose work affects product quality and those personnel whose work does not affect product quality. Why would you not want all your personnel to be competent? It makes no sense! In principle, everyone's work affects the quality of the products and services supplied by the organization, some directly, others indirectly.

The additional automotive requirements in Clause 6.2.2.2 for personnel to be qualified with particular attention to the satisfaction of customer requirements means that personnel are deemed capable of applying the mathematics, science or other discipline required by the customer.

Why is this necessary?

This requirement responds to the Leadership and Involvement of People Principles.

Traditionally personnel have been selected on the basis of certificated evidence of qualifications, training and experience rather than achievement of results. Here are some examples showing the inadequacy of this method:

- A person may have received training but not have had the opportunity to apply the knowledge and skills required for the job.
- A person may have practiced the acquired skills but not reached a level of proficiency to work unsupervised.
- A person may possess the knowledge and skills required for a job but may be temporarily or permanently incapacitated. (A professional footballer with a broken leg is not competent to play the game until his leg has healed and he is back on top form.)
- A person may have qualified as a chemist 30 years ago but not applied the knowledge since.
- Airline pilots who spend years flying one type of aircraft will require some period in the flight simulator before flying another type because they are not deemed competent until they have demonstrated competency.
- A person may have been competent in maintaining particular process equipment but has not had occasion to apply the skills in the last 12 months. The engineers need to demonstrate competence before being assigned to maintain a particular piece of equipment because process capability is at risk.

The above examples illustrate why the possession of qualifications, training certificates and years of experience are not necessarily adequate proof of competence. With QS-9000, all that was needed was to demonstrate that personnel had the qualifications, training and experience required for the work – not that they were competent to perform it although in many cases, the staff appraisal system was offered as proof that staff were suitable for the job. However, staff appraisal systems are notoriously inadequate because the standards are not measurable and performance is based on the subjective and sometimes biased judgement of the managers and not the measured performance and behaviours of their staff.

How is this implemented?

In any organization there are positions that people occupy, jobs they carry out and roles they perform in these positions. For each of these certain outcomes are required from the occupants. The starting point is therefore to define the outcomes required from a job, and then define what makes those performing it successful and agree these with the role or job holder. However, having set these standards, they become the basis for competence assessment and competence development. The education and training provided should be consistent with enabling the individual concerned to achieve the agreed standards.

The outcomes required for a role or job has two dimensions. There are the hard results such as products and decisions, and the soft results such as behaviours, influence and stamina. The outcomes are also dependent on the conditions or context in which the role or job is performed. For instance a manager of a large enterprise may produce the same outcomes as the manager of a small enterprise but under entirely different conditions. It follows therefore that a competent manager in one context may not be competent in another.

For instance, a person who occupies the position of production manager in a glass factory performs the role of a manager and therefore needs to be competent in achieving the required results through use of appropriate physical and human resources. This person also performs several different jobs concerned with the production of glass and therefore may need to be competent to negotiate with suppliers, use computers, produce process specifications, blow glass, drive trucks, test chemicals, administer first aid, etc. depending on the scale of the operations being managed. Each job comprises a number of tasks that are required to deliver a particular result. Having a forklift truck-driving certificate is not a measure of competence. All this does is prove that the person can drive a forklift truck. What the organization may need is someone who can move 4 tonnes of glass from point A to point B safely in 5 minutes using a forklift truck. The management role may occupy some people 100% of the time and therefore the number of competencies needed is less than those who perform many different types of jobs in addition to management. For this reason there is no standard set of competencies for any particular position because each will vary but there are national schemes for assessing competence relative to specific occupations. In the UK, the National Vocational Qualification (NVQ) scheme has been operating since 1986. The central feature of NVQs is the National Occupational Standards (NOS) on which they are based. NOS are statements of performance standards that describe what competent people in a particular occupation are expected to be able to do. They cover all the main aspects of an occupation, including current best practice, the ability to adapt to future requirements, and the knowledge and understanding that underpin competent performance. NVQs are work-related, competence-based qualifications. They reflect the skills and knowledge needed to do a job effectively and represent national standards recognized by employers throughout the country. Five levels of competence are defined covering knowledge, complexity, responsibility, autonomy and relationships.

While ISO/TS 16949 does not require vocational qualifications, such as the UK NVQ scheme, the implication of this requirement for competence is that managers will need to select personnel on the basis of their ability to deliver the outcomes required. Selecting a person simply because they are related to the boss would not be appropriate unless of course they can also deliver the required outcomes!

The standard requires:

- personnel to be competent,
- the necessary competence to be determined,
- actions to be taken to satisfy these needs,
- the evaluation of the effectiveness of actions taken.

The standard is therefore implying that organizations should employ competence-based assessment techniques but as indicated previously, it is not explicit nor is it clarified in ISO 9004 whether competence in this context is based on what people know or what they can do. The current method of selection may be on the basis of past performance but without performance standards in place and a sound basis for measurement, this method is not capable of delivering competent people to the workplace.

You will need to maintain documentary evidence that personnel are competent to perform the jobs assigned to them and to do this you need to identify the competence needed and demonstrate that a competence assessment process is employed to validate competence. This is addressed in the next section. Fletcher[1] provides useful guidance on designing competence-based assessment programmes. Competence-based assessment has a number of uses:

- Assessment for certification (not required for ISO/TS 16949 but regulations or customers may require this; welders and personnel performing non-destructive testing are some examples).
- Performance appraisal (indirectly required by ISO/TS 16949 from the requirement for an evaluation of actions taken to satisfy needed competence).
- Identification of training needs (required by ISO/TS 16949).
- Skills audit (indirectly required by ISO/TS 16949 from the requirement to ensure the organization has the ability to meet product requirements and from the requirement for selecting competent personnel).
- Accreditation of prior learning (not required by ISO/TS 16949).
- Selection and recruitment (required by ISO/TS 16949).
- Evaluating training (required by ISO/TS 16949).

Training, awareness and competence (6.2.2)

Determining competence necessary (6.2.2a, 6.2.2.1 and 6.2.2.2)

The standard requires the organization *to determine the necessary competence for personnel performing work affecting product quality and to maintain documented procedures for identifying training needs and achieving competence.* The additional automotive requirements require *personnel with design responsibility to be competent to achieve design requirements and be skilled in applicable tools and techniques.*

What does this mean?

Individual competence is concerned with the ability of a person to achieve a result whereas training is concerned with the acquisition of skills to perform a task and education is concerned with the acquisition of knowledge. It is therefore not a question of whether a person has the skills and knowledge to do a job but on whether a person is able to achieve the desired outcome. This is known as the competence-based approach. Academic qualifications tend to focus on theory or application of theory to work situations. In contrast, the competence-based approach focuses on the results the individuals are achieving. People are either competent or not yet competent. There are no grades, percentages or ratings. People are deemed competent when they have demonstrated performance that meets all the requirements standards. A person who is appropriately educated, trained and experienced is competent only if they have the ability to produce the desired results when required. If for some reason a competent person became incapacitated they would no longer be deemed competent to perform the job they were performing prior to the incapacity.

Why is this necessary?

This requirement responds to the Leadership and Involvement of People Principles.

When assigning responsibility to people we often expect that they will determine what is needed to produce a good result and perform the job right first time. We are often disappointed. Sometimes it is our fault because we did not adequately explain what we wanted or more likely, we failed to select a person that was competent to do the job. We naturally assumed that because the person had a college degree, had been trained in the job and had spent the last 2 years in the post, that they would be competent. But we would be mistaken, primarily because we had not determined the necessary competence for the job and assessed whether the person had reached that level of competence. In theory we should select only personnel who are competent to do a job but in practice we select the personnel we have available and compensate for their weaknesses either by close supervision or by providing the means to detect and correct their failures.

The competence-based movement developed in the 1960s out of a demand from businesses for greater accountability and more effective means of measuring and managing performance. This led to research into what makes people effective and what constitutes a competent worker. Two distinct competence-based systems have emerged. The British model focuses on standards of occupational performance and the American model focuses on competency development. In the UK, the standards reflect the outcomes of workplace performance. In the USA, the standards reflect the personal attributes of individuals who have been recognized as excellent performers[2] but what individuals achieved in the past is not necessarily and indication of what they achieve in the future they age, they forget, their eyesight deteriorates and they may not be as agile both physically and mentally as they once were.

Competence is particularly important in the professions because the outputs result from an intellectual process rather than an industrial process. We put our trust in professionals and expect them to be competent but methods of setting standards of competence and their evaluation have only been developed over the last 15 years. It was believed that education, training and experience were enough, but the recent cases of malpractice particularly in the medical profession have caused the various health authorities to look again at clinical competence.

How is this implemented?
Determining the competence necessary for performing a job is a matter of determining the outcomes required of a job, the performance criteria or standards to be achieved, the evidence required and the method of obtaining it. It is important that the individuals whose performance is to be assessed are involved in the setting of these standards.

The jobs that people perform must be related to the organization's objectives and as these objectives are achieved through processes, these jobs must contribute to achievement of the process objectives. In the decomposition from the system level where the business processes are identified through to work processes and subprocesses you will arrive at a level where the results are produced by a single person. The objectives for these processes or subprocesses describe outcomes "What must be achieved?". If you then ask, "What must be done for this to be achieved?". These are termed *units of competence*. Several units of competence will be necessary to achieve a given outcome. For example, a font line operator's primary output is conforming product. The operator needs to possess several competences for conforming product to be produced consistently.

An operator might need the ability to:

- understand and interpret technical specifications,
- set up equipment,
- operate the equipment so as to produce the required output,
- undertake accurate measurements,
- apply variation theory to the identification of problems,
- apply problem-solving methods to maintain control of the process.

Simply possessing the ability to operate a machine is not a mark of competence.

Table 6.2 *Competence-based assessment*

Key question	What this is called	Example
What must be achieved?	Outcome	Conforming product
What must be done for this to be achieved?	Unit of competence	The ability to apply variation theory to the identification of problems (this is one of several)
How well must this be achieved?	Performance criteria	Distinguishes special cause problems from common cause problems (this is one of several)
How should assessment be conducted?	Assessment method	Observation of performance
What evidence should be collected?	Evidence requirement	Run charts indicating upper and lower control limits with action taken only on special causes (this is one of several)

Table 6.2 shows the key questions to be asked, the terms used and one example. It should be noted that there may be several performance criteria and a range of methods used to collect the evidence. It should also be noted that the evidence should be against the unit of competence not against each performance criteria because it is competence to deliver the specified outcome that is required not an ability to produce discrete items of evidence.[3] Terminology in this area is not yet standardized and therefore there are some differences in the terms between the British and American competence-based systems. For instance, performance criteria seem to be clustered into Elements of Competence in the British system.

When considering the introduction of a competence-based assessment system Fletcher provides a useful checklist.[4]

Is the proposed system:

- Based on the use of explicit statements of performance?
- Focused on the assessment of outputs or outcomes of performance?
- Independent of any specified learning programme?
- Based on a requirement in which evidence of performance is collected from observation and questioning of actual performance as the main assessment method?
- One which provides individualized assessment?
- One which contains clear guidance to assessors regarding the quality of evidence to be collected.
- One which contains clear guidelines and procedures for quality assurance?

The determination of competence requires that we have defined a standard for competence, measured performance and acquired evidence of attainment. We therefore need to ask:

- What are the key results or outcomes for which the person is responsible? (The units of competence.)
- What are the principal tasks the individual is expected to perform and the expected behaviours the individual is required to exhibit to achieve these outcomes? (The elements of competence.)
- What evidence is required to demonstrate competence?
- What method of measurement will be used to obtain the evidence?

The methods for setting competence standards are quite complex therefore the reader should consult the various references in Appendix B. A documented procedure for identifying competence or training needs would be written around the topics addressed above.

The purpose in determining competence is to identify the requirements for the job. Requirements for new competencies arise in several ways as a result of the following:

- Job specifications.
- Process specifications, maintenance specifications, operating instructions, etc.
- Development plans for introducing new technologies.
- Project plans for introducing new equipment, services, operations, etc.
- Marketing plans for launching into new markets, new countries, new products and services.
- Contracts where the customer will only permit trained personnel to operate customer owned equipment.
- Corporate plans covering new legislation, sales, marketing, quality management, etc.
- An analysis of nonconformities, customer complaints and other problems.
- Developing design skills, problem-solving skills or statistical skills.
- Introducing a quality system thus requiring awareness of the topics covered by ISO 9000, the quality policies, objectives and training in the implementation of quality system procedures, standards, guides, etc.

You have a choice as to the form in which the competences are defined. Some organizations produce a skills' matrix which shows the skills each person has in a department. This is not strictly a list of competences because it is not linked to outputs. It is little more than a training record but could have been developed from observation by supervisors. The problem is that it is people oriented not process oriented. A better way is for a list of competences to be generated for each process based on the required process outputs. This could be a table with the process name and objective at the top and two columns:

- *Column 1*: Process activity output. (What must each activity achieve – the intended and unintended outputs.)
- *Column 2*: Unit of competence. (What must be done to achieve the outcome – these are the individual tasks.)

In identifying the units of competence, you are in effect identifying all the tools, techniques, methods and practices used by the organization, such as design of experiments, computer-aided design (CAD), SPC, etc. This will satisfy the automotive requirement for personnel with product design responsibility to be skilled in applicable tools and techniques.

Once the competency requirements have been specified, managers should plan the development needed for their staff.

Providing for training (6.2.2b and 6.2.2.3)

The standard requires the organization *to provide training including on-the-job training for personnel in any new or modified job affecting product quality, or take other actions to satisfy these needs.*

What does this mean?
Having identified the competence needs, this requirement addresses the competence gap but this gap is only established after assessing competence. Therefore, there are two types of actions needed to satisfy these needs: *competence assessment* and *competence development*.

Why is this necessary?
This requirement responds to the Leadership and Involvement of People Principles.

Having identified the competence needed to achieve defined outcomes, it is necessary to determine the current level of competence and provide the means to develop the competence of personnel where it is found that staff are not yet competent in some areas of their job.

How is this implemented?
Competence assessment
To operate a competence-based approach to staff selection, development and assessment, it is necessary to:

- set criteria for the required performance,
- collect evidence of competence,

- match evidence to standards,
- plan development for areas in which a "not yet competent" decision has been made.

A number of questions arise when considering the collection and assessment of evidence:

- What do we want to assess?
- Why do we want to assess it?
- Who will perform the assessment?
- How will we ensure the integrity of the assessment?
- What evidence should be collected?
- Where will the evidence come from?
- How much evidence will be needed?
- When should the assessment commence?
- Where should the assessment take place?
- How will we conduct the assessment?
- How will we record and report the findings?

Answers to these questions can be found in Fletcher's book on Competence based Assessment Techniques.

Bridging the competence gap

Once the results of the competence assessment are known, the gap may be bridged by a number of related experiences:

- Training courses where an individual undertakes an internal or external course.
- Mentoring where a more senior person acts as a point of contact to give guidance and support.
- Coaching on-the-job where a more experienced person transfers knowledge and skill.
- Job rotation where a person is temporarily moved into a complementary job to gain experience or relieve boredom.
- Special assignments where a person is given a project that provides new experiences.
- Action learning where a group of individuals work on their own but share advice with others and assist in solving each other's problems.
- On-the-job learning where the individual explores new theories and matches these with organizational experience.

On-the-job training

In many cases, formal assessment may not be required simply because the gap is glaringly obvious. A person is competent to produce a particular result and the result required or the method of achieving it is changed. A need therefore arises for additional training or instruction. After completing the training or instruction the records need to be updated indicating the person's current competence. This would apply equally to contractors or agency personnel.

External training

Beware of training courses that are no more than talk and chalk sessions where the tutor lectures the students, runs through hundreds of slides and asks a few questions! There is little practical gain from these kinds of courses. A course that enables the participants to learn by doing, to learn by self-discovery and insight is a *training* course. The participants come away having had an experience. Just look back on your life, and count the lessons you have learnt by listening and watching, and compare that number with those you have learnt by doing. The latter will undoubtedly out-number the former.

Encourage your staff to make their mistakes in the classroom not on the job or if this is not practical, provide close supervision on the job. Don't reprimand staff under training as anyone can make mistakes. An environment in which staff are free to learn is far better than one in which they are frightened of doing something wrong.

Training aids

If training is necessary to improve skills involving the operation or maintenance of tools or equipment, you need to ensure that any practical aids used during training:

* represent the equipment that is in use on the production line;
* adequately simulate the range of operations of the production equipment;
* are designated as training equipment and only used for that purpose;
* are recorded and maintained indicating their serviceability and their design standards including records of repairs and modifications.

Students undertaking training may inadvertently damage equipment. It may also be necessary to simulate or inject fault conditions so as to teach diagnostic skills. Training activities may degrade the performance, reliability and safety of training equipment, and so it should be subject to inspection before and after training exercises. The degree of inspection required would depend on whether the equipment has been designated for use only as training equipment or whether it will be used either as test equipment or to provide operational spares. If it is to be used as test or operational equipment, it will need to be re-certified after each training session. During the training sessions, records will need to be maintained of any fault conditions injected, parts removed and any other act which may invalidate the verification status of the equipment. In some cases it may be necessary to refurbish the equipment and subject it to the full range of acceptance tests and inspections before its serviceability can be assured. Certification can only be maintained while the equipment remains under controlled conditions. As soon as it passes into a state where uncontrolled activities are being carried out, its certification is immediately invalidated. It is for such reasons that it is often more economical to allocate equipment solely for training purposes.

Evaluating the effectiveness of personnel development activities (6.2.2c)

The standard requires *the organization to evaluate the effectiveness of the actions taken.*

What does this mean?

All education, training, experience or behavioural development should be carried out to achieve a certain objective. The effectiveness of the means employed to improve competence is determined by the results achieved by the individual doing the job. Regardless of how well the education, training or behavioural development has been designed; the result is wholly dependent on the intellectual and physical capability of the individual. Some people learn quickly while others learn slowly, and therefore personnel development is incomplete until the person has acquired the appropriate competence, i.e. is delivering the desired outcomes.

Why is this necessary?

This requirement responds to the Factual Approach and Involvement of People Principles.

The mere delivery of education or training is not proof that it has been effective. Many people attend school only to leave without actually gaining an education. They may pass the examinations but are not educated because they are often unable to apply their knowledge in a practical way except to prescribed examples. The same applies with training. A person may attend a training course and pass the course examination but may not have acquired the necessary proficiency – hence the necessity to evaluate the effectiveness of the actions taken.

How is this implemented?

Competence is assessed from observed performance and behaviours in the workplace not from an examination of education and training programmes far removed from the workplace.

Having established and agreed standards of competence for each role and job in the organization, those who can demonstrate attainment of these standards are competent. Thus competent personnel have the opportunity to prove their competence.

Fletcher[5] identifies the following key features of a competence-based assessing system:

- Focus on outcomes
- Individualized assessment
- No percentage rating

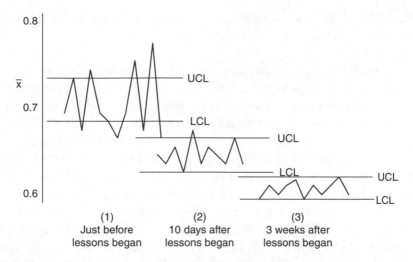

Figure 6.3 *Average daily scores for a patient learning to walk after an operation*

- No comparison with other individual's results
- All standards must be met
- Ongoing process
- Only "competent" or "not yet competent" judgments made.

There are three parts to the evaluation:

1 An evaluation of the personnel performance activity before development.
2 An evaluation of the personnel performance immediately on completion of development activity.
3 An evaluation of the personnel development activity within weeks of its completion.

Deming illustrates this as a run chart[6] an example of which is shown in Figure 6.3.

Development activity evaluation (the initial stage)

Activity evaluation by the students themselves can only indicate how much they felt motivated by the event. It is not effective in evaluating what has been learnt. This is more likely to be revealed by examination at the end of the event or periodically throughout the development period. However, the type of examination is important in measuring the effectiveness of the personnel development, e.g. a written examination for a practical course may test the theories behind the skills but not the practical mastery of the skills themselves. A person may fail an examination by not having read the question correctly, so examination by itself cannot be a valid measure of training effectiveness. You need to examine the course yourself before sending your staff on it. If you want information to be conveyed to your staff, a lecture with accompanying slide show may suffice. Slide shows are good for creating awareness but not for skill training. Skills cannot be acquired by any other means than by doing.

Development activity effectiveness short term (the intermediate stage)

We often think of training as a course away from work. We go on training courses. But the most effective training is performed on the job. Training should be primarily about learning new skills not acquiring knowledge – that is education. On returning to work or normal duties after a course it is important that the skills and knowledge learnt are put to good effect as soon as possible. A lapse of weeks or months before the skills are used will certainly reduce the effectiveness. Little or no knowledge or skill may have been retained. Training is not about doing something once and once only. It is about doing something several times at frequent intervals. One never forgets how to ride a bicycle or drive a car regardless of the time-lapse between each attempt, because the skill was embedded by frequency of opportunities to put the skill into practice in the early stages. Therefore to ensure effectiveness of training you ideally need to provide opportunities to put into practice the newly acquired skills as soon as possible. The person's supervisor should then examine the students' performance through sampling work pieces, reading documents he or she produces and observing the person doing the job. If you have experts in the particular skills then in addition to appraisals by the supervisor, the expert should also be involved in appraising the person's performance. Pay particular attention to the person's understanding of customer requirements. Get this wrong and you could end up in trouble with your customer!

Development activity effectiveness long term (the final stage)

After several months of doing a job and applying the new skills, a person will acquire techniques and habits. The techniques shown may not only demonstrate the skills learnt but also those being developed through self-training. The habits may indicate that some essential aspects of the training have not been understood and that some re-orientation is necessary. It is also likely that the person may have regressed to the old way of doing things and this may be due to matters outside his or her control. The environment in which people work and the attitudes of the people they work with can have both a motivating and de-motivating effect on an individual. Again the supervisor should observe the person's performance and engage the expert to calibrate his or her judgement. Pay particular attention to customer requirements and whether the trainee really understands them. If there are significant signs of regression you will need to examine the cause and take corrective action.

Increasing sensitivity to the impact of activities (6.2.2d and 6.2.2.4)

The standard requires the organization *to have a process that ensures its personnel are aware of the relevance and importance of their activities, and how they contribute to the achievement of the quality objective and to measure the effectiveness of this process.*

What does this mean?

Every activity an individual is required to perform should serve the organization's objectives either directly or indirectly. All activities impact the organization in some way and the quality of results depends on how they are perceived by the person performing them. In the absence of clear direction, personnel use intuition, instinct, knowledge and experience to the select activities they perform and how they should behave. Awareness of the relevance of an activity means that individuals are more able to select the right activities to perform in a given context. Awareness of the importance of an activity means that individuals are able to approach the activity with the appropriate behaviour. Some activities make a significant contribution to the achievement of objectives and others make less of a contribution but all make a contribution. Awareness of this contribution means that individuals are able to apportion their effort accordingly.

Measurement of the effectiveness of the awareness process means determining if the awareness activities achieved the specified objectives.

Why is this necessary?

This requirement responds to the Customer Focus and Involvement of People Principles.

Other than those at the front end of the business, personnel often don't know why they do things, why they don't do other things, why they should behave in a certain way and why they should or should not put a lot of effort into a task. Some people may work very hard but on activities that are not important, not relevant or not valued by the organization. Working smart is much better and is more highly valued, and why awareness of the relevance and importance of activities and their contribution to the organization's objectives is essential for enabling an organization to function effectively.

Awareness of contribution also puts a value on the activity to the organization and therefore awareness of the contribution that other people make puts a job in perspective, overcomes grievances and discontent. Personnel can sometimes get carried away with their sense of self-importance that may be based on a false premise. When managers make their personnel aware of the context in which activities should be performed, it helps redress the balance and explain why some jobs are paid more than others, or more highly valued than others.

There are perhaps thousands of activities that contribute to the development and supply of products and services some of which create features that are visible to the customer or are perceived by the customer as important. They may be associated with the appearance, sound, function or feel of a product where the activity that creates such features is focused on a small component within the product the customer purchases. They may also be associated with the actions, appearance or behaviour of other personnel where the impact is immediate because the personnel come face to face with the customer. However, close to or remote from the customer and seemingly insignificant, the result of an activity will impact customer satisfaction. Explaining the relationship between what people do and its effect on customers can have a remarkable impact on how personnel approach the work they perform. Awareness creates pride, a correct sense of importance and serves to focus everyone on the organization's objectives.

How is this implemented?
Designing the awareness process
As with any other process, the place to start is by defining what it is that you want to achieve; in other words, What are the process objectives? Now, establish how you will measure whether these objectives have been achieved. If staff are aware of the relevance and importance of their activities, and how they contribute to the achievement of the quality objective, they will be doing certain things and not doing other things, what you need to do is define what these are. Then you need to define the activities that will be carried out to achieve these objectives as measured.

An obvious way of implementing this requirement is for managers to advise staff before an event of the types of actions and behaviours that are considered appropriate. Also managers should advise staff during or immediately after an event that their action or behaviour is inappropriate and explain the reasons for this. If done in a sympathetic and sensitive manner the effect can be productive and people will learn. If done insensitively, abruptly and in a condescending manner, the effect can be counterproductive and people will not learn. However, the success of this method depends on managers being on the scene of action to observe intended or actual behaviour.

There are other ways of building awareness:

- Induction to a new job
- Training for a new or changed job
- Product briefings
- Chart displays and warning notices
- Performance results briefings
- Videos showing activities in context, where components are used, safety incidents, etc.
- Coaching of personnel by demonstrating appropriate behaviour that they may follow.

A source of information is the FMEA carried out on the product and process as a means of identifying preventive action. In this analysis the possible modes of failure are anticipated and measures taken to eliminate, reduce or control the effect. Such measures may include staff training and awareness as to the consequence of failure or nonconformity. However, it is sometimes not enough to explain the consequences of failure, you may need to enable them to see for themselves the effect using simulations, prototypes or case studies. Staff may have no idea of the function performed by the part they are producing, where it fits, how important it is. This education is vital to increasing sensitivity. In many organizations this sensitivity is low. The manager's task is to heighten sensitivity so that everyone is in no doubt what effect nonconformity has on the customer.

You can take a horse to water but you can't make it drink so the saying goes! It is the same with people! Making personnel aware of the quality issues and how important they are to the business and to themselves and the customer may not motivate certain individuals. The intention should be to build an understanding of the collective advantages of adopting a certain style of behaviour. It is therefore more important to modify behaviour than promote awareness.

Measuring effectiveness
Measuring employee understanding of appropriate quality objectives is a subjective process but measuring the outputs employees produce is not. Competence assessment would therefore be an effective means of measuring the effectiveness of the awareness process. In this way you don't have to measure it twice. Competence assessment serves to indicate whether staff have the ability to do the job and also serves to demonstrate the awareness process has or has not been effective.

If you don't employ competence assessment techniques there is an alternative. Through the data analysis carried out to meet the requirements of Clause 8.4 you will have produced metrics that indicate whether your quality objectives are being achieved. If they are being achieved you could either assume your employees understand the quality objectives or you could conclude that it doesn't matter. However, it does matter, as the standard requires a measurement. Results alone are insufficient evidence, i.e. you need to know how the results were produced. The results may have been achieved by pure chance and in 6 months time your performance may have declined significantly. The only way to test understanding is to check the decisions people make. This can be done with a questionnaire but is more effective if one checks decision made in the workplace. Is their judgment in line with your objectives or do you have to repeatedly adjust their behaviour? Take a walk around the plant and observe what people do, how they behave, what they are wearing, where they are walking and what they are not managing. You might conclude:

- A person not wearing eye protection obviously does not understand the safety objectives.
- A person throwing good product into a bin obviously does not understand the product handling policy.

- A person operating a machine equipped with unauthorized fittings obviously does not understand either the safety objectives, the control plan or the process instructions.
- An untidy yard with evidence of coolant running down the public drains indicates a lack of understanding of the environmental policy.

For each quality objective you should have a plan that defines the processes involved in its achievement. Assess these processes and determine where critical decisions are made and who is assigned to make them. Audit the decisions and ascertain whether they were contrary to the objectives. A simple example is where you have an objective of decreasing dependence on inspection. By examining corrective actions taken to prevent recurrence of nonconformities you can detect whether a person decided to increase the level of inspection in order to catch the nonconformities or considered alternatives. Any person found making such a decision has clearly not understood the quality objective.

Maintaining training records (6.2.2e)

The standard requires the organization *to maintain records of education, training, skills and experience.*

What does this mean?

As the requirement references Clause 4.2.4, the records referred to are records that provide:

- Evidence of the extent to which a person's abilities fulfil certain competence requirements.
- Evidence of activities performed to specify, develop or verify the abilities of a person who is intended to fulfil certain competence requirements.

Such records will include a personal development plan indicating the actions to be taken by the organization and the individual in meeting competence requirements as well as records of the actions taken and records of any measurement and verification of competence. Therefore the records required need to extend beyond lists of training courses, academic qualifications and periods of experience because these only record actions taken and not whether they were planned or whether they achieved the desired result.

Referring to these records as training records becomes misleading because they will contain evidence of other activities as well as training. They are part of the personnel records but do not constitute all the personnel records because these will undoubtedly include confidential information, such as promotions, grading, disciplinary records, etc. A more suitable label for the records that contain the results of competence assessment would be Personnel Competence Records.

Why is this necessary?

This requirement responds to the Factual Approach Principle.

The justification for records is provided in Chapter 4.

How is this implemented?

The standard is somewhat inconsistent in this requirement. Previously it required personnel to be competent, next it required the necessary competence to be determined but instead of requiring records of competence it reverts back to requiring only records of education, skills, training and experience. It would be more useful to generate records of the competence assessments.

Comparing product records with personnel records can be a useful way to determine the information required in competence records (Table 6.3).

The Job Specification identifies the competence needed and the Personnel Development Plan (PDP) identifies the education, training and behavioural development required to bridge the gap in terms of courses of study, training and development together with dates. Re-verification therefore provides evidence of education, training and behavioural development undertaken together with dates completed. The Certification of Competence provides evidence that the actions were effective. However, certification of competence is not required unless it is necessary for regulatory purposes. With this method you

Table 6.3 *Contrasting product records and personnel records*

Product record	Personnel record
Identity of product	Identity of person
Product specification reference	Job specification reference
Required characteristics	Required competences
Product verification stages	Competence assessment stages
Product verification method	Competence assessment method
Inspector or tester	Competence assessor
Product verification results	Assessment results
Nonconformity reports	Opportunities for improvement
Remedial action plan	Personal development plan
Re-verification results	Re-assessment results
Certification of conformity	Certification of competence
Certifying authority	Certifying authority

will also need to maintain separately in the personnel records, historical records of education, training and experience to provide a database of capability that can be tapped when searching for potential candidates for new positions.

Whenever any personnel development is carried out you should record on the individual's personal file, details of the course taken, the dates, duration and examination results (if one was taken). Copies of the certificate should be retained on file as evidence of training but these are not necessarily evidence of competence. You may find it useful to issue each individual with a PDP that includes personal development log, but do not rely on this being maintained or retained by the person in question. Often personnel development records are held at some distance away from an individual's workplace and in certain cases, especially for certificated personnel performing special processes, individuals should carry some identification of their proficiency so as to avoid conflict if challenged.

The records should indicate whether the prescribed level of competence has been attained. In order to record competence, formal training needs to be followed by on-the-job assessment. The records should also indicate who has conducted the education, training or behavioural development, and there should be evidence that this person or organization has been assessed as competent to deliver and evaluate such activities.

Competence records should contain evidence that the effectiveness of action taken has been evaluated and this may be accomplished by a signature and date from the assessor against the stages of evaluation.

Periodic reviews of Competence records should be undertaken to clearly identify personnel development needs.

You will need two types of competence records – those records relating to a particular individual and those relating to particular activities. The former is used to identify an individual's competence and the latter to select competent people for specific assignments.

Awareness of the consequences of nonconformity (6.2.2.3)

The standard requires personnel whose work can affect quality *to be informed about the consequences to the customer of nonconformity to quality requirements*.

What does this mean?

If a nonconforming product is allowed to be shipped to the customer either on its own or as part of an assembly the nonconformity might disrupt the customers production programme or cause other problems detrimental to customer supplier relations. As a result this will have consequences not only for the customer but also for the organization. Any deviation from the product acceptance criteria contained in the approved specifications and drawings can potentially impact the customer if not detected before shipment.

Why is this necessary?

If the processes were truly capable, nonconforming product would not be produced never mind shipped. However, this is based on the assumption that the controls in place will detect 100% of all errors, which is not possible. Even at six sigma performance levels there will be 3.4 defects/million and someone will have produced one of those defective products.

How is this implemented?

This requirement is tougher than you might think but you can make it easier. You have produced the design FMEA and the process FMEA, and in these two documents you have the basic information you need to inform your staff. The FMEA should have identified the sources and causes of failure. Make your staff aware of these documents but also provide other information that enables them to see the effect a part failure has at system level or on the complete vehicle. Staff may have no idea of the function performed by the part they are producing, where it fits, how important it is. This education is vital to increasing sensitivity. In many organizations this sensitivity is low. The manager's task is to heighten sensitivity so that everyone is in no doubt what effect nonconformity has on the customer.

Employee motivation and empowerment (6.2.2.4)

The standard requires the organization *to have a process to motivate employees to achieve quality objectives, make continual improvements and create an environment to promote innovation.* It also requires this process *to include the promotion of quality and technological awareness throughout the whole organizations.*

What does this mean?

Motivation has been defined as an inner mental state that prompts a direction, intensity and persistence in behaviour.[7] It is therefore a driving force within an individual that prompts him or her to achieve some goal. The process referred to in this requirement is therefore one that creates this driving force in every employee in the organization.

Everyone is motivated but not all are motivated to achieve their organization's goals. Many may be more interested in achieving their personal goals. Motivation is key to performance.

Why is this necessary?

Everything achieved in or by an organization ultimately depends on the activities of its workforce. It is therefore imperative that the organization is staffed by people who are motivated to achieve its goals. The performance of a task is almost always a function of three factors: Environment, Ability and Motivation. To maximize performance of a task, personnel have not only to have the necessary ability or competence to perform it but also need to be in the right surroundings and have the motivation to perform it.[8] Motivation comes from within. A manager cannot alter employees at will despite what they may believe is possible.

How is this implemented?

Motivation

There is a motivation process – not an organizational process but a process operating inside to the individual. This process is illustrated in Figure 6.4.

From this diagram it will be observed that motivation comes from satisfying personal needs and expectations of work, therefore the motivation to achieve quality objectives must be triggered by the expectation that achievement of objectives will lead to a reward that satisfies a need of some sort. This does not mean that you can motivate personnel solely by extrinsic rewards such as financial incentives. It requires a good understanding of an individual's pattern of needs. People desire psychological rewards from the work experience or like to feel a part of an organization or team. People can be motivated by having their efforts recognized and appreciated or included in discussions. However, this will only occur if the conditions they experience allow them to feel this way.

One way of motivating people is to set up process improvement teams comprised of personnel who perform activities within the process. Arrange presentations from others involved in interfacing processes so that there is an awareness of the impact of inputs or outputs that do not meet the acceptance criteria.

Encourage the teams to:

- Study the process objectives and challenge the measures by which success will be determined.
- Measure performance of the process by the chosen methods.
- Produce maps, charts and graphs showing the performance in ways that gets attention.
- Challenge the methods by which performance is measured.
- Identify barriers preventing the objectives being achieved.
- Propose innovative solutions for improving the process by better control.

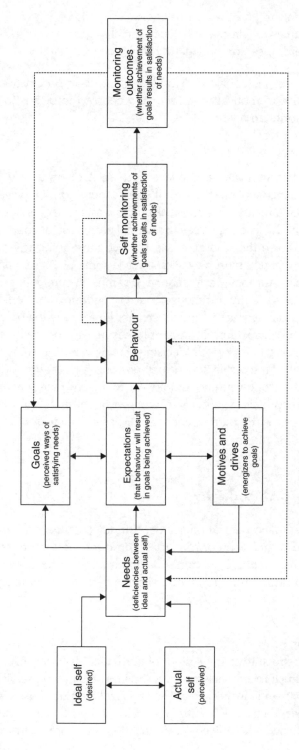

Figure 6.4 *Motivation process. Rollinson et al. (1998)*

- Conduct experiments to find the best solution.
- Estimate the resources required to introduce the changes.
- Present to the board or senior management their chosen solutions.

Then, if the solutions are viable, give the go ahead, i.e. don't procrastinate. There is nothing more destructive to motivation than giving people the means to change and withdrawing their authority on a whim.

Empowerment

Empowerment is said to motivate employees because it offers a way of obtaining higher level of performance without the use of strict supervision. However, it is more theory and rhetoric than a reality. To *empower* employees, managers not only have to delegate authority but to release resources for employees to use as they see fit and to trust their employees to use the resources wisely. If you are going to empower your employees, remember that you must be willing to cede some of your authority but also as you remain responsible for their performance, you must ensure your employees are able to handle their new authority. Employees have to be trained not only to perform tasks but also need a certain degree of experience in order to make the right judgements and therefore need to be competent. Some employees may acknowledge that they are willing to accept responsibility for certain decisions but beware, they may not be ready to be held accountable for the results when they go sour. It is also important that any changes arising from the empowering of employees to improve the process, be undertaken under controlled conditions. However, empowerment does not mean that you should give these individuals the right to change policies or practices that affect others without due process.

Infrastructure (6.3, 6.3.2)

The standard requires the organization *to determine, provide and maintain the infrastructure needed to achieve conformity to product requirements* and *to prepare contingency plans to satisfy customer requirements in the event of an emergency, such as utility interruptions, labour shortages, key equipment failure and field returns.*

What does this mean?

ISO 9000 defines infrastructure as the system of permanent facilities and equipment of an organization. Infrastructure also includes basic facilities, equipment, services and installations needed for growth and functioning of the organization. Such basics would include the buildings and utilities, such as electricity, gas, water and telecommunications. Within the buildings it would include the office accommodation, furniture, fixtures and fittings, computers, networks, dining areas, medical facilities, laboratories, plant, machinery and on the site it would include the access roads and transport. In fact everything an organization needs to operate other than the financial, human and consumable resources. In many organizations infrastructure is classified under the heading of *capital expenditure* because it is not order-driven, i.e. it does not change on receipt of an order.

The identification and provision of the infrastructure needs no explanation but in maintaining the infrastructure the implications go beyond the maintenance of what exists. Maintenance is more to do with maintaining the capability the infrastructure provides. Plant and facilities can be relatively easily maintained, but maintaining their capability means continually providing a capability even when the existing plant and facilities are no longer serviceable. Such situations can arise due to man-made and natural disasters. Maintaining the infrastructure means maintaining output when there is a power cut, a fire, a computer virus, a flood or a gas explosion. Maintaining the infrastructure therefore means making provision for disaster recovery and therefore maintaining business continuity.

The emphasis in this requirement is on infrastructure needed to achieve conformity to product requirements. As the conforming product is the organization's output, it follows that most of the infrastructure exists for this purpose. However, there will be areas, buildings, facilities, etc. that may not be dedicated to this purpose but to meeting requirements of interested parties other than the customer of the organization's products. The requirement is not implying that these other facilities do not need to be identified, provided and maintained, but that such provision is not essential to meet ISO/TS 16949. As with determining resources previously ask: "Why would we want to exclude a particular resource from the management system? What business benefit is derived from doing so?"

Why is this necessary?

This requirement responds to the Process Approach Principle.

The design, development and supply of products and services do not exist in a vacuum. There is always an infrastructure within which these processes are carried out and on which these processes depend for their results. Without an appropriate infrastructure and maintenance of that infrastructure the desired results will not be achieved. A malfunction in the infrastructure can directly affect results.

How is this implemented?

Identifying infrastructure

In identifying the resources needed to implement the quality system, some are product, project, contract, order specific, others are needed for maintenance and growth of the organization. These are likely to be classified as capital assets. The management of the infrastructure is a combination of asset management (knowing what assets you have, where they are, how they are depreciating and what value they could realize) and of facilities management (identifying, acquiring, installing and maintaining the facilities).

As the infrastructure is a critical factor in the organization's capability to meet customer requirements, and ability to continually meet customer requirements, its management is vital to the organization's success. Within the resource management process there are therefore several subprocesses related to the management of the infrastructure. It would be impractical to put in place one process because the processes will differ depending on the services required. Based on the generic model for resource management (see Figure 6.1) several planning processes will be needed for identifying and planning the acquisition, deployment, maintenance and disposal of the various assets. In describing these processes, you need to cover the aspects addressed in Chapter 4, under the heading *Process descriptions* and in doing so, identify the impact of failure on the organization's ability to achieve conforming product.

Providing infrastructure

Providing the infrastructure is associated with the acquisition and deployment of resources and therefore processes addressing the acquisition and deployment of buildings, utilities, computers, plant, transport, etc. need to be put in place. Many will use the purchasing process but some require special versions of this process because provision will include installation and commissioning and all the attendant architectural and civil engineering services. Where the new facility is required to provide additional capability so that new processes or products can be developed, the time to market becomes dependent on the infrastructure being in place for production to commence. Careful planning is often required because orders for new products may well be taken on the basis of projected completion dates and any delays can adversely affect achievement of these goals and result in dissatisfied customers. A model facility maintenance process flow is illustrated in Figure 6.5.

Maintaining infrastructure

There are two aspects to maintenance as addressed previously. Maintaining the buildings, utilities and facilities in operational condition is the domain of planned preventive and corrective maintenance. Maintaining the capability is the domain of contingency plans, disaster recovery plans and business continuity provisions. In some industries there is no obligation to continue operations as a result of *force majeure*, i.e. an event, circumstance or effect that cannot be reasonably anticipated or controlled including natural disasters caused by weather and land movement, war, riots, labour stoppage, illness, disruption in utility supply by service providers, etc. However, in other industries, provisions have to be made to continue operations albeit at a lower level of performance in spite of force majeure.

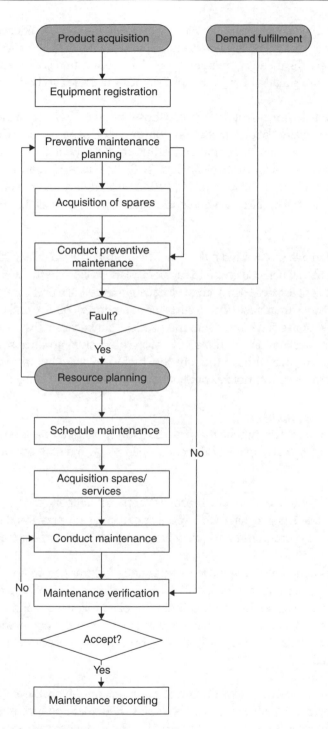

Figure 6.5 *Model facility maintenance process flow*

Although such events cannot be prevented, their effects can be reduced and in some cases eliminated. Contingency plans should therefore cover those events that can be anticipated where the means to minimize the effects are within your control. What may be a force majeure situation for your suppliers does not need to be the same for you.

Start by doing a risk assessment and identify those things on which continuity of business depends: power, water, labour, materials, components, services, etc. Determine what could cause a termination of supply and estimate the probability of occurrence. For those with a relatively high probability (1 in 100) find ways to reduce the probability. For those with lower probability (1 in 10,000 chance) determine the action needed to minimize the effect. The FMEA technique works for this as well as for products and processes.

If you are located near a river and it floods in the winter, Can you claim it to be an event outside your control when you chose to site your plant so close to the river? (OK the land was cheap – you got a special deal with the local authority – but was it wise?) You may have chosen to outsource manufacture to a supplier in a poorer country and now depend on them for your supplies. They may ship the product but because it is seized by pirates it doesn't reach its destination – you may therefore need an alternative source of supply! A few years ago we would have thought this highly unlikely, but after 300 years and equipped with modern technology pirates have returned to the seas once again.

Maintenance of equipment

In a manufacturing environment the process plant, machinery and any other equipment on which process capability depends need to be maintained, and for this you will need:

- A list of the equipment on which process capability depends.
- Defined maintenance requirements specifying maintenance tasks and their frequency.
- A maintenance programme which schedules each of the maintenance tasks on a calendar.
- Procedures defining how specific maintenance tasks are to be conducted.
- Procedures governing the decommissioning of plant prior to planned maintenance.
- Procedures governing the commissioning of plant following planned maintenance.
- Procedures dealing with the actions required in the event of equipment malfunction.
- Maintenance logs that record both the preventive and corrective maintenance work carried out.

In a service environment if there is any equipment on which the capability of your service depends, this equipment should be maintained. Maintenance may often be subcontracted to specialists but nevertheless needs to be under your control. If you are able to maintain process capability by bringing in spare equipment or using other available equipment, your maintenance procedures can be simple. You merely need to ensure you have an operational spare at all times. Where this is not possible you can still rely on the Call-out Service if you can be assured that the anticipated down time will not reduce your capability below that which you have been contracted to maintain.

The requirement does not mean that you need to validate all your word-processing software or any other special aids you use. Maintenance means retaining in an operational condition and you can do this by following some simple rules.

There are several types of maintenance:

Planned maintenance is maintenance carried out with forethought as to what is to be checked, adjusted, replaced, etc.

Preventive maintenance is maintenance carried out at predetermined intervals to reduce the probability of failure or performance degradation. An effective maintenance system should be one that achieves its objectives in minimizing down time, i.e. the period of time in which the equipment is not in a condition to perform its function.

Corrective maintenance is maintenance carried out after a failure has occurred and is intended to restore an item to a state in which it can perform its required function.

Predictive maintenance is part of planned preventive maintenance. In order to determine the frequency of checks you need to predict when failure may occur. Will failure occur at some future time, after a certain number of operating hours, when being operated under certain conditions or some other time? An example of predictive maintenance is vibration analysis. Sensors can be installed to monitor vibration and thus give a signal when normal vibration levels have been exceeded. This can signal tool wear and wear in other parts of the machine in advance of the stage where nonconforming product will be generated.

The manuals provided by the equipment manufacturer's should indicate the recommended preventive maintenance tasks and the frequency they should be performed covering aspects, such as cleaning, adjustments, lubrication, replacement of filters and seals, inspections for wear, corrosion, leakage, damage, etc.

Another source of data is from your own operations. Monitoring tool wear, corrective maintenance, analysing cutting fluids and incident reports from operators you can obtain a better picture of a machine's performance, and predict more accurately the frequency of checks, adjustments and replacements. For this to be effective you need a reporting mechanism that causes operators to alert maintenance staff to situations where suspect malfunctions is observed. In performing such monitoring you cannot wait until the end of the production run to verify whether the tools are still producing conforming product. If you do you will have no data to show when the tool started producing nonconforming product and will need to inspect the whole batch.

An effective maintenance system depends on it being adequately resourced. Maintenance resources include people with appropriate skills, replacement parts and materials with the funds to purchase these material, and access to support from original equipment manufacturers (OEMs) when needed. If the OEM no longer supports the equipment, you may need to cannibalize old machines or manufacture the parts yourself. This can be a problem because you may not have a new part from which to take measurements. At some point you need to decide whether it is more economical to maintain the old equipment than to buy new. Your inventory control system needs to account for equipment spares and to adjust spares holding based on usage.

For the system to be effective there also has to be control of documentation, maintenance operations, equipment and spare parts. Manuals for the equipment should be brought under document control. Tools and equipment used to maintain the operational equipment should be brought under calibration and verification control. Spare parts should be brought under identity control and the locations for the items brought under storage control. The maintenance operations should be controlled to the extent that maintenance staff should know what to do, know what they are doing and be able to change their performance if the objectives and requirements are not being met. While the focus should be on preventive maintenance, one must not forget corrective maintenance. The maintenance crew should be able to respond to equipment failures promptly and restore equipment to full operational condition in minimum time. The function needs resourcing to meet both the preventive and corrective demands because, it is down time that will have most impact on production schedules.

The exact nature of the controls should be appropriate to the item concerned, the emphasis being placed on that which is necessary to minimize operational equipment down time. It would be far better to produce separate procedures for these tasks rather than force fit the operational procedures to maintenance applications.

Jigs, tools and fixtures
Drawings should be provided for jigs, fixtures, templates and other hardware devices, and they should be verified as conforming with these drawings prior to use. They should also be proven to control the dimensions required by checking the first-off to be produced from such devices. Once these devices have been proven they need checking periodically to detect signs of wear or deterioration. The frequency of such checks should be dependent on usage and the environment in which they are used. Tools which form characteristics, such as crimping tools, punches, press tools, etc. should be checked prior to first use

to confirm they produce the correct characteristics and then periodically to detect wear and deterioration. Tools that need to maintain certain temperatures, pressures, loads, etc. in order to produce the correct characteristics in materials should be checked to verify that they would operate within the required limits.

Steel rules, tapes and other indicators of length should be checked periodically for wear and damage and although accuracy of greater than 1 millimetre is not normally expected, the loss of material from the end of a rule may result in inaccuracies that affect product quality.

While you may not rely entirely on these tools to accept product, the periodic calibration or verification of these tools may help prevent unnecessary costs and production delays. While usage and environment may assist in determining the frequency of verification hardware checks, these factors do not affect software. Any bugs in software have always been there or were introduced when it was last modified. Software therefore needs to be checked prior to use and after any modifications have been carried out, so you cannot predetermine the interval of such checks.

Plant, facility and equipment planning (6.3.1)

Plant layouts (6.3.1)

The standard requires the organization *to use multidisciplinary approach for developing plant, facility and equipment plans,* and for plant layouts *to optimize material travel, handling and value added use of floor space with synchronous material flow.*

What does this mean?
The multidisciplinary approach is addressed in Chapter 7 against Clause 7.3.1.1. A layout is the physical disposition of equipment, storage, space and services for a particular process or group of processes. The term *plant* relates to the collection of buildings and equipment designed for a particular industrial purpose whereas, the term *facility* is a smaller collection of machines, equipment and tools within a plant to facilitate particular operations or processes.

Why is this necessary?
The layout of plant and facilities impact process output. A poor layout might impede the flow of material and product; create hazards to personnel and damage to product. A good layout enables smooth flow, minimizes hazards and damage.

How is this implemented?
For some types of production, the facility housing the equipment and machinery needed to create, move and store product has to be designed. The pathways for raw materials

and semi-finished product need to be thought out and the gangways for staff and material movement need careful planning to prevent hazard, and ensure smooth and timely throughput.

Plant design cannot commence until conceptual design of the product and the process flow have been worked out and the planned production targets determined. You need to know what quantities of material and product need to be stored and moved plus the limitations on your current layout, and determine whether changes are needed to minimize material travel and handling. This can cause some difficult decisions since plant design decisions tend to be long term and cannot be implemented without considerable disruption to existing facilities. This is one advantage of cellular manufacturing where a complete cell can be redesigned without affecting other cells.

You should develop a documented procedure for the facility planning activity that will ensure the provision of adequate information on which to base plant design decisions. The procedures should provide for a separate development plan with allocation of responsibilities for the various tasks to be undertaken and cover the layout, specification, procurement, installation and commissioning of the new or revised plant.

Effectiveness monitoring methods (6.3.1)

The standard requires the organization *to develop and implement methods of monitoring the effectiveness of existing operations.*

What does this mean?
This requirement refers to how well the plant layout facilitates synchronous material flow.

Why is this necessary?
Clearly some layouts are better than others and therefore an indication of the effectiveness of the layouts is needed to determine whether it is at an optimum.

How is this implemented?
Your procedures should detail the plant evaluation methods and require consideration to be given to the overall plan of the plant, automation, ergonomics, operator and line balance, inventory levels, and value added labour content. Reports of the evaluation should be required so that they facilitate analysis by management and auditors.

The layout of your plant and facilities should be documented to facilitate its analysis. Detail installation drawings and commissioning procedures should be prepared in order to ensure completion on time. Plant dimensions and movement times should be recorded, and the constraints imposed by handling equipment, safety and environmental regulations registered. The documents should be brought under document control, as their revision is necessary whenever the layout is changed. Out-of-date plans will hinder future planning activities so their maintenance is a preventive action.

You may have chosen to outsource manufacture to a supplier in a poorer country and now depend on them for your supplies. They may ship the product but because it is stopped at customs due to a change in government of the country concerned, it doesn't reach its destination; hence you may need an alternative source of supply.

Work environment (6.4)

The standard requires the organization *to identify and manage the work environment needed to achieve conformity of product.*

What does this mean?

The word environment is very topical at the dawn of the 21st century. The protests in the 1980s and 1990s made us all more aware of the damage being done to the natural environment by human activity. The boom and bust years from the 1970s onwards made us aware of the fragile nature of the business environment and the decline of life-time employment so prevalent in the first half of the 20th century. However, trade union protests in the 1970s and 1980s also made us aware of the environment in which people work and the breakdown in relations between management and workers. With the 1990s came the growth in violence in the home, a sharp rise in single parent families, homelessness and issues within the home environment that spilled over into class-rooms and urban areas. These four types of environment are important and although only work environment is addressed directly in ISO/TS 16949, they are all interrelated. They all involve people and it's the same people that populate the work environment.

The natural environment describes the physical conditions that affect the development of the living organisms on the planet. In the context of environmental management systems, it is the natural environment that is being managed. The natural environment is addressed through legislation and therefore covered by Clause 5.1 of ISO/TS 16949.

The home environment is the physical, human and economic conditions that affect the development of the family unit. It is in the home environment where family values are developed, the code of ethics, the difference between right and wrong, and the religious and racial beliefs that the individual takes into the workplace. The home is also an environment where abuse, violence and poverty can exist that affect an individual's attitude towards the treatment of others. Although outside the scope of ISO/TS 16949, we cannot ignore the home environment when dealing with interpersonal relationships because it is often a source of problems that an individual brings into the workplace and affects his or her performance.

The business environment is the economic, political and market conditions under which an organization operates. Unlike the natural and the home environment, there is little that organizations can do to change the economic, political and market factors. Even protests, pickets and lobbying members of parliament have little immediate effect – certainly not sufficient to prevent economic failure of an organization. Organizations have to learn to predict these conditions, turn them into a force for good, manoeuvre the organization around them or adapt. The business environment needs to be taken into account when formulating the quality policy and the quality objectives in Clauses 5.3 and 5.4 of ISO/TS 16949 so that the organization adapts to the environment in which it operates. In essence here we are concerned about the environmental impact on the organization. Although these factors are outside the organization's control, smart tactics can minimize their impact.

ISO 9000 defines *the work environment* is a set of conditions under which people operates and include physical, social and psychological environmental factors. The social and psychological factors can be considered to be human factors as both are related to human behaviour. However, there are two dimensions to the work environment, namely organizational design and job design. A job exists within an organization and there are factors that arise from performing the job, such as physical movement, the man–machine interface, the physical environment and the psychological factors of the job. These form the basis of ergonomics (see BS 3138:1992). Even when the ergonomics of the job have been optimized, performance can still be adversely affected by wider influence of factors inherent in the organization caused by the way it functions or the way things get done, i.e. its culture and climate (see below).

In essence, the organization impacts the natural and the work environments and both the business and home environments impact the organization.

Physical factors of the work environment include space, temperature, noise, light, humidity, hazards, cleanliness, vibration, pollution, accessibility, physical stress and airflow. In addition to visible light, other types of radiation across the whole spectrum impact the physical environment.

Social factors of the work environment are those that arise from interactions between people and include the impact of an individual's family, education, religion and peer pressure as well as the impact of the organization's ethics, culture and climate.

Psychological factors of the work environment are those that arise from an individuals inner needs and external influences, and include recognition, responsibility, achievement, advancement, reward, job security, interpersonal relations, leadership, affiliation, self-esteem, and occupational stress. They tend to affect or shape the emotions, feelings, personality, loyalty and attitudes of people, and therefore the motivation of people towards the job to which they have been assigned.

While this grouping serves to identify related factors, it is by no means comprehensive or exclusive. Each has an influence on the other to some extent. Relating the identification of such factors to the achievement of conformity of product tends to imply that there are factors of the work environment that do not affect conformity of product. Whether people produce products directly or indirectly, their behaviour affects their actions and decisions, and consequently the results of what they do. It is therefore difficult to exclude any factor on the basis that it is not needed to achieve conformity of product in some way or other.

Managing such factors means creating, monitoring and adjusting conditions in which the physical and human factors of the work environment act positively towards achievement of the planned results. Some of the factors affecting the work environment are constraints rather than objectives, i.e. they exist only because we have an objective to achieve. Noise in the workplace occurs because we need to run machines to produce product that satisfies the customer. If we didn't need to run the machines, there would be no noise. However, some of the constraints are of our own making. If the style of management created an environment that was more conducive to good industrial relations, the workforce would be more productive.

Why is this necessary?
This requirement responds to the Leadership and Involvement of People Principles.

The work environment is critical to worker performance and extends beyond the visible and audible factors commonly observed in a workplace. All the above factors influence individual behaviour, which has a direct impact on organizational performance and consequently product quality. It is the duty of management to control the physical factors firstly within the levels required by law and as necessary for people to perform their jobs as efficiently and effectively as possible. It is also the task of management to create conditions in which personnel are motivated to achieve the results for which they are responsible, and therefore remove or contain any de-motivating elements such as friction and conflict in the workplace.

The physical factors of the work environment influence individual behaviour by causing fatigue, distraction, accidents and a series of health problems. There are laws governing many of the physical factors, such as noise, air pollution, space and safety. There are also laws related to the employment of disabled people that impact the physical environment in terms of access and ergonomics. The social interactions in the workplace influence interpersonal relationships, such as the worker–boss, worker–subordinate, worker–colleague and worker–peer relationship. The social factors if disregarded cause unpredictable effects and some of these are the subject of legislation, such as discrimination on the basis of religion, gender, race and disability. The issue is not whether product will be affected directly, but whether performance will be affected. It requires no more than common sense (rather than scientific evidence) to deduce that intimidation, sexual harassment, invasion of privacy and similarly unfair treatment by employees and employers will adversely affect the performance of people and consequently the quality of their output. Social factors can have a psychological effect on employees causing de-motivation and mental stress. This is not to say that employees have to be mollycoddled, but it is necessary to remove the negative forces in the work environment if productivity is to be maximized and business continuity maintained.

The information produced to support ISO 9000 from the standards and certification bodies seems to restrict the work environment to those aspects that directly affect the product, such as cleanliness, temperature and regulations relating to the product. This narrow view overlooks the principal factors that affect product quality – that of human behaviour and adds little to the movement to enhance customer satisfaction. A more progressive organization would realize that the influence of the physical factors of the environment is small in comparison with the human factors and place their improvement effort on this. It is clear from Figure 8.2 that General Motors in 1993 were looking far beyond the physical factors of the work environment at determinants of customer satisfaction.

How is this implemented?

Previously the work environment was limited to its effect on the product and although not explicit in ISO 9001, it is clear from ISO 9004-1 that this was the intent. The implication here is that personnel safety, job security, recognition, relationships, responsibility, etc. influence the performance of people in an organization and that managers have to manage these factors to provide an environment in which personnel will be motivated. This is a wide subject and beyond the scope of this book but it is important that the reader appreciates the breadth and scope of the factors that influence the working environment in preparing for further study.

For a solution we can use a similar approach to that taken towards the natural environment. Environmental management is the control over activities, products and processes that cause or could cause environmental impacts. The approach taken is based on the management of cause and effect where the activities, products and processes are the causes or 'aspects', and the resulting effects or potential effects on the environment are the impacts. All effort is focused on minimizing or eliminating the impacts. In the context of the work environment, the causes or aspects would be the physical and human factors, and the impacts would be the changes in working conditions. However, unlike management of the natural environment, the effort would not all be focused on reducing or eliminating impacts where a state of zero impact is ideal. In the working environment, the effort should be focused on eliminating negative impact and creating positive or beneficial impacts that also lead to an improvement in performance.

Dealing with the physical factors

The physical factors are more easily dealt with than the human factors primarily because they are more tangible, measurable and controllable. To manage the physical factors they firstly need to be identified and this requires a study of the work environment to be made relative to its influence on the worker. We are not necessarily dealing only with safety issues although these are very important. The noise levels do not need to cause harm for them to be a factor that adversely affects worker performance. Libraries are places of silence simply to provide the best environment in which people can concentrate on reading. No harm arises if the silence is broken!

In dealing with physical factors there is a series of steps you can take to identify and manage these factors:

- Use an intuitive method such as brainstorming to discover the safety-related and none safety-related factors of the environment, such as noise, pollution, humidity, temperature, light, space, accessibility, etc.
- Research legislation and associated guidance literature to identify those factors that could exist in the work environment due to the operation of certain processes, use of certain products or equipment. We do X therefore from historical and scientific evidence there will be Y impact. (VDUs, RSI, airborne particles, machinery, etc.)
- Determine the standard for each factor that needs to be maintained to provide the appropriate environment.
- Establish whether the standard can be achieved by workspace design, by worker control or by management control or whether protection from the environmental impact is needed (protection of ears, eyes, lungs, limbs, torso or skin).
- Establish what could fail that would breach the agreed standard using FMEA or Hazards Analysis; identify the cause and the effect on worker performance.
- Determine the provisions necessary to eliminate, reduce or control the impact.
- Put in place the measures that have been determined.
- Measure and monitor the working environment for compliance with the standards and implementation of the provisions defined.
- Periodically repeat the previous steps to identify any changes that would affect the standards or the provisions currently in place. Ask: Is the standard still relevant? Are there better methods now available for dealing with this environmental impact?

Dealing with the human factors

Managers are often accused of ignoring the human factors but such factors are not easily identified or managed. With physical factors you can measure the light level and adjust it if it's too bright or too dim. You can't measure ethics, culture, climate, occupational stress – all you see are its effects and the primary effect is employee motivation.

Managers need to understand and analyse human behaviour and provide conditions in which employees are motivated to achieve the organization's objectives.

Ethics   Ethics concern a person's behaviour towards others and therefore within a particular work environment there will be some accepted norms of right and wrong. These values or standards vary from group to group and culture to culture. For example, bribery it is an accepted norm in some countries but in others it is illegal. When the work environment includes employees of different races and religions, conflict can arise simply due to an individual behaving in a way that to them is normal but to another in the group is unethical.

Ethical standards therefore vary and change with time. Family, religion, and education will influence an individual's code of ethics, and this may conflict with that of the work environment that the person joins. Unfortunately it is often not until a situation arises that challenges the ethical standards of the individual or the group that the conflict becomes apparent. People may be content to abide by the unwritten code of ethics under normal conditions but when an important prize is within reach, the temptation to put the principles to one side is too great and some will succumb to the pressure and put self-interest or profits first, causing harm to the interests of others. Employees may be easily led by other less ethical employees in a desire to follow the pack and those that do challenge their peers and their managers get accused of 'rocking the boat' and being 'troublemakers'. Management may turn a blind eye to unethical practices if in doing so they deliver the goods and no one appears to be harmed thus strengthening the beliefs of the instigator of such practices. Sometimes managers are simply unaware of the impact of their decisions. A one-off instruction to let a slightly defective product be despatched because it was needed urgently, gets interpreted by the employees as permission to deviate from requirements. A one-off instruction to take the previous test results as evidence of conformity instead of waiting until the test equipment has been repaired, gets interpreted as permission to do this every time. Employees naturally take the lead from the leader and can easily misread the signals. They can also be led by a manager who does not share the same ethical values and under threat of dismissal; an otherwise law-abiding citizen can be forced into falsifying evidence.

Questions of right and wrong in the work environment can arise between two individuals or between an individual and the group or the top management. There is also the wider perspective of the relationship between the organization and society. Social responsibility is becoming more dominant in the boardroom because investors look for "green" companies and those that do not support apartheid or purchase from producers that use child labour. While the organization's behaviour with respect to its interested parties is outside the work environment the internal and external relationships cannot be isolated. The behaviour of the organization externally will undoubtedly influence internal behaviour.

The moral maze created by attempting to satisfy all interested parties within an environment in which there are so many variables calls for a code of ethics. Some companies have ethical policies that are intended to guide employees in dealing with a wide range of issues and assist managers to manage the work environment. However, writing down the code of ethics is fraught with problems:

- They have to be followed in all circumstances – there are no exceptions.
- If an employee 'blows the whistle' on the company for unethical behaviour, it must not reprimand that individual but accept the situation as a consequence of its unethical behaviour.
- It cannot put 'gagging' conditions in contracts of employment because this itself would be indicative of unethical behaviour.

Culture   If we ask people to describe what it is like to work for a particular organization, they often reply in terms of their feelings and emotions that are their perceptions of the essential atmosphere in the organization. This atmosphere is encompassed by two concepts: culture and climate.[9] Rollinson, Broadfield and Edwards observe that Schein conceptualizes culture as a layered phenomenon that has three interrelated levels of meaning. The observations of Schein, Peters, Waterman and Ouchi have been integrated to illustrate a more diverse range of cultural characteristics.

Artefacts and creations
- Rites and ceremonies (e.g. morning exercises and celebrations of success)
- Symbols (e.g. status symbols)
- Taboos (e.g. addressing the boss by his or her first name and parking in the manager's parking space)
- Myths and stories (e.g. how the founder grew the company)
- Language (e.g. how managers address subordinates and vice versa, jargon and verbal signals)
- Norms (e.g. behaviour that is accepted as being normal for various levels)

Values and beliefs
- Honesty
- Basis of reward and punishment
- Effort
- Trust
- Commitment to employees

- Evaluation of employees (e.g. evaluation on quantitative or qualitative criteria)
- Career paths
- Employee control (e.g. rules, procedures or mentoring and coaching)
- Decision-making (e.g. decisions by individuals or by groups, based on fact or intuition, consensus or dictate)
- Concern for people
- Management contact (e.g. managers work behind closed doors or walk about and make contact)
- Autonomy (e.g. small autonomous units rather than large bureaucracy)
- Customer focus not self-interest

Basic assumptions
- Respect for the individual
- Responsibility for actions and decisions
- Internal cooperation
- Freedom

Culture evolves and can usually be traced back to the organization's founder. The founder gathers around people of like mind and values, and these become the role models for new entrants. Culture has a strong influence on people's behaviour but is not easily changed. It is an invisible force that consists of deeply held beliefs, values and assumptions that are so ingrained in the fabric of the organization that many people might not be conscious that they hold them. Unless the recruitment process recognizes the importance of matching people with the culture, mavericks may well enter the organization and either cause havoc in the work environment or be totally ineffective due to a lack of cultural awareness. People who are oblivious to the rites, symbols, customs, norms, language, etc. may not advance and will become demotivated. There is however, no evidence to suggest a right or wrong culture. What is important is that the culture actually helps an organization to achieve its goals – that it is pervasive and a positive force for good.

The role of management relative to culture is to set a good example and ensure that personnel are likely to fit in with the existing culture before being offered employment. It is also important for management not to reprimand staff for failing to observe the protocols when they may not understand them and to take time to induct new staff in a way that reduces any anxieties associated with the new job. New entrants often don't know what to ask. It doesn't occur to them to ask about the culture, what the rites and customs are and in effect discover information that will be vital to their performance. They are often more concerned about the "Who, What, Where, When, and How" of the job. Even if it is explained on induction, it will not become significant to the individual until something happens that reinforces the culture. For many people, the culture in an organization has to be discovered – it is not something that is necessarily articulated by the managers. Statements of the vision and values help but the rites, symbols and taboos are all learnt mainly by observation.

Climate   Climate is allied to culture and although people experience both, climate tends to be something of which there is more awareness. Culture provides a code of conduct that defines acceptable behaviour whereas climate tends to result in a set of conditions to which people react. Culture is more permanent whereas climate is temporary and is thought of as a phase the organization passes through. In this context therefore, the work environment will be affected by a change in the organizational climate. Several external forces cause changes in the climate, such as economic factors, political factors and market factors. These can result in feelings of optimism or pessimism, security or insecurity, complacency or anxiety. There are also several internal characteristics of the organization that impact the climate:

- *Job design*: Do these limit freedom, stifle initiative and innovation?
- *Technology*: High manual input and personal control or extensive automation and little personal involvement.
- *Management philosophies*: The roles and functions of subordinates, employee performance measurement.
- *Authority*: Autocratic, democratic, strong delegation or little delegation, multiple approvals.
- *Local practices*: Supervision manipulates policies to serve their own purpose, varying rules to exert power.
- *Accessibility*: Supervision being accessible or inaccessible for support, help, discussion.
- *Fear*: Workers frightened of reporting problems, stopping the process, owning up to mistakes.

These and probably many more climatic factors influence the behaviour of people in the work environment, and therefore it is important to avoid those factors that lead to poor performance and employee dissatisfaction.

Ergonomics   An employee's body movement in performing a job has important implications on the work environment.

The study of the relationship between a person and his or her job is referred as Ergonomics (see BS 3138) and it deals principally with the relationship between a person and his or her job, equipment and environment and particularly the anatomical, physiological and psychological aspects arising thereon. The layout of the workplace, the distances involved the areas of reach, seating, frequency and type of movement all impact the performance of the worker. These factors require study to establish the optimum conditions that minimize fatigue, meet the safety standards while increasing productivity.

Where people are an integral part of a mechanized process the man–machine interface is of vital importance and has to be carefully considered in process design. The information on display panels should be clear and relevant to the task. The positioning of instruments, input, output and monitoring devices should allow the operator to easily access information without abnormal movement. The emergency controls should be within easy reach and the operating instructions accessible at the workstation. Legislation and national standards cover many of these aspects.

Managing the human factors

Identifying the barriers   The role of the manager in enabling a person to be motivated is that of removing barriers to work motivation. There are two types of barriers that cause the motivation process to break down. The first barrier is job-related, i.e. there is something about the job itself that prevents the person performing it from being motivated (e.g. boring and monotonous work in mass production assembly lines). The second barrier is goal-related, i.e. attainment of the goals is thwarted in some way (e.g. unrealistic goals, insufficient resources and insufficient time for preparation). When targets are set without any regard for the capability of processes this often results in frustration and a decline in motivation (see also Chapter 5 under the heading *Expressing quality objectives*).

Common barriers are:

- fear of failure, reprisals, rejection, losing, conflict, humiliation and exploitation;
- distrust of management, favouritism and discrimination;
- work is not challenging or interesting;
- little recognition, respect and reward for a job well done;
- no authority and responsibility.

Measuring employee satisfaction   Within the work environment the bottom line is whether the objectives are being achieved and the employees satisfied. It is not enough just to achieve objectives. The first 200 years of the industrial revolution did that but in response to worker exploitation the labour unions were born and thus commenced 100 years of strife. The very idea that employees should be satisfied at work is a comparatively recent notion but clearly employee dissatisfaction leads to lower productivity. The measurement of employee satisfaction together with the achievement of the organization's objectives would therefore provide an indication of the quality of the work environment, i.e. whether the environment fulfils its purpose.

Many companies carry out employee surveys in an attempt to establish their needs and expectations, and whether they are being satisfied. It is a fact that unsatisfied employees may not perform at the optimum level and consequently product quality may deteriorate. As with customer satisfaction surveys, employee satisfaction surveys are prone to bias. If the survey hits the employee's desk following a reprimand from a manager, the result is likely to be negatively biased. The results of employee satisfaction surveys are also often disbelieved by management. Management believe their decisions are always in their employees best interests whereas the employees may not believe management if its track record has not been all that great. Employee satisfaction has less to do with product quality and more to do with relationships. However, employee relationships can begin to adversely affect product quality if no action is taken.

You will need a procedure for measuring employee satisfaction but design the survey with great care and treat the results with an open mind because they cannot be calibrated. A common method for measuring satisfaction is to ask questions that require respondents to check the appropriate box on a scale from 'strongly agree' to 'strongly disagree'. An alternative is for an outsider appointed by management to conduct a series of interviews. In this way you will obtain a more candid impression of employee satisfaction. The interviewer needs some knowledge of the management style, the efforts management has actually made to motivate their workforce – not the rhetoric they have displayed through newsletters, briefings, etc. On hearing what management has actually done, the employees may react differently. They also have short memories and are often reacting to immediate circumstances forgetting the changes that were made some time ago. The interviewer is also able to discover whether the employee has done anything about the feelings of dissatisfaction. It could be that a supervisor or middle manager is blocking communication. Whatever the method, management needs unbiased information of the level of employee satisfaction to do the job.

The forgoing treatment of the work environment may appear at odds with the current interpretation of the requirement in ISO/TS 16949. But to claim that the human factors of the work environment have nothing to do with quality would be a spurious argument. Most of Deming's 14 points focus on the human factors, the constancy of purpose, leadership, driving out fear, removing barriers, etc. If you are serious about quality, you cannot ignore the human factors – they are key to your success. If you want to adopt the minimalist approach and do only what lies in the words of the standard, you will miss the point completely. You will not succeed as an organization and be constantly trying to reduce variation as though each deviation from the norm has a unique cause. When you have eliminated the impossible, whatever remains, however improbable, must be the truth (Arthur Conan Doyle). The improbable is the human factor.

Personnel safety to achieve product quality (6.4.1)

The standard requires the organization *to address product safety and potential risks to employees in the manufacturing process activities.*

What does this mean?

Product safety as an inherent characteristic of the product is addressed under the heading *Design and development* in Chapter 7 therefore, what this requirement refers to is the protection of the product for safety reasons and of the people who come into contact with the product as well as protection of the people operating the processes that make the product. If the requirement applied to safety in design, the heading to this requirement would have reflected this: as it is, the heading only refers to personnel.

Why is this necessary?

Suppliers in the automotive sector need to demonstrate they care about personnel safety. This requirement indicates that risks to personnel have to be minimized regardless of there being safety legislation in the country of origin.

How is this implemented?

Some of the topics your safety management practices should address are as follows:

- Methods for assessing the health and safety hazards present in your organization, its products and its operations.
- Safety objectives and targets based on the results of the safety assessment.
- A programme for achieving the safety objectives.
- Methods for making staff aware of their safety responsibilities, the benefits of compliance and consequences of a failure to comply.
- Methods for alerting staff to hazardous situations.
- Methods for creating controlled conditions in which safety hazards are a minimum.
- Methods for dealing with accidents, incidents and emergency situations, investigating their cause and preventing recurrence.
- Methods of measuring the achievement of safety objectives and targets.

Instructions concerning safety issues should be integrated into the control and operating procedures such that the instructions are given at the stage in the process when they apply. In this way staff do not have to consult several documents and the chance of error is reduced.

Personnel safety is also addressed in other parts of this book.

Cleanliness of premises (6.4.2)

The standard requires the organization *to maintain its premises in a state of cleanliness, order and repair consistent with the product and manufacturing process needs.*

What does this mean?

It means that you need to keep the working areas free of any particles or fluids that could adversely affect product quality should they come into contact with them.

For auditors it means that they do not need to find evidence that product has been affected by the working conditions; they need only proof that the conditions are not appropriate and that product may be affected in due course.

Why is this necessary?

This requirement should not be necessary as Clause 6.4 addresses working environment but it emphasizes that poor housekeeping and maintenance can affect product quality. However, contamination can arise from people handling product, air currents or being picked up on machinery, equipment, tooling, etc.

How is this implemented?

The level of cleanliness that is maintained in working areas needs to be commensurate with the needs of the product. There are several aspects to consider:

- Efficient removal of waste that is produced from the process. Modern machinery has extraction equipment and after machining, components can be put through washing cycles to remove swarf.
- Regular cleaning cycles, clearing floors of spillages, swarf, etc.
- Demarcation between working areas and walkways for through traffic. Often a strict code is enforced that prohibits visitors crossing into the working area thus carrying into that area contamination.
- Regular plant inspections for dilapidation of roofs, doors and drainage systems.

Summary

In this chapter we have examined the requirements contained in Section 6 of ISO/TS 16949. Although still a relatively short set of requirement, they are among the most important, for without adequate resources no organization will fulfil its purpose and mission. We have discovered that we cannot exclude any of the organization's resources as all either directly or indirectly affects our ability to satisfy the needs and expectations of the interested parties. We have learnt that there is a significant difference between qualified personnel and competent personnel. We now know we need people who can deliver results and not people who can provide evidence of certain academic qualifications, training and experience. Recognizing that we need to take a new approach to the selection and development of our human resources. We have examined some proven methods used to identify competence needs and assess competence. We have learnt that the infrastructure is important in sustaining the organization's capability and that customer satisfaction depends on maintaining continuity of supply when disaster strikes. Finally we have looked deeply into the work environment and discovered that this is not simply a question of managing physical conditions of the workplace but of managing the culture and climate within the organization. We have shown that product quality depends on maintaining a motivated workforce and that motivation is affected by both the human and physical factors that can be identified and managed. Effective management of human resources therefore arises from deploying competent personnel into a work environment in which they are motivated to achieve the organization's goals.

We have dwelt on competence and culture rather more than most technical subjects in this book primarily because they are so often overlooked. The achievement of quality depends more on people than it does on technology for it is people who create and use the technology. If the people are not competent to use the technology and work in an environment in which they are frightened to own up to mistakes, the most modern and sophisticated technology and the most detailed of specifications, plans and procedures won't produce the outputs that management require to satisfy the customers.

Resource management requirements checklist

These are the topics that the requirements of ISO/TS 16949 address consecutively. Topics 42–84 appeared in Chapter 5.

Provision of resources
85   Determination and provision of resources to implement the QMS
86   Determination and provision of resources to enhance customer satisfaction

Human resources
General
87   Competence of personnel

Competence, awareness and training
88   Determination of competence
89   Provision of training
90   Evaluating the effectiveness of training
91   Awareness of impact on the achievement of quality objectives
92   Records of education, training, skills and experience

Product design skills
93   Qualification of design personnel
94   Identification of design tools and techniques

Training
95   Establishing procedures for identifying training needs
96   Qualification of personnel
97   Training in specific customer requirements

Training on the job
98   Job training for new or modified jobs
99   Informing personnel about the consequence of nonconformity

Employee motivation and empowerment
100 Employee motivation process
101 Promotion of quality awareness
102 Measurement of employee awareness

Infrastructure
103 Provision and maintenance of infrastructure

Plant, facility and equipment planning
104 Developing plant, facility and equipment plans
105 Optimizing plant layouts
106 Facilitating synchronous flow
107 Evaluating the effectiveness of existing operations

Contingency plans
108   Preparation of contingency plans

Work environment
109   Determination and management of work environment

Personnel safety to achieve product quality
110   Addressing product safety and risk to employees

Cleanliness of premises
111   Cleanliness of premises

Resource management – Food for thought

1 Are there any activities in an organization that require no human or financial resources?
2 What business benefit is derived from excluding particular resources from the management system?
3 Why are the costs of maintaining the management system different from those of maintaining the organization?
4 Are functional budgets justified when in reality the organization's objectives are achieved through a series of interconnected processes?
5 If your staff are your most important asset, why are staff development programmes a prime target for cost reduction?
6 Do you select people on what they demonstrate they know or what they demonstrate they can do?
7 How do you know that your staff are competent to achieve the objectives you have agreed with them?
8 From where does the evidence come from to assess staff competence?
9 There are seven ways of developing staff. Before you send them on a training course have you eliminated the other six ways as being unsuitable for the particular situation?
10 What action was taken the last time your staff returned from an external training course that resulted in an improvement in their performance?
11 How do you ensure that staff work on those things that add value to the organization?
12 What's the second thing you do on receipt of a customer complaint?
13 How much information is contained in your personnel training records to help you make decisions on staff development needs?
14 Do you know of all the ways in which an infrastructure failure might impact customer satisfaction?
15 How much of what affects individual performance depends on the relationship between management and staff?
16 In describing what it is like to work in your organization, would you identify any factors that were detrimental to overall performance?
17 Would anyone in your organization take an action or make a decision that might be considered unethical in your society?
18 What actions do managers take to create conditions in which their staff are motivated to achieve their objectives?
19 When did you last examine the organization's culture for its relevance to the current conditions in which the organization operates?
20 When did you last recognize or appreciate the efforts of your staff in contributing towards improved performance?
21 When did you last reprimand a member of staff for something that was symptomatic of the natural variation inherent in the process?
22 If every employee were to follow the examples set by the management, in what way would the organization's performance change?

23   Do you wait until you have no option but to take drastic action to restore financial stability or do you involve your workforce in seeking ways in which to lessen the impact?

24   When did last calculate the time lost by the inappropriate location of tools, information, equipment and facilities?

25   If an employee was dissatisfied with the working environment, could you be sure that he or she would approach the manager with confidence that the matter would be dealt with objectively and sympathetically?

Chapter 7

Product realization

Quality must be built into each design and each process. It cannot be created
through inspection.
Kaoru Ishikawa

Summary of requirements

Product Realization as expressed in Section 7 of ISO/TS 16949 is the Demand Fulfilment Process referred to previously that has interfaces with Resource Management and Demand Creation processes. However the Product Realization requirements include requirements for purchasing, a process that could fit as comfortably under resource management because it is not limited to the acquisition of components, but is a process that is used for acquiring all physical resources including services. Section 7 also includes requirements for control of measuring devices, which would fit more comfortably into Section 8, but it omits the control of nonconforming product which is more to do with handling product than measurement. Product realization does not address demand creation or marketing. The demand has already been created when the customer approaches the organization with either an order or invitation to tender. Note that Demand Creation is addressed by the standard only through Clause 5.2 and that product design is located in Section 7 simply because, it refers to the design of customer-specific products. If the products were designed in order to create a demand this work process would be part of Demand Creation.

If we link the requirements together in a cycle (indicating the headings from ISO/TS 16949 in bold italics type), having marketed the organization's capability and attracted a customer, the cycle commences by the need to **communicate with customers** and **determine the requirements** of customers, of regulators and of the organization relative to the product or service to be supplied. This will undoubtedly involve more *customer communication* and once requirements have been determined we need to *review the requirements* to ensure they are understood and confirm we have the capability to achieve them. If we have identified a need for new products and services, we would then need to *plan product realization* and in doing so use *preventive action* methods to ensure the success of the project and take care of any *customer property* on loan to us. We would undertake product *design and development* and in doing so we would probably need to *identify product, purchase* materials, components and services, build prototypes using the process of *production provision* and *validate* new *processes*. After *design validation* we would release product information into the market to attract customers and undertake more *customer communication*. As customers enquire about our offerings we would once more *determine the requirements* in order to match customer needs with product offerings and our ability to supply.

On receipt of a production contract, we would *review the requirements* and review the *manufacturing feasibility* to ensure we had the capability to supply the product in the quantities and to the delivery schedule required before entering into a commitment to supply. We would then proceed to *plan product realization* once again and undertake *production or service provision*. As part of this work we would initiate the *purchasing process* and ensure we used *customer-approved sources*. Where necessary we would undertake *supplier development* and *supplier monitoring* to ensure suppliers developed and maintained the capability required, and carry out *incoming product quality checks* as necessary to ensure product entering the production line was of the correct standard. During production we would perform the operations to produce the product in the sequence and as specified in the *control plan* and would *monitor and measure processes* and *monitor and measure products* using the *measurement and monitoring* tools, techniques and control methods that were also stated in the control plan. The control plan would alert us to the *special characteristics* and the actions to take when results were outside specified tolerance. We would have carried out *measurement system analysis* as part of developing process capability and will use and *control the measuring and monitoring devices* in facilities that meet international standards. We would maintain *traceability* of the product to help us trace and eliminate the root cause of problems, and if we found unacceptable variations in the product we would undertake the *control of nonconforming product* and *analyse data* to facilitate *corrective action*. Throughout production we would seek the *preservation of product* and take care of *customer property*. In the event of changes in the product or the process we would initiate the *change control* process to ensure the impact of performance was managed effectively and the customer informed. Once we had undertaken all the *product verification* and *preserved* the product for delivery, we would ship the product to the customer or complete the service transaction. To complete the cycle *customer communication* would be initiated once more to obtain feedback on our performance.

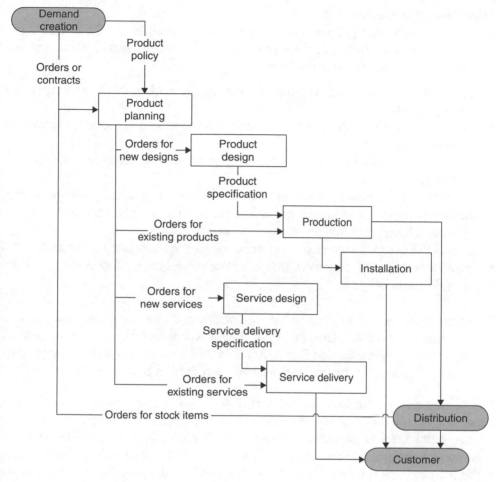

Figure 7.1 *Relationship between processes in product realization*

Here we have linked together all the clauses in Section 7 and many in Section 8 of the standard because the two cannot be separated. The relationship between the primary product realization processes is illustrated in Figure 7.1.

Planning product realization processes (7.1)

Planning and developing product realization processes (7.1)

The standard requires the organization to *plan and develop the processes required for product realization.*

What does this mean?

The product realization processes are the processes needed to specify, develop, produce and supply the product or service required, and in the context of the automotive industry would include those needed to:

- specify the products and services required by the organization's customers (the sales process);
- plan the provision of the identified products and services (the project-, contract- or order-planning process);
- design the identified products and services so as to meet customer requirements (the design process);
- procure the materials, components, services needed to accomplish the design and/or generate or deliver the product or service (the procurement process);
- generate the product (the production process);
- supply the product or service (the distribution or service delivery processes);
- provide support to customers (the after sales, technical support or customer-support process).

These processes are all product or service specific and take the input from the customer through a chain of related processes that deliver acceptable products or services to customers. However, Section 7 of ISO/TS 16949 does not cover all of the product realization processes – some of them are placed in Section 8.

Product design may not have formed part of the organization's management system under QS-9000 if the products were not designed for specific customers but were proprietary designs (to the organization's own specifications). However, ISO 9001:2000 removes this constraint and requires the processes to realize the product to be included in the management system whether or not the products are provided to the organization's specification or to a specification defined by a customer.

Planning these processes means identifying the processes required for a specific project, contract or order and determining their sequence and interrelation. In many cases, the work processes will have been designed and will form part of the management system. However, the nature and complexity of specific projects, contracts or orders may require these work processes to be developed, i.e. tailored or enhanced to suit particular needs. However, let us not confuse business processes with manufacturing processes. The Demand Fulfilment process is likely to enable the organization to deliver conforming product regardless of the specific features and characteristics of the product. The Demand Fulfilment process will generate product-specific control plans, work instructions, flow charts, etc.; therefore the business process won't change.

As the nature of planning will vary significantly from organization to organization, the generic term for this type of planning is *product realization planning* thus distinguishing it from specific planning activities such as design planning, production planning, installation planning, etc.; product realization planning is therefore the overall planning activities needed to meet all requirements for a project, product, contact or order.

Why is this necessary?

This requirement responds to the Process Approach Principle.

In designing the management system the work processes needed to produce the organization's products and services should have been developed so that planning to meet specific orders does not commence with a blank sheet of paper. These work processes provide a framework that aids the planners in deciding on the specific processes, actions and resources required for specific projects, contracts or orders. The process descriptions may not contain details of specific products, dates, equipment, personnel or product characteristics. These may need to be determined individually for each product, hence the need to plan and develop processes for product realization.

How is this implemented?

Planning product realization processes

There are too many variations in the product realization process to provide much more than an overview.

For customer-specific products, product planning is driven by customer enquires, orders or contracts. This may require a project-planning process in order to provide the customer with a viable proposal. On receipt of a contract or order, an order-processing process is then needed to confirm and agree customer requirements and the terms and conditions for the supply of product or service. Once the contract or order has been agreed, a project- or order-planning process is needed to establish the provisions needed to meet the contract or order requirements.

For proprietary products, product planning is driven by the demand creation process which searches for new opportunities that will result in the development of new products and service, and thus its output will lead into project planning which subsequently supplies the demand fulfilment process with proven products and service to sell.

Product planning can therefore be driven by customer-specific requirements as well as requirements determined by the organization that will reflect or create customer needs and expectations. The product realization-planning process therefore includes the sales process and the product-planning process. The sales process may not require tailoring for specific enquiries but with major projects for the large procurement agencies, this process often varies depending on the nature of the project and may require careful planning for the organization to be successful. The product-planning process will often require tailoring for specific projects or contracts as the nature of the work involved will vary. Where the organization takes orders for existing products or services which the customer selects from a catalogue, no special planning may be needed other than the creation of work orders.

In planning product realization, there are several factors involved: task, timing, responsibility, resources, constraints, dependencies and sequence. The flow charts for each process that were developed in establishing the management system identify the tasks. The planners job is to establish whether these tasks, their sequence and the process characteristics in terms of throughput, resources, capacity and capability require any modification to meet the requirements of a particular project, contract or order. A typical product-planning process is illustrated in Figure 7.2.

Tools often used in product realization planning are Gantt and PERT charts. The Gantt chart depicts the tasks and responsibilities on a timescale showing when the tasks are to commence and when they are to be complete. PERT charts display the same information but show the relationship between the tasks. These tools are useful in analysing a programme of work, determining resources and determining whether the work can be completed by the required end date using the allocated resources. It is not the purpose here to elaborate on project-planning techniques, but merely to indicate the scope of the requirement and what is needed to implement it.

A common method of planning projects is to prepare a Project Plan that includes the following depending on whether the project includes new product development or the development of new processes to deliver high-volume production:

1   Project objectives
2   System requirements
3   Project strategy
4   Critical success factors and success criteria
5   Project milestones
6   Project timeline
7   Project organization (chart and team responsibilities and authority)
8   Work breakdown structure (major tasks, work packages and deliverables)
9   Resource provision (in terms of space, development tools and equipment)
10 Supplier control plan
11 Information system (strategy, tools and their development)
12 Communication plan (strategy, methods and tasks)
13 Personnel development plan (strategy, education and training for those engaged on the project)
14 Evaluation plan (audits, design reviews and assessments)
15 Project reviews (strategy, project reviews and team reviews)
16 Contract management.

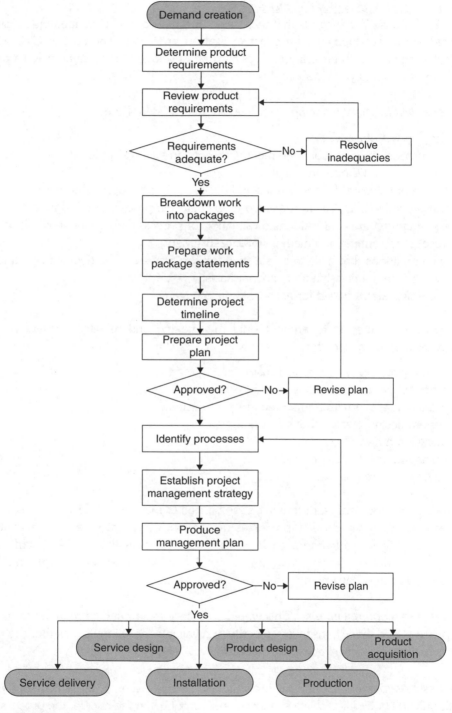

Figure 7.2 *Product planning process flow*

Developing product realization processes

Note 2 of Clause 7.1 suggests that the design and development requirements may be applied to the development of product realization processes. The note in Clause 7.3 basically requires manufacturing processes to be developed in the same way as products. So how would you go about this?

Process development planning   This would involve identifying:

- the process objectives;
- the process specification (parameters, targets and success criteria);
- the process development stages;
- the responsibilities for the actions and decisions in developing the process;
- the procedures, instructions needed to execute the process development tasks;
- the stages where verification and validation of process and product take place, the acceptance criteria and methods of measurement;
- the conditions that have to be satisfied for the process to be deemed operational (who make the decision and on what basis);
- the timescales required for each task in each stage.

The processes referred to are not only the making and moving processes. New processes may be needed for:

- communicating with overseas customers;
- managing a multinational project;
- managing a centralized procurement programme;
- system design involving new technologies;
- installing a new IT system;
- managing major subcontractors;
- setting up a remote servicing unit.

A process development team should be established to manage the development of any new processes. If several new processes are to be developed, several teams will be needed. By building a team for each process you will focus the efforts of staff more clearly than loading several new jobs onto the same individuals, but if you lack resources you may have no option.

Process development inputs   This would consist of the input data required to design the process, and the actions taken to verify and validate these inputs (whether the right inputs had been received and were correct).

Process development   Curiously the standard does not address the activity of design itself, only the boundary activities of input and output but in the middle, the process has to be designed and this requires several important steps.

It would involve determining:

- the process stages, their sequence and interrelation, the inputs and outputs and their destination;
- the procurement and/or construction activities (make or buy);
- the methods for performing the tasks in the process;
- the means of conveying product and information through the process;
- the means of storing product or data awaiting processing;
- the means of disposing of process waste;
- the potential failure modes and eliminating, reducing or controlling the effects including error-proofing measures;
- the resources required (equipment, tooling, facilities, personnel competency, finance, physical and human environment);
- the measures necessary to maintain the equipment, facilities and environment;
- the installation and commissioning activities including layout and access.

Process development outputs   This would consist of the process development deliverables (the specifications, floor plans, layouts, procedures, set-up and operating instructions, forms, notices, guides, standards, certificates, competency requirements, handling requirements, samples needed to operate and maintain the process).

Process development reviews   This would be the review stages where process development is assessed, i.e. planning, construction, verification and validation in terms of risks, costs, utilization, lead times, critical paths, etc.

Process verification   This would be the action needed to verify that the process is being fed with the correct inputs and resources, and that it is producing the required results at each stage. Verification is also a period when parameters are optimized and special cause variation is removed. This would also involve the monitoring of resource consumption in terms of time, materials and labour, verification of error-proofing provisions, measurement system capability, packaging, health, safety and environmental requirements, and other regulations that apply to the process.

Process validation   This would be the action needed to determine the capability of the process to produce consistent output that meets the requirements in terms of quality, cost and delivery. In some cases, this may require product approval by the customer or the designer.

Process change control   This would be the action needed to propose, review, evaluate, approve and implement changes in the process design.

Creating consistency in process planning (7.1)

The standard requires planning of the realization processes to *be consistent with the other requirements of the organization's quality management system.*

What does this mean?
This means that the processes employed for specific products and services should either be those that form part of the management system, are developments of those that form part of the management system or are new processes that fit into the set of management system processes and meet the same organizational objectives.

Why is this necessary?
This requirement responds to the Process Approach Principle.

The management system developed to meet the requirements of ISO/TS 16949 is likely to be a generic system, not specific to any particular product, project or contract other than the range of products and services which your organization supplies. By implementing the processes of the management system, product-, project- or contract-specific plans, procedures, specification, etc. are generated. However, specific variants, modifications or new activities may be required for particular projects, contracts or orders. It is therefore essential that the provisions made for any particular product, service, project or contract do not conflict with the authorized policies and practices so that the integrity of the system is maintained. Also, if staff are familiar with one way of working, by receiving conflicting instructions staff may apply the incorrect policies and practices to the project.

How is this implemented?
There is often a temptation when planning for specific contracts to change the policies and work processes where they are inflexible, invent new forms, change responsibilities, by-pass known bottlenecks, etc. You need to be careful not to develop a mutant management system for specific contracts. If the changes needed are good for the business as a whole, they should be made using the prescribed management system change procedures.

Quality objectives and requirement for product (7.1a)

The standard requires the organization to *determine the quality objectives and requirements for the product.*

What does this mean?
This means that for every product or service that is to be supplied there has to be a specification of requirements which if met will deliver a product or service that meets customer requirements.

The quality objectives for the product are those inherent characteristics of the product or service that aims to satisfy customers. For products these would include objectives for functional performance and physical attributes, reliability, maintainability, durability, etc. Those for services would include accessibility, responsiveness, promptness, reliability, etc. Quality requirements for the product are the inherent characteristics that are required to be met and may equal the quality objectives, but quality objectives may aim higher than the requirements either in an attempt to delight customers or simply to ensure requirements are met. In some cases, the characteristic may be prone to variation due to factors that are not easily controllable and therefore targets are set higher to be certain of meeting the requirement. In other cases, the objective may be stated in subjective or non-specific terms such as high reliability or relevant EU Directives governing the Automotive Industry whereas the requirement will be specific and quantifiable such as requiring a mean time between failure (MTBF) of 100,000 hours or requiring compliance with EU Directive 72/306/EEC. These quality objectives are for specific products whereas the quality objectives referred to in Clause 5.4.1 relate to the whole organization.

Why is this necessary?

This requirement responds to the Leadership Principle.

The objective of product realization processes is to deliver product or service that meets requirements. It is therefore necessary to establish exactly what requirements the product must satisfy in order to determine whether the processes to be employed are fit for their purpose, i.e. are capable of delivering a product that meets the requirements.

How is this implemented?

A specification should be produced or supplied for each product or service that is designed, produced and delivered. This specification should not only define the characteristics of the product, but also should define its purpose or function so that a product possessing the stated characteristics can be verified as being fit for its purpose. It is of little use for a product to meet its specification if the specification does not accurately reflect customer needs.

Determining the need for specific processes (7.1b)

The standard requires the organization to *determine the need to establish processes specific to the product.*

What does this mean?

As the standard does not define processes other than something that transforms inputs into outputs, the notion of a hierarchy of processes is overlooked. With this definition of a process, the organization itself is a customer-oriented process; i.e. it takes inputs from customers and transforms them into products that hopefully satisfy the inputs. So with the requirement to *determine the need to establish processes specific to the product*, we are not talking about business processes or even work processes, but the processes needed to execute a task such as to make a body panel or assemble an engine where the inputs are materials and the output is a finished or semi-finished product.

Why is this necessary?
This requirement responds to the Process Approach Principle.

When planning for specific products, it is necessary to determine whether the intended product characteristics are within the design limits of the existing processes. If the product is similar to existing products no change to the processes may be needed. If the nature of the product is different or if the performance required is beyond the capability of existing processes, new or modified processes will be required. Many problems arise where managers load product into processes without being aware of the process's limitations. Often because people are so flexible, it is assumed that because they were successful at producing the previous product they will be successful with any other products. It is only when the differences are so great as to be glaringly obvious that they stop and think.

How is this implemented?
In planning for a contract or new product or service, the existing processes need to be reviewed against the customer or market requirements. One can then identify whether the system provides an adequate degree of control. Search for unusual requirements and risks to establish whether any adjustment to processes is necessary. This may require you to introduce new activities or provide additional verification stages and feedback loops or prepare contingency plans.

If you have a process hierarchy such as that in the example of Figure 4.3 (repeated in Figure 7.3) new 'processes' might be needed at the task level. In the example, the tasks vary depending on what parts are being made so in effect the new process is reflected in the Control Plan. In order to make the parts, after machining the materials might need heat treatment, plating or any number of processes.

Determining the need for documentation (7.1b)

The standard requires the organization to *determine the need to establish documents specific to the product.*

What does this mean?
Documentation specific to the product is any documentation that is used or generated by the product realization processes. Such documents include specifications, drawings, plans, standards, datasheets, manuals, handbooks, procedures, instructions, records, reports, etc. that refer to the product or some aspect of the product. Determining the need to establish documentation means that in planning product realization you need

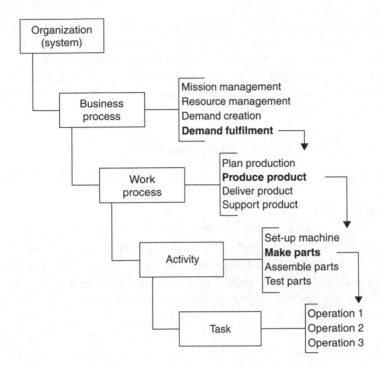

Figure 7.3 *System decomposition*

to determine the information carriers that will feed each of the processes and be generated by each of the processes.

Why is this necessary?
This requirement responds to the Process Approach Principle.

The process descriptions will specify the types of information required to operate the process and required to be generated by the process. However, depending on the nature of the product, contract or project, these may need to be customized for the specific product so that they carry the required information to the points of implementation. Information required for one project may not be required for another project. Product configuration, organization structure and locations may all be different, and require specific documents that are not used on other projects.

How is this implemented?
A common method for project work is to establish a Work Breakdown Structure (WBS) that identifies all the major packages of work to be carried out. For each major task a work statement is produced that defines the inputs, tasks and outputs required. The outputs are described together with a series of deliverables. Some of these will be documents,

particularly in the design and planning phases. For less complex projects a list of deliverables may be all that is required, identifying the document by name, the author and delivery date. Another method is APQP. The *APQP Manual* identifies all the documents requirement from identification of need through to launch into production as indicated in Table 7.1 but there are a few omissions, such as a design verification plan and design validation plan.

Determining the need for resources (7.1b)

The standard requires the organization to *provide resources specific to the product.*

What does this mean?
Resources are an available supply of equipment, environment, machines, materials, processes, labour, information and utilities such as heat, light, water and power, etc. that can be drawn on when needed. Therefore, this requires detailed planning and logistics management, and may require many lists and sub-plans so that the resources are available when required. Inventory Management and Information Technology are an element of such planning.

Why is this necessary?
This requirement responds to the Process Approach Principle.

All businesses are constrained by their resources. No organization has an unlimited capability. It is therefore necessary when planning new or modified products to determine what resources will be required to design, develop, produce and supply the product or service. Even when the requirement is for existing products, the quantity or delivery required might strain existing resources to an extent where failure to deliver becomes inevitable.

How is this implemented?
Successful implementation of this requirement depends on managers having current details of the capability of the process at their disposal. At the higher levels of management, a decision will be made as to whether the organization has the inherent capability to meet the specific requirements. At the lower levels, resource planning focuses on the detail, identifying specific equipment, people, materials, capacity and most important, the time required. A common approach is to use a project-planning tool such as Microsoft Project that facilitates the development of Gantt and PERT charts, and the ability to predict resources levels in terms of manpower and programme time. Other planning tools will be needed to predict process throughput and capability. The type of resources to be determined might include any of the following:

- Special equipment tools, test software and test or measuring equipment.
- Equipment to capture, record and transmit information internally or between the organization and its customers.

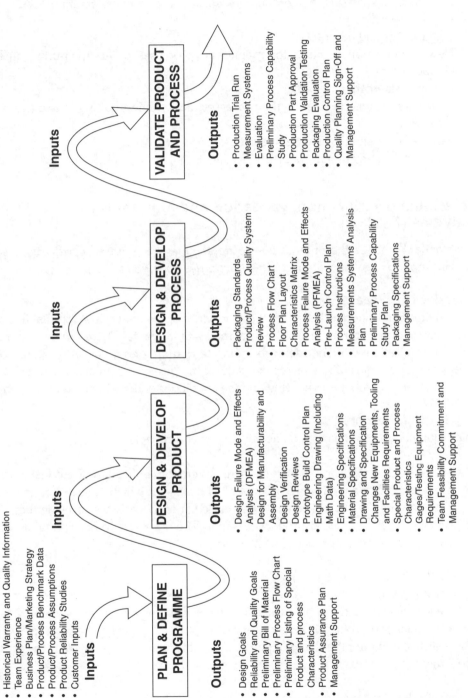

Inputs
- Market Research
- Historical Warranty and Quality Information
- Team Experience
- Business Plan/Marketing Strategy
- Product/Process Benchmark Data
- Product/Process Assumptions
- Product Reliability Studies
- Customer Inputs

Inputs

Inputs

Inputs

PLAN & DEFINE PROGRAMME

DESIGN & DEVELOP PRODUCT

DESIGN & DEVELOP PROCESS

VALIDATE PRODUCT AND PROCESS

Outputs
- Design Goals
- Reliability and Quality Goals
- Preliminary Bill of Material
- Preliminary Process Flow Chart
- Preliminary Listing of Special Product and process Characteristics
- Product Assurance Plan
- Management Support

Outputs
- Design Failure Mode and Effects Analysis (DFMEA)
- Design for Manufacturability and Assembly
- Design Verification
- Design Reviews
- Prototype Build Control Plan
- Engineering Drawing (Including Math Data)
- Engineering Specifications
- Material Specifications
- Drawing and Specification Changes New Equipments, Tooling and Facilities Requirements
- Special Product and Process Characteristics
- Gages/Testing Equipment Requirements
- Team Feasibility Commitment and Management Support

Outputs
- Packaging Standards
- Product/Process Quality System Review
- Process Flow Chart
- Floor Plan Layout
- Characteristics Matrix
- Process Failure Mode and Effects Analysis (PFMEA)
- Pre-Launch Control Plan
- Process Instructions
- Measurements Systems Analysis Plan
- Preliminary Process Capability Study Plan
- Packaging Specifications
- Management Support

Outputs
- Production Trial Run
- Measurement Systems
- Evaluation
- Preliminary Process Capability Study
- Production Part Approval
- Production Validation Testing
- Packaging Evaluation
- Production Control Plan
- Quality Planning Sign-Off and Management Support

Table 7.1 *Advanced product quality planning*

- New technologies such as computer-aided design and manufacturing (CAD and CAM, respectively).
- Fixtures, jigs and other tools.
- New instrumentation either for monitoring processes or for measuring quality characteristics.
- New measurement capabilities.
- New skills required to operate the processes, design new equipment and perform new roles.
- New research and development facilities.
- New handling equipment, plant and facilities.

Determining verification, validation and monitoring activities (7.1c)

The standard requires the organization to *determine the required verification, validation, monitoring, inspection and test activities specific to the product.*

What does this mean?

The required verification activities are those activities necessary to establish that a product meets or is meeting the defined and agreed requirements. There are several methods of verification that include inspection, test, monitoring, analysis, simulation, observation or demonstration each serving the same purpose, but each being different in the manner in which it is conducted and the conditions under which it is appropriate.

Why is this necessary?

This requirement responds to the Factual Approach Principle.

Provision should be made in the management system for verification of product at various stages through the realization processes. The stage at which verification needs to be performed and the characteristics to be verified at each stage are dependent on the requirements for the particular product. It is therefore necessary to determine the verification required for each product and process.

How is this implemented?

Product verification

If all the key features and characteristics of your product or service can be verified by a simple examination on final inspection or at the point of delivery, the requirement is easily satisfied. On the other hand, if you can't do this, whilst the principle is the same, it becomes more complex.

Generically there are two types of requirements: *defining requirements* and *verification requirements*. Defining requirements specify the features and characteristics required of a product, process or service. These may be wholly specified by the customer or by the organization or a mixture of the two. *Verification requirements* specify the requirements for verifying that the defining requirements have been achieved and again may be wholly specified by the customer or by the organization or a mixture of the two. With verification requirements, however, other factors need to be taken into consideration depending on what you are supplying and to whom you are supplying it. In a contractual situation, the customer may specify what he wants to be verified and how he wants it verified. In a non-contractual situation, there may be statutory legal requirements, compliance with which is essential to avoid prosecution. Many of the national and international standards specify the tests that products must pass rather than performance or design requirements, so identifying the verification requirements can be quite a complex issue. It is likely to be a combination of:

- What your customer wants to be verified to meet the need for confidence. (The customer may not demand you demonstrate compliance with all customer requirements, only those which are judged critical.)
- What you need to verify to demonstrate that you are meeting all your customer's defining requirements. (You may have a choice as to how you do this so it is not as onerous as it appears.)
- What you need to verify to demonstrate that you are meeting your own defining requirements. (Where your customer defines the product or service in performance terms, you will need to define in more detail the features and characteristics that will deliver the specified performance and these will need to be verified.)
- What you need to verify to demonstrate that you are complying with the law. (Product safety, personnel health and safety, conservation, environmental and other legislation.)
- What you need to verify to obtain confidence that your suppliers are meeting your requirements.

Verification requirements are not limited to product or service features and characteristics. One may need to consider who carries out the verification, where and when it is carried out and under what conditions and on what quantity (sample or 100%) and standard of product (prototype or production models).

You may find that the only way you can put your product on the market is by having it tested by an independent test authority. You may need a licence to manufacture it, to supply it to certain countries and this may only be granted after independent certification. Some verification requirements only apply to the type of product or service, others to the process or each batch of product and others to each product or service delivery. Some requirements can only be verified under actual conditions of use. Others can be verified by analysis or similarity with other products that have been thoroughly tested. The range is so widespread it is not possible in this book to explore all examples, but as you can see, this requirement contains a minefield unless you have a simple product or unless the customer has specified everything you need to verify.

There are a number of ways of documenting verification requirements:

- By producing defining specifications which prescribe requirements for products or services and also the means by which these requirements are to be verified in-house in terms of the inspections, tests, analyses, audits, reviews, evaluations and other means of verification.
- By producing separate verification specifications which define which features and characteristics of the product or service are to be verified and the means by which such verification is to be carried out.
- By producing a quality plan or a verification plan that identifies the verification stages from product conception to delivery and further as appropriate, and refers to other documents that define the specific requirements at each stage.
- By control plan referencing drawings and specifications.
- By inspection and test instructions specific to a production line, product or range of products.

In fact you may need to employ one or more of the above techniques to identify all the verification requirements.

Document verification
It is necessary to verify that all the documentation needed to produce and install the product is compatible; that you haven't a situation where the design documentation requires one thing and the production documents require another or that details in the design specification conflict with the details in the test specification. Incompatibilities can arise in a contract that has been compiled by different groups. For example, the contract requires one thing in one clause and the opposite in another. Many of the standards invoked in the contract may not be applicable to the product or service required. Production processes may not be qualified for the material specified in the design – the designer may have specified materials that are unavailable!

In order to ensure compatibility of these procedures, quality-planning reviews need to be planned and performed as the new documentation is produced. Depending on the type of contract, several quality-planning reviews may be necessary, each scheduled to occur prior to commencing subsequent stages of development, production, installation or servicing. The quality-planning reviews during product development can be held in conjunction with the design stage reviews required in Section 7.3.4 of ISO/TS 16949. At these reviews the technical and programme requirements should be examined to determine whether the existing provisions are adequate, compatible and suitable to achieve the requirements, and if necessary additional provisions put in place.

Determining the criteria for product acceptance (7.1c)

The standard requires the organization to *determine the criteria for product acceptance.*

What does this mean?

The criteria for product acceptability are those characteristics that the product or service needs to exhibit for it to be deemed acceptable to the customer or the regulator. These are those standards, references and other means used for judging compliance with defined requirements. In some cases, the requirements can be verified directly such as when a measurable dimension is stated. In other cases, the measurements to be made have to be derived, such as engineering the product so as to comply with specific EU Directives thus deeming it to be safe. Another example is in traffic management systems where speed limits are imposed for certain roads because they have been proved to represent a safe driving speed under normal conditions. The requirement is for people to drive safely but this is open to too much interpretation consequently measurable standards are imposed for effective communication and to ensure consistency in the application of the law.

Why is this necessary?

This requirement responds to the Factual Approach Principle.

In order to verify that the products or services meet the specified requirements there needs to be unambiguous standards for making acceptance decisions. These standards need to be expressed in terms that are not open to interpretation so that any qualified person using them would reach the same decision when verifying the same characteristics in the same environment using the same equipment. In some cases, the requirement is expressed definitively and in other cases subjectively. It is therefore necessary to establish how reliable is 'reliable', how safe is 'safe', how clean is 'clean' and how good is 'good quality'. Specifications often contain subjective statements such as good commercial quality, smooth finish, etc., and require further clarification in order that an acceptable standard can be attained.

How is this implemented?

A common method of determining acceptance criteria is to analyse each requirement and establish measures that will indicate that the requirement has been achieved. In some cases, national or international standards exist for use in demonstrating acceptable performance. The secret is to read the statement then ask yourself, "Can I verify we have achieved this?" If not, select a standard that is attainable, unambiguous and acceptable to both customer and supplier, that if achieved will be deemed as satisfying the intent of the requirement.

The results of some processes cannot be directly measured using gauges, tools, test and measuring equipment, and so an alternative means has to be found of determining what is conforming product. The term given to such means is 'Workmanship Criteria', criteria that will enable producers and inspectors to gain a common understanding of what is acceptable and unacceptable. Situations where this may apply in manufacturing are soldering, welding, brazing, riveting, deburring, etc. It may also include criteria for finishes, blemishes and many others. Samples indicating the acceptable range of colour, grain and texture may be needed and if not provided by your customer, those that you provide will need customer approval.

The criteria need to be defined by documented standards or by samples and models that clearly and precisely define the distinguishing features that represent both conforming and nonconforming product. In order to provide adequate understanding it may be necessary to show various examples of workmanship from acceptable to unacceptable so that the producer or inspector doesn't strive for perfection or rework product unnecessarily. These standards like any others need to be controlled. Documented standards should be governed by the document control provisions. Samples and models need to be governed by the provision for controlling measuring devices and be subject to periodic examination to detect deterioration and damage. They should be certified as authentic workmanship samples and measures taken to preserve their integrity. Ideally they should be under the control of the inspection authority or someone other than the person responsible for using them so that there is no opportunity for them to be altered without authorization. The samples represent your company's standards, they do not belong to any individual, and if used by more than one person you need to ensure consistent interpretation by training the users.

Determining the need for records (7.1d)

The standard requires the organization to *determine the records needed to provide evidence that the realization processes and resulting product meet requirements.*

What does this mean?

Countless verification activities will be carried out at various levels of product and service development, production and delivery. These activities will generate data and this data needs to be collected in a form that can be used to demonstrate that processes and products fulfil requirements. This does not mean that *every* activity needs to be recorded, but the manner in which the data is recorded and when it is to be recorded should be determined as part of the planning activity.

Why is this necessary?

This requirement responds to the Factual Approach Principle.

Without records indicating the results that have been obtained from product and process verification, compliance cannot be demonstrated to those on the scene of the action such as customers, managers and analysts. When investigating failures and plotting performance trends, records are also needed for reference purposes.

While procedures should define the records that are to be produced, these are the records that will be produced if these procedures are used. On particular contracts, only those procedures that are relevant will be applied and therefore the records to be produced will vary from contract to contract. Special conditions in the contract may make it necessary for additional records to be produced.

How is this implemented?
There are two parts to this requirement: one concerning product records and the other concerning process records.

Product records
By assessing the product requirements and identifying the stage in the process where these requirements will be verified, the type of records needed to capture the results should be determined. In some cases, common records used for a variety of products may suffice but in others, product-specific records may be needed that prescribe the characteristics to be recorded and the corresponding acceptance criteria to be used to indicate pass or fail conditions.

Process records
The records required for demonstrating process performance should be identified during process development (see previously). Continued operation of the process should generate further records that confirm that the process is functioning properly, i.e. meeting the requirements for which it was designed.

Process records should indicate the process objectives and exhibit performance data showing the extent to which these objectives are being achieved. These may be in the form of bar charts, graphs, pie charts, etc. From the process hierarchy of Figure 7.3, you will observe that there would be process records for business processes, work processes, and for activities and tasks.

Documenting product realization planning (7.1 and 7.1.1)

The standard requires the product realization planning to *be in a form suitable for the organization's method of operations* and the automotive additions require *references to customer requirements and technical specifications to be included as a component of quality plans as part of planning for product realization*.

What does this mean?
The output of planning can be in a variety of forms depending on the nature of the product, project, contract or service, and its complexity. For simple products, the planning

output may be a single document. For complex products, the planning output may take the form of a project plan and several supplementary plans and other documents. A typical example is the APQP approach which generates six documented outputs from the planning stage of the product realization process (Table 7.1). A quality plan might be label given to the consolidation of these outputs.

Why is this necessary?

This requirement responds to the Process Approach Principle.

The standard does not impose a particular format for the output of the planning activity or insist that such information carriers are given specific labels. Each product is different and, therefore, the planning outputs need to match the input requirements of the processes they feed.

How is this implemented?

Discrete plans are needed when the work to be carried out requires detailed planning beyond that already planned for by the management system. The system will not specify everything you need to do for every job. It will usually specify only general provisions that apply in the majority of situations. You will need to define the specific activities to be performed, the documentation to be produced and resources to be employed. The contract may specify particular standards or requirements that you must meet and these may require additional provisions to those defined in the documented management system.

The disadvantage in giving any document a label with the word *quality* in the title is that it can sometimes be thought of as a document that serves only the Quality Department rather than a document that defines the provisions for managing the various processes that will be utilized on the project. A useful rule to adopt is to avoid giving documents a title that reflect the name of a department wherever possible.

Acceptance criteria (7.1.2)

The standard requires the organization *to define acceptance criteria and confine acceptance levels for attribute data to zero defects.*

What does this mean?

With attribute data the product either has or has not the ascribed attribute – it can therefore either pass or fail the test. There are no grey areas. Attributes are measured on a go or no-go basis. With variables the product can be evaluated on a scale of measurement. However, with inspection by attributes we sometimes use an acceptable quality level (AQL) that allowed us to ship a certain per cent defective in a large batch of product – probably no more than 10 in 1000, but to the automobile industry that is not good enough. The standard imposes a strict requirement on characteristics that are measured by attributes. There shall be no AQL, there shall be zero per cent defective in the sample selected for inspection, otherwise the batch shall be rejected.

Why is this necessary?
Your customer does not want to be supplied any defective products. With automated assembly equipment, any defects in fasteners, washers and other components will produce defective products that have to be withdrawn from the line and put through a separate rework process then is clearly uneconomical. It is far better to ensure that the only components that feed the assembly tools are conforming products.

How is this implemented?
The way to ensure conformity in this instance is to make the process that produces the attribute incapable of producing nonconforming product. The process can be designed so that errors are detected automatically and rejects ejected so that they are separated from the batch that is inspected.

For inspection by variables the acceptance criteria has to be specified and the place to specify it is the Control Plan that is submitted to your customer for approval.

Confidentiality (7.1.3)

The standard requires the organization *to ensure the confidentiality of products, projects and related product information contracted by customers.*

What does this mean?
Quite simply, this requirement implies that information about products being produced for a customer are not be conveyed to anyone not authorized to receive such information.

Why is this necessary?
A problem that may face many suppliers to the automotive industry is that of having multiple customers that are competitors, thus creating a need to preserve confidentiality. Customers are naturally concerned that their information or product does not reach their competitors.

How is this implemented?
In responding to this requirement you need to define how you intend to ensure confidentiality. How you do this is not as easy as getting everyone to sign a declaration. The

declaration is useful in a prosecution but that will be after confidentiality has been breached! Things you can do to minimize a breach in confidentiality are as follows:

- Employ a classification system for identifying information that requires difference security measures.
- Code the identity of the customer on classified information.
- Control filing or storing of customer data.
- Identify customer data with a customer code rather than a name.
- Control photocopying machines where access to customer data can be obtained.
- Destroy data by shredding and secure disposal.
- Remove labels from obsolete product before disposal.
- Escort and record visitors on site.
- Bar visitors from carrying mobile phones into the plant.
- Control the possession of cameras.
- Create project offices for new product development, installing controlled access if necessary.
- Advise staff never to discuss company matters in a public place.

Change control (7.1.4)

The standard requires the organization *to have a process to control and react to changes in product and manufacturing processes that impact product realization.*

What does this mean?
Although change control is addressed in several other clauses of the standard, this requirement focuses on technical change only. It is therefore concerned with what is referred to in the aerospace sector as Configuration Management – management of the functional and physical description of a product or process. It applies to both custom and proprietary product. This requirement could just as easily been placed under Clause 7.3.7, but the editors chose a separate clause so as to give greater impact and scope.

Why is this necessary?
The complexity and speed of innovation in the automotive sector requires a change management process that will maintain control. The traditional keeping pace with change by reacting to problems is not an effective solution. What is needed is a more co-ordinated approach that ensures all components of a product or process remain compatible when a change to one of them is introduced.

How is this implemented?
The ISO 9001 requirements on change control are documentation changes (Clause 4.2.3), system changes (Clause 5.4.2), contract changes (Clause 7.2.2), design changes (Clause 7.3.7), control plans (Clause 7.5.1.1) and process changes (Clause 8.2.3.1).

However, the requirement for continual improvement implies that there needs to be a change management process in place for any type of change. The requirement in ISO/TS 16949 focuses on technical change and is limited to the product realization processes, but this does not diminish the general ISO 9001 requirement for a change management process. Document changes are often the result of a change process that is triggered by other situations. Therefore, there is a process that has the key stages of identifying the need for change, evaluating the potential change, defining the change, validating the change, implementing and verifying the change. These stages are triggered by the following questions:

- What product or process do we need to change?
- Why do we want to change this product or process?
- What will be the cost, benefit and impact of the change?
- What are the competing or alternative solutions?
- Which is the best solution?
- How do we intend to prove the solution is effective before we make the change?
- How will the change be made to the products and processes affected?
- How will the cost, benefits and impact of the change be measured and reviewed?

To control change there needs to be a defined baseline to which all change is referred. For each version of a product or process there would be a hierarchy of its components along with their associated specifications and versions of these specifications. The list itself would be subject to version control. One list could cover several versions of the product simply by laying the information out in columns with the version code heading each column.

When a component needs to be changed, it needs to be clear from the list what the parent and child relationships are so that one can see which other components might be affected. However, this would not be the only way of determining the impact of the change. A study of interface specifications, assembly drawings, wiring lists, etc. will reveal the items that could be affected by the change and who the suppliers are.

Once the components affected have been identified, a thorough evaluation is needed to determine the impact on form, fit and function, and to assess the cost. More detail is given under the heading *Identification and recording of design changes* later in this chapter.

A form would be needed to collect this information and an example is given in Figure 7.4.

Customer-related processes (7.2)

Determination of requirements related to the product (7.2.1)

This heading implies there are other requirements that do not relate to the product that may form part of the customer requirements. However, ISO 9000 defines a product as

DESIGN CHANGE PROPOSAL						DCP/	
ORIGIN						Class	
Name	Company	Dept		Tel:	Date	Priority	
System		Subsystem			Item		
SPECIFICATIONS CONCERNED							
Specification		Title			Date		Revision

REASON FOR APPLICATION

[ ] Hazards [ ] Incompatibility [ ] Potential failure	[ ] Production costs [ ] Maintenance costs [ ] Disposal costs	[ ] Non-compliance [ ] Other [ ] Other

CHANGE DESCRIPTION

COST ESTIMATE (+Additions or −Savings)

Phase	Hardware	Software	Tools/test equipment	Publications	Planning
Development					
Production					
Operation					
Maintenance					

EFFECT OF CHANGE

On customers and suppliers

On other systems/equipments/parts

On programmes

On transportation and distribution networks

DECISION of ENGINEERING CHANGE BOARD

Accepted [ ]	Rejected [ ]	Hold [ ]
Reasons		
Authorized by		Date

Figure 7.4 *Sample change proposal*

the result of a process and includes services among these. It is therefore difficult to imagine any aspect of customer requirements that would not relate to the product or service that is being provided. Requirements related to the product or service could include:

- Characteristics that the product is required to exhibit, i.e. the inherent characteristics.
- Price and delivery requirements.
- Procurement requirements that constrain the source of certain components, materials or the conditions under which personnel may work.
- Management requirements related to the manner in which the project will be managed, the product developed, produced and supplied.
- Security requirements relating to the protection of information.
- Financial arrangements for the deposit of bonds, payment conditions, invoicing, etc.
- Commercial requirements such as intellectual property, proprietary rights, labelling, warranty, resale, copyright, etc.
- Personnel arrangements such as access to the organization's facilities by customer personnel and vice versa.

A process for determining product requirements should be designed so that it takes as its input the identified need for a product and passes this through several stages where requirements from various sources are determined, balanced and confirmed as the definitive requirements that form the basis for product realization. The input can either be a customer-specific requirement or the market specification that results from market research (see Chapter 5 under the heading *Customer focus*) or a sales order for an existing product. However, this is not blue-sky stuff; remember the product realization process is triggered by a customer placing a demand on the organization either because of what the organization has to sell or offer. The output may indeed be presented in several documents: the product requirement specification containing the hardware and software requirements, and the service requirement specification containing the service requirements. Alternatively where service is secondary, the requirement may be contained in the contract.

Products requirements specified by the customer (7.2.1a)

The standard requires the organization to *determine requirements specified by the customer including requirements for delivery and post-delivery activities.*

What does this mean?
ISO 9000 defines the customer as the organization that receives the product; however, this is easily taken out of context by referring to internal and external customers. The term customer in ISO/TS 16949 is reserved only for the external customer because the organization that is the subject of the standard is the whole organization not its component

parts. Customers are also consumers, clients, end users, retailers, purchasers and benefici-aries; therefore, requirements specified by the customer need not be limited to the organi-zation that is purchasing the product or the service. ISO 9000 also defines requirements as needs or expectations that are stated, generally implied or obligatory and therefore any information that is expressed by the customer as a need or expectation, whether in writing or verbally is a requirement. To determine such requirements means that the needs and expectations that are either stated verbally, or in writing, implied or obligatory have to be resolved, pinned-down and defined so that neither party is in any doubt as to what is required. The requirements for delivery mean requirements pertaining to the ship-ment, transportation, transmission or other means for conveying the product or service to the customer in a specified condition. Similarly with post-delivery requirements, these are the requirements pertaining to the support the customer requires from the organization to maintain, service, assist, or otherwise retain the product or service in a serviceable state.

Why is this necessary?
This requirement responds to the Customer Focus Principle.

The purpose of the requirements is to ensure that you have established the requirements you are obliged to meet before you commence work. This is one of the most import-ant requirements of the standard. The majority of problems downstream can be traced either to a misunderstanding of customer requirements or insufficient attention being paid to the resources required to meet customer requirements. Get these two things right and you are half way towards satisfying your customer needs and expectations.

How is this implemented?
Customers will convey their requirements in various forms. Many organizations do business through purchase orders or simply order over the telephone or by electronic or surface mail. Some customer prefer written contracts others prefer a handshake or a verbal telephone agreement. However, a contract does not need to be written and signed by both parties to be a binding agreement. Any undertaking given by one party to another for the provision of products or services is a contract, whether written or not. The requirement for these provisions to be determined rather than documented, places the onus on the organization to understand customer needs and expectations, not simply react to what the customer has transmitted. It is therefore necessary in all but simple transactions to enter into a dialogue with the customer in order to understand what is required. Through this dialogue, assisted by checklists that cover your product and service offerings, you can tease out of the customer all the requirements that relate to the product. Sometimes the customer wants one of your products or services but in fact needs another but has failed to realize it. Customers wants are not needs unless the two coincide. It is not until you establish needs that you can be certain that you can sat-isfy the customer. There may be situations when you won't be able to satisfy customers

needs because the customer simply does not have sufficient funds to pay you for what is necessary!

Many customer requirements will go beyond end product or service requirements. They will address delivery, quantity, warranty, payment and other legal obligations. With every product one provides a service; for instance, one may provide delivery to destination, invoices for payment, credit services (if they don't pay on delivery they are using your credit services), enquiry services, warranty services, etc., and the principal product may not be the only product either – there may be packaging, brochures, handbooks, specifications, etc. With services there may also be products such as brochures, replacement parts and consumables, reports, certificates, etc.

In ensuring that contract requirements are adequately defined, you should establish where applicable that:

* there is a clear definition of the purpose of the product or service you are being contracted to supply;
* the conditions of use are clearly specified;
* the requirements are specified in terms of the features and characteristics that will make the product or service fit for its intended purpose;
* the quantity, price and delivery are specified;
* the contractual requirements are specified including: warranty, payment conditions, acceptance conditions, customer-supplied material, financial liability, legal matters, penalties, subcontracting, licences and design rights;
* the management requirements are specified such as points of contact, programme plans, work breakdown structure, progress reporting, meetings, reviews and interfaces;
* the quality assurance (QA) requirements are specified such as quality system standards, quality plans, reports, customer approvals and surveillance, product approval procedures and concessions.

It is wise to have the requirement documented in case of a dispute later. The document also acts as a reminder as to what was agreed, but it is vital when either of the parties that made agreement move on to pastures new, leaving their successors to continue the relationship. This becomes very difficult if the agreements were not recorded particularly if your customer's representative moves on before you have submitted your first invoice. The document needs to carry an identity and if subject to change, an issue status. In the simple case this is the serial numbered invoice and in more complicated transactions, it will be a multipage contract with official contract number, date and signatures of both parties.

Product requirements not specified by the customer (7.2.1b)

The standard requires the organization to *determine product requirements not specified by the customer but necessary for known intended use.*

What does this mean?

There are two ways of looking at this requirement:

1 From the viewpoint of an identified market need.
2 From the viewpoint of a specific contract or order.

Market need

The process of identifying future customer needs and expectations was addressed under the heading *Customer focus* in Chapter 5. The output of this process will be in the form of a market research report that contains information from which a new product requirement can be developed.

Specific contract or order

The customer is not likely to be an expert in your field. The customer may not know much about the inner workings of your product and service offerings, and may therefore specify the requirements only in performance terms. In such cases, the onus is on the organization to determine the requirements that are necessary for the product or service to fulfil its intended use. For example, if a customer requires an electronic product to operate in the engine compartment, the electronics will need to be screened to prevent radiation from other components affecting its performance. The customer may not know that this is necessary but during your dialogue, you establish the conditions of use and as a result identify several other requirements that need to be met. These are requirements not specified by the customer but necessary for known intended use.

Why is this necessary?

This requirement responds to the Customer Focus Principle.

It is necessary to convert the results of market research into definitive product requirements so as to form a basis for new product development.

In the case of specific orders it is important to identify requirements necessary for intended use. For instance, subsequent to delivery of a product, a customer could inform you that your product does not function properly and you establish that it is being used in an environment that is outside its design specification. You would not have a viable case if the customer had informed you that it was going to be placed in the engine compartment and you took no action.

How is this implemented?

Careful examination of customer needs and expectations is needed in order to identify all the essential product requirements. A useful approach is to maintain a checklist or datasheet of the products and services offered which indicate the key characteristics but also indicate the limitations, what it can't be used for, what your processes are not capable of. Of course such data needs to be kept within reasonable bounds, but it is interesting to note that a manufacturer of refrigerators was successfully sued under product liability legislation for not providing a warning notice that the item was not safe for a person to stand on. It is therefore important to establish what the customer intends to use the product for, where and how they intend to use it and for how long they expect it to remain serviceable. With proprietary products many of these aspects can be clarified in the product literature supplied with the goods or displayed close to the point of service delivery. With custom designed products and services, a dialogue with the customer is vital to understand exactly what the product will be used for through its design life.

Statutory and regulatory requirements (7.2.1c)

The standard requires the organization to *determine statutory and regulatory requirements related to the product.*

What does this mean?

Almost all products are governed by regulations that constrain or prohibit certain inherent characteristics. Many of the regulations apply to human safety but some also apply to equipment safety such as those pertaining to electromagnetic radiation. There are also regulations that apply to the import and export of goods and environmental regulations that apply to pollution. While there may be no pollution from using the product, there may be pollution from making, moving or disposing of the product and therefore regulations that apply to production processes are indeed product related.

Why is this necessary?

This requirement responds to the Customer Focus Principle.

The customer may not be aware of all regulations that apply but will expect the supplier of the products and services required to be fully aware and have complied with all of them without exception. It is necessary to be fully aware of such statutes and regulations for the following reasons:

- A failure to observe government health and safety regulations could close a factory for a period and suspend your ability to supply customers.
- Health and safety hazards could result in injury or illness and place key personnel out of action for a period.
- Environmental claims made by your customers regarding conservation of natural resources, recycling, etc. may be compromised if environmental inspections of your organization show a disregard for such regulations.
- The unregulated discharge of waste gases, effluent and solids may result in public concern in the local community and enforce closure of the plant by the authorities.

- A failure to take adequate personnel safety precautions may put product at risk.
- A failure to dispose of hazardous materials safely and observe fire precautions could put plant at risk.
- A failure to provide safe-working conditions for personnel may result in public concern and local and national enquiries that may harm the reputation of the organization.

It is therefore necessary to maintain an awareness of all regulations that apply regardless of the extent to which they may or may not relate to the product.

How is this implemented?

In order to determine the applicable statutes and regulations you will need a process for scanning the environment, identifying those that are relevant and capturing them in your management system. The legislators don't know what is relevant to your organization, only you know that; so a dialogue with legal experts may be necessary to identify all those regulations that apply.

There are lots of regulations and no guarantees of finding them all. However, you can now search through libraries on the Internet and consult bureaus, trade associations and government departments to discover those that apply to you. Ignorance of the law they say is no excuse.

The requirement also applies to products you purchase that are resold under the original manufacturer's label or re-badged under your label or incorporated into your product. Regulations that would apply to your products apply to products you have purchased. There may be regulations that only apply to products you have purchased because of their particular form, function or material properties and may not apply to your other products. It pays therefore to be vigilant when releasing purchase orders.

It is not simply the product that may have to meet regulations but the materials used in making the product. These materials are a direct consequence of the chosen design solution, therefore in general:

- the product has to be safe during use, storage and disposal;
- the product has to present minimum risk to the environment during use and disposal;
- the materials used in the manufacture of the product have to be safe during use, storage and disposal;
- the materials used in manufacture of the product have to present minimum risk to the environment during use and disposal.

Although you may not have specified a dangerous substance in the product specification, the characteristics you have specified in the product specification may be such than can only be produced by using a dangerous substance.

Organization's product requirement (7.2.1d)

The standard requires *any additional requirements determined by the organization for the product to be determined.*

What does this mean?

In addition to the requirements specified by the customer and the regulations that apply, there may be requirements imposed by the organization's policies that impinge on the particular products or services that are to be supplied. The product policy may impose certain style, appearance, reliability and maintainability requirements or prohibit use of certain technologies or materials. Other requirements may serve to aid production or distribution that are of no consequence to the customer but necessary for the efficient and effective realization and supply of the product.

Why is this necessary?

This requirement responds to the Leadership Principle.

The requirement is necessary in order that relevant organizational policies and objectives are deployed through the product and service offerings. A failure to identify such constraints at the requirement definition stage could lead to abortive design work or if left undetected, the supply of products or services that harm the organization's reputation. Often, an organization is faced with the task of balancing customer needs with those of other interested parties. It may therefore be appropriate in some circumstances for the organization to decline to meet certain customer requirements on the grounds that they conflict with the needs of certain stakeholders.

How is this implemented?

The organization's requirements should be defined in technical manuals that are used by designers, production and distribution staff. These will often apply to all the organization's products and services but will, however, need to be reviewed to identify the specific requirements that apply to particular products.

Customer-designated special characteristics (7.2.1.1)

The standard requires the organization *to demonstrate conformity to customer requirements for the designation, documentation and control of special characteristics.*

What does this mean?

In addition to the requirements of Clause 7.3.2.3, any requirements relating to the designation of special characteristics will be specified by the customer in the contract. These requirements simply require the organization to demonstrate that it conforms to these additional requirements.

Why is this necessary?
This requirement was unnecessary simply because the organization should be able to demonstrate conformity with any of the requirement not just those relating to special characteristics.

How is this implemented?
Although you might have your own symbols for special characteristics, if customer symbols are different, it would be prudent to use customer symbols in drawings and control plans to avoid any confusion especially as these documents need to be submitted to the customer in the product or process approval submission.

The control plans should make provision for any specific controls required by the customer and these must be implemented. Evidence is required to show that all the controls specified in the control plan have been implemented and a way of doing this is to make provision for recording verification of conformity against the relevant requirement in production records.

Review of requirements related to the product (7.2.2 and 7.2.2.1)

Conducting the review (7.2.2)

The standard requires the organization to *review the requirements related to the product and for the automotive sector, this requirement can only be waived with customer authorization.*

What does this mean?
A review of the requirements related to the product means that all the requirements that have been identified through the requirement determination process should be examined together preferably by someone other than those who gathered the information. The review may be quite independent of any order or contract but may need to be repeated should an order or contract for the product be received.

Why is this necessary?
This requirement responds to the Factual Approach Principle.

The process of determining the requirements that relate to the product consists of several stages culminating in a definitive statement of the product requirements. All processes should contain review stages as a means of establishing that the process output is correct and that the process is effective. The review referred to in this requirement is therefore necessary to establish that the output of the requirement determination process is correct.

How is this implemented?

The information gathered as a result of determining the various product requirements should be consolidated in the form of a specification, contract or order and then subject to review. The personnel who should review these requirements depend on their complexity and there are three situations that you need to consider:

1   Development of new product to satisfy identified market needs – new product development.
2   Sales against the organization's requirements – proprietary sales.
3   Sales against specific customer requirements – custom sales.

New product development

In setting out to develop a new product there may not be any customer orders – the need for the product may have been identified as a result of market research and from the data gathered a definitive product requirement is developed. The product requirement review is performed to confirm that the requirements do reflect a product that will satisfy the identified needs and expectations of customers. At the end-product level, this review may be the same as the design input review, but there other outputs from market research such as the predicted quantities, the manner of distribution, packaging and promotion considerations. The review should be carried out by those functions representing the customer, design and development, production, service delivery and in-service support so that all views are considered.

Proprietary sales

In a proprietary sales situation, you may simply have a catalogue of products and services advertising material and a sales office taking orders over the telephone or over the counter. There are two aspects to the review of requirements. The first is the initial review of the requirements and advertising material before they are made available for potential customer to view and the second is where the sales person reviews the customer's request against the catalogue to determine if the particular product is available and can be supplied in the quantity required. We could call these reviews: Requirement Review and Transaction Review. As a customer may query particular features, access to the full product specification or a technical specialist may be necessary to answer such queries.

Custom sales

In custom sales situation the product or service is being produced or customized for a specific customer and with several departments of the organization having an input to the contract and its acceptability. These activities need co-ordinating so that you ensure all are working with the same set of information. You will need to collect the contributions of those involved and ensure they are properly represented at meetings. Those who negotiate contracts on behalf of the company, carry a great responsibility.

One aspect of a contract often overlooked is shipment of finished goods. You have ascertained the delivery schedule, the place of delivery, but how do you intend to ship it (by road, rail and ship or by air). It makes a lot of difference to the costs. Also delivery dates often mean the date on which the shipment arrives not the date it leaves. You therefore need to build into your schedules an appropriate lead time for shipping by the means agreed to. If you are late you may need to employ speedier means but that will incur a premium for which you may not be paid. Your financial staff will therefore need to be involved in the requirement review.

Having agreed the requirements, you need to convey them to their point of implementation in sufficient time for resources to be acquired and put to work.

Timing of review (7.2.2)

The standard requires the review to *be conducted prior to the decision or commitment to supply a product to the customer (e.g. submission of a tender, acceptance of a contract or order).*

What does this mean?

A tender is an offer made to a potential customer in response to an invitation. The acceptance of a contract is a binding agreement on both sides to honour commitments. Therefore, the period before the submission of a tender or acceptance of a contract or order is a time when neither side is under any commitment and presents an opportunity to take another look at the requirements before legal commitments are made.

Why is this necessary?

This requirement responds to the Factual Approach Principle.

The purpose of the requirement review is to ensure that the requirements are complete, unambiguous and attainable by the organization. It is therefore necessary to conduct such reviews before a commitment to supply is made so that any errors or omissions can be corrected in time. There may not be opportunities to change the agreement after a contract has been signed without incurring penalties. Customers will not be pleased by organizations that have underestimated the cost, time and work required to meet their requirements, and may insist that organizations honour their commitments – after all an agreement is a promise and organizations that break their promises do not survive for long in the marketplace.

How is this implemented?

The simplest method of implementing this requirement is to make provision in the requirement determination process for a requirement review to take place before tenders are submitted, contracts are signed or orders accepted. In order to ensure this happens staff need to be educated and trained to react in an appropriate manner to situations in which the organization will be committed to subsequently honouring its obligations. At one level this means that staff stop, think and check before accepting an order. At another level, this means that staff seek out someone else to perform the checks so that there is another pair of eyes focused on the requirements. At a high level, this means that a review panel is assembled and the requirements debated, and all issues resolved before the authorized signatory signs the contract. One means of helping staff to react in an appropriate manner is to provide forms with provision for a requirement review box that has to be checked or signed and dated before the process may continue. With computer-based systems, provision can also be made to prevent the transaction being completed until the correct data has been entered. This process is needed also for any amendments to the contract or order so that the organization takes the opportunity to review its capability with each change.

Ensuring that product requirements are defined (7.2.2a)

The standard requires that the review of requirements *ensure product requirements are defined.*

What does this mean?
This means that the review should verify that all the requirements specified by the customer, the regulators and the organization have been defined, there are no omissions, no errors, no ambiguities and no misunderstandings.

Why is this necessary?
This requirement responds to the Factual Approach Principle.

During the requirement determination process there are many variables. The time allowed, the competence of the personnel involved, the knowledge of the customer of what is needed and the accessibility of information. A deficiency in any one of these can result in the inadequate determination of requirements. It is therefore necessary to subject these requirements to review to ensure they are correct before a commitment to supply is made.

How is this implemented?
Some organizations deal with orders that are so predictable that a formal documented review before acceptance adds no value. But however predictable the order, it is prudent to establish that it is what you believe it to be before acceptance. Many have been caught out by the small print in contracts or sales agreements such as the following wording: "This agreement takes precedence over any conditions of sale offered by the supplier" or "Invoices must refer to the order reference otherwise they will be rejected".

If the customer is choosing from a catalogue or selecting from a shelf of products, you need to ensure that the products offered for sale are properly described. Such descriptions must not be unrepresentative of the product otherwise you may be in breech of national laws and statutes. In other situations you need some means of establishing that the customer requirements are adequate.

One means of doing this is to use checklists that prompt the reviewers to give proper consideration to important aspects before accepting contracts. Another method is to subject the requirements to an independent review by experts in their field, thus ensuring a second pair of eyes scans the requirements for omissions, ambiguities and errors.

Resolving differences (7.2.2b)

The standard requires the review to *ensure that contract or order requirements differing from those previously expressed are resolved*.

What does this mean?
Previously expressed requirements are those that may have been included in an invitation to tender issued by the customer. Whether or not you have submitted a formal tender, any offer you make in response to a requirement is a kind of tender. Where a customers needs are stated and you offer your product, you are implying that it responds to your customers stated needs. You need to ensure that your 'tender' is compatible with your customers needs otherwise the customer may claim you have sold a product that is not 'fit for purpose'.

Why is this necessary?
This requirement responds to the Customer Focus Principle.

In situations where the organization has responded to an invitation to tender for a contract, it is possible that the contract when it arrives may differ from the draft conditions against which the tender was submitted. It is therefore necessary to check whether any changes have been made that will affect the validity of the tender. Customers should indicate the changes that have been made but they often don't – they would if there was a mutually beneficial relationship in place.

How is this implemented?
On receipt of a contract that has been the subject of an invitation to tender, or the subject of an unsolicited offer of product or service, it is prudent to check that what you are now being asked to provide is the same as that which you offered. If the product or service you offer is in any way different than the requirement, you need to point this out to your customer and reach agreement before you accept the order. Try and get the contract changed, but if this is not possible, record the differences in your response to the contract. Don't rely on verbal agreements because they can be conveniently forgotten when it suits one party or the other, or as is more common, the person you conversed with moves on and the new person is unable to act without written agreement – such is the world of contracting!

Ensuring that the organization has the ability to meet defined requirements (7.2.2c and 7.2.2.2)

The standard requires that the review *ensure that the organization has the ability to meet defined requirement* and *to investigate, confirm and document the manufacturing feasibility including the analysis of risks of the proposed product when reviewing requirements related to the product.*

What does this mean?

The organization needs to be able to honour its obligations made to its customers. Checks therefore need to be made to ensure that the necessary resources including plant, equipment, facilities, technology, personnel, competency and time are available or will be available to discharge these obligations when required.

Often the contract for design and development and the contract for production are two separate contracts. They may be placed on the same organization but it is not unusual for the production contract to be awarded to an organization that did not design the product for cost reasons. This requirement is concerned with business capability rather than process capability and addresses the question, do we as a business have the capability to make this product in the quantity required and deliver it in the condition required to the destination required over the time period required and within the price required.

Why is this necessary?

This requirement responds to the Leadership Principle.

You must surely determine that you have the necessary capability before accepting the contract as to find out afterwards that you haven't the capability to honour your obligations could land you in deep trouble. There may be penalty clauses in the contract or the nature of the work may be such that the organization's reputation could be irrevocably damaged as a result. This requirement demands an analytical approach and objective evidence to show that it's feasible to meet the requirements. When the customer awards such a contract, there is a commitment on both sides and if the supplier is later found to be unable to deliver, the customer suffers – it puts the programme in jeopardy.

How is this implemented?

Prior to accepting a contact to manufacture a product of proven design, a business risk assessment is necessary as a safeguard against programme failure. The risk assessment should answer the following questions:

- What new technical capabilities will be needed for us to make this product?
- Will we be able to develop the additional capabilities within the timescales permitted?
- Can we make this product in the quantities required and in the timescales required?

- In consideration of the timescales and quantities required, can we make this product at a price that will provide an acceptable profit?
- Do we have the slack in our capacity to accommodate this programme at this time?
- Do we have the human resources required?
- If we require additional human resources can we acquire them and train them to the level of competence required in the timescales?
- Is there a sufficient supply of materials and components available in order to resource the production processes?
- Do we (or our partners) have the capability to transport this product to the required destination and protect it throughout the journey?

If you don't have any of the above, you will need to determine the feasibility of acquiring the relevant licence, the skills, the technology, etc. within the timescale. Many organizations do not need staff waiting for the next contract. It is a common practice for companies to bid for work for which they do not have the necessary numbers of staff. However, what they need to ascertain is from where and how quickly they can obtain the appropriate staff. If a contract requires specialist skills or technologies that you don't already possess, it is highly probable that you will not be able to acquire them in the timescale. It is also likely that your customer will want an assurance that you have the necessary skills and technologies before the contract is placed. No organization can expect to hire extraordinary people at short notice; in fact, all you can expect to be available are average people and you may well have no choice than to accept less than average people. With good management skills and a good working environment you may be able to get these average people to do extraordinary things but it is not guaranteed!

A sales person who promises a short delivery to win an order invariably places an impossible burden on the company. A company's capability is not increased by accepting contracts beyond its current level of capability. You need to ensure that your sales personnel are provided with reliable data on the capability of the organization do not exceed their authority and always obtain the agreement of those who will execute the contractual conditions before their acceptance.

In telephone sales transactions or transactions made by sales personnel alone, the sales personnel need to be provided with current details of the products and services available, the delivery times, prices and procedures for varying the conditions.

Maintaining records of product requirement reviews (7.2.2)

The standard requires *the results of the review and actions as a consequence of the review to be recorded (see Clause 4.2.4).*

What does this mean?
A requirement review is an action that generates an outcome and this requirement means that the outcome of the review should be recorded. The outcome may be a decision in

which case, a record of the decision is all that is necessary. However, the outcome could be a list of actions to be executed to correct the definition of requirement, or a list of concerns that need to be addressed. If the review is conducted with customer's representatives present, records of the review could include modifications, interpretations and correction of errors that may be held back until the first contract amendment. In such cases the review records act as an extension to any contract.

Why is this necessary?
This requirement responds to the Factual Approach Principle.

For both new product development and order processing, records of the review are necessary as a means of recalling accurately what took place or as a means of reference in the event of a dispute or to distribute to those having responsibility for any actions that have been agreed. During the processing of orders and contracts, records of the requirement review indicate the stage in the process that has been reached and are useful if the process is interrupted for any reason.

How is this implemented?
There should be some evidence that a person with the authority to do so has accepted each product requirement, order or contract. This may be by signature or by exchange of letters or e-mails. You should also maintain a register of all contracts or orders and in the register indicate which were accepted and which declined for use when assessing the effectiveness of the sales process. If you prescribe in your contract acquisition procedures the criteria for accepting a contract; the signature of the contract or order together with this register can be adequate evidence of requirement review. If requirement reviews require the participation of several departments in the organization, their comments on the contract, minutes of meetings and any records of contract negotiations with the customer will represent the records of product requirement review. It is important, however, to be able to demonstrate that the requirement being executed was reviewed for adequacy, differences in the tender and for supplier capability, before work commenced. The minimum you can have is a signature accepting an assignment to do work or supply goods but you must ensure that those signing the document know what they are signing for. Criteria for accepting orders or contracts can be included in the appropriate procedures. It cannot be stressed too strongly the importance of these actions. Most problems are caused by the poor understanding or poor definition of requirements.

Handling undocumented statements of requirements (7.2.2)

The standard requires *that where the customer provides no documented statement of requirement, the customer requirements are to be confirmed by the organization before acceptance.*

What does this mean?

Customers often place orders by telephone or in face-to-face transactions where no paperwork passes from the customer to the organization. Confirmation of customer requirements is an expression of the organization's understanding of the obligations it has committed to honour.

Why is this necessary?

This requirement responds to the Customer Focus Principle.

Confirmation is necessary because when two people talk, it is not uncommon to find that although they use the same words, they each interpret the words differently. Confirming an understanding will avoid problems later. Either party to the agreement could move leaving the successors to interpret the agreement in a different way.

How is this implemented?

The only way to implement this requirement is for the organization to send a written acknowledgement to the customer confirming the requirements that form the basis of the agreement. In this way there should be no ambiguity, but if later the customer appears to be requiring something different, you can point to the letter of confirmation. If you normally use e-mail for correspondence, obtain an e-mail receipt that it has been read (not merely *received* as it could be overlooked) otherwise always send confirmation by post as e-mails can easily be inadvertently lost or deleted. Keep a copy of the e-mail and the letter, and bring them under records control. Saving specific e-mails as text files in an appropriate directory on the server is better than simply keeping them on your local hard drive as a message in Outlook Express or similar program.

Changes to product requirements (7.2.2)

The standard requires that *where product requirements are changed, the organization ensure that relevant documents are amended and that relevant personnel are made aware of the changed requirements.*

What does this mean?

Product requirements may be changed by the customer, by the regulators or by the organization itself, and this may be made verbally or by changing the affected product requirement documents. This requirement means that all documents affected by the change are amended and that the changes are transmitted to those who need to know.

Why is this necessary?

This requirement responds to the Factual Approach Principle.

When changes are made to product requirements, the documents defining these requirements need to be changed otherwise those using them will not be aware of the changes. Also, changes to one document may have an impact on other related documents and unless these too are changed, the users will be working with obsolete information. It is therefore necessary to promulgate changes in a way that users are able to achieve the desired results.

How is this implemented?

In some organizations *product requirement change control* is referred to as *configuration management.*[1] Once a baseline set of requirements has been agreed, any changes to the baseline need to be controlled such that accepted changes are promptly made, and rejected changes are prevented from being implemented. If there is only one product specification and no related information, configuration management is similar to document control (see Chapter 4). When there are many specifications and related information, configuration management introduces a further dimension of having to control the compatibility between all the pieces of information. Document control is concerned with controlling the information carriers, whereas configuration management is concerned with controlling the information itself. If a system parameter changes there may be a knock-on effect through the subsystems, equipments and components. The task is to identify all the items affected and as each item will have a product specification, this task will result in a list of affected specifications, drawings, etc. While the list looks like a list of documents, it is really a list of items that are affected by the change. The requirement of ISO/TS 16949 makes it appear a simple process but before documents are amended, the impact on each item may vary and have to be costed before being implemented. Major redesign may be necessary; tooling, handling equipment, distribution methods, etc. may be affected; and therefore, the process of not only communicating the change, but also communicating the effects and the decisions relating to the change is an essential part of configuration management. It is not so much as who should be informed as what is affected. Identify what is affected and you should be able to identify who should be informed.

Customer communication (7.2.3 and 7.2.3.1)

Providing product information (7.2.3a)

The standard requires the organization to *determine and implement effective arrangements for communicating with customers relating to product information.*

What does this mean?

Product information could be in the form of advertising material, catalogues, a web site, specifications or any medium for promoting the organization's products and services. Effective arrangements would be the processes that identified, planned, produced and distributed information that accurately describes the product.

Why is this necessary?
This requirement responds to the Customer Focus Principle.

Customers are only aware of product information that is accessible but whether they receive it or retrieve it, it must accurately represent the products and services offered otherwise, it is open to misrepresentation and liable to prosecution in certain countries. It is therefore necessary to employ an effective process for communicating production information.

How is this implemented?
This requirement is concerned with the quality of information available to customers and has two dimensions. There is the misleading of customers into believing a product or service provides benefits that it cannot deliver (accuracy) and there is the relationship between information available to customers and information as would need to be to properly represent the product (compatibility).

Accuracy depends on getting the balance right between imaginative marketing and reality. Organizations naturally desire to present their products and services in the best light – emphasizing the strong points and playing down or omitting the weak points. Providing the omissions are not misleading to the customer, this is legitimate. What is needed is a product advertising process that ensures product information accurately represents the product and does not infringe advertising regulations and sale of goods laws.

Compatibility depends on maintaining the product information once it has been released. Product information takes many forms and keeping all of it compatible is not an easy task. An information control process is therefore needed to ensure that information compatibility is maintained when changes are made.

Handling enquiries (7.2.3a)

The standard requires the organization to *determine and implement effective arrangements for communicating with customers relating to enquiries.*

What does this mean?
Customer enquiries are the result of the effectiveness of the marketing process. If this has been successful, customers will be making contact with the organization to seek more information, clarify price, specification or delivery or request tenders, proposals or quotations.

Why is this necessary?
This requirement responds to the Customer Focus Principle.

If the personnel receiving a customer enquiry are uninformed or not competent to deal effectively with enquiries, a customer may not receive the treatment intended by the organization and either go elsewhere or be misled. Both situations may result in lost business and dissatisfied customers.

How is this implemented?

An enquiry-handling process is needed as part of the sales process that ensures customers are fed correct information and treated in a manner that maximizes the opportunity of a sale. Enquiries should be passed through a process that will convert the enquiry into a sale. As the person dealing with the enquiry could be the first contact the customer has with the organization, it is vital that they are competent to do the job. Frequent training and monitoring is therefore necessary to prevent customer dissatisfaction. This has become more apparent with telephone sales where a recorded message informs the customer that the conversation may be monitored for quality assurance purposes. As with all processes, you need to establish what you are trying to achieve, what affects your ability to get it right and how you will measure your success. The potential for error is great, whether customers are dealing with someone in person or cycling through a menu during a telephone transaction or on the Internet. Both human and electronic enquiry-handling processes need to be validated regularly to ensure their continued effectiveness. A typical enquiry conversion process flow is illustrated in Figure 7.5.

Handling contracts and orders (7.2.3b)

The standard requires the organization to *determine and implement effective arrangements for communicating with customers relating to contracts or order handling.*

What does this mean?

When an order or contract is received, several activities need to be performed in addition to the determination and review of product requirements and in each of these activities, communication with the customer may be necessary to develop an understanding that will secure an effective relationship.

Why is this necessary?

This requirement responds to the Customer Focus Principle.

Customer enquiries may or may not result in orders. However, when an order or a contract is received, it is necessary to pass it through an effective process that will ensure both parties are in no doubt as to the expectations under the contract.

How is this implemented?

A process should be established for handling orders and contracts with the objective of ensuring both parties are in no doubt as to the expectations under the contract before work commences. A typical order-processing process is illustrated in Figure 7.6.

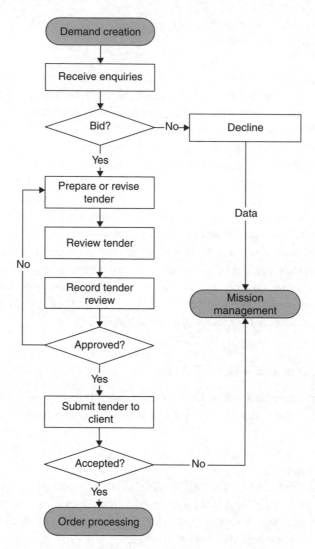

Figure 7.5 *Enquiry conversion process flow*

Apart from the requirement determination and reviews stages, there will be:

- order or contract registration (recording its receipt);
- order or contract acknowledgment (informing the customer the order has been received and that the organization intends or does not intend to offer a bid or supply a product or service);
- requirement determination (as addressed previously);
- requirement review (as addressed previously);

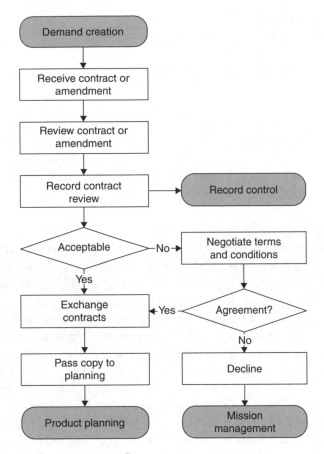

Figure 7.6 *Order-processing process flow*

- order or contract negotiation;
- order or contract acceptance;
- order or contract communication.

Handling contract amendments (7.2.3b)

The standard requires the organization to *determine and implement effective arrangements for communicating with customers relating to amendments.*

What does this mean?
An order or contract amendment is a change that corrects errors, rectifies ambiguities or otherwise makes improvements.

Why is this necessary?

This requirement responds to the Customer Focus Principle.

The need for an amendment can arise at any time and be initiated by either party to an agreement.

As orders and contracts are primarily a source of reference, it is necessary to ensure that only agreed amendments are made and that any provisional amendments or disagreed amendments are not acted on or communicated as though they were approved. Otherwise, the basis of the agreement becomes invalid and may result in a dissatisfied customer or an organization that cannot recover its costs.

How is this implemented?

There may be several reasons why a customer needs to amend the original contract – customer needs may change, your customer's customer may change the requirement or details unknown at the time of contract may be brought to light. Whatever the reasons you need to provide a process for amending existing contracts under controlled conditions. On contracts where direct liaison with the customer is permitted between several individuals, e.g. a project manager, contract manager, design manager, procurement manager, manufacturing manager, quality assurance manager, it is essential to establish ground rules for amending contracts, otherwise your company may unwittingly be held liable for meeting requirements beyond the funding that was originally predicted. It is often necessary to stipulate that only those changes to contract that are received in writing from the contract authority of either party will be legally binding. Any other changes proposed, suggested or otherwise communicated should be regarded as being invalid. Agreement between members of either project team should be followed by an official communication from the contract authority before binding either side to the agreement.

Having officially made the change to the contract, a means has to be devised to communicate the change to those who will be affected by it. You will need to establish a distribution list for each contract and ensure that any amendments are issued on the same distribution list. The distribution list should be determined by establishing who acts on information in the contract and may include the managers of the various functions that are involved with meeting the contract or order requirements. Once established the distribution list needs to be under control because the effect of not being informed of a change to contract may well jeopardize delivery of conforming product.

Customer feedback (7.2.2c)

The standard requires the organization to *determine and implement effective arrangements for communicating with customers relating to customer feedback including customer complaints.*

What does this mean?

Customer feedback is any information conveyed by the customer in relation to the quality of the products or services provided. Sometimes this may be positive in the form of compliments, praise, gifts or tips and other times the feedback could be negative in the form of a complaint or an expression of disapproval.

Why is this necessary?

This requirement responds to the Customer Focus Principle.

Without an effective process for capturing customer feedback the organization would be missing opportunities for improving its performance as it is perceived to be by its customers. While not an accurate measure of customer satisfaction, customer feedback provides objective evidence that can be used in such an assessment.

How is this implemented?

You can only handle effectively the customer feedback that you receive and record. Customers may complain about your products and services but not go to the extent of writing a formal complaint. They may also compliment you on your products and services but again not bother to put it in writing. Compliments and complaints may arise in conversation between the customer and your sales and service staff, and this is where you need to instil discipline and ensure they are captured. The primary difference between compliments and complaints is that compliments deserve a thank you and complaints deserve action; therefore, the processes for dealing with compliments and complaints will differ.

The complaint-handling process should cover the following aspects to be effective:

- A definition of when a message from a customer can be classified as a complaint.
- The method of capturing the customer complaints from all interface channels with the customer.
- The behaviour expected from those on the receiving end of the complaint.
- The registration of complaints in order that you can account for them and monitor progress.
- A form on which to record details of the complaint, the date, customer name, etc.
- A method for acknowledging the complaint in order that the customer knows you care.
- A method for investigating the nature and cause of the complaint.
- A method for replacing product, repeating the service or for compensating the customer.
- A link with other processes to trigger improvements that will prevent a recurrence of the complaint.

The compliment-handling process should cover the following:

- A definition of when a message from a customer can be classified as a compliment.
- The method of capturing the compliments from all interface channels with the customer.
- The behaviour expected from those on the receiving end of the compliment.
- The registration of compliments in order that you can account for those you can use in your promotional literature.
- A method for keeping staff informing of the compliments made by customers.
- A method of rewarding staff when compliments result in further business.

Language and format (7.2.3.1)

The standard requires the organization *to possess the ability to communicate the necessary information in a customer-specified language and format.*

What does this mean?

Although the requirement does not specify that the data format be electronic, much of the information automotive customers require is required to be transmitted electronically so that they can manipulate it, import into reports and retrieve it quickly through search engines. Format also applies to the layout of reports such as Control Plans and FMEA as well as the software programs used to generate these documents. Language applies not only to the spoken language of the supplier and customer but also to computer software.

Why is this necessary?

The pressure on customers to get product to market ahead of their competitors makes paper-based communication, storage and retrieval systems no longer practical. The increase in speed that electronic data import and export functions has created makes electronic transmission essential.

How is this implemented?

For electronic data transmission to work effectively, there has to be compatibility between customer and supplier computer systems in terms of size, connectivity, software and version control.

Even with common software programs, problems can arise that causes a breakdown in the interface. It is not uncommon to find advanced organizations using versions of word processing and spreadsheet software that are several years behind those of much smaller organizations. The customer may not be using the latest software simply because of the cost of upgrading is far greater proportionally to the small business.

You need to agree file naming conventions and version control conventions. When saving files after making modifications it is often best to change the file name otherwise, the recipient might overwrite a file when he or she intended to preserve the previous version.

If the spoken language of the customer is different to that of the supplier the engineering staff need to speak the customer's language but at some stage, translation will be necessary so that work instructions can be in the native tongue. Customer requirements need to be translated so that they do not lose their intent and accuracy at the point of implementation. This may require a translation control function through which important communications are processed or an assurance function that audits the accuracy of important communications.

Design and development (7.3)

Design and development control (7.3.1)

The standard requires the organization to *control design and development of the product and of manufacturing processes*.

What does this mean?

With the addition of manufacturing processes, this requirement now has two dimensions: product design and process design, and while both have many similarities, there are also some differences. It is not intended that the word process replace the word product in Section 7.3. Those requirements of Section 7.3 that apply to process design are indicated separately, so if it is not stated that the requirement refers to product design or process design, assume it refers to product design.

Product design can be as simple as replacing the motor in an existing vehicle with one of a different specification, or as complex as the design of a new automobile or any of its subsystems. Product design can be of hardware, software (or a mixture of both) and can be of new services or modified services. Before design commences there is either a requirement or simply an idea. Product design is a creative process that creates something tangible out of an idea or a requirement. The controls specified in the standard apply to the product and process design process. There are no requirements that will inhibit creativity or innovation. In order to succeed, the process of converting an idea into a design that can be put into production or service has to be controlled. Product design is often a process which strives to set new levels of performance, new standards or create new wants and as such can be a journey into the unknown. On such a journey we can encounter obstacles we haven't predicted which may cause us to change our course but our objective remains constant. Design control is a method of keeping the design on course towards its objectives and as such will comprise all the factors that may prevent the design from achieving its objectives. It controls the process not the designer, i.e. the inputs, the outputs, the selection of components, standards, materials, processes, techniques and technologies.

To control any design activity be it either product or process design, there are 10 primary steps you need to take in the design process:

1 Establish the customer needs.
2 Convert the customer needs into a definitive specification of the requirements.
3 Plan for meeting the requirements.
4 Organize resources and materials for meeting the requirements.
5 Conduct a feasibility study to discover whether accomplishment of the requirements is feasible.
6 Conduct a project definition study to discover which of the many possible solutions will be the most suitable.
7 Develop a specification which details all the features and characteristics of the product or service.
8 Produce a prototype or model of the proposed design.
9 Conduct extensive trials to discover whether the product, service or process which has been developed meets the design requirements and customer needs.
10 Feed data back into the design and repeat the process until the product, service or process is proven to be fit for the task.

Control of design and development does not mean controlling the creativity of the designers – it means controlling the process through which new or modified designs are produced so that the resultant design is one that truly reflects customer needs.

Why is this necessary?
This requirement responds to the Process Approach Principle.

Without control over the design and development process several possibilities may occur:

- Design will commence without an agreed requirement.
- Costs will escalate as designers pursue solutions that go beyond what the customer really needs.
- Costs will escalate as suggestions get incorporated into the design without due consideration of the impact on development time and cost.
- Designs will be released without adequate verification and validation.
- Designs will be expressed in terms that cannot be implemented economically in production or use.

The bigger the project, the greater the risk that the design will overrun budget and timescale. Design control aims to keep the design effort on course so that the right design is released on time and within budget.

How is this implemented?

Control of the design and development requires the application of the same principles as any other process. The standard actually identifies the controls that need to be applied to each design but there are other controls that need to be applied to the design process in order to apply the requirements of ISO/TS 16949 Clause 4.1. A typical product design process flow chart is illustrated in Figure 7.7 with process design illustrated in Figure 7.8.

The design process is a key process in enabling the organization to achieve its objectives. These objectives should include goals that apply to the design process (see Chapter 5 – Clause 5.4.1). Consequently there need to be:

- objectives for the design process;
- measures for indicating achievement of these objectives;
- a defined sequence of subprocesses or tasks that transform the design inputs into design outputs;
- links with the resource management process so that human and physical resources are made available to the design process when required;
- review stages for establishing that the process is achieving its objectives;
- processes for improving the effectiveness of the design process.

Design and development planning (7.3.1)

Preparing the plans (7.3.1)

The standard requires the organization to *plan design and development of the product.*

What does this mean?

Planning the design and development of a product means determining the design objectives and the design strategy, the design stages, timescales, costs, resources and responsibilities needed to accomplish them. Sometimes the activity of design itself is considered to be a planning activity but what is being planned is not the design but the product.

Why is this necessary?

This requirement responds to the Leadership Principle.

The purpose of planning is to determine the provisions needed to achieve an objective. In most cases, these objectives include not only a requirement for a new or modified product but also requirements governing the costs and product introduction timescales (Quality, Cost and Delivery or QCD). Remove these constraints and planning becomes less important but there are few situations when cost and time are not the constraints. It is therefore

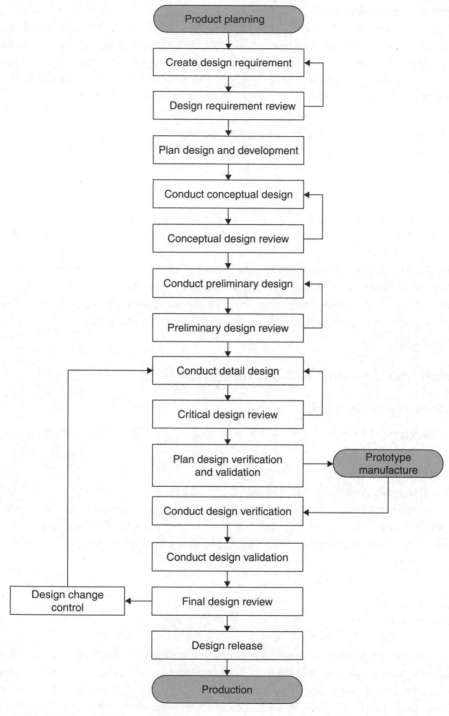

Figure 7.7 *Product design process flow*

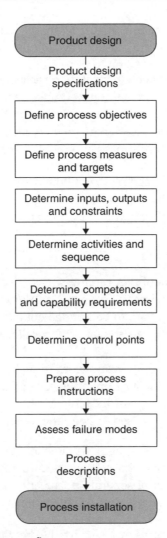

Figure 7.8 *Process design process flow*

necessary to work out in advance whether the objective can be achieved within the budget and timescale. One problem with design is that it is often a journey into the unknown and the cost and time it will take cannot always be predicted. Without a best guess some projects would not get underway so planning is a vital first step to get the funding and secondly to define the known and unknown so that risks can be assessed and quantified.

How is this implemented?
You should prepare a design and development plan for each new design and also for any modification of an existing design that radically changes the performance of the

product or service. For modifications that marginally change performance, control of the changes required may be accomplished through the design change process.

Design and development plans need to identify the activities to be performed, by whom they will be performed and when they should commence and should be completed. One good technique is to use a network chart (often called a PERT chart), which links all the activities together. Alternatively a bar chart may be adequate. There does need to be some narrative in addition as charts in isolation rarely convey everything required.

Design and development is not complete until the design has been proven as meeting the design requirements, so in drawing up a design and development plan you will need to cover the planning of design verification and validation activities. The plans should identify as a minimum:

- The design requirements.
- The design and development programme showing activities against time.
- The work packages and names of those who will execute them. (Work packages are the parcels of work that are to be handed out either internally or to suppliers.)
- The work breakdown structure showing the relationship between all the parcels of work.
- The reviews to be held for authorizing work to proceed from stage to stage.
- The resources in terms of finance, manpower and facilities.
- The risks to success and the plans to minimize them.
- The controls that will be exercised to keep the design on course.

Planning for all phases at once can be difficult as information for subsequent phases will not be available until earlier phases have been completed. So, your design and development plans may consist of separate documents, one for each phase and each containing some detail of the plans you have made for subsequent phases.

Your design and development plans may also need to be subdivided into plans for special aspects of the design such as reliability plans, safety plans, electromagnetic compatibility plans and configuration management plans. With simple designs there may be only one person carrying out the design activities. As the design and development plan needs to identify all design and development activities, even in this situation you will need to identify who carries out the design, who will review the design and who will verify the design. The same person may perform both the design and the design verification activities; however, it is good practice to allocate design verification to another person or organization because it will reveal problems overlooked by the designer. On larger design projects you may need to employ staff of various disciplines such as mechanical engineers, electronic engineers, reliability engineers, etc. The responsibilities of all these people or groups need to be identified and a useful way of parcelling up the work is to use work packages that list all the activities to be performed by a particular group. If you subcontract any of the design activities, the supplier's plans need to be integrated with your plans and your plan should identify which activities are the supplier's responsibility. While purchasing is dealt with in Clause 7.4 of the standard, the requirements also apply to design activities.

Stages of design and development process (7.3.1a)

The standard requires *the stages of design and development to be determined*.

What does this mean?

A stage in design and development is a point at which the design reaches a phase of maturity. There are several common stages in a design process. The names may vary but the intent remains the same:

- *Feasibility stage*: The stage during which studies are made of a proposed object-ive to determine whether practical solutions can be developed within time and cost constraints. This stage usually terminates with a design brief.
- *Conceptual design stage*: The stage during which ideas are conceived and theor-ies tested. This stage usually terminates with a preferred solution in the form of a design requirement.
- *Design definition stage*: The stage during which the architecture or layout takes form and the risks assessed and any uncertainty resolved. This stage usually terminates with definitive design specifications for the components comprising the product.
- *Detail design stage*: The stage during which final detail characteristics are deter-mined and method of production or delivery established. This stage usually termin-ates with a set of specifications for the construction of prototypes. Process design and development will also commence as soon as engineering drawings and tooling requirements are released.
- *Development stage*: The stage during which the prototype is proven using models or simulations and refined. This stage usually terminates with a set of approved specifications for the construction, installation and operation of the product. Process validation also occurs during this phase.

Why is this necessary?

This requirement responds to the Process Approach Principle.

Any endeavour is more easily accomplished when undertaken in small stages. By pro-cessing a design through several iterative stages, a more robust solution will emerge than if the design is attempted in one cycle.

How is this implemented?

In drawing up your design and development plans you need to identify the principal activities and a good place to start is with the list of ten steps detailed previously of which the last five are explained further above. Any more detail will in all probability be a breakdown of each of these stages initially for the complete design and subsequently for each element of it. If dealing with a system you should break it down into subsystems, and the subsystems into equipments and equipments into assemblies, and so on. It is most important that you agree the system hierarchy and associated terminology early on in the development programme otherwise you may well cause both technical and organizational problems at the interfaces.

Planning review, verification and validation activities (7.3.1b)

The standard requires *the review, verification and validation activities appropriate to each design and development stage to be determined.*

What does this mean?

Each design stage is a process that takes inputs from the previous process and delivers outputs to the next stage. Within each process are verification, validation and review points that feedback information into the process to produce a further iteration of the design. At the end of each stage the output needs to be verified, validated and reviewed before being passed on the next stage. The further along the design cycle, the more rigorous and complex the verification, validation and review stages will need to be. The verification stages are those stages where design output of a stage is checked against the design input for that stage to ensure the output is correct. The validation stages occur sequentially or in parallel to confirm that the output is the right output by comparing it with the design brief or requirement. The review stages are points at which the results of verification and validation are reviewed to confirm the design solution, recommend change and authorize or halt further development.

Why is this necessary?

This requirement responds to the Factual Approach Principle.

The checks necessary to select and confirm the design solution need to be built into the design process so that they take place when they will have the most beneficial effect on the design. Waiting until the design is complete before verification, validation or review will in all probability result in extensive rework and abortive effort.

How is this implemented?

The stages of verification, validation and review should be identified in the design and development plan, but at each stage there may need to be supplementary plans to contain more detail of the specific activities to be performed. This may result in a need for a separate design verification plan.

The design verification plan should be constructed so that every design requirement is verified and the simplest way of confirming this is to produce a verification matrix of requirement against verification methods. Another matrix in a similar form is a Quality Function Deployment (QFD) chart. You need to cover all the requirements, those that can be verified by test, by inspection, by analysis, by simulation or demonstration or simply by validation of product records. For those requirements to be verified by test, a test specification will need to be produced. The test specification should specify which characteristics are to be measured in terms of parameters, limits and the conditions under which they are to be measured.

The verification plan needs to cover some or all of the following details as appropriate:

- A definition of the product design standard that is being verified.
- The objectives of the plan. (You may need several plans covering different aspects of the requirement.)
- Definition of the specifications and procedures to be employed for determining that each requirement has been achieved.
- Definition of the stages in the development phase at which verification can most economically be carried out.
- The identity of the various models that will be used to demonstrate achievement of design requirements. (Some models may be simple space models, others laboratory standard or production standard depending on the need.)
- Definition of the verification activities that are to be performed to qualify or validate the design and those which need to be performed on every product in production as a means of ensuring that the qualified design standard has been maintained.
- Definition of the test equipment, support equipment and facilities needed to carry out the verification activities.
- Definition of the timescales for the verification activities in the sequence in which the activities are to be carried out.
- Identification of the venue for the verification activities.
- Identification of the organization responsible for conducting each of the verification activities.
- Reference to the controls to be exercised over the verification activities in terms of the procedures, specifications and records to be produced, the reviews to be conducted during the programme and the criteria for commencing, suspending and completing the verification operations. (Provision should also be included for dealing with failures, their remedy, investigation and action on design modifications.)

As part of the verification plan, you should include an activity plan that lists all the planned activities in the sequence they are to be conducted and use this plan to progressively record completion and conformance. The activity plan should make provision for planned and actual dates for each activity and for recording comments such as recovery plans when the programme does not proceed exactly as planned. It is also good practice to conduct test reviews before and after each series of tests so that corrective measures can be taken before continuing with abortive tests (see also under the heading *Design validation*).

The designers and those performing the verification activities should approve the verification plan. Following approval the document should be brought under document control. Design verification is often a very costly activity and so any changes in the plan should be examined for their effect on cost and timescale. Changes in the specification can put back the programme by months whilst new facilities are acquired, new jigs, cables, etc. procured. However small your design, the planning of its verification is vital to the future of the product. Lack of attention to detail can rebound months (or even years) later during production.

Determining responsibilities and authority for design and development activities (7.3.1c)

The standard requires *the responsibilities and authorities for design and development activities to be determined.*

What does this mean?
To cause the activities in the design and development plan to happen, they have to be assigned to either a person or an organization. Once assigned and agreed by both parties, the assignee becomes responsible for delivering the required result. The authority delegated in each assignment conveys a right to the assignee to make decisions affecting the output. The assignee becomes the design authority for the items designed but this authority does not extend to changing the design requirement – this authority is vested in the organization that delegated or sponsored the design for the item.

Why is this necessary?
This requirement responds to the Leadership Principle.

Responsibility for design activities needs to be defined so that there is no doubt as to who has the right to take which actions and decisions. Authority for design activities needs to be delegated so that those responsible have the right to control their own output. Also the authority responsible for the requirements at each level of the design needs defining so that there is a body to which requests for change can be routed. Without such a hierarchy, there would be anarchy resulting in a design that failed to fulfil its requirements.

How is this implemented?
Within the design and development plan the activities need to be assigned to a person, group or organization equipped with the resources to execute them. Initially the feasibility study may be performed by one person or one group, but as the design takes

shape, other personnel are required and possibly other external organizations may be required to undertake particular tasks or products.

One way of assigning responsibilities is to use the work package technique. With this approach you can specify not only what is to be done but estimate the required hours, days, months or years to do it and then obtain the group's acceptance and consequently commitment to the task.

One of the difficulties with assigning design work is ensuring that those to whom the work is assigned understand the boundary conditions, i.e. what is included and what is excluded (see below under the heading *Organizational interfaces*).

You also need to be careful that work is not delegated or subcontracted to parties about whom you have little knowledge. In subcontracts, clauses that prohibit subcontracting without your approval need to be inserted, thereby enabling you to retain control.

Managing organizational interfaces (7.3.1)

The standard requires *the interfaces between different groups involved in design and development process to be managed to ensure effective communication and clarity of responsibilities.*

What does this mean?
Where there are many different groups of people working on a design they need to work together to produce an output that meets the overall requirement when all outputs are brought together. To achieve this each party needs to know how the design work has been allocated and to which requirements each party is working so that if there are problems, the right people can be brought together.

Why is this necessary?
This requirement responds to the Leadership Principle.

If the interfaces between design groups are not properly managed, there are likely to be technical problems arising from groups changing interface requirement without communicating the changes to those affected, political problems arising from groups assuming the right to do work or make decisions that have been allocated to other groups and cost overruns arising from groups not communicating their difficulties when they are encountered.

How is this implemented?
You should identify where work passes from one organization to another and the means used to convey the requirements such as work instructions, work package descriptions or contracts. Often in design work, the product requirements are analysed to identify further

requirements for constituent parts. These may be passed on to other groups as input requirements for them to produce a design solution. In doing so these groups may in fact generate further requirements in the form of development specifications to be passed to other groups and so on. For example, the systems engineer generates the system speci-fication and subsystem specifications, and passes the latter to the subsystems engineers. These engineers design the subsystem and generate equipment specifications to pass on to the equipment engineers. To meet the equipment specification new parts may be neces-sary and so these engineers generate part specifications and pass these to the parts engi-neers. Some of these transactions may be in-house but some might be subcontracted. Some organizations only possess certain design capabilities and subcontract other design activities. In this way they concentrate on the business they are good at and get the best specialist support through competitive tenders. These situations create organizational interfaces that require effective information control processes.

In managing the organizational interfaces you will need to:

- define the customer and the supplier in the relationship;
- define the product requirements that the supplier is to meet (the objectives and outputs);
- define the work that the supplier is to carry out with the budget and time constraints;
- define the responsibility and authority of this work (who does what, who approved what);
- define the process used for conveying information and receiving feedback;
- define the reporting and review requirements for monitoring the work;
- conduct regular interface review meetings to check progress and resolve concerns;
- periodically review the interface control process for its effectiveness.

One mechanism of transmitting technical interface information is to establish and promulgate a set of baseline requirements that are to be used at commencement of design for a particular phase. Any change to these requirements should be processed by a *change control board or* similar body and following approval a change to the base-line is made. This baseline listing becomes a source of reference and if managed prop-erly ensures that no designer is without the current design and interface information.

Interfaces should be reviewed along with other aspects of the design at regular design reviews scheduled prior to the completion of each phase or more often if warranted. Where several large organizations are working together to produce a design, an interface control board or similar body may need to be created to review and approve changes to technical interfaces. Interface control is especially difficult with complex projects. Once underway an organization like a large ship gains momentum and takes some time to stop. The project manager may not know of everything that is happening. Control is largely by information and it can often have a tendency to be historical information by the time it reaches its destination. So it is important to control changes to the interfaces. If one small change goes unreported, it may cause months of delay correcting the error.

Ensuring that plans are updated as the design progresses (7.3.1)

The standard requires planning output to *be updated, as appropriate, as the design and development progresses.*

What does this mean?
Planning commences before work is performed but as work progresses and the unknown becomes the known, its direction may need to change and therefore the plans need to change.

Why is this necessary?
This requirement responds to the Factual Approach Principle.

The design and development plan is a source of reference for those on the project – it communicates the work to be performed, who is to perform it and when. It should form the basis on which the design and development costs have been estimated and the work is proceeding. Therefore, if the basis for the plan changes, the plan needs to change so that it reflects the design that is being produced and provides legitimacy for the actions and decisions being carried out. If the plan is not updated, those in possession of it may waste valuable effort in performing work that is no longer required or may not be able to provide the resources when needed and therefore additional costs may be incurred.

How is this implemented?
Some design planning needs to be carried out before any design commences, but it is an iterative process and therefore the design plans may be completed progressively as more design detail emerges. It is not unusual for plans to be produced and as design gets underway, problems are encountered which require a change in direction. When this occurs the original plans should be changed. The design and development plan should be placed under control after it has been approved. When a change in the plan is necessary you should use the document change request mechanism to change the design and development plan and not implement the change until the request has been approved. In this way you remain in control.

Multidisciplinary approach (7.3.1.1)

The standard requires the organization *to use a multidisciplinary approach to pre-pare for product realization.*

What does this mean?

A multidisciplinary approach is another term for a *cross-functional team* or a *project team*. Such teams comprise representatives from each line and staff department so that decisions are taken close to the development work by those who will need to implement the decisions or verify their implementation. Such teams facilitate communication and overcome delays that often occur when reliant on line-staff relationships.

Why is this necessary?

The project organization has been used for several decades as an effective means of organizing knowledge-based staff, pooling ideas, obtaining consensus, and making decisions that don't need to be sold to the line departments since they are usually well represented. They do have some disadvantages as several project teams may call on a single resource at the same time and this is where upper management need to prioritize projects. Also if standards for each project differ, errors can occur as staff juggle with different requirements for the same piece of work.

How is this implemented?

Most organizations are structured in the form of functional components each representing a primary discipline such as Marketing, Engineering, Purchasing, Production, Servicing, Quality, Human Relations, etc. within each function there is planning, organizing, doing, checking and acting (PDCA). Some do more planning than others and some do more checking than others; in fact, some check on the work that others do, or plan the work that others do with consequential overlap, conflict, duplication, omission and waste. This is the fate of functional structures but they suite many organizations. To overcome these weaknesses, the project structure or multidisciplinary team can be formed in which representatives from each of the function come together to work on a project. They come together not to do what they did in their functional departments but to contribute their knowledge and skills collectively to the resolution of problems. So in this regard, the requirements, specifications, solutions, tests, trials and changes would each have a champion supported by specialists who provide inputs, perform evaluations, set up experiments and then review, evaluate and critique the results.

The benefits of a multidisciplinary approach won't arise if the people come together, talk then leave to return to their functional silos. The people have to form a team that is focused on a common goal that has priority over departmental goals. A cross-functional approach sometimes makes the project goal subservient and thus fails to work.

Design and development input (7.3.2 and 7.3.2.1)

Determining and recording design inputs (7.3.2)

The standard requires inputs relating to product requirements to *be determined and records maintained (see 4.2.4). The automotive additions in Clause 7.3.2.1 also require design inputs to be reviewed and to include customer requirements, use of information from competitor analysis, supplier feedback and field data. Design inputs are also required to include targets for product quality, reliability, durability, maintainability, timing and cost.*

What does this mean?

The design inputs are the requirements governing the design of the intended product. They include all the requirements determined from an analysis of customer and regulatory requirements and the organization's requirements. It may appear that this requirement duplicates those addressed in Clause 7.2 of the standard, but as the design is decomposed into subsystems, equipments, components, materials and processes, design inputs are the inputs into the design of each of these levels and will therefore become more specific through the hierarchy. The records to be maintained are the resultant specifications that describe these requirements.

Why is this necessary?

This requirement responds to the Factual Approach Principle.

The design input requirements constitute the basis for the design without which there is no criteria to judge the acceptability of the design output.

How is this implemented?

Design inputs should reflect the customer, regulator and organization's needs and be produced or available before any design commences.

To identify design input requirements you need to identify the following factors:

- The purpose of the product or service.
- The conditions under which it will be used, stored and transported.
- The skills and category of those who will use and maintain the product or service.
- The countries to which it will be sold and the related regulations governing sale and use of products.
- The special features and performance characteristics which the customer requires the product or service to exhibit including life, reliability, durability and maintainability (see Chapter 1 for list of other typical features and characteristics).
- The constraints in terms of timescale, operating environment, cost, size, weight or other factors.
- The standards with which the product or service needs to comply.
- The products or services with which it will directly and indirectly interface, and their features and characteristics.
- The documentation required of the design output necessary to manufacture, procure, inspect, test, install, operate and maintain a product or service.

Organizations have a responsibility to establish their customer requirements and expectations. If you do not determine conditions that may be detrimental to the product and you supply the product as meeting the customer needs and it subsequently fails, the failure is your liability. If the customer did not provide reasonable opportunity for you to establish the requirements, the failure may be the customer's liability. If you think you may need some extra information in order to design a product that meets the customer needs, you must obtain it or declare your assumptions. A nil response is often taken as acceptance in full.

In addition to customer requirements there may be industry practices, national standards, company standards, the experience gained from previous designs and other sources of input to the design input requirements to be taken into account. The result of competitive analysis should be used to ensure product design requirements are not putting the product at a distinct disadvantage even before design commences. You should provide design guides or codes of practice that will assist designers identify the design input requirements that are typical of your business.

The design output has to reflect a product that is producible or a service that is deliverable. The design input requirements may have been specified by the customer and consequently not have taken into account your production capability. The product of the design may therefore need to be producible within your current production capability using your existing technologies, tooling, production processes, material-handling equipment, etc.

Having identified the design input requirements, you need to document them in a specification that when approved is brought under document control. The requirements should not contain any solutions at this stage so as to provide freedom and flexibility to the designers. If the design is to be subcontracted, it makes for fair competition and removes from you the responsibility for the solution. Where specifications contain solutions, the supplier is being given no choice and if there are delays and problems the supplier may have a legitimate claim against you.

Defining functional and performance requirements (7.3.2a)

The standard requires design inputs to *include functional and performance requirements*.

What does this mean?
Functional requirements are those related to actions that the product is required to perform with or without external stimulus. Performance requirements relate to the results or behaviours required by such actions under stated conditions. Normally a product's characteristics are stated in physical, functional and performance terms rather than

functional and performance but no matter, the intent of the requirement is that all characteristics that the product is required to exhibit should be included in the design input requirements and expressed in terms that are measurable.

Why is this necessary?
This requirement responds to the Process Approach Principle.

All the characteristics need to be stated otherwise the resultant design may not reflect a product that fulfils the conditions for intended use. Two products may possess the same physical and functional characteristics but perform differently due to the individual arrangement of their component parts and the materials and processes used in their construction.

How is this implemented?
From the statement of product purpose, the conditions of use and the skills of those who will use it, the most obvious characteristics can be derived and divided into physical, functional and performance requirements. The physical characteristics might include size, mass, appearance and material properties. Functional characteristics might include speed, power, capacity and a wide range of characteristics that give the product distinctive features. Performance characteristics might include reliability, maintainability, durability, flammability, portability, safety, etc. Sustainability may be a customer requirement; therefore, material and component selection criteria may need to be specified.

Defining statutory and regulatory requirements (7.3.2b)

The standard requires design inputs to *include applicable statutory and regulatory requirements.*

What does this mean?
At the end-product level, the applicable statutory and regulatory requirements are those addressed by Clause 7.2.1c. However, as the design unfolds additional statutory and regulatory requirements may become applicable as specific subsystems, equipments, components, materials and processes are identified.

Why is this necessary?
This requirement responds to the Process Approach Principle.

The statutory and regulatory requirements that apply are dependent on a range of factors that emerge during the design process once it is known what type of device is required. These regulations need to be identified in the design input so that the resultant design is proven to meet them before commitment to production is granted.

How is this implemented?

Statutory and regulatory requirements are those that apply in the country to which the product or service is to be supplied. Whilst some customers have the foresight to specify these, others often don't. Just because such requirements are not specified in the contract doesn't mean you don't need to meet them.

Statutory requirements may apply to the prohibition of items from certain countries, power supply ratings, security provisions, markings and certain notices.

Regulatory requirements may apply to health, safety and environmental emissions, electromagnetic compatibility, and these often require accompanying certification of compliance.

If you intend exporting the product or service, it would be prudent to determine the regulations that would apply before you complete the design requirement. Failure to meet some of these requirements can result in no export licence being granted as a minimum and imprisonment in certain cases if found to be subsequently non-compliant.

Defining information from previous designs (7.3.2c)

The standard requires the design inputs to *include applicable information derived from previous similar designs.*

What does this mean?

Most designs are a development of that which was designed previously. It is rare for a design to be completely new. Even if the product concept is new, it may contain design solutions used previously. The history of these previous designs contains a wealth of information that may be applicable to the application that is currently being considered.

Why is this necessary?

This requirement responds to the Continual Improvement Principle.

Using the lessons learnt from previous designs is corrective action – preventing the recurrence of problems that have occurred in the past. If previous design history is not utilized, the problems may recur.

How is this implemented?

In principle the design history of a product should be archived and made available to future designers. Design history can be placed in a database or library that is accessible to future designers. A rather old way of doing this was for companies to create design manuals containing data sheets, fact sheets and general information sheets on design topics – a sort of design guide that captured experience. Companies should still be doing this but many will by now have converted to electronic storage medium with the added advantage of a search engine. Information will also be available from trade associations, libraries and learned societies. Often professional journals, published literature and even newspapers can contain useful information for designers. In your model of the design process you need to install a research process that is initiated prior to commencing design of a system, subsystem, equipment or component. The research process needs to commence with an enquiry such as "Have we done this or used this before? Has anyone done this or used this before?". The questions should initiate a search for information but to make this structured approach, the database or libraries need to structure the information in a way that it will be used. One advantage of submitting the design to a review by those not involved in the design is that they bring their experience to the review and identify approaches that did not work in the past, or put forward more effective ways of doing such things in the future. A case for keeping the old designers on tap, if not on top!

Within the design input requirements, such information would appear as either as preferred solutions or non-preferred solutions, either directly or by reference to learned papers, standards, guide, etc.

Identifying other essential requirements (7.3.2d)

The standard requires design inputs to *include any other requirements essential for design and development.*

What does this mean?
In addition to the requirements identified there may be requirements that are dictated because of the organizational policies, national and international politics as was addressed under Clause 7.2.1d.

Why is this necessary?
This requirement responds to the Process Approach Principle.

The organization may wish to maintain a certain profile or reputation through its designs and therefore may impose requirements that may impact the design input requirements.

How is this implemented?
One specific series of requirement that may not emerge from the forgoing are technical interface requirements. Some of these may need to be written around a particular supplier. However, within each development specification the technical interfaces between systems, subsystems, equipments, etc. should be specified so that when all these components are integrated they function properly. In some situations it may be necessary to generate separate interface specifications defining requirements that are common to all components of the system. In a large complex design, minor details of a component may be extremely important in the design of another component.

Instead of providing designers with specifications of all the components, it may be more economical (as well as more controllable) if the features and characteristics at the interface between components are detailed in separate interface specifications.

Reviewing design input requirements (7.3.2)

The standard requires design inputs to *be reviewed for adequacy* and for the requirements *to be complete, unambiguous and not in conflict with other requirements.*

What does this mean?

Adequacy in this context means that the design input requirements are a true reflection of the customer needs.

Why is this necessary?

This requirement responds to the Factual Approach Principle.

The determination of design inputs results in information that needs to be reviewed prior to its release otherwise incorrect information may enter the design process. It is prudent to obtain customer agreement to the design requirements before commencing the design. In this way, you will establish whether you have correctly understood and translated customer needs. It is advisable also to hold an internal design review at this stage so that you may benefit from the experience of other staff in the organization.

How is this implemented?

The review of the design input requirements needs to be a systematic review, not a superficial glance. Design work will commence on the basis of what is conveyed in the requirements or the brief, although you should ensure there is a mechanism in place to change the information should it become necessary later. In fact, such a mechanism should be agreed at the same time as agreement to the requirement is reached.

In order to detect incomplete requirements you either need experts on tap or checklists to refer to. It is often easy to comment on what has been included but difficult to imagine what has been excluded. It is also important to remove subjective statements.

Ambiguities arise where statements imply one thing but the context implies another. You may also find cross-references to be ambiguous or in conflict. To detect the ambiguities and conflicts you need to read statements and examine diagrams very carefully. Items shown on one diagram may be shown differently in another. There are many other aspects you need to check before being satisfied they are fit for use. Any inconsistencies you find should be conveyed to the appropriate person with a request for action. Any changes to correct the errors should be self-evident so that you do not need to review all the information again.

Manufacturing process design input (7.3.2.2)

The standard requires the organization *to identify, document and review the manufacturing process design input requirements.*

What does this mean?

As with product design where there is a specification detailing the design requirements there needs to be a specification detailing the process design requirements – what process is required to deliver in terms of the intended and unintended outputs.

Why is this necessary?

If process requirements are not defined in advance, there is no basis from which to design a process and if the process is constructed, installed or selected without a definitive specification there will also be no legitimate basis for judging its subsequent performance. You will get what the configuration of components gives you and this might be nowhere near the performance you need to meet customer requirements.

How is this implemented?

The processes we are concerned with here are those needed to produce the product that has been designed. From the outset you need to be clear about what you want the process to deliver. You may want 100% conforming product at a rate of 1000 parts/hour but this may not be feasible. The inherent variation in the materials, the environment, the personnel and the equipment will limit the performance of the process. So, process performance is generally derived from a historical perspective. What is the known performance of such a process and is this good enough to meet this customers requirements or do we need a breakthrough in technology to achieve higher levels of performance? All this might only be with respect to intended outputs. Consider emissions, hazards, waste and other unintended outputs or outcomes.

The process design requirements will therefore be a combination of historical performance, customer requirements and improvement potential in the particular technology.

Special characteristics (7.3.2.3)

The standard requires the organization *to identify special characteristics and include them within the control plan.*

What does this mean?

Special characteristics are those characteristics of products and processes designated by the customer and/or selected by the supplier through knowledge of the product and the process. These characteristics are special because they can affect the safe functioning of the vehicle and compliance with government regulations such as flammability, occupant protection, steering control, braking, emissions, noise, EMC, etc.

Why is this necessary?

All characteristics are important and need to be controlled. However, some need special attention as excessive variation may affect product safety, compliance with government regulations, fit, form, function, appearance or the quality of subsequent operations. Designating such characteristics with special symbols alerts planners and operators to take particular care. It also alerts those responsible for dispositioning nonconforming product to exercise due care when reaching their decisions. By bringing special characteristics to the attention of production personnel through the control plan, it prompts the staff to take the appropriate measures when problems arise relative to these characteristics.

How is this implemented?

During the planning phase, a preliminary list of special product characteristics should be produced. During the product design and development phase, the list should be refined, reviewed and consensus reached. The output should be documented in the prototype control plan. During process design and development, the list should be converted into a matrix which displays the relationship between the process parameters and the manufacturing stations, and this documented in the production control plan.

The standard also requires documents such as FMEA, control plans, etc. to be marked with the customer's-specific symbols to indicate those process steps that affect special characteristics. As the characteristics in question will be specified within documents, the required symbols should be applied where the characteristic is mentioned rather than on the face of the document. For drawings the symbol should be applied close to the appropriate dimension or item. Alternatively, where a document specifies processes that affect a special characteristic, the appropriate symbol should be denoted against the particular stage in the process which affects that characteristic. The symbols therefore need to be applied during document preparation and not to copies of the document. The instructions to apply these symbols should be included within the procedures that govern the preparation of the documents concerned.

Design and development output (7.3.3 and 7.3.3.1)

Documenting the design and development output (7.3.3)

The standard requires that *the outputs of design and development be provided in a form that enables verification against the design and development inputs.* The automotive additions also require *the outputs to be expressed in terms that can be verified and validated against product design input requirements and for the outputs to include such things as design FMEA, reliability studies, error-proofing, etc.*

What does this mean?

Design output is the product of the design process and will therefore comprise information and/or models and specimens that describe the design in all its detail, the calculations, assumptions and the rationale for the chosen solution. It is not simply the specifications or drawings because should the design need to be changed, the designer may need to revisit the design data to modify parameters and assumptions. By requiring the design output to be in a form that enables verification, the characteristics of the product need to be expressed in measurable terms. One would therefore expect form, fit and function to be specified in units of measure with allowable tolerances or models and specimens to be capable of use as comparative references.

It is interesting to note that the requirement omits validation. This is because design outputs are verified against design inputs whereas, the design is validated against the original product requirement using a product or simulation that accurately reflects the design, thereby by-passing the design input and output as illustrated in Figure 7.9.

Why is this necessary?

This requirement responds to the Process Approach Principle.

Unless the design output is expressed in a form that enables verification, it will not be possible to verify the design with any certainty.

How is this implemented?

The design input requirements should have been expressed in a way that would allow a number of possible solutions. The design output requirements should therefore be expressed as *all* the inherent features and characteristics of the design that reflect a product that will satisfy these requirements. It should therefore fulfil the stated or implied needs, i.e. be fit for purpose.

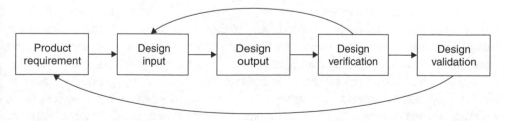

Figure 7.9 *Relationship between design verification and design validation*

Product specifications

Product specifications should specify requirements for the manufacture, assembly and installation of the product in a manner that provides acceptance criteria for inspection and test. They may be written or CAD-generated specifications, engineering drawings, diagrams, inspection and test specifications and schematics. With complex products you may need a hierarchy of documents from system drawings showing the system installation to component drawings for piece-part manufacture. Where there are several documents that make up the product specification there should be an overall listing that relates documents to one another.

Service specifications should provide a clear description of the manner in which the service is to be delivered, the criteria for its acceptability, the resources required including the numbers and skills of the personnel required, the numbers and types of facilities and equipment necessary, and the interfaces with other services and suppliers.

In addition to the documents that serve product manufacture and installation or service delivery, documents may also be required for maintenance and operation. The product descriptions, handbooks, operating manuals, user guides and other documents which support the product or service in use are as much a part of the design as the other product requirements. Unlike the manufacturing data, the support documents may be published either generally or supplied with the product to the customer. The design of such documentation is critical to the success of the product as poorly constructed handbooks can be detrimental to sales.

The requirements within the product specification need to be expressed in terms that can be verified. You should therefore avoid subjective terms such as "good quality components", "high reliability", "commercial standard parts", etc., as these requirements are not sufficiently definitive to be verified in a consistent manner.

Design calculations

Throughout the design process, calculations will need to be made to size components and determine characteristics and tolerances. These calculations should be recorded and retained together with the other design documentation but may not be issued. In performing design calculations it is important that the status of the design on which the calculations are based is recorded. When there are changes in the design these calculations may need to be repeated. The validity of the calculations should also be examined as part of the design verification activity. One method of recording calculations is in a designer's logbook that may contain all manner of things and so the calculations may not be readily retrievable when needed. Recording the calculations in separate reports or in separate files along with the computer data will improve retrieval.

Design analyses

Analyses are types of calculations but may be in the form of comparative studies, predictions and estimations. Examples are stress analysis, reliability analysis, failure modes analysis and hazard analysis. Analyses are often performed to detect whether the design has any inherent modes of failure and to predict the probability of occurrence. FMEA is addressed in Chapter 8 as it is a preventive measure. The analyses assist in design improvement and the prevention of failure, hazard, deterioration and other adverse conditions. Analyses may need to be conducted because the end-use conditions may not be reproducible in the factory. Assumptions may need to be made about the interfaces, the environment, the actions of users, etc. and analysis of such conditions assists in determining characteristics as well as verifying the inherent characteristics.

Design is an iterative process therefore the above analyses are not performed once but after each iteration of the design so that by the end of the design process any inherent weaknesses will have either been eliminated, reduced or contained by component redundancy, derating, error-proofing or warning notices on the product or in the user documentation.

Ensuring that design output meets design input requirements (7.3.3a)

The standard requires that design and development output *meets the design and development input requirements.*

What does this mean?
The characteristics of the resultant design should be directly or indirectly traceable to the design input requirements. In some cases, a dimension may be stated in the design input which is easily verified when examining the design specifications, drawings, etc. In other cases, the input requirement may be stated in performance terms that is translated into a number of functions which when energized provide the required result. In other cases, a parameter may be specified above or below the design input requirement so as to allow for production variation.

Why is this necessary?
This requirement responds to the Factual Approach Principle and needs no explanation.

How is this implemented?
The techniques of design verification can be used to verify that the design output meets the design input requirements. However, design verification is often an iterative process. As features are determined, their compliance with the requirements should be checked by calculation, analysis or test on development models. Your development

plan should identify the stages at which each requirement will be verified so as to give warning of non-compliance as early as possible.

Providing information for purchasing, production and service provision (7.3.3b)

The standard requires that *design and development output provide appropriate information for purchasing, production and service provision.*

What does this mean?
A design is a set of instructions that are necessary to construct a product. Therefore, the inherent characteristics of the product that facilitates procurement, production and servicing need to be defined. Tooling for production is considered to be part of the production process but information within the design output is needed to enable tooling to be designed.

The instructions needed to produce, inspect, test, install and maintain the product may be produced by the designers but are strictly outputs of the production, installation and servicing processes that are derived from inputs that comprise the design description.

Why is this necessary?
This requirement responds to the Systems Approach Principle.

A design description alone will not result in its realization unless information is provided for procuring the materials and components, preparing the product for production and maintaining the product in service.

How is this implemented?
Products should be designed to facilitate procurement, manufacture, installation and servicing and therefore additional characteristics to those required for end use may be necessary. Examples include geometric tolerances, specific part numbers, part marking, assembly aids, error-proofing, lifting points, transportation and storage protection. Techniques used to identify such design provisions are as follows:

- Failure Mode and Effects Analysis (see Chapter 8)
- Producibility Analysis
- Testability Analysis
- Maintainability Analysis.

Defining acceptance criteria (7.3.3c)

The standard requires design and development output to *contain or reference product acceptance criteria.*

What does this mean?

Acceptance criteria are the requirements that, if met, will deem the product acceptable. It means that characteristics should be specified in measurable terms with tolerances or limits. These limits should enable all production versions to perform to the product specification, providing such limits are well within the limits to which the design has been tested. It means that every requirement should be stated in such a way that it can be verified – that there is no doubt as to what will be acceptable and what will be unacceptable.

Why is this necessary?

This requirement responds to the Factual Approach Principle.

Where product characteristics are specified in terms that are not measurable or are subjective, they lend themselves to misinterpretation and variation such that no two products produced from the design will be the same and will exhibit consistent performance.

How is this implemented?

A common method used to ensure characteristics are stated in terms of acceptance criteria is to define them by reference to product standards. These standards maybe developed by the organization or maybe of national or international status. Standards are employed to enable interchangeability, repeatability and to reduce variety.

Where there are common standards for certain features, these may be contained in a standards manual. Where this method is used it is still necessary to reference the standards in the particular specifications to ensure that the producers are always given full criteria. Some organizations omit common standards from their specifications. This makes it difficult to specify different standards or to subcontract the manufacture of the product or operation of a service without handing over proprietary information.

Specifying essential characteristics (7.3.3d)

The standard requires design and development output to *define the characteristics of the product that are essential to its safe and proper use.*

What does this mean?

Certain characteristics will be critical to the safe operation of the product. These can be divided into two types. Those characteristics that the product needs to exhibit in order to function correctly and those characteristics that are exhibited when the product is put together, used or maintained incorrectly.

Why is this necessary?

This requirement responds to the Customer Focus Principle.

Alerting assemblers, users and maintainers to safety-critical characteristics increases their sensitivity, provides the awareness to plan preventive measures and thus reduces the probability of an incident or accident.

How is this implemented?

The design output data should identify by use of symbols or codes those characteristics that are safety critical under normal operating conditions. This will enable the manufacturers to determine the measures needed to ensure no variation from specification when the characteristics are initially produced and ensure no alteration of these characteristics during subsequent processing.

Drawings should indicate the warning notices required, where such notices should be placed and how they should be affixed. Examples that indicate improper function or potential danger are red lines on tachometers to indicate safe limits for engines, audible warnings that lights are on when opening a car door or low oil warning lights. In some cases, it may be necessary to mark dimensions or other characteristics on drawings to indicate that they are critical and employ special procedures for dealing with any variations. In passenger vehicle component design, certain parts are regarded as safety critical because they carry load or need to behave in a certain manner under stress. Others are not critical because they carry virtually no load so there can be a greater tolerance on deviations from specification.

Failure Mode and Effects Analysis and Hazard Analysis are techniques that aid the identification of characteristics crucial to the safe and proper functioning of the product.

Approval of design outputs (7.3.3)

The standard requires design and development outputs to *be approved prior to release*.

What does this mean?

Although the requirement for design outputs to be approved prior to release may appear to duplicate the requirement of Clause 4.2.3 on document control, there is a subtle difference. Document approval is not the same as design approval and design release is not the same as document issue. When a design is approved it is the description of that design in whatever form that is approved. Design approval therefore applies to all the documents, models and specimens, etc. that constitute the design description, not as separate entities but as a whole.

Why is this necessary?

This requirement responds to the Factual Approach Principle.

While design documents should pass into the document control process for issue, several iterations may be needed before the design output is complete and ready for release. It is important that these iterations are under control otherwise the full impact of changes in one component may not be reflected in other components. When the design is ready for verification it is released.

How is this implemented?

The requirements in QS-9000 and both the 1999 and 2002 versions of ISO/TS 16949 do not recognize that design reviews are not only concerned with reviewing a design but are also concerned with granting design approval. This oversight resulted in design approval being planted under design outputs and the inevitable consequence that it would be interpreted as document approval.

There are sound reasons for separating design approval from document approval. Design approval should proceed through three stages:

1   Design information should be approved before being presented to a design review – in this way the reviewers only work with information that has been checked and found acceptable.
2   Designs should be approved before the design is subject to verification – in this way prototypes are produced, or simulated using a complete set of design information that has been found acceptable.
3   Designs should be verified before being subject to validation – in this way trials are only conducted on models representative of those that will enter production.

The design output may consist of many documents each of which fulfils a certain purpose. It is important that these documents are reviewed and verified as being fit for their purpose before release using the documentation controls developed for meeting Section 4.5 of ISO/TS 16949. In the software industry, where documentation provides the only way of inspecting the product prior to installation, document inspections called *Fagan Inspections* (after Michael Fagan of IBM) are carried out not only to identify the errors, but to collect data on the type of error and the frequency of occurrence. By analysing this data using statistical techniques the results assist in error removal and prevention.

Design documentation reviews can be made effective by providing data requirements for each type of document as part of the design and development planning process. The data requirement can be used both as an input to the design process and as acceptance criteria for the design output documentation review. The data requirements would specify the input documents and the scope, content and format required for the output document.

As design documents are often produced at various stages in the design process, they should be reviewed against the input requirements to verify that no requirements have been overlooked and that the requirements have been satisfied.

Manufacturing process design output (7.3.3.2)

The standard requires manufacturing process design output *to be expressed in terms that can be verified against manufacturing process design input requirements and validated and for the process design output to include such things as process FMEA, work instructions, control plans, etc.*

What does this mean?

There are two requirements here one implies that the results of process design are defined and documented, and the other implies that there has to be link between what is defined and documented and the originating process design requirements. Therefore, if the process is required to achieve certain productivity targets, the process design outputs need to include data indicating the level of productivity that the specific process design will achieve. The standard requires the process design output to include several different types of documents. Collectively these can be termed a *Process Design Disclosure*.

Why is this necessary?

Unless the process design output is expressed in a form that enables verification, it will not be possible to verify the process design with any certainty.

How is this implemented?
Specifications and drawings
Process specifications define the features and characteristics of the process and would include such aspects as follows:

- *Outputs characteristics*: Characteristics of the output that the process is designed to produce. It may be a component machined to a particular accuracy, the finish with a particular smoothness, a bond with a particular strength, etc.
- *Performance characteristics*: Throughput, yield or productivity, setup time, shutdown time, reliability and maintainability.
- *Maintenance characteristics*: Daily, weekly, monthly and annual maintenance routines required to maintain performance.
- *Safety characteristics*: The provisions designed in to safeguard health and safety such as guards, trips, warnings, alerts and sensors.
- *Measuring and monitoring*: The built-in measuring and monitoring provisions for verifying the quality of the product and the stability of the process.

- *Error-proofing*: The provisions made to prevent errors in setup, operation and shutdown.
- *Environmental provisions*: The provisions built into remove waste, control emissions, effluent and discharges.

Much of this data might be included in the equipment manufacturers specifications, but the specific configuration and process inputs requirements will modify such parameters.

Process flow charts

Process flow charts are no more than a sequence of operations turned into a diagram. However, they do have the benefit of showing where inputs come from and output go. It is not uncommon for Control Plans to include a flow chart on the left side of the page instead of simply text in cells within a table. When the sequence has no interruptions there appears to be no advantage other than emphasizing the sequential nature of the operations. With processes that involve several intermediate stages with some running in parallel the control plan would not be an appropriate location for such flow charts and separate documents would be needed. The flow chart is also a convenient way of showing the relationship between major stages in a process such as might be illustrated in an engine production process that commences with receipt of materials and components and ends with delivery of engines to the vehicle maker. Each stage might warrant its own flow chart and be deemed to be a separate process but on their own they are only stages as the output is not a commercially viable entity. It is the completed engine that is delivered not a machined cylinder block except as a spare.

Process layouts

While the flow chart indicates the sequence and relationship between the various operations, the layout shows the physical disposition of the plant, equipment, tooling and holding areas. There will be an optimum configuration but individual sites may have variations due to the space available. Process layouts are needed to evaluate material flow (see under the heading *Plant, facility and equipment planning* in Chapter 6).

Process FMEA

Process FMEA is a vital tool in determining process reliability. As a technique, it is identical to Product FMEA and this topic is addressed in Chapter 8 under the heading *Preventive action*. A Process FMEA would use the format shown in Figure 8.7.

Control plan

The control plan is addressed in detail in Clause 7.5.1.1 and later in this chapter. The point being made here is that the Control Plan is produced as part of the process design disclosure not simply as an add-on to production.

Work instructions

Work instructions are also required under Clause 7.5.1.2, but they are the same work instructions that form part of the process design disclosure.

Process approval acceptance criteria

The process approval acceptance criteria are the success measures that if achieved, indicate that the process is capable of delivering the requirement outputs in the quantity, cost and timescale required. The criteria may be expressed as a result that is to be obtained when the process output is measured in a certain way.

Quality, reliability, maintainability and measurability data

This data will be the results of analytical studies performed to predict:

- the process yield and throughput;
- the mean time between process failure relative to the acceptance criteria;
- the mean time to restore performance following failure;
- the accuracy and precision with which the monitoring and measuring devices will monitor and measure process and product characteristics.

Error-proofing data

The requirement for the process design outputs to include the results of error-proofing is in the wrong place as it related to design validation. The design specification would include information on error-proofing, the reliability data should address the probability that the error-proofing measures to fail and the Process FMEA would look at the potential for failure in error-proofing measures.

Nonconformity detection and feedback

The degree to which nonconformities would be detected and alerts sounded depends on the extent of built-in monitoring and measurement facilities. Clearly, if product characteristics were measured automatically, channels could be provided along which nonconforming product would pass. For example, in a bearing plant, in-process measurements are taken and rejected components diverted to a quarantine area alongside the machine. In other cases, a co-ordinate measuring sensor might scan the components before being released and provisions installed feed the results into a computer for on-line analysis by skilled personnel.

Design reviews (7.3.4)

Planning design reviews (7.3.4)

The standard requires that *at suitable stages, systematic reviews of design and development be performed in accordance with planned arrangements.*

What does this mean?

ISO 9000 defines a review as an activity undertaken to determine the suitability, adequacy and effectiveness of the subject matter to achieve established objectives. This is the way the term should be interpreted in the context of ISO/TS 16949. When the term review is used in another context, it simply means to have another look at something. The ISO 9000 definition implies that a design review is an activity undertaken to determine the suitability, adequacy and effectiveness of a design to meet the design requirement. Suitability means "Has an appropriate design solution been developed?". Adequacy means "Does the design solution meet all the design requirements?" and effectiveness means "Have we got the right design objective?". Design reviews are therefore *not* document reviews.

Systematic reviews are those that cover the complete design from the high level down to the smallest component and all the associated requirements in a logical manner. The review has to be stage-by-stage, methodical with purpose. Systematic reviews probe the design solution and the interfaces between all components for design weaknesses and delves into the detail to explore how requirement are fulfilled.

Suitable stages are at the transition between the various phases of design maturity in the design process. In simple terms designs begin with a conceptual phase, proceed through a definition phase and end with a detail design phase. Development commences with a detail design and proceeds through several iterations involving verification and validation, and may continue through several enhancements before the design becomes obsolete and a new design idea is conceived.

The planned arrangements are the stages in the design and development cycle that reviews have been planned and the expected inputs and outputs, and the acceptance criteria for proceeding to the next stage.

Why is this necessary?

This requirement responds to the Leadership Principle.

A design represents a considerable investment by the organization. There is therefore a need for a formal mechanism for management and the customer (if the customer is sponsoring the design) to evaluate designs at major milestones. The purpose of the review is to determine whether the proposed design solution is compliant with the design requirement and should continue or should be changed before proceeding to the next phase. It should also determine whether the documentation for the next phase is adequate before further resources are committed. Design review is that part of the design control process which measures design performance, compares it with pre-defined requirements and provides feedback so that deficiencies may be corrected before the design is released to the next phase.

How is this implemented?

Review schedules

A schedule of design reviews should be established for each product or service being developed. In some cases there will need to be only one design review. After completion of all design verification activities, but depending on the complexity of the design and the risks, you may need to review the design at some or all of the following intervals:

- *Design Requirement Review*: To establish that the design requirements can be met and reflect the needs of the customer before commencement of design.
- *Conceptual Design Review*: To establish that the design concept fulfils the requirements before project definition commences.
- *Preliminary Design Review*: To establish that all risks have been resolved and development specifications have been produced for each sub-element of the product or service before detail design commences.
- *Critical Design Review*: To establish that the detail design for each sub-element of the product or service complies with its development specification and that product specifications have been produced before manufacture of the prototypes.
- *Design Validation Readiness Review*: To establish the configuration of the baseline design and readiness before commencement of design validation.
- *Final Design Review*: To establish that the design fulfils the requirements of its development specification before preparation for its production.

Design review input data

The input data for the review should be distributed and examined by the review team well in advance of the time when a decision on the design has to be made. A design review is not a meeting. However, a meeting will often be necessary to reach a conclusion and to answer questions of the participants. Often analysis may need to be performed on the input data by the participants in order for them to determine whether the design solution is the most practical and cost-effective way of meeting the requirements.

Conducting design reviews (7.3.4a and 7.3.4b)

The standard requires design reviews to *be conducted to evaluate the ability of the results of the design and development to fulfil requirements, identify problems and propose required actions.*

What does this mean?

Design reviews occur at the end of a design phase when there are results to review. This means that every phase needs an objective, the achievement of which is evaluated at the review. The results of the design may be concepts, models, calculations, drawings, specifications or any output which describes the maturity of the design at a particular stage. During the initial phases, the key performance characteristics will be evaluated and at subsequent design reviews further definition enables the design to be evaluated against more definitive requirements until all requirements are fulfilled. Each review may reveal design weaknesses that need to be resolved before proceeding to the next phase.

Why is this necessary?
This requirement responds to the Factual Approach Principle.

It would be folly to proceed with a design that possesses significant weaknesses and therefore the design review provides an opportunity to identify these weaknesses early on and take action to eliminate them before compounding the errors.

How is this implemented?
Although design documents may have been through a vetting process, the purpose of the design review is not to review documents but to subject the design to an independent board of experts for its judgement as to whether the most satisfactory design solution has been chosen. By using a design review methodology, flaws in the design may be revealed before it becomes too costly to correct them. Design reviews also serve to discipline designers by requiring them to document the design logic and the process by which they reached their conclusions, particularly the options chosen and the reasons for rejecting other options.

The experiences of previous designs provide a wealth of information of use to designers that can alert them to potential problems. In compiling this information designers can feed of the experience of others not only in the same organization, but also in different organizations and industries. By using technical data available from professional institutions, associations, research papers, etc., checklists can be complied that aid the evaluation of designs.

Participants at design reviews (7.3.4)

The standard requires participants in design reviews to *include representatives of functions concerned with the design and development stage(s) being reviewed.*

What does this mean?
Representatives of functions concerned with a design include not only the designers but those sponsoring the design such as the customers, marketing personnel or upper management, those that will be responsible for transforming the design into a product or service, those responsible for maintaining the product, using the product or disposing of the product in fact any party that has an interest in the quality of the design solution.

Why is this necessary?
This requirement responds to the Involvement of People Principle.

Design reviews should be performed in order to release a design to the next phase of development. A review is another look at something. The designer has had one look at the design and when satisfied presents the design to an impartial body of experts so as to seek approval and permission to go-ahead with the next phase. Designers are often not the budget holders, or the sponsors. They often work for others. Even in situations where there is no specific customer or sponsor or third party, it is good practice to have someone else look at the design. A designer may become too close to the design to spot errors or omissions and so will be biased towards the standard of his or her own performance. The designer may welcome the opinion of someone else because it may confirm that the right solution has been found or that the requirements can't be achieved with the present state of the art. If a design is inadequate and the inadequacies are not detected before production commences the consequences may well be disastrous. A poor design can lose a customer, a market or even a business so the advice of independent experts should be valued.

How is this implemented?
The review team should have a collective competency greater than that of the designer of the design being reviewed. For a design review to be effective it has to be conducted by someone other than the designer. The requirement for participants to include representatives of all functions concerned with the design stage means those who have an interest in the results.

The review team should comprise as appropriate, representatives of the purchasing, manufacturing, servicing, marketing, inspection, test, reliability, quality assurance authorities, etc., as a means of gathering sufficient practical experience to provide advance warning of potential problems with implementing the design. The number of people attending the design review is unimportant and could be as few as the designer and his or her supervisor provided that the supervisor is able to impart sufficient practical experience and there are no other personnel involved at that particular design stage. There is no advantage gained in staff attending design reviews that add no value in terms of their relevant experience, regardless of what positions they hold in the company. The representation at each review stage may well be different: i.e. may be being just the designer and his or her supervisor at the conceptual review, and may be representation from manufacturing, servicing, etc. at the final review.

The chairman of the review team should be the authority responsible for placing the development requirement and should make the decision as to whether design should proceed to the next phase based on the evidence substantiated by the review team.

Design review records (7.3.4)

The standard requires *the results of the reviews and actions arising from the review to be recorded.*

What does this mean?
The results of the design review are not simply minutes of a meeting but all the evidence that has been accumulated in evaluating a particular design, identifying problems and determining actions required to resolve them.

Why is this necessary?
This requirement responds to the Factual Approach Principle.

The results of the design review should be documented in a report rather than minutes of a meeting because they represent objective evidence that may be required later to determine product compliance with requirements, investigate design problems and compare similar designs. Even when no problems are found, the records of the review provide a baseline that can be referred to when making subsequent changes.

How is this implemented?
The report should have the agreement of the full review team and should include:

- The criteria against which the design has been reviewed.
- A list of the documentation that describes the design being reviewed and any evidence presented which purports to demonstrate that the design meets the requirements.
- The decision on whether the design is to proceed to the next stage.
- The basis on which confidence has been placed in the design.
- A record of any uncompleted corrective actions from previous reviews.
- The recommendations and reasons for corrective action, if any.
- The members of the review team and their roles.

Monitoring (7.3.4.1)

The standard requires *measurements at specific stages of design and development to be defined, analysed and reported with summary results as an input to management review.*

What does this mean?
Each design and development stage produces an output and these outputs should be measured in order to verify that they meet the expectations for the design and development process.

Why is this necessary?
The intent of this requirement is to provide a means for assuring management of performance and alerting them to potential and actual problems in the design and development process for specific projects. Management need to know whether projects are proceeding on course and hence periodic reporting is necessary to provide management with factual data on which to make decisions.

How is this implemented?

The specified stages might include the following:

- Project launch
- Programme approval
- Start and end of product design
- Start and end of process design
- Start and end of prototype manufacture
- Start and end of product verification
- Start and end of process verification
- Product approval.

The kinds of measurements you can make at each stage are as follows:

- The extent to which planned tasks are being completed on time.
- The degree of slippage or slack in the programme.
- The critical paths and changes in criticality.
- Lead times and effect of changes on advanced procurement.
- Resource utilization.
- Spend versus budget.
- Estimated spend to completion.
- Quality risks in terms of potential and actual failures that affect critical tasks.

The precise staging of the reviews will depend on the nature of the project. However, the principle is to hold a review prior to a major decision that dictates the direction of the project or at a stage in a project where the nature of work changes (see Figure 7.10). Alternatively, reviews can be held monthly providing a project review precedes the change in phase of work.

Project reviews are not design reviews. Project reviews assess performance of the project and take into account timing, costs, organization, work assignments, subcontracts, etc. Design reviews look back at the technical aspects of design and look forward to the technical aspects of the design tasks ahead.

Design verification (7.3.5)

Performing design verification (7.3.5)

The standard requires design and development verification to *be performed in accordance with planned arrangements to ensure the output meets the design and development inputs.*

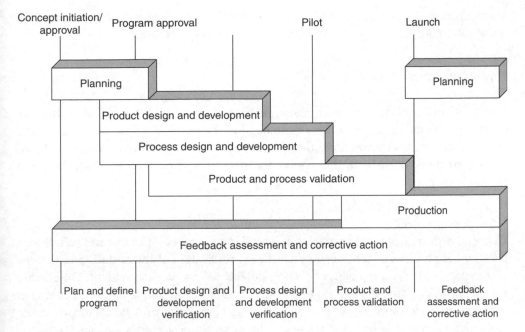

Figure 7.10 *Project planning timing chart*

What does this mean?

ISO 9000 defines verification as confirmation, through the provision of objective evidence that specified requirements have been fulfilled. There are two types of verification, those verification activities performed during design and on the component parts to verify conformance to specification and those verification activities performed on the completed design to verify performance against the design input. When designing a system there should be design requirements for each subsystem, each item of equipment and each unit, and so on down to component and raw material level. Each of these design requirements represents acceptance criteria for verifying the design output of each stage. Verification may take the form of a document review, laboratory tests, alternative calculations, similarity analyses or tests and demonstrations on representative samples, prototypes, etc. In all these cases, the purpose is to prove that the design is right, i.e. it meets the requirements. Validation on the other hand serves to confirm that it is the right design, i.e. that the requirements were the right requirements for a specific application.

The reference to planned arrangements again means that verification plans should be adhered to.

Why is this necessary?

This requirement responds to the Factual Approach Principle.

Verification is fundamental to any process and unless the design is verified, there will be no assurance that the resultant design meets the requirements.

How is this implemented?
Timing
The standard does not state when design verification is to be performed although verification of the design after launch of product into production would not be appropriate. The stages of verification will therefore mirror the design review schedule but may include additional stages. Design verification needs to be performed when there is a verifiable output.

Verification process
During the design process many assumptions may have been made and will require proving before commitment of resources to the replication of the design. Some of the requirements such as reliability and maintainability will be time dependent. Others may not be verifiable without stressing the product beyond its design limits. With computer systems, the wide range of possible variables is so great that proving total compliance would take years. It is however necessary to subject a design to a series of tests and examinations in order to verify that all the requirements have been achieved.

The design verification process should provide for the following:

- Test specifications to be produced that define the features and characteristics that are to be verified for design qualification and acceptance.
- Test plans to be produced that define the sequence of tests, the responsibilities for their conduct, the location of the tests and test procedures to be used.
- Test procedures to be produced that describe how the tests specified in the test specification are to be conducted together with the tools and test equipment to be used and the data to be recorded.
- All measuring equipment to be within calibration during the tests.
- The test sample to have successfully passed all planned in-process and assembly inspections and tests prior to commencing qualification tests.
- The configuration of the product in terms of its design standard, deviations, non-conformities and design changes to be recorded prior to and subsequent to the tests.
- Test reviews to be held before tests commence to ensure that the product, the facilities, tools, documentation and personnel are in a state of operational readiness for verification.
- Test activities to be conducted in accordance with the prescribed specifications, plans and procedures.
- The results of all tests and the conditions under which they were obtained to be recorded.

- Deviations to be recorded, remedial action taken and the product subject to re-verification prior to continuing with the tests.
- Test reviews to be performed following qualification tests to confirm that sufficient objective evidence has been obtained to demonstrate that the product fulfils the requirements of the test specification.

Development tests

Where tests are needed to verify conformance with the design specification, development test specifications will be needed to specify the test parameters, limits and operating conditions. For each development test specification there should be a corresponding development test procedure that defines how the parameters will be measured using particular test equipment and taking into account any uncertainty of measurement (see Measurement uncertainty under the heading *Defining measuring processes*). Test specifications should be prepared for each testable item. Whilst it may be possible to test whole units, equipment or subsystems you need to consider the procurement and maintenance strategies for the product when deciding which items should be governed by a test specification. Two principal factors to consider are:

- testable items sold as spare parts,
- testable items the design and/or manufacture of which are subcontracted.

If you conduct trials on parts and materials to prove reliability or durability, these can be considered to be verification tests. For example, you may conduct tests in the laboratory on metals for corrosion resistance or on hinges for reliability and then conduct validation tests under actual operating conditions when these items are installed in the final product.

Verifying compliance with regulations

Having designed the product to meet the applicable statutes and regulations you need to plan for verifying that they have been met. Verification of compliance can be accomplished through discrete checks combined with other tests, inspections and analyses. However, it may be more difficult to demonstrate compliance through the records alone. In some cases tests such as pollution tests, safety tests, proof-loading tests, electromagnetic compatibility tests, pressure vessel tests, etc. are so significant that separate tests and test specifications are the most effective method of verifying compliance with regulations.

Alternative design calculations

Verification of some characteristics may only be possible by calculation rather than by test, inspection or demonstration. In such cases, the design calculations should be checked either by being repeated by someone else or by performing the calculations by an alternative method. When this form of verification is used the margins of error permitted should be specified in the verification plan.

Comparing similar designs

Design verification can be a costly exercise. One way of avoiding unnecessary costs is to compare the design with a similar one that has been proven to meet the same requirements. This approach is often used with designs that use a modular construction. Modules used in previous designs need not be subject to the range of tests and examinations necessary if their performance has been verified either as part of a proven design or has been subject to such in-service use that will demonstrate achievement of the requirements. Care has to be taken when using this verification method that the requirements are the same and that evidence of compliance is available to demonstrate compliance with the requirements. Marginal differences in the environmental conditions and operating loads can cause the design to fail if it was operating at its design limit when used in the previous design.

Recording design verification results (7.3.5)

The standard requires *the results of the verification and any required actions to be recorded.*

What does this mean?

The results of design verification comprise:

- the criteria used to determine acceptability;
- data testifying the standard of the design being subject to verification;
- the verification methods;
- data testifying the conditions, facilities and equipment used to conduct the verification;
- the measurements;
- analysis of the differences between planned and achieved results;
- actions to be taken on the differences.

Why is this necessary?

This requirement responds to the Factual Approach Principle.

Any decision to proceed either to the next stage of development or into production or operations needs to be based on fact and the records of design verification provide the facts. As the definition of verification explains, it is the provision of objective evidence that requirements have been fulfilled and the records of verification constitute this objective evidence.

How is this implemented?

In planning design verification, consideration needs to be given to the output, its format and content. The basic content is governed by the design specification but the data to be recorded both before, during and after verification need to be prescribed. Some data may be generated electronically and other data may be collected from observation. Often there are lots of different pieces of evidence that need to be collected, collated and assembled into a dossier in a secure format. These factors need to be sorted out before commencing verification so that all the necessary information is gathered at the time. After verification a report of the activities may also be necessary to explain the results, possible causes of any variation and recommendations for action for presentation at a design review.

Design validation (7.3.6 and 7.3.6.1)

Performing validation (7.3.6)

The standard requires design and development validation to *be performed in accordance with planned arrangements to confirm that resulting product is capable of fulfiling the requirements for the specified application or intended use where known.*

What does this mean?
ISO 9000 defines validation as confirmation through the provision of objective evidence that requirements for a specific intended use or application have been fulfilled. Specified requirements are often an imperfect definition of needs and expectations and therefore to overcome inadequacies in the manner in which requirements can be specified, the resultant design needs to be validated against intended use or application.

Design validation is a process of evaluating a design to establish that it fulfils the intended use requirements. It goes further than design verification in that validation tests and trials may stress the product of such a design beyond operating conditions in order to establish design margins of safety and performance. Design validation can also be performed on mature designs in order to establish whether they will fulfil different user requirements to the original design input requirements. An example is where software designed for one application can be proven fit for use in a different application or where a component designed for one environment

> **The case of the "copper-cooled engine"**
>
> Between 1918 and 1923, General Motors (GM) experimented with an air-cooled engine that would eliminate the radiator and plumbing associated with a water-cooled engine. Even though the engine seemed to work in the laboratory, it failed in service. Production was commenced before the design had been validated. The engine had pre-ignition problems and showed a loss of compression and power when hot. As a result many cars with the engine were scrapped. Apart from the technical problems GM experienced with its development, it did prove to be a turning point in GM's development strategy, probably resulting in what is now their approach to product quality planning.

can be shown to possess a capability that would enable it to be used in a different environment. Multiple validations may therefore be performed to qualify a design for different applications.

Why is this necessary?

This requirement responds to the Factual Approach Principle.

Merely requiring that the design output meets the design input would not produce a quality product or service unless the input requirements were a true reflection of the customer needs and this is not always possible. If the input is inadequate the output will be inadequate: "garbage-in–garbage-out" to use a common software expression. Let's face it; if design validation were performed effectively, all the product recalls would be down to errors in manufacturing.

How is this implemented?

Design validation may take the form of qualification tests which stress the product up to and beyond design limits – beta tests where products are supplied to several typical users on trial in order to gather operational performance data, performance trials and reliability and maintainability trials where products are put on test for prolonged periods to simulate usage conditions.

Road trials on test tracks are validation tests as are the customer trials conducted over several weeks or months under actual operating conditions on pre-production models. Sometimes the trials are not successful as was the experience at General Motors in 1920[2] (see box in page 471).

As the cost of testing vast quantities of equipment would be too great and take too long, qualification tests, particularly on hardware, are usually performed on a small sample. The test levels are varied to take account of design assumptions, variations in production processes and the operating environment.

Products may not be put to their design limits for some time after their launch into service, probably far beyond the warranty period. Customer complaints may appear years after product launch. When investigated this may be traced back to a design fault which was not tested for during the verification programme. Such things as corrosion, insulation, resistance to wear, chemicals, climatic conditions, etc. need to be verified as being within the design limits.

Prototype programmes (7.3.6.2)

The standard requires the organization *to have a prototype programme when required by the customer and wherever possible, use the same suppliers, tooling and manufacturing processes as will be used in production*. It also requires performance testing *to be monitored and if the services are outsourced the organization is to take technical leadership.*

Many different types of models may be needed to aid product development, test theories, experiment with solutions, etc. However, when the design is complete, prototype models representative in all their physical and functional characteristics to the production models may need to be produced to ensure that:

- the product functions as intended under normal operating conditions;
- the product does not exhibit any unexpected characteristics when malfunctioning;
- the margins of safety relative to static and dynamic loading are adequate under worse-case conditions.

There will be situations where the customer requires a prototype programme but when no such requirement has been stated it does not mean you should not produce prototypes. Prototypes will not normally be required when the design is similar to a previously proven design or standard, or the design is so simple that sufficient evidence can be obtained during the production trail run.

If design is proven on uncontrolled models then it is likely that there will be little traceability to the production models. Production models may therefore contain features and characteristics that have not been proven. The only verification that needs to be performed on production models is for those features and characteristics that are subject to change due to the variability in manufacturing, either of raw materials or of assembly processes.

When building prototypes, the same materials, locations, suppliers, tooling and processes should be used as will be used in actual production so as to minimize the variation. With early prototypes, this might not be possible in all cases as perhaps the tooling or eventual suppliers might not be known. What is important is that the prototype used in any trails is representative of production in the features and characteristics being tested during those trails.

Development tests will not yield valid results if obtained using uncontrolled measuring equipment; therefore, the requirements of Clause 7.6 on measuring devices apply to the design process. Design is not complete until the criteria for accepting production versions has been established. Products need to be designed so as to be testable during production using the available production facilities. The proving of production acceptance criteria is therefore very much part of design verification.

Tracking performance testing (7.3.6.2)
As part of the verification plan discussed previously, you should include an activity plan that lists all the planned activities in the sequence they are to be conducted and use this

plan to progressively record completion and conformance. The activity plan should make provision for planned and actual dates for each activity and for recording comments such as recovery plans when the programme does not proceed exactly as planned. It is also good practice to conduct test reviews before and after each series of tests so that corrective measures can be taken before continuing with abortive tests (see also under the heading *Design validation*).

Outsourcing prototyping services (7.3.6.2)

Where you do not possess the necessary facilities for building prototypes or conducting design verification and validation, these activities may be outsourced. However, ISO/TS 16949 requires that you exercise technical leadership in such matters. This means that you need to enter into a formal contract with the supplier, apply the controls you established to meet Clause 7.4 and manage the test programme. You should require the suppliers to submit test plans and procedures for your approval prior to commencement of the test unless you are providing this information yourself. You need to be confident that the tests will produce valid data so the test setup, test equipment, test environment and monitoring methods need to be periodically reviewed. You should have a representative present during test, and retain authority for starting and stopping the test.

Demonstrations

Tests exercise the functional properties of the product. Demonstrations, on the other hand, serve to exhibit usage characteristics such as access, maintainability including interchangeability, repairability and serviceability. Demonstrations can be used to prove safety features such as the crash tests filmed at high speed that when played at normal speed, the crumpling of the steel and movement of the dummy against the air bag show up characteristics that prove whether the safety features behave as intended. However, one of the most important characteristics that need to be demonstrated is producibility. Can you actually make the product economically in the quantities required? Does production yield a profit or do you need to produce 50 to yield 10 good ones? The demonstrations should establish whether the design is robust. Designers may be selecting components at the outer limits of their capability. A worst-case analysis should have been performed to verify that under worst-case conditions, i.e. when all the components fitted are at the extreme of their tolerance range, the product would perform to specification. Analysis may be more costly to carry out than a test and by assembling the product with components at their tolerance limits you may be able to demonstrate economically the robustness of the design.

Timing of validation (7.3.6)

The standard requires validation to *be completed wherever applicable prior to the delivery or implementation of the product*. The automotive additions require *design*

and development validation to be performed in accordance with customer require-
ments including programme timing.

What does this mean?

As indicated previously, validation trials may take some considerable time but until confidence in the capability of a design to fulfil intended use requirements is known, any
decision to launch into production or into operation involves risk. The requirement,
however, recognizes that it may not be possible or practical to hold production until all
the results of validation have been obtained and assessed.

Why is this necessary?

This requirement responds to the Systems Approach Principle.

There are some characteristics such as safety and reliability that need to be demonstrated before launching into production otherwise unsafe or unreliable products might
be put onto the market. One has only to scan the recall programmes accessible on the
Internet to notice that many products are indeed launched into production with major
faults (see Appendix B). Some failures may be due to the quality of conformity but
there are also some that are due to design weaknesses that should have been detected
in the verification and validation programmes.

It is thought that in the USA alone, there are 30 million product recalls every year.
Probably the biggest recall of all time occurred in April 1996 when Ford USA recalled
up to 9 million vehicles that may have been equipped with a faulty ignition switch. In
July 1999, General Motors USA recalled 1.1 million vehicles that may have had anti-
lock brake problems. Launching into production without sufficient evidence that the
decision satisfies all interested parties can therefore be very costly.

How is this implemented?

The simplest approach is to wait until all the evidence from verification and validation
trials has been assessed before launching into production or going operational. In practice, it depends on knowing what the risks are and therefore is a balance between risk
and the impact any delay in production launch or going operational may have. It would
therefore be prudent to conduct a risk assessment in such circumstances. However, it
should be noted that there is no mean time between failure (MTBF) until you actually
have a failure, so you need to keep on testing until you know anything meaningful
about the product's reliability.

Recording results (7.3.6)

The standard requires *the results of the validation and subsequent follow-up actions*
to be recorded.

What does this mean?
The results of validation are similar to those required for verification except that duration of testing and trials is important in quantifying the evidence. The results should not only indicate that the product meets intended use requirements but also satisfies market need.

Why is this necessary?
This requirement responds to the Factual Approach Principle.

The reasons for recording validation results are the same as for verification results but the decision based on the results is far more significant. Going into production or going operational with the wrong product will be disastrous. There have been several examples of this over the years. The Ford Edsel was a classic example as was the Sinclair C5 motorized tricycle. There was no market for such a product and sales did not materialize in the quantities anticipated.

How is this implemented?
As with design verification, consideration of the output, its format and content needs to be given early in the design phase so that the correct data is captured during validation trials.

Product approval process (7.3.6.3)

The standard requires the organization *to conform to a product and manufacturing process approval procedure recognized by the customer*.

What does this mean?
The product approval process or PPAP is intended to validate that products made from production materials, tools and processes meet the customer's engineering requirements and that the production process has the potential to produce product meeting these requirements during an actual production run at the quoted production rate.

Why is this necessary?
If you don't use a product approval process that conforms to customer requirements your part approval submissions will not be accepted. The customer needs certain information in order to carry out part evaluation. Without this information, the customer is unable to judge whether production parts will possess the same characteristics as the part being evaluated.

When one considers the potential risk involved in assembling unapproved products into production vehicles, it is hardly surprising that the customers impose such stringent requirements. The process is similar in other industries but more refined and regulated in mass production where the risks are greater.

How is this implemented?

The approval process commences following design and process verification during which a production trial run using production standard tooling, suppliers, materials, etc. produces the information needed to make a submission for product approval. Until approval is granted, shipment of production product will not be authorized. If any of the processes change then a new submission is required. Shipment of parts produced to the modified specifications or from modified processes should not be authorized until customer approval is granted.

The requirements for product approval are defined in the reference manuals. You may not need to prepare product approval submissions for *all* the parts you supply. The applicability of product approval procedures is affected by several factors so definitive solutions cannot be offered. The fundamental requirement is that if you supply product to the automotive customers you need a product approval procedure in place to gain ISO/TS 16949 registration. If you have been supplying parts for some time without product approval then you should confirm with your customer that you may continue to do so.

The documentation required varies but is likely to include the following:

- A production part submission warrant – a form that captures essential information about the part and contains a declaration about the samples represented by the warrant.
- Appearance approval report – a form that captures essential information about the appearance characteristics of the part.
- Design records including specifications, drawings and CAD/CAM mathematical data.
- Engineering change orders not yet incorporated into the design data but embodied in the part.
- Dimensional results using a pro forma or a marked up print.
- Test results.
- Process flow diagrams.
- Process FMEA.
- Design FMEA where applicable.
- Control plans.
- Process capability study report.
- Measurement systems analysis report.

The data on which the product approval submission is based should be generated during the process verification phase.

Approval of supplied parts

Your suppliers may not need to supply product approval submissions for all parts they supply but there are situations where supplier's product approval submissions are required. For example, General Motors requires product approval of all commodities supplied by suppliers to first-tier suppliers. The standard does point out that suppliers are responsible for supplier's material and services so if your submission relies on your supplier operating capable processes, you should be requesting a product approval submission from them.

Control of design and development changes (7.3.7 and 8.2.3.1)

Identification and recording of design changes (7.3.7)

The standard requires design and development changes to *be identified and records maintained. The note in Section 7.3 require this requirement to apply to product realization processes as is also indicated in Clause 8.2.3.1 where it requires records of effective dates of process changes to be maintained.*

What does this mean?

This clause covers two different requirements involving two quite different control processes. Design changes are simply changes to the design and can occur at any stage in the design process from the stage at which the requirement is agreed to the final certification that the product or process design is proven. Development changes can occur at any time in the life cycle of the design that extends until the product or process is obsolete. Following design certification, i.e. when all design verification has been completed and the product launched into production (or in the case of processes), the process begins producing deliverable product, changes to the product or process to incorporate design changes are generally classed as "modifications".

Changes to design documents are not design changes unless the characteristics of the product are altered. Changes in the presentation of design information or to the system of measurement (imperial units to metric units) are not design or development changes.

Why is this necessary?

This requirement responds to the Process Approach Principle.

You need to control design changes to permit desirable changes to be made and to prohibit undesirable changes from being made. Change control during the design process is a good method of controlling costs and timescales because once the design process has commenced every change will cost time and effort to address. This will cause delays whilst the necessary changes are implemented and provides an opportunity for additional errors to creep into the design. "If it's not broke don't fix it!" is a good maxim to adopt during design. In other words, don't change the design unless it already fails to meet the requirements or you have discovered the requirements to be wrong. Designers are creative people who love to add the latest devices, the latest technologies, to stretch performance and to go on enhancing the design regardless of the timescales or costs.

One reason for controlling design changes is to restrain the otherwise limitless creativity of designers in order to keep the design within the budget and timescale.

How is this implemented?

Product design changes

The imposition of change control is often a difficult concept for designers to accept. They would prefer change control to commence after they have completed their design rather than before they have started. They may argue that until they have finished there is no design to control. They would be mistaken. Designs proceed through a number of stages (as described previously under the heading *Design reviews*). Once the design requirements have been agreed, any changes in the requirements should be subject to formal procedures. When a particular design solution is complete and has been found to meet the requirements at a design review, it should be brought under *change control*. Between the design reviews, the designers should be given complete freedom to derive solutions to the requirements. Between the design reviews there should be no change control on incomplete solutions.

Design changes will result in changes to documentation but not all design documentation changes are design changes. This is why design change control should be treated separately from document control. You may need to correct errors in the design documentation and none of these may materially affect the product. The mechanisms you employ for such changes should be different from those you employ to make changes that do affect the design. By keeping the two types of change separate you avoid bottlenecks in the design change loop and only present the design authorities with changes that require *their* expert judgement.

Identifying and recording product design changes

The documentation for design changes should comprise the change proposal, the results of the evaluation, the instructions for change and traceability in the changed documents to the source and nature of the change. You will therefore need:

- A *Change Request Form* which contains the reason for change and the results of the evaluation – this is used to initiate the change and obtain approval before being implemented.
- A *Change Notice* that provides instructions defining what has to be changed – this is issued following approval of the change as instructions to the owners of the various documents that are affected by the change. A change notice is probably unnecessary for process changes.
- A *Change Record* that describes what has been changed – this usually forms part of the document that has been changed and can be either in the form of a box at the side of the sheet (as with drawings) or in the form of a table on a separate sheet (as

with specifications). For processes, the change record could be incorporated into the document that describes the process be it a specification, flow chart or control plan.

Where the evaluation of the change requires further design work and possibly experimentation and testing, the results for such activities should be documented to form part of the change documentation.

At each design review a design baseline should be established which identifies the design documentation that has been approved. The baseline should be recorded and change control procedures employed to deal with any changes. These change procedures should provide a means for formally requesting or proposing changes to the design. For complex designs you may prefer to separate proposals from instructions and have one form for proposing design changes and another form for promulgating design changes after approval. You will need a central registry to collect all proposed changes and provide a means for screening those that are not suitable to go before the review board (either because they duplicate proposals already made or because they may not satisfy certain acceptance criteria which you have prescribed). On receipt, the *change proposals* should be identified with a unique number that can be used on all related documentation that is subsequently produced. The change proposal needs to:

- identify the product of which the design is to be changed;
- state the nature of the proposed change;
- identify the principal requirements, specifications, drawings or other design documents which are affected by the change;
- state the reasons for the change either directly or by reference to failure reports, nonconformity reports, customer requests or other sources;
- provide for the results of the evaluation, review and decision to be recorded.

Identifying and recording product modifications

As *modifications* are changes to products resulting from design changes, the identity of modifications needs to be visible on the product that has been modified. If the issue status of the product specification changes, you will need a means of determining whether the product should also be changed. Not all changes to design documentation are design changes that result in product changes and not all product changes are modifications. (Nonconformities may be accepted which change the product but not the design.) Changes to the drawings or specifications that do not affect the form, fit or function of the product are usually called "alterations" and those which affect form, fit or function are "modifications". Alterations should come under "document control" whereas design changes should come under "configuration control". You will therefore need a mechanism for relating the modification status of products to the corresponding drawings and specifications. Following commencement of production the first design change to be incorporated into the product will usually be denoted by a number such as Mod 1 for hardware and by Version or Release number for software. The practices for software differ in that versions can be incremented by points such as 1.1, 1.2, etc., where the second digit denotes a minor change and the first digit a major change. This modification notation relates to the product whereas, issue notation relates to the documentation that describes the product. You will need a modification procedure that describes the notation to be used for hardware and software.

Within the design documentation you will need to provide for the attachment of modification plates on which to denote the modification status of the product.

Prior to commencement of production, design changes do not require any modification documentation, the design changes being incorporated in prototypes by rework or rebuild. However, when product is in production, instructions will need to be provided so that the modification can be embodied in the product. These modification instructions should detail:

- The products that are affected by part number and serial number.
- The new parts that are required.
- The work to be carried out to remove obsolete items and fit new items or the work to be carried out to salvage existing items and render them suitable for modification.
- The markings to be applied to the product and its modification label.
- The tests and inspections to be performed to verify that the product is serviceable.
- The records to be produced as evidence that the modification has been embodied.

Modification instructions should be produced after approval for the change that has been granted and should be submitted to the change control board or design authority for approval before release.

Process design changes

Process design changes can be treated in a similar way to product design changes. In some cases processes are not replicated like products, but when laying down an assembly line, more than one may be required; perhaps not in the same plant but when setting up plants in different countries, process replication is being performed but not to the extent that one would hope products are replicated.

As with product design, there should be a process specification that describes the process characteristics not simply the flow chart. The resources required, the set-up procedures, alignments and tests, etc. There may be other documents such as the process FMEA, process capability studies, layout diagrams, work instructions, process acceptance criteria, etc. All the documents representing the process configuration should be listed in a Process Design Record that is subject to version control. The version given this document is the version of the process design. Thus, any changes to the process should be reflected in changes to the Process Design Record. The record should be constructed so as to identify the versions of the listed documents relative to the version of the process design so that, for instance, one can trace the version of the

FMEA that related to a specific version of the process design. It is often the case that there is no common link between these documents. Changes are driven either by problems or review date so that it is not uncommon for the relationship between them to get out of sync. The Process Design Record is intended to prevent this.

Review and evaluation of changes (7.3.7)

The standard requires *the changes to be reviewed and approved before implementation, including the evaluation of the effect of changes on constituent parts and delivered product.*

What does this mean?

A change to a design that has not proceeded beyond a design review or verification stage is still in progress and therefore requires no approval. When a design is reviewed or verified it means that any change to the information on which that decision was taken needs to be evaluated for its effect on the design and any product produced from that design which may be in production or in service.

Why is this necessary?

This requirement responds to the Factual Approach Principle.

By controlling change you control cost so it is a vital organ of the business and should be run efficiently. The requirement for changes to be approved before their implementation emphasizes the importance of this control mechanism. The requirement for evaluation of changes for impact on product in service is necessary to transfer any benefits from the change to customers. This is especially important if the reason for change was to improve safety, reliability or regulatory compliance.

How is this implemented?

Following the commencement of design you will need to set up a change control board or panel comprising those personnel responsible for funding the design, administering the contract and for accepting the product. All change proposals should be submitted to such a body for evaluation and subsequent approval or disapproval before the changes are implemented. Such a mechanism will give you control of all design changes. By providing a two-tier system you can also submit all design documentation changes through such a body. They can filter the alterations from the modifications and the minor changes from the major changes.

The change proposals need to be evaluated:

- to validate the reason for change;
- to determine whether the proposed change is feasible;
- to judge whether the change is desirable;

- to determine the effects on performance, costs and timescales;
- to determine the impact of the change on other designs with which it interfaces and in which it is used;
- to examine the documentation affected by the change and consequently pro-gramme their revision;
- to determine the stage at which the change should be embodied.

The evaluation may need to be carried out by a review team, by suppliers or by the original proposer, however, regardless of who carries out the evaluation, the results should be presented to the change control board for a decision.

During development there are two decisions the board will need to make:

1   whether to accept or reject the change;
2   when to implement the change in the design documentation.

If the board accepts the change, the changes to the design documentation can either be submitted to the change control board or processed through your document control procedures. With CAD systems there is no reason why changes cannot be incorpor-ated immediately following their approval. One does not need to accumulate design changes for incorporation into the design when design validation has been completed.

During production the change control board will need to make four decisions:

1   Whether to accept or reject the change.
2   When to implement the change in the design documentation.
3   When to implement the modification in new product.
4   What to do with existing product in production, in store and in service.

The decision to implement the modification will depend on when the design documen-tation will be changed and when new parts and modification instructions are available. The modification instructions can either be submitted to the change control board or through your document control procedures. The primary concern of the change control board is not so much the detail of the change but its effects, its costs and the logistics in its embodiment. If the design change has been made for safety or environmental rea-sons you may need to recall product in order to embody the modification. Your modifi-cation procedures need to provide for all such cases. With safety issues, there may be regulatory procedures that need to be implemented to notify customers, recall product, implement modifications and to release modified product back into service.

Verification and validation of design changes (7.3.7)

The standard requires design changes to *be verified and validated as appropriate before implementation.*

What does this mean?
Design changes are no different than original designs in that they need to be verified and validated.

Why is this important?
This requirement responds to the Factual Approach Principle.

When a change is made to an approved design, any previous review, verification and validation may be invalidated by the change. Re-verification and validation may therefore be necessary.

How is this implemented?
Depending on the nature of the change, the verification may range from a review of calculations to a repeat of the full design verification programme. The changes may occur before the design has reached the validation phase and therefore not warrant any change to the validation programme. It is therefore necessary when evaluating a design change to determine the extent of any verification and validation that may need to be repeated. Some design changes warrant being treated as projects in their own right, recycling the full design process. Other changes may warrant verification on samples only or verification may be possible by an analysis of the differences with a proven design.

In some cases, the need for a design change may be recognized during production tests and in order to define the changes required you might wish to carry out trial modifications or experiments. Any changes to the product during production should be carried out under controlled conditions, hence the requirement that approval of design changes be given before their implementation. To allow such activities as trial modifications and experiments to proceed you will need a means of controlling these events. If the modification can be removed in a way that will render the production item in no way degraded, you can impose simple controls for the removal of the modification. If the item will be rendered unserviceable by removing the modification, alternative means may need to be determined otherwise you will sacrifice the product. It is for this reason that organizations provide development models on which to try out modifications.

Purchasing (7.4)

Purchasing process (7.4.1)

Ensuring purchased product conforms to specified requirements (7.4.1)

The standard requires the organization to *ensure that purchased product conforms to specified purchase requirements*.

What does this mean?

ISO 9000 defines a supplier as an organization or person that provides a product and in ISO/TS 16949 a product can be services, hardware, software or processed materials. A supplier may therefore be a producer, distributor, retailer, vendor, contractor, subcontractor or service provider. Purchased product is any product or service that is purchased rather than freely given or otherwise acquired and applies to any product or service that affect compliance with customer requirements. Specified purchase requirements are those requirements that are specified by the customer, the organization or by statutes and regulations that apply to purchased product. This would include any requirements limiting the conditions or the source of supply.

Why is this necessary?

This requirement responds to the Factual Approach Principle.

All organizations have suppliers of one form or another in order to provide products and service to their customers. Some of them directly or indirectly impact the product being supplied to the organization's customers and others may have no impact at all such as office supplies. From the scope of the standard we draw the conclusion that the requirement is not intended to apply to products and service that have no impact on the customer but why would you not want to manage such purchasing activities as effectively as other purchasing activities? This is not to say that one should apply the same controls and constraints to all products – one needs to apply those controls and constraints that are appropriate to their use.

How is this implemented?

Once the make or buy decision has been made, control of any purchasing activity follows a common series of activities.

There are four key processes in the procurement cycle for which you should prepare procedures:

1 The *specification process* which starts once the need has been identified and ends with a request to purchase.
2 The *evaluation process* which starts with the request to purchase and ends with the placement of the order or contract.
3 The *surveillance process* which starts with placement of order or contract and ends on delivery of supplies.
4 The *acceptance process* which starts with delivery of supplies and ends with entry of supplies onto the inventory and/or payment of invoice.

Whatever you purchase the processes will be very similar although there will be variations for purchased services such as subcontract labour, computer maintenance, consultancy services, etc. Where the purchasing process is relatively simple, one route may suffice but where the process varies you may need separate routes so as to avoid *all* purchases, regardless of value and risk, going through the same process and incurring unnecessary costs and delay. A typical procurement process flow is illustrated in Figure 7.11.

Control of suppliers (7.4.1)

The standard requires *the type and extent of control applied to the supplier and the purchased product to be dependent on the effect of the purchased product upon subsequent product realization or the final product.*

What does this mean?
The requirement contains two quite separate requirements: one applying to the product and the other applying to the supplier.

Regarding the product, the type of control refers to whether the controls should act before, during or after receipt of product. The extent of control refers to whether it is remote or on the suppliers' premises and whether product is accepted on the basis of supplier data or is to be evaluated before authorizing delivery.

Regarding the supplier the type of control refers to whether or not to qualify the supplier and the extent of control refers to the degree to which the organization is involved with the supplier in managing the purchase.

Purchased product can have varying degrees of impact on the processes and products of the organization ranging from no impact to critical impact. A product with a critical impact would warrant stringent control over its purchase, whereas a product with negligible or no impact may warrant no more than a simple visual check on delivery to verify receipt of the right product.

Why is this necessary?
This requirement responds to the Mutually Beneficial Supplier Relationships Principle.

As it would not be prudent to exercise no control over suppliers, it would also be counterproductive to impose rigorous controls over every supplier and purchased product. A balance has to be made on the basis of risk to the processes in which the purchased product is to be used and the final product into which the purchased product may be installed.

How is this implemented?
Selecting the degree of control
You need some means of verifying that the supplier has met the requirements of your order, and the more unusual and complex the requirements, the more control will be required. The degree of control you need to exercise over your suppliers depends on the

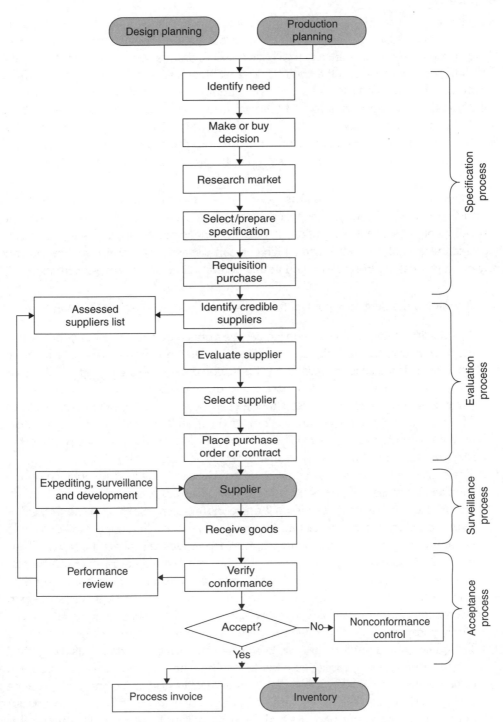

Figure 7.11 *Procurement process flow*

confidence you have in their ability to meet your requirements. If you have high confidence in a particular supplier you can concentrate on the areas where failure is more likely. If you have no confidence you will need to exercise rigorous control until you gain sufficient confidence to relax the controls. The fact that a supplier has gained ISO 9000 registration for the products and service you require should increase your confidence, but if you have no previous history of their performance it does not mean they will be any better than the supplier you have used for years which is not registered to ISO 9000. Your supplier control procedures need to provide the criteria for selecting the appropriate degree of control and for selecting the activities you need to perform.

With suppliers of proprietary products, your choices are often limited because you have no privileges. Control over your suppliers is therefore exercised by the results of receipt inspection or subsequent verification. If your confidence in a supplier is low you can increase the level of verification and if high you can dispense with receipt verification and rely on in-process controls to alert you to any deterioration in supplier performance.

In determining the degree of control to be exercised you need to establish whether:

- The quality of the product or service can be verified on receipt using your normal inspection and test techniques. Sampling inspection on receipt should be used when statistical data is unavailable to you or you don't have the confidence for permitting ship-to-line. (This is the least costly of methods and usually applies where achievement of the requirements is measurable by examination of the end product.)
- The quality of the product can be verified on receipt providing you acquire additional equipment or facilities. (This is more costly than the previous method but may be economic if there is high utilization of the equipment.)
- The quality of the product can be verified by witnessing the final acceptance on the suppliers' premises. (If you don't possess the necessary equipment or skill to carry out product verification, this method is an economic compromise and should yield as much confidence in the product as the previous methods. You do, however, need to recognize that your presence on the supplier's premises may affect the results. They may omit tests that are problematical or your presence may cause them to be particularly diligent, a stance that may not be maintained when you are not present.)
- The verification of the product could be contracted to a third party such as a part evaluation laboratory. (This can be very costly and is usually only applied with highly complex products and where safety is of paramount importance.)
- The quality of the product can only be verified by the supplier during its design and manufacture. (In such cases you need to rely on what the contractor tells you and to gain sufficient confidence you can impose quality system requirements, require certain design, manufacturing, verification documents to be submitted to you for approval and carry out periodic audit and surveillance activities. This method is usually applied to one-off systems or small quantities when the stability of a long

production run cannot be achieved to resolve problems or when re-qualifying a supplier after production problems have been detected).

In order to relate the degree of inspection to the importance of the item, you should categorize purchases as follows:

- If the subsequent discovery of nonconformity will *not* cause design, production, installation or operational problems of any nature, a simple identity, carton quantity and damage check may suffice. An example of this would be stationery.
- If the subsequent discovery of nonconformity will cause *minor* design, production, installation or operational problems, you should examine the features and characteristics of the item on a sampling basis. An example of this would be electrical, electronic or mechanical components.
- If the subsequent discovery of nonconformity will cause *major* design, production, installation or operational problems, then you should subject the item to a complete test to verify compliance with all prescribed requirements. An example of this would be an electronic unit.

These criteria would need to be varied depending on whether the items being supplied were in batches or separate. However, these are the kind of decisions you need to take in order to apply practical receipt verification procedures.

Defining supplier controls

When carrying out supplier surveillance you will need a plan which indicates what you intend to do and when you intend to do it. You will also need to agree the plan with your supplier. If you intend witnessing certain tests the supplier will need to give you advanced warning of commencing such tests so that you may attend.

The quality plan would be a logical place for such controls to be defined. Some companies produce a Quality Assurance Requirement Specification to supplement ISO 9001 and also produce a Supplier Surveillance Plan. In most other cases, the controls may be defined on the reverse side of the purchase order as standard conditions coded and selected for individual purchases. However, don't impose onerous requirements on simple purchases. Requiring test samples for literature you purchase or ISO 9000 certification from a bookseller is rather ludicrous. Make provision for the relevant conditions to be selected by the buyer otherwise you run the risk of suppliers ignoring requirements that might be relevant.

Evaluation and selection of suppliers (7.4.1)

The standard requires the organization to *evaluate and select suppliers based on their ability to supply product in accordance with the organization's requirements and to establish criteria for selection, evaluation and re-evaluation.*

What does this mean?

In searching for a supplier you need to be confident that the supplier can provide the product or service you require. This means that the decision to select a supplier should be based on knowledge about that supplier's capability to meet your requirements. The decision should be based on facts gathered as a result of an evaluation against criteria that you have established. Note 2 in ISO/TS 16949 Clause 7.4.1 advises that continuity of supply should be verified following mergers, acquisitions or other affiliations.

Why is this necessary?

This requirement responds to the Mutually Beneficial Supplier Relationships Principle.

It would be foolish to select a supplier without first verifying that it was able to meet your requirements in some way or other. Failure to check out the supplier and its products may result in late delivery of the wrong product. It may also mean that you might not know immediately that the product does not meet your requirement and discover much later that it seriously impacts your commitments to your customer.

How is this implemented?

Selection process

The process for selection of suppliers varies depending on the nature of the products and services to be procured. The more complex the product or service, the more complex the process. You either purchase products and services to your specification (custom) or to the supplier's (proprietary). For example, you would normally procure stationery, fasteners or materials to the *supplier's specification* but procure an oil filter, brake pads or body panel to *your specification*. There are grey areas where proprietary products can be tailored to suit your needs and custom made products or services that primarily consist of proprietary products configured to suit your needs. There is no generic model, each industry seems to have developed a process to match its own needs. However, we can treat the process as a number of stages some of which do not apply to simple purchases as shown in Table 7.2. At each stage the number of potential suppliers is whittled down to end with the selection of what is hoped to be the most suitable that meets the requirements. With "custom" procurement this procurement cycle may be exercised several times. For instance, there may be a competition for each phase of project feasibility, project definition, development and production. Each phase may be funded by the customer. On the other hand, a supplier may be invited to tender on the basis of previously demonstrated capability but has to execute project feasibility, project definition and development of a new version of a product at its own cost. Supplier capability will differ in each phase. Some suppliers have good design capability but lack the capacity for quantity production; others have good research capability but lack development capability.

Table 7.2 *Supplier evaluation and selection stages*

Stage	Purpose	Proprietary	Tailored	Custom
Preliminary Supplier Assessment	To select credible suppliers	✓	✓	✓
Pre-qualification of Suppliers	To select capable bidders		✓	✓
Qualification of Suppliers	To qualify capable bidders			✓
Request for Quotation (RFQ)	To obtain prices for products or services	✓	✓	
Invitation to Tender (ITT)	To establish what bidders can offer			✓
Tender or Quote Evaluation	To select a supplier	✓	✓	✓
Contract Negotiation	To agree terms and conditions	✓	✓	✓

You need to develop a supplier evaluation and selection process, and in certain cases this may result in several closely related procedures for use when certain conditions apply. Do not try to force every purchase through the same selection process. Having purchasing policies that require three quotations for every purchase regardless of past performance of the current supplier is placing price before quality. Provide flexibility so that process complexity matches the risks anticipated. Going out to tender for a few standard nuts and bolts would seem uneconomical. Likewise, placing an order for £1 million of equipment based solely on the results of a third-party ISO 9000 certification would seem unwise.

Preliminary supplier assessment
The purpose of the preliminary supplier assessment is to select a *credible* supplier and not necessarily to select a supplier for a specific purchase. There are millions of suppliers in the world, some of which would be happy to relieve you of your wealth given half a chance, and others that take pride in their service to customers and are a pleasure to have as partners. You need a process for gathering intelligence on potential suppliers and for eliminating unsuitable suppliers so that the buyers do not need to go through the whole process from scratch with each purchase. The first step is to establish the type of products and services you require to support your business, and then search for suppliers that claim to provide such products and services. In making your choice, look at what the supplier says it will do and what it has done in the past. Is it

the sort of firm that does what it says it does or is it the sort of firm that says what you want to hear and then conducts its business differently? Some of the checks needed to establish the credibility of suppliers are time consuming and would delay the selection process if undertaken only when you have a specific purchase in mind. You will need to develop your own criteria but typically unsuitable suppliers may be those that:

- are unlikely to deliver what you want in the quantities you may require;
- are unable to meet your potential delivery requirements;
- cannot provide the after-sales support needed;
- are unethical;
- do not comply with the health and safety standards of your industry;
- do not comply with the relevant environmental regulations;
- do not have a system to assure the quality of supplies;
- are not committed to continuous improvement;
- are financially unstable.

You may also discriminate suppliers on political grounds such as a preference for supplies from certain countries or a requirement to exclude supplies from certain countries.

The supplier assessment will therefore need to be in several parts:

- *Technical assessment*: This would check the products, processes or services to establish they are what the supplier claims them to be. Assessment of design and production capability may be carried out at this stage or be held until the pre-qualification stage when specific contracts are being considered.
- *Quality system assessment*: This would check the certification status of the quality system, verifying that any certification was properly accredited. For non-ISO 9000 registered suppliers, a quality system assessment may be carried out at this stage either to ISO 9000 or the customer's standards.
- *Financial assessment*: This would check the credit rating, insurance risk, stability, etc.
- *Ethical assessment*: This would check probity, conformance with professional standards and codes.

These assessments do not need to be carried out on the supplier's premises. Much of the data needed can be accumulated from a supplier questionnaire and searches through directories and registers of companies, and you can choose to rely on assessments carried out by accredited third parties to provide the necessary level of confidence. (The Directories of Companies of Assessed Capability that are maintained by the Accreditation Agencies can be a good place to start.) The assessments may yield suppliers over a wide range and you may find it beneficial to classify suppliers (see Table 7.3, for reference).

Caution is advised on the name you give to this list of suppliers. All have been assessed (even if it is limited to a literature survey) but all may not have been visited or used.

Table 7.3 *Classification of suppliers*

Class A	ISO 9000 certified and demonstrated capability. This is the class of those certified suppliers with which you have done business for a long time and gathered historical evidence that proves their capability.
Class B	Demonstrated capability. This is the class of those suppliers you have done business with for a long time and warrant continued patronage on the basis that its better to deal with those suppliers you know than those you do not. They may not even be contemplating ISO 9000 certification, but *you* get a good product, a good service and no hassle. If these suppliers were providing parts to your specification you would need to encourage these suppliers to seek ISO 9000 certification in order to meet your commitment to ISO/TS 16949. If you are purchasing proprietary parts the quality of which can be assured on receipt, ISO 9001 certification is unnecessary.
Class C	ISO 9000 certified and no demonstrated capability. This is the class of those certified suppliers with whom you have done no business. This may appear a contradiction because ISO 9000 certification is obtained on the basis of demonstrated capability, but you have not established their capability to meet *your* requirements. These suppliers may be customer-designated suppliers.
Class D	Capable with additional assurance. This is the class of first-time suppliers with which you have not done sufficient business to put in Class B and where you may need to impose ISO 9000 requirements or similar to gain the confidence you need
Class E	Unacceptable performance that can be neutralized. This class is for those cases where you may be able to compensate for poor performance, if they are sole suppliers of the product or service. Unfortunately, with ISO/TS 16949, you will still need to impose ISO 9001 on these suppliers if they provide parts to your specification.
Class F	No demonstrated capability. This is the class of those suppliers you have not used before and therefore have no historical data. It is unlikely you will classify these but it is a category for placing those suppliers that produce the type of products you purchase and so might become eligible when they become eligible for Class C status
Class G	Demonstrated unacceptable performance. This is the class of those suppliers that have clearly demonstrated that their products and services are unacceptable, and it is uneconomic to compensate for their deficiencies.

Some organizations refer to it as an Approved Suppliers List (ASL) or Approved Vendor List (AVL) or Assessed Vendor List (AVL), but if you include unacceptable suppliers you cannot call it an Approved Vendor List. If it is in paper form, two lists may be preferable. Some organizations use coloured paper to distinguish between approved and unapproved suppliers. If the data is stored electronically, the fields can be protected to prevent selection of unacceptable suppliers.

If your requirements vary from product to product, suppliers approved for one product may not be approved for others. If your procurement requirements do not vary from product to product, you may well be able to maintain an AVL. Most will meet your minimum criteria for doing business with your company but may not be capable of meeting specific product or service requirements. Others you will include simply because they do supply the type of product or service you require but their credibility is too low at present to warrant preferred status. In the process you have eliminated the 'cowboys' or 'rogues' – there is no point in adding these to the list because you have established that they won't change in the foreseeable future.

Pre-qualification of suppliers

Pre-qualification is a process for selecting suppliers for known future work. The design will have proceeded to a stage where an outline specification of the essential parameters has been developed. You know roughly what you want but not in detail. Pre-qualification is undertaken to select those suppliers that can demonstrate they have the capability to meet your specific requirements on quality, quantity, price and delivery. A supplier may have the capability to meet quality, quantity and price requirements, but not have the capacity available when you need the product or service. One that has the capacity may not offer the best price and one that meets the other criteria may not be able to supply product in the quantity you require.

A list of potential bidders can be generated from the Assessed Suppliers List by searching for suppliers that match given input criteria specific to the particular procurement. However, the evidence you gathered to place suppliers on your Assessed Suppliers List may now be obsolete. Their capability may have changed and therefore you need a sorting process for specific purchases. If candidates are selected that have not been assessed, an assessment should be carried out before proceeding any further.

Once the list is generated a Request for Quotation (RFQ) or Invitation to Tender (ITT) can be issued depending on what is required. RFQs are normally used where price only is required. This enables you to disqualify bidders offering a price well outside your budget. ITTs are normally used to seek a line-by-line response to technical, commercial and managerial requirements. At this stage you may select a number of potential suppliers and require each to demonstrate its capability. You know what they do but you need to know if they have the capability of producing a product with specific characteristics and can control its quality.

When choosing a bidder you also need to be confident that continuity of supply can be assured. One of the benefits of ISO 9001 certification is that it should demonstrate that the supplier has the capability to supply certain types of products and services. However, it is not a guarantee that the supplier has the capability to meet *your* specific requirements. Suppliers that have not gained ISO 9001 registration may be just as good.

(There is no evidence to demonstrate that ISO 9001 certified suppliers perform significantly better than non-certificated suppliers.) If the product or service you require can only be obtained from a non-registered supplier you will need to encourage that supplier to seek ISO 9001 certification to win your business or seek a concession from your customer. Using an ISO 9001 registered supplier should enable you to reduce your supplier controls.

Depending on the nature of the work you may require space models, prototypes, process capability studies, samples of work as evidence of capability. You may also make a preliminary visit to each potential bidder but would not send out an evaluation team until the qualification stage.

Qualification of suppliers

Of those potential bidders that are capable, some may be more capable than others. Qualification is a stage executed to compile a short list of bidders following pre-qualification. A detail specification is available at this stage and production standard models may be required to qualify the design. Some customers may require a demonstration of process capability to grant production part approval.

During this stage of procurement a series of meetings may be held depending on the nature of the purchase. A pre-bid meeting may be held on the customers premises to enable the customer to clarify the requirements with the bidders. A mid-bid meeting or pre-award assessment may be held on the supplier's premises at which the customer's Supplier Evaluation Team carries out a capability assessment on site. This assessment may cover:

- an evaluation of the product;
- an audit of design and production plans to establish that, if followed, they will result in compliant product;
- an audit of operations to verify that the approved plans are being followed;
- an audit of processes to verify their capability;
- an inspection and test of product (on- or off-site) to verify that it meets the specification.

The result of supplier qualification is a list of capable suppliers that will be invited to bid for specific work.

ISO 9001 certification was supposed to reduce the amount of supplier assessments by customers and it has in certain sectors. However, the ISO 9001 certification whilst focused on a specific scope of registration is often not precise enough to give confidence to customers for specific purchases.

The evaluation may qualify two or three suppliers for a specific purchase. The tendering process will yield only one winner but the other suppliers are equally suitable and should not be disqualified because they may be needed if the chosen supplier fails to deliver.

Invitation to tender
Once the bidders have been selected, an ITT needs to be prepared to provide a fixed baseline against which unbiased competitive bids may be made. The technical, commercial and managerial requirements should be finalized and subject to review and approval prior to release. It is important that all functions with responsibilities in the procurement process review the tender documentation. The ITT will form the basis of any subsequent contract.

The requirements you pass to your bidders need to include as appropriate:

* The tender conditions, date, format, content, etc.
* The terms and conditions of the subsequent contract.
* A specification of the product or service that you require that transmits all of the relevant requirements of the main contract (see under the heading *Purchasing specifications*).
* A specification of the means by which the requirements are to be demonstrated (see under the heading *Purchasing specifications*).
* A statement of work which you require the supplier to perform – it might be design, development, management or verification work and will include a list of required deliverables such as project plans, quality plans, production plans, drawings, test data, etc. (You need to be clear as to the interfaces both organizationally and technically.)
* A specification of the requirements which will give you an assurance of quality – this might be a simple reference to ISO 9001, but as this standard does not give you any rights or flow down your customers requirements you will probably need to amplify the requirements (see under the section *Quality management system requirements*).

In the tendering phase each of the potential suppliers is in competition, so observe the basic rule that what you give *one* must be given to *all!* It is at this stage that your supplier conducts the tender review defined in Clause 7.2.2 of ISO/TS 16949.

Tender or quote evaluation
On the due date when the tenders should have been received, record those that have been submitted and discard any submitted after the deadline. Conduct an evaluation to determine the winner – the supplier that can meet all your requirements (including confidence) for the lowest price. The evaluation phase should involve all your staff that were involved with the specification of requirements. You need to develop scoring criteria so that the result is based on objective evidence of compliance.

The standard does not require that you only purchase from "approved suppliers". It does require that you maintain records of the results of supplier evaluations but does not prohibit you from selecting suppliers that do not fully meet your purchasing requirements. There will be some suppliers that fully meet your requirements and others that provide a product with the right functions but quality, price and delivery may be less than you require. If the demonstrated capability is lacking in some respects you can adjust your controls to compensate for the deficiencies.

In some cases, your choice may be limited to a single source because no other supplier may market what you need. On other occasions you may be spoilt for choice. With some proprietary products you are able to select particular options so as to tailor the product or service to your requirements. It remains a proprietary product because the supplier has not changed anything just for you. The majority of products and services you will purchase from suppliers, however, is likely to be from catalogues. The designer may have already selected the item and quoted the part number in the specification. Quite often you are buying from a distributor rather than the manufacturer and so need to ensure that *both* the manufacturer and the distributor will meet your requirements.

Contract negotiation

After selecting a "winner" you may need to enter contract negotiations in order to draw up a formal subcontract. It is most important that none of the requirements are changed without the supplier being informed and given the opportunity to adjust the quotation. It is at this stage that your supplier conducts the requirement review defined in Clause 7.2.2 of ISO/TS 16949. It is pointless negotiating the price of products and services that do not meet your needs. You will just be buying a heap of trouble! Driving down the price may also result in the supplier selling their services to the highest bidder later and leaving you high and dry!

Satisfying regulatory requirements

The first step in meeting this requirement is to establish a process that will identify all current regulatory requirements pertaining to the part or material. You need to identify the regulations that apply in the country of manufacture and the country of sale. This may result in two different sets of requirements. For example, a part may be manufactured in Mexico and sold in California or made in UK and sold in India. In one case, the regulations on recycling materials may be tougher in the country of sale and in the other case, there may be restrictions prohibiting sale of vehicles containing materials from a particular country. It is difficult to keep track of changes in import and export regulations but using the services of a legal department or agency will ease the burden.

In order to ensure compliance with this requirement you need to impose on your suppliers through the purchase order, the relevant regulations and through examination of specifications, products and by on-site assessment, verify that these regulations are being met. It is not sufficient to merely impose the requirement on your supplier through the purchase order. You can use the certified statements of authorized independent inspectors as proof of compliance instead of conducting the assessment yourself. However, such inspections may not extend to the product being supplied and therefore a thorough examination by your technical staff will be needed. Once deemed compliant, you need to impose change controls in the contract that prohibit the supplier changing the process or the product without your approval. This may not be possible when dealing with suppliers supplying product to their specification or when using offshore suppliers where the system of law enforcement cannot be relied on. In such cases, you will need to accurately define the product required and carry out periodic verification for continued compliance.

Criteria for periodic evaluation

For one-off purchases periodic re-evaluation would not be necessary. Where a commitment from both parties is made to supply products and services continually until terminated, some means of re-evaluation is necessary as a safeguard against deteriorating standards.

The re-evaluation may be based on supplier performance, duration of supply, quantity, risk or changes in requirements and conducted in addition to any product verification that may be carried out. Suppliers are no different than customers in that their performance varies over time. People, organizations and technologies change and may impact the quality of the service obtained from suppliers. *In addition the effect of recent mergers, acquisitions and affiliations on the effectiveness of the quality system should be verified.*

Results of supplier evaluation

The standard requires *the records of evaluations and any necessary follow-up actions to be maintained.*

What does this mean?

Records of evaluations are documents containing the results of the evaluation. This is not the ASL or AVL used to select suppliers but the objective evidence that was used to make the decision as to whether a supplier should be listed in such a document.

Why is this necessary?

This requirement responds to the Factual Approach Principle.

Records of supplier evaluation are necessary in order to select suppliers on the basis of facts rather than opinion. They are also necessary for comparisons between competing suppliers as a mere listing provides little information on which to judge acceptability.

How is this implemented?

Although records of evaluations are not the same as a list of approved suppliers, there is a need for both. The list identifies which suppliers have been or not been evaluated and the records support the decision to include suppliers in the list or qualify suppliers in different categories.

Evaluation records

Evaluation records can be classified in three groups:

1 Initial evaluation for supplier selection.
2 Supplier performance monitoring.
3 Re-evaluation to confirm approval status.

The initial evaluation records would include the evaluation criteria, the method used, the results obtained and the conclusions. They may also include information relevant to the supplier such as supplier history, advertising literature, catalogues and approvals. These records may not contain actions and recommendations because the evaluation may have been carried out under a competitive tender. The actions come later, when re-evaluations are performed and continued supply is decided.

You should monitor the performance of all your suppliers and classify each according to prescribed guidelines. Supplier performance will be evident from audit reports, surveillance visit reports and receipt inspections carried out by you or the third party if one has been employed. You need to examine these documents for evidence that the supplier's quality system is controlling the quality of the products and services supplied. You can determine the effectiveness of these controls by periodic review of the supplier's performance. What some firms call "vendor rating". By collecting data on the performance of suppliers over a long period you can measure their effectiveness and rate them on a scale from excellent to poor. In such cases you should measure at least three characteristics, quality, delivery and service. Quality would be measured by the ratio of defective products to conforming products. Delivery would be measured by the number of days early or late and service would be measured by the responsiveness to actions requested by you on scale of excellent to poor. The output of these reviews should be in the form of updates to the list of assessed suppliers.

Re-evaluation records would include all the same information as the initial evaluation but in addition contain follow-up actions and recommendations, the supplier's response and evidence that any problems have been resolved.

Listing suppliers

It is important that you record those suppliers that should not be used due to previously demonstrated poor performance so that you don't repeat the mistakes of the past.

Assessing suppliers is a costly operation. Having established that a supplier has or hasn't the capability of meeting your requirements you should enter their details in a database. The database should be made available to the purchasing authority thereby avoiding the necessity of re-assessments each time you wish to place an order. The database of assessed suppliers should not only identify the name and address of the company but also provide details of the products and service that have been evaluated. This is important because the evaluation performed to place suppliers on the list will have only covered particular products and services. Other products and services offered by the supplier may not have been acceptable. Some firms operate several production lines each to different standards. Just because one production line met your requirement doesn't mean that the other production line will also meet your requirements. Calling it a List of Assessed Suppliers does not imply that it only lists approved firms – it allows you to include records of all firms with which you have done business and classify them accordingly. By linking purchases with the List of Assessed Suppliers you can indicate usage status, e.g. current, dormant or unused.

You will need a procedure for generating and managing the database of Assessed Suppliers adding new suppliers, changing data and reclassifying suppliers that no longer meet your criteria.

Regulatory compliance (7.4.1.1)

The standard requires *all purchased products or materials used in product to conform to applicable regulatory requirements.*

What does this mean?
From whichever country the products are purchased, they will need to meet the prevailing health, safety and environmental regulations of the country or state into which they will be supplied. Such regulations might apply not only to the use, handling or installation of the product but its storage and disposal.

Why is this necessary?
This is so that products may pass through the relevant customs controls without a problem and be installed in vehicles without hazard to operators or end users and detriment to the natural environment.

How is this implemented?
Parts that are designed by the vehicle maker should not pose a problem because the specifications for materials and processes are known, and will have addressed the applicable regulations. Proprietary parts pose the biggest problem because their composition and specification might be unknown.

It will be necessary to firstly determine the regulations that apply in the country where the intermediate and end product will be used, stored and disposed and establish whether the parts being purchased satisfy these regulations. Assurances might be given in writing by the supplier and confirmed later by testing samples. This should form part of the component approval process.

Supplier quality management system development (7.4.1.2)

The standard requires the organization *to perform supplier quality management system development with the goal of supplier conformity with ISO/TS 16949 and conformity with ISO 90001:2000 being the first step in achieving this goal.*

What does this mean?
The increasing trend for customers to develop partnerships with suppliers has led to supplier development programmes where customers work with suppliers to develop their capability to improve process capability, delivery schedules or reduce avoidable costs. These programmes replace re-evaluations because they are ongoing and any deterioration in standards is quickly detected.

This requirement means that as a minimum, all of your suppliers should be registered to ISO 9001:2000 by an accredited certification body. First-tier suppliers should also develop their suppliers (second-tier suppliers) to achieve conformity with ISO/TS 16949. There is no requirement for second-tier suppliers to be registered to ISO/TS 16949: 2002 unless stated otherwise in the contract between first- and second-tier suppliers. This equally applies down the supply chain.

Why is this necessary?
The systems in place need to be tuned so as to sing to the same hymn sheet and this tuning may require the co-operation and collaboration of the several parties involved. Hence, the quality management systems of your suppliers need to be tuned to deliver product that meets your specific requirements just as your systems need to be tuned to deliver to your customer's-specific requirements. The standard recognizes that the products being produced are principally, designed for a specific vehicle, meeting specific customer requirements. Unlike proprietary products that are totally the responsibility of the supplier, custom products represent a shared responsibility.

How is this implemented?
So if your contract does not specifically require ISO/TS 16949:2002 certification by an accredited third party, you don't have to seek ISO/TS 16949 certification nor do you have to impose ISO/TS 16949 certification on your suppliers. If you choose this approach and your organization is a first-tier supplier, you may elect to perform your own

supplier assessments against ISO/TS 16949 or simply to require third-party certification to ISO/TS 16949. Whether or not you impose ISO/TS 16949 certification, your suppliers need to be registered to ISO 9001:2000 to begin with.

The requirement implies that supplier development may involve developing a supplier to achieve ISO 9001:2000 certification and then advancing to ISO/TS 16949 compliance or certification. You need to work with your suppliers in enabling them to meet these requirements. It is not a case of imposing the requirement and letting the third party do your work for you. The requirement for supplier development rests on your shoulders and not on those of the certification body.

As part of your development effort you might perform the following as appropriate:

*   Initial assessment to ISO 9001:2000 or ISO/TS 16949 to size the gap in compliance and effectiveness.
*   Awareness sessions developing an understanding of the requirements, their importance and impact.
*   Coaching with respect to the development of solutions to the requirements or specific improvement programmes.
*   Full capability assessment to ISO/TS 16949:2002.
*   Semi-annual confirmation of capability.
*   Re-confirmation of capability after taking corrective action.

You can't develop all your suppliers and hence you should prioritize those for development based on performance and importance of product or service supplied.

Customer-approved sources (7.4.1.3)

The standard requires the organization *to purchase products, materials or services from approved sources when specified by the contract.*

What does this mean?
Where the product your organization is required to supply to an automotive customer incorporate parts produced by other suppliers, your customer will instruct you to purchase such parts from designated sources.

The approved sources will be specified in the contract so it should not be difficult to establish contact with them. However, just because the customer has designated these suppliers, the quality of the products they supply remains the responsibility of your organization. It might appear that your customer has removed your rights but they have not. When the customer says you are responsible for the quality of these products it means that you do have the authority, delegated by your customer to take whatever action is necessary to enforce the standards those parts are required to meet and the standards under which those parts are required to be produced.

Why is this necessary?

The customer is often the design authority either for the product your organization suppliers or the assembly into which your organization's product will be integrated. In order to maintain the integrity of the assembly, the parts need to be supplied by the same suppliers that were granted part approval for parts within that assembly. Also, your customer may have invested heavily in the development of that supplier and will not wish to sacrifice that investment.

How is this implemented?

Quite simply you place orders on the suppliers designated in your contract. You need to take care to ensure that the conditions contained in the purchase order are consistent with those your customer imposes on your organization otherwise the supplier may have justification for deviation. Apart from this, manage them just the same as you manage your other suppliers. Report problems with these suppliers promptly to your customer and do not wait for your customer to react – take the action you believe is necessary. Just be careful not to breach the conditions of the purchase order.

In some cases, the division of responsibility might not be straightforward. The customer might be on one country, the approved source in another and your organization in a third. You may have inherited a programme from another supplier, a merger, takeover; anything could have brought about the situation in which your organization now finds itself. A documented agreement may need to be produced defining the division of responsibility between your organization, your customer and any other third party that may be involved. It is essential that this agreement be developed from the outset otherwise, months or even years into the programme, disputes will be a recurring feature and cause deterioration in performance, delay payment from your customer and ultimately cause failures in service.

Purchasing information (7.4.2)

Describing products to be purchased

The standard requires purchasing information to *describe the product to be purchased*.

What does this mean?

Purchasing information is the information that identifies the product or service to be purchased and which is used to make purchasing decisions. Not all of this information may be conveyed to the supplier. Some information may be needed by buyers to select the correct product or service required.

Why is this necessary?
This requirement responds to the Process Approach Principle.

The supplier needs to know what the organization requires before it can satisfy the need and although the standard does not specifically require the information to be recorded, you need to document purchasing requirements so that you have a record of what you ordered. This can then be used when the goods and the invoice arrive to confirm that you have received what you ordered. The absence of such a record may prevent you from returning unwanted or unsatisfactory goods.

How is this implemented?
The essential purchasing information must be communicated to suppliers so that they know what you require, but it is not essential to submit your purchasing documents to your suppliers. In fact many purchases will be made from catalogues by telephone, quoting reference number and quantity required. Providing you have a record and can compare this with the goods received and the invoice, you are protected against paying for goods you didn't order.

Product identification
The product or service identification should be sufficiently precise as to avoid confusion with other similar products or services. The supplier may produce several versions of the same product and denote the difference by suffixes to the main part number. To ensure you receive the product you require you need to carefully consult the literature provided and specify the product in the same manner as specified in the literature or as otherwise advised by the supplier.

Purchasing specifications
If you are procuring the services of a supplier to design and/or manufacture a product or design and/or deliver a service, you will need specifications which detail all the features and characteristics which the product or service is to exhibit. The reference number and issue status of the specifications need to be specified in the event that they change after placement of the purchase order. This is also a safeguard against the repetition of problems with previous supplies. These specifications should also specify the means by which the requirements are to be verified so that you have confidence in any certificates of compliance that are supplied. For characteristics that are achieved using special processes you need to ensure that the supplier employs qualified personnel and equipment. Products required for particular applications need to be qualified for such applications and so your purchasing documents will need to specify what qualification tests are required.

Quality management system requirements
Management system requirements are only necessary inclusions in purchasing information when the quality of the product cannot be verified on receipt or when confidence

in the product and the supplier is needed to permit the supplier to ship direct into stock or onto the production line. This might appear at odds with the requirement for all suppliers to be certified to ISO 9001 but this requirement applies to suppliers providing products to your specification. If specifying ISO 9001 on the purchase order will not increase your level of confidence then it would be foolish to do so. There is little point in imposing ISO 9000 on non-registered suppliers when ordering from a catalogue. It only makes sense when the supplier is prepared to make special arrangements for your particular order – arrangements which may well cost you more for no added value. Remember that the requirements in Clause 7.4.2 apply only where appropriate.

Adequacy of purchasing requirements

The standard *requires* the organization to *ensure the adequacy of specified purchase requirements prior to their communication to the supplier.*

What does this mean?

The adequacy of purchasing information is judged by the extent to which it accurately reflects the requirements of the organization for the products concerned. Communication of such requirements to the supplier can be verbal or through documentation and processed by post or electronically.

Why is this necessary?

This requirement responds to the Factual Approach Principle.

The acceptance of an order by a supplier places it under an obligation to accept product or service that meets the stated requirements. It is therefore important that such information is deemed adequate before being released to the supplier.

How is this implemented?

Prior to orders being placed the purchasing information should be checked to verify that it is fit for its purpose. The extent to which you carry out this activity should be on the basis of risk and if you choose not to review and approve all purchasing information, your procedures should provide the rationale for your decision. In some cases, orders are produced using a computer and transmitted to the supplier directly without any evidence that the order has been reviewed or approved. The purchase order does not have to be the only purchasing document. If you enter purchasing data onto a database, a simple code used on a purchase order can provide traceability to the approved purchasing documents.

You can control the adequacy of the purchasing data in four ways:

1 Provide the criteria for staff to operate under self-control.
2 Check everything they do.

3   Select those orders that need to be checked on a sample basis.
4   Classify orders depending on risk, and only review and approve those that present
    a certain risk.

A situation where staff operate under self-control would be in the case of telephone
orders where there is little documentary evidence that a transaction has taken place.
There may be an entry on a computer database showing that an order has been placed
with a particular supplier. So how would you ensure the adequacy of purchasing
requirements in such circumstances? There follows a number of steps you can take:

* Provide buyers with read-only access to approved purchasing data in the database.
* Provide buyers with read-only access to a list of approved suppliers in the database.
* Provide a computer file containing details of purchasing transactions with read and
  write access.
* Provide a procedure that defines the activities, responsibilities and authority of all
  staff involved in the process.
* Train the buyers in the use of the database.
* Route purchase requisitions only to trained buyers for processing.

The above approach is suitable for processing routine orders, however, where there are
non-standard conditions a more variable process needs to be developed.

Verification of purchased product (7.4.3 and 7.4.3.1)

Ensuring purchased product meets requirements (7.4.3)

The standard requires the organization to *establish and implement the inspection or
other activities necessary for ensuring that purchased product meets specified purchase requirements. The automotive additions also require the organization to
have a process to assure the quality of purchased product utilizing one or more of
five defined methods.*

What does this mean?

Verification is one of the fundamental elements of the control loop and in this case the
verification serves to ensure that the output from the purchasing process meets the
purchasing requirement. Verification may be achieved by several means, inspection,
test, analysis before or after product is delivered or by building confidence in the source
of supply so that product may enter the organization without any physical inspection.
The requirement does not state when such verification should be performed and clearly
it can be before, during and after receipt of the product. The standard leaves it to the
organization's discretion to choose the timing that is appropriate to its operations.

Why is this necessary?

This requirement responds to the Factual Approach Principle.

When we purchase items as individuals it is a natural act to inspect what you have purchased before you use it. To neglect to do this may result in you forfeiting your rights to return it later if found defective or nonconforming. When we purchase items on behalf of our employers we may not be as tenacious, so the company has to enforce its own verification policy as a way of protecting itself from the mistakes of its suppliers. Another reason for product verification is that it is often the case that characteristics are not accessible for inspection or test after subsequent processing. Characteristics that have not been verified prior to or on receipt may never be verified.

How is this implemented?

There are several ways of verifying that purchased product meets requirements and these were outlined previously under the heading *Selecting the degree of control*. Assessments by third parties alone would not give sufficient confidence to remove all receiving inspection for deliveries from a particular supplier. You need to examine product as well as the system until you have gained the confidence to reduce inspection and eventually remove it.

Timing of verification activities

If you have verified that product conforms to the specified requirements before it arrives you can receive product into your company and straight onto the production line. An example of this is where you have performed acceptance tests or witnessed tests on the supplier's premises. You may also have obtained sufficient confidence in your supplier that you can operate a "just-in-time" arrangement, but you must be able to show that you have a continuous monitoring programme that informs you of the supplier's performance.

If you have not verified that product meets requirements before it arrives you need to install a "gate" through which only conforming items may pass. You need to register the receipt of items and then pass them to an inspection station equipped to determine conformance with your purchasing requirements. If items would normally pass into stores following inspection, as a safeguard you should also make provision for the store-person to check that all items received have been through inspection, rejecting any that have not. By use of labels attached to items you can make this a painless routine. If some items are routed directly to the user, you need a means of obtaining written confirmation that the items conform to the prescribed requirements so that at receipt inspection you can provide evidence that:

- nothing comes into the company without being passed through inspection;
- nothing can come out of inspection without it being verified as conforming.

If the user is unable to verify that requirements have been met, you will need to provide either evidence that it has passed your receipt inspection or has been certified by the vendor.

Receiving inspection

The verification plans should prescribe the acceptance criteria for carrying out receipt inspection. The main aspects to cover are as follows:

- Define how the receipt inspection personnel obtain current purchasing requirements.
- Categorize all items that you purchase so that you can assign levels of receipt inspection based on given criteria.
- For each level of inspection, define the checks that are to be carried out and the acceptance criteria to be applied.
- Where dimensional and functional checks are necessary, define how the receipt inspection personnel obtain the acceptance criteria and how they are to conduct the inspections and tests.
- Define the action to be taken when the product, the packaging or the documentation is found to be acceptable.
- Define the action to be taken when the product, the packaging or the documentation is found to be unacceptable.
- Define the records to be maintained.

Evaluation of supplier's statistical data

If the supplier supplies statistical data from the manufacturing process that indicates that quality is being controlled, then an analysis of this data based on assurances you have obtained through site evaluation can provide sufficient confidence in part quality to permit release into the organization.

Where the supplier provides a Product Approval Warrant the evidence provided needs to be evaluated against customer acceptance criteria and subsequent batches traceable to the Product Approval Warrant.

Where you have required your suppliers to send a certificate of conformity (C of C) testifying the consignment's conformity with the order, you cannot omit all receiving checks. Once supplier capability has been verified, the C of C allows you to reduce the frequency of incoming checks but not to eliminate them. The C of C may need to be supported with test results therefore you would need to impose this requirement in your purchasing documents. However, take care to specify exactly what test results you require and in what format you require them presented because you could be provided with attribute data when you really want variables data.

Purchased labour

This requirement poses something of a dilemma when purchasing subcontract labour because clearly it cannot be treated the same as product. You still need to ensure, however, that the labour conforms with your requirements before deployment to the job.

Such checks will include verification that the personnel provided could demonstrate competence and they are who they say they are. These checks can be made on the documentary evidence provided such as certificates, but you will probably wish to monitor their performance because it is the effort you have purchased not the people. You will not be able to verify whether they are entirely suitable until you have evaluated their performance so you need to keep records of the personnel and their performance during the tenure of the contract.

Dealing with product audits on supplier's premises

Within your procedures you need to provide a means of identifying which items have been subject to inspection at the supplier's premises and the receipt inspection action to be taken depending on the level of that inspection. In one case, your representative on the supplier's premises may have accepted the product. In another case, your representative may have accepted a product from the same batch but not the batch that has been delivered. Alternatively your representative may have only performed a quality audit to gain a level of confidence. You need to specify the inspection to be carried out in all such cases. Even if someone has performed inspection at the supplier's premises, if there is no evidence of conformance the inspections are of little value. The fact that an inspection was carried out is insufficient. There has to be a statement of what was checked, what results were obtained and a decision as to whether conformance had been achieved. Without such evidence you may need to repeat some of the inspections carried out on the supplier's premises.

Third-party assessments or part evaluation

In cases where you don't have the skills or the resources to verify products on supplier sites it can be outsourced to a competent organization such as a part evaluation laboratory. This is what is known as *Third-Party Assessment*. You could use the third party to undertake specialist assessments that support to your own incoming inspection so as to give confidence that the components you receive are built under adequately controlled conditions.

Legitimizing verification on supplier's premises (7.4.3)

The standard requires that *where the organization or its customer intends to perform verification activities at the supplier's premises, the organization is to state the intended verification arrangements and method of product release in the purchasing information.*

What does this mean?

If you choose a verification method other than receipt inspection that involves a visit to the supplier's premises, the supplier has a right to know and the proper vehicle for doing this is through the purchasing information such as a contract or order.

Why is this necessary?
This requirement responds to the Factual Approach Principle.

The supplier needs to know if you or your customers intend to enter its premises to verify product before shipment so that they may make the necessary arrangements and establish that the proposed methods are acceptable to them.

How is this implemented?
Verification by the organization
The acceptance methods need to be specified at the tendering stage so that the supplier can make provision in the quotation to support any of your activities on site. When you visit a supplier you enter its premises only with their permission. The product remains their property until you have paid for it and therefore you need to be very careful how you behave. The contract or order is likely to only give access rights to products and areas related to your order and not to other products or areas. You cannot dictate the methods the supplier should use unless they are specified in the contract. It is the results in which you should be interested not the particular practices unless you have evidence to demonstrate that the steps they are taking will affect the results.

Verification by the customer
In cases where your customer requires access to your suppliers to verify the quality of supplies, you will need to transmit this requirement to your supplier in the purchasing information and obtain agreement. Where a firm's business is wholly that of contracting to customer requirements, a clause giving their customers certain rights will be written into their standard purchasing conditions. If this is an unusual occurrence, you need to identify the need early in the contract and ensure it is passed on to those responsible for preparing subcontracts. You may also wish to impose on your customer a requirement that you are given advanced notice of any such visits so that you may arrange an escort. Unless you know your customer's representative very well it is unwise to allow unaccompanied visits to your suppliers. You may for instance have changed, for good reasons, the requirements that were imposed on you as the main contractor when you prepared the subcontract and in ignorance your customer could inadvertently state that these altered requirements are unnecessary.

When customers visit your suppliers or inspect product on receipt, they have the right to reserve judgement on the final acceptance of the product because it is not under their direct control, and they may not be able to carry out all the test and inspections that are required to gain sufficient confidence. Customer visits are to gain confidence and not to accept product. The same rules apply to you when you visit your suppliers. The final decision is the one made on receipt or some time later when the product is integrated with your equipment and you can test it thoroughly in its operating environment or equivalent.

Supplier monitoring (7.4.3.2)

The standard requires *supplier performance to be monitored using several specified indicators.*

What does this mean?

Monitoring supplier performance means checking periodically and systematically how well a supplier performs against certain prescribed criteria. It implies that the monitoring is of results rather than actions so is carried out from information gathered within the organization rather that on supplier's premises.

Why is this necessary?

Wherever supplier performance is not to be monitored, a deterioration in standards might well occur without being detected until it reached a level from which recovery was impossible within the timescales required. When degradation in supplier performance is detected on the production line it is already too late, as it is already affecting production and potentially the organization's performance in satisfying its customers.

How is this implemented?

It is interesting to observe that the list of indicators given for supplier monitoring is the same as those given for customer satisfaction, thereby recognizing that customer satisfaction and supplier satisfaction are intertwined and mutually dependent.

For each supplier you will need to set up a database that has provision for recording:

* variance in quantity supplied,
* variance in quality supplied,
* stage at which defects were detected,
* variance in delivery date,
* disruptions caused by variance in either delivery or quality,
* notifications of unsatisfactory performance, i.e. when prescribed targets have not been met.

Some of these data will not be collected on receipt of product but after the product is first used and therefore you will need a reporting mechanism that feeds data from the assembly line into the supplier database.

In the same way, that your customer expects you to monitor the same indicators relative to your performance, you need to encourage your suppliers to do the same. For them to do so, they need feedback from your organization. Similar mechanisms used by your customers could be used for passing this information to your suppliers. The database that contains supplier performance data could be accessible by your suppliers through a secure

Internet connection. They would have read-only access but it gives them immediate access to the information and thus no excuse for not taking action well before performance drops outside the agreed levels.

One of the problems with any monitoring data is dispute over the method of collection and its interpretation. So that there is no dispute when the time for action arises, methods, criteria and presentation format need to be agreed between your organization and your suppliers from the outset.

Production and service provision (7.5)

Control of production and service provision (7.5.1)

Planning production and service provision (7.5.1)

The standard requires the organization to *plan and carry out production and service provision under controlled conditions*.

What does this mean?
The process referred to in this section of the standard is the result-producing process, the process of implementing or replicating the design. It is the process that is cycled repeatedly to generate product or to deliver service. It differs from the design process in that it is arranged to reproduce product or service to the same standard each and every time. The design process is a journey into the unknown whereas the production process is a journey along a proven path with a predictable outcome. The design process requires control to keep it on course towards an objective; the production process requires control to maintain a prescribed standard.

There are two ways in which product quality can be controlled: by controlling the product that emerges from the producing processes or by controlling the processes through which the product passes. Process control relies on control of the elements that drive the process, whereas product control relies on verification of the product as it emerges from the process. In practice it is a combination of these that yields products of consistent quality. If you concentrate on the process output to the exclusion of all else, you might find there is a high level of rework of the end product. If you concentrate on the process using the results of the product verification, you will gradually reduce rework until all output products are of consistent quality. It will therefore be possible to reduce dependence on output verification.

Controlled conditions are conditions under which the outputs are predictable and are capable of being changed by a measurable degree. If the factors that affect process outputs could not be identified and changed, the process is not under control. There are usually eight factors that affect the control of any process:

1   The quality of the people: competence to do the job with the required proficiency when required. If you can't identify the competency of personnel required and choose those with the required competence, you are not in control of the process unless you can compensate by changing other parameters (Clause 6.2.1).

2   The quality of the physical resources: capability of plant, machinery, equipment and tools. If you can't identify the physical resources required and choose or change them so that capability is improved, you are not in control of the process (Clause 7.5.1c).

3   The quality of the physical environment: level of temperature, cleanliness and vibration. If you can't identify the physical environment required or change it, you are not in control of the process (Clause 6.4).

4   The quality of the human environment: degree of physical stress, physiological stress and motivation. If you can't identify the human factors required in the work environment required and alter their effect, you are not in control of the process (Clause 6.4).

5   The quality of the information: degree of accuracy, currency, completeness, usability and validity. If you can't identify the information required and improve it, you are not in control of the process (Clause 7.5.1a and b).

6   The quality of materials: adequacy of physical properties, their consistency and purity. If you can't identify the materials required and change their properties, you are not in control of the process (Clause 7.4).

7   The quantity of resources: time, money, information, people, materials, components, equipment, etc. If you can't identify the quantities required and adjust them to suit the demand, you are not in control of the process (Clause 6.1).

8   The quality of measurement: units, values, timing and integrity. If you can't identify the measurements to be taken, the units of measure, the target values, and when to take them and control the integrity of measurement you are not in control of the process (Clauses 7.5.1d and e and 7.6).

Only three of these factors are addressed by Clause 7.5.2 of the standard. The others are addressed by other clauses of the standard meaning that the list of items in Clause 7.5.1 is not all that you would need to do to provide controlled conditions. Although the process you are controlling may not necessarily provide the resources needed, to remain in control you need the ability to verify you are getting what you need to produce the requirement process outputs.

The use of the term *provision* in the requirement is not significant to its meaning. The term *process* would have been more consistent with the principles on which the standard is based.

There is a conflict between the requirements in Section 7.3 and those in Section 7.5.1. If the manufacturing processes have been designed and developed to a stage of being approved, the control plans, work instructions, equipment maintenance and tooling arrangements required by Clauses 7.5.1.1, 7.5.1.2, 7.5.1.4 and 7.5.1.5 would already exist in order to achieve process approval. Therefore, in all other respects the production planning referred to in Section 7.5.1 is not that concerned with design and development of manufacturing processes but with resourcing the previously validated processes, scheduling production and setting up these processes to deliver the quantities required by the customer. It is like planning a trip in a car that you have just driven out of the showroom. We will therefore address Clauses 7.5.1.1, 7.5.1.2, 7.5.1.4 and 7.5.1.5 as though the work is being performed during process design and development.

Why is this necessary?
This requirement responds to the Leadership Principle.

Controlled conditions enable the organization to achieve its objectives. If operations were carried out in conditions in which there were no controls, the outputs would be the result of chance and totally unpredictable.

How is this implemented?
The planning of manufacturing processes was accomplished when the processes received process approval. What is needed now is a Production Schedule detailing the quantity of product and delivery target. The labour and materials required to meet the schedule need to be specified. By studying the product specifications and Bill of Materials (BOM) you will be able to calculate the materials, bought out components and manpower required to product the finished product based on various shift patterns. The output here is likely to be a resource plan.

To ensure that the processes are carried out under controlled conditions the production plans need to:

- identify the product in terms of the specification reference and its issue status;
- define the quantity required;
- define which section is to perform the work;
- define each stage of manufacture and assembly;
- provide for progress through the various processes to be recorded so that you know what stage the product has reached at any one time;
- define the special tools, processing equipment, jigs, fixtures and other equipment required to produce the product (general-purpose tools and equipment need not be specified because your staff should be trained to select the right tool for the job);
- define the methods to be used to produce the product either directly or by reference to separate instructions;
- define the environment to be maintained during production of the product if anything other than ambient conditions;

- define the process specifications and workmanship standards to be achieved;
- define the stages at which inspections and test are to be performed and the methods to be used;
- define any special handling, packaging, marking requirements to be met;
- define any precautions to be observed to protect health, safety and environment.

These plans create a basis for ensuring that work is carried out under controlled conditions, but the staff, equipment, materials, processes and documentation must be up to the task before work commences. Many of the above activities will be addressed through Control Plans, Work Instructions and Process Specifications. A simple production process is illustrated in Figure 7.12. The shaded boxes indicate interfaces external to the production process. The variables are too numerous to illustrate the intermediate steps.

Availability of information that describes the product (7.5.1a)

The standard requires the organization to *control production and service provision through the availability of information that describes the characteristics of the product.*

What does this mean?
This information tells you what to make or provide and the criteria the output must meet for it to be fit for its purpose. The information is the input to the production or service delivery process usually coming out of the design process but may be direct from customers. It may take the form of definitive specifications, drawings, layouts or any information that specifies the physical and functional characteristics that the product or service is required to meet.

Why is this necessary?
This requirement responds to the Process Approach Principle.

Information is one of the key factors needed to control processes. Without information specifying the product to be produced or the service to be delivered, there is no basis on which to commence production.

How is this implemented?
In order to ensure the right information is available, there needs to be a communication channel opened between the product design and production process or the service design and the service delivery process as appropriate. Along this channel needs to pass all the information required to produce and accept the product or deliver the service. Provision also needs to be made for transmitting changes to this information in such a manner that the recipients can readily determine what the changes are and why they have been made. Often the design information is reissued, identified with a revision

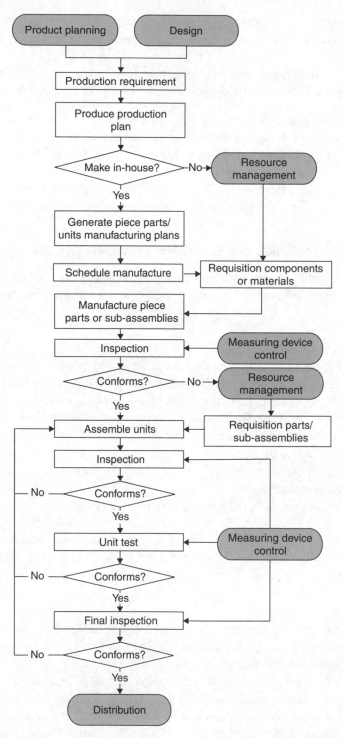

Figure 7.12 *Production process flow*

code leaving the recipient to work out what has changed and whether it affects what has gone before. The change management process therefore needs to take into account the factors that affect the effectiveness of the interfacing processes.

Availability of work instructions (7.5.1b and 7.5.1.2)

The standard requires the organization to *control production and service provision through the availability of work instructions where necessary*. The standard also requires the organization *to prepare documented work instructions that are derived from appropriate sources for all employees having responsibility for the operation of the processes that impact product quality*.

What does this mean?

Information is one of the key factors needed to control processes. In addition to information specifying the product characteristics to be achieved, information relating to how, when and where the activities required to convert the inputs into useable outputs are to be performed will be needed. This type of information is often termed as work instructions. There are two forms of work instructions – instructions that inform people what work to do and when to do it, and instructions that inform people how to do work – the latter are often called procedures. This topic is addressed in more detail in Chapter 4.

The requirement implies that work instructions may not always be required. While it is true that documented work instructions may not always be necessary, it is hard to imagine any production or service work proceeding without there being either an instruction or an identified need. Some people take instructions from their managers; others take instruction from the situation, the equipment or other signals they receive.

The requirement for instructions to be derived from sources such as the quality plan, the control plan and product realization process means that all instructions should be traceable to one or more of these documents.

Why is this necessary?

This requirement responds to the Process Approach Principle.

As people are not normally mind readers, they are unlikely to know what is required unless the work instructions come direct from the customer or the supplier. Once they know what is required they may need further instructions on how to carry out the work because it is not intuitive or learnt through education and training. Work instructions may also be necessary to ensure consistent results. In fact any operation that requires tasks to be carried out in a certain sequence to obtain consistent results should either be specified through work instructions or procedures or developed into a habit.

How is this implemented?

Work instructions can take many forms:

- Schedules indicating when work should be complete.
- Plans indicating what work is to be performed and what resources are available.
- Specifications indicating the results the work should achieve.
- Process descriptions indicating the stages through which work must pass.
- Procedures indicating how work should be performed and verified.

There are instructions for specific activities and instructions for specific individuals – whether they are contractors or employees is not important – the same requirements apply. As each employee may perform different jobs, they may each have a different set of instructions that direct them to specific sources of information. Therefore, it is unnecessary to combine all instructions into one document although they could all be placed in the same binder for easy access.

By imposing formal controls you safeguard against informality that may prevent you from operating consistent, reliable and predictable processes. The operators and their supervisors may know the tricks and tips for getting the equipment or the process to operate smoothly. You should discourage informal instructions because you cannot rely on them being used by others when those who know them are absent. If the tips or tricks are important, encourage those who know them to bring them to the process owner's attention so that changes can be made to make the process run smoothly all the time.

If you have a manufacturing process that relies on skill and training then instructions at the workstation are unnecessary. For example, if fixing a tool in a tool holder on a lathe is a skill, learnt during basic training, you don't need to provide instructions at each workstation where normal tool changes take place. However, if the alignment of the tool is critical and requires knowledge of a setting up procedure, then either documented instruction or training is necessary. Even for basic skills you can still provide standard machinery data books that are accessible near the workstation. There is merit in not providing basic textbooks to operators because the information is soon outdated and operators relying on such data instead of consulting the authorized data may inadvertently induce variation into the process.

Use of suitable equipment (7.5.1c)

The standard requires the organization to *control production and service provision through the use of suitable equipment.*

What does this mean?

Equipment is one of the key factors needed to control processes. Suitable equipment is equipment of the right type and capability to fulfil the requirements for which it is needed. Although the term equipment is used, the intent is to imply any physical resource that is needed to achieve the process objectives. It means that the resources have to be serviceable and capable of the performance, accuracy and precision required, i.e. maintained, calibrated, qualified, verified or otherwise approved as appropriate.

Why is this necessary?

This requirement responds to the Process Approach Principle.

Process outputs cannot be achieved unless the physical resources that are essential to perform the work are fit for their purpose. In any other state, the human resources would be used to compensate for the inadequacies of the equipment – a state that can be sustained in some circumstances but not for long without degrading the quality of the work.

How is this implemented?

The equipment should be selected during the planning process. In selecting such equipment you should determine whether it is capable of producing, maintaining or handling conforming product in a consistent manner. You also need to ensure that the equipment is capable of achieving the specified dimensions within the stated tolerances. Process capability studies can reveal deficiencies with equipment that are not immediately apparent from inspection of the first off.

There may be documentation available from the supplier of the equipment that adequately demonstrates its capability; otherwise you may need to carry out qualification and capability tests to your own satisfaction. In the process industries the plant is specially designed and so needs to be commissioned and qualified by the user. Your procedures need to provide for such activities and for records of the tests to be maintained.

When equipment or plant is taken out of service either for maintenance or for repair, it should not be re-introduced into service without being subject to formal acceptance tests which are designed to verify that it meets your declared standard operating conditions. Your procedures need to provide for such activities and for records of the tests to be maintained.

Use of monitoring and measuring devices (7.5.1d)

The standard requires the organization to *control production and service provision through the availability and use of measuring and monitoring devices.*

What does this mean?

Measurement is one of the key factors needed to control processes. This means providing the devices needed to measure product features and monitor process performance, and also providing adequate training and instruction for these devices to be used as intended.

Why is this necessary?
This requirement responds to the Process Approach Principle.

Product quality can only be determined if the devices needed to measure and monitor product and process characteristics are available and used.

How is this implemented?
When designing the process for producing product or delivering service you should have provided stages at which product or service features are verified and/or installed monitoring devices that indicate when the standard operating conditions have been achieved and whether they are being maintained. The devices used to perform measurements need to be available where the measurements are to be performed. The monitoring devices need to be accessible to process operators for information on the performance of the process to be obtained. The monitoring devices may be located in inaccessible places providing the signals are transmitted to the operators controlling the process.

Implementation of monitoring and measurement activities (7.5.1e)

The standard requires the organization to *control production and service provision through the implementation of monitoring and measurement.*

What does this mean?
Measurement is one of the key factors needed to control processes. Measurement and monitoring is the means by which product and process characteristics are determined. The specifications define the target values and the process description or plan defines when measurements should be taken to ascertain whether the targets have been met.

Why is this necessary?
This requirement responds to the Factual Approach Principle.

In order to control product quality the achieved characteristics need to be measured and the process operating conditions need to be monitored. All controls need a verification stage and a feedback loop. You cannot control production processes without performing some kind of verification.

How is this implemented?
Controlled conditions in production include in-process monitoring and in-process inspection and test. In a service delivery process they may include inspections of information, personnel and facilities as well as a review of process outputs.

The production of some products can be controlled simply by inspection after the product has been produced. In other cases, you may need to monitor certain process parameters to be sure of producing conforming product. By observing the variability of certain parameters using control charts, you can determine whether the process is under control within the specified limits.

The purpose of monitoring the process is firstly to establish its capability of producing product correctly and consistently, and secondly to alert the process operators to conditions that indicate that the process is becoming incapable of producing the product correctly and consistently.

Process monitoring can be achieved by observing sensors installed in the production process that measure key process parameters or, samples can be taken at discrete intervals and prescribed measurements taken. In both cases, the measurements should be recorded for subsequent analysis and any decision made to allow the process to continue or to stop should also be recorded together with the reasons for the decision. The data to be recorded should be specified in advance on the forms or computer screens provided at the workstation. This will give personnel a clear indication of what to record, when and where to record it. It also simplifies auditing if data is required in all boxes on a form or computer screen. A blank box would then indicate an unusual occurrence that should be checked. The forms should also indicate the accept or reject limits so that the operator can easily judge when the process is out of control.

Operators should be trained to both operate the plant and control the process. As added assurance you should take samples periodically and subject them to a thorough examination. The sampling plan should be defined and documented, and operators trained to determine what causes the results they observe. *Operators should therefore understand what causes the dots on the chart to vary.* Process control comes about by operators knowing what results to achieve, by knowing what results are being achieved and by being able to correct performance should the results not be as required. They need to understand what is happening during processing to cause any change in the results as they are being monitored. In the process specification you will need to define the parameters to be observed and recorded, and the limits within which the process is to be controlled.

Release processes (7.5.1f)

The standard requires *the implementation of release activities.*

What does this mean?

Release activities are decision points where process output is confirmed as complete and moved onto the next stage in a process or to another process. These are sometimes called "gates" through which product has to pass before being deemed acceptable for further processing or delivery.

Why is this necessary?
This requirement responds to the Process Approach Principle.

Within the chain of processes from customer requirement to delivery of product there are many interfaces between processes. Wherever there to be no decision points between processes, errors may pass from one process to another resulting in customer dissatisfaction.

How is this implemented?
In the production process there are several release points:

* Release from component storage, where components and materials are held pending completion of the input requirement.
* Release from job set-up, where work is held pending verification of job set-up.
* Release from in-process inspection, where product is held pending completion or rework, sample tests, repair, etc.
* Release from final inspection, where product is held pending completion of all inspection stages, sampling, etc.
* Release from finished product storage, where product is held pending receipt of customer order or customer instruction.

Release from storage areas
The content of storage areas should be known at all time in order that you can be confident that only that which is in storage areas is of a known condition. Storage areas containing conforming items should be separate from those containing nonconforming items. It follows therefore that when an item is taken from a storage area the person taking it should be able to rely on it conforming unless otherwise stated on the label. If free access is given to add and remove items in such areas, this confidence is lost. If at any time the controls are relaxed, the whole stock becomes suspect.

There is often a need to supply items as free issue because the inadvertent loss of small value items is less than the cost of the controls to prevent such loss. This practice can be adopted only if the quality of the items can be determined wholly by visual inspection by the person using them.

There are, however, issues other than quality that will govern the control of items in stock. Inventory control is a vital part of any business. Stock ties up capital, so the less stock that is held the more capital the firm has available to apply to producing output.

A common solution that satisfies both inventory control and quality control is to institute a stock requisition system. Authorization of requisitions may be given by a person's supervisor or can be provided via a work order. If someone has been authorized to carry out a particular job, this should authorize the person to requisition the items needed. Again for inventory control reasons you may wish to impose a limit on such authority, requiring the person to seek higher authority for additional items above a certain value. If an operator is requesting additional items you need to know why as the process was intended to deliver output from supplied inputs.

Held product
In continuous production, product is inspected by taking samples from the line that are then examined whilst the line continues producing product. In such cases, you will need a means of holding product produced between sampling points until the results of the tests and inspections are available. You will also need a means of releasing product when the results indicate that the product is acceptable. So a Product Release Procedure, or Held Product Procedure, may be necessary.

Release from job set-ups
See below under the heading *Verification of job set-ups*.

Release from inspection
Every verification is a stage where product is verified as either conforming or noncon-forming. Provisions are needed for signifying when product is ready for release either to the next stage in the process, into quarantine store awaiting decision or back into the process for rework or completion. Often this takes the form of an inspection label appropriately annotated.

Delivery processes (7.5.1f)

The standard requires *the implementation of delivery activities*.

What does this mean?
Delivery is an activity that serves the shipment or transmission of product to the customer and is one part of the distribution process. Delivery may include preparation for delivery, such as packing, notification, transportation, customs, arrival at destination and unpacking on customer premises.

Why is this necessary?
This requirement responds to the Process Approach Principle.

The process of moving goods from producers to customers is an important process in the management system. Although good product design, economic production and effective promotion are vital for success, these are useless if the customer cannot buy the product and take ownership. It is necessary to control delivery activities because conforming product may be degraded by the manner in which it is protected during transit and it may be delayed by the manner in which it is transported. You may be under an obligation to supply product by certain dates or within so many days of order and as a consequence control of the delivery process is vital to honour these obligations.

How is this implemented?

A typical distribution process is illustrated in Figure 7.13 indicating that like any process it needs to be designed and that the origin of the process inputs is the Marketing Process.

The distribution or marketing channel promotes the physical flow of goods and services along with ownership title, from producer to consumer or business user.[3] Often the logistics for moving goods to outlets where consumers are able to purchase them is a business in its own right, but nevertheless starts out in the marketing process when determining the distribution strategy. There are several different distribution channels depending on the type of goods and the market into which they are to be sold as illustrated in Figure 7.14.

Delivery takes place between each of the parties in the distribution chain and for each party there are several aspects to the delivery process:

- Preparation of product such as cleaning and preservation
- Packing of product
- User information
- Product certification
- Labelling and transit information
- Handling
- Customer notification
- Transportation
- Tracking.

Preparation and packing of product is addressed under the heading *Preservation of product* later in this chapter as the methods also apply to internal processing. However, within the delivery process there will be specific packing stages that are different in nature to internal packing stages.

Sometimes delivery is made electronically using a modem and telephone line. The product may be a software package, a document stored in electronic form or a facsimile. Protection of the product is still required but takes a different form. You need to protect the product against loss and corruption during transmission.

Customers may require product certificates testifying the fulfilment of contracts or order requirement. Customs may require certain legal information on the outside of the

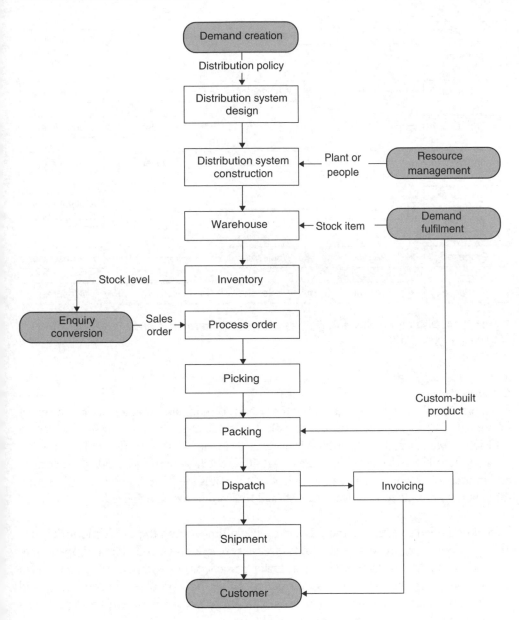

Figure 7.13 *Distribution process flow*

package otherwise the consignment will be held at the port of entry and customers will be none too pleased.

The type of transport employed is a key factor in getting shipments to customers on time.

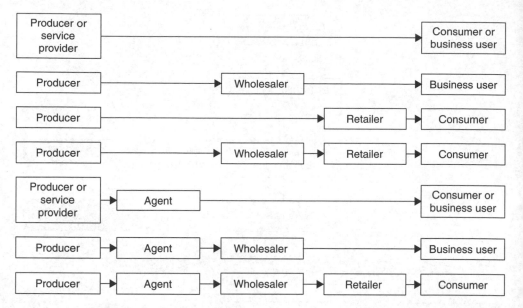

Figure 7.14 *Distribution channels (Source: Adapted from Boonze and Kurtz, Contemporary Marketing Figure 13.2)*

On-time delivery

To guarantee shipment on time, you either need to maintain an adequate inventory of finished goods, for shipment on demand or utilize only predictable processes and obtain sufficient advanced order information from your customer. Without sufficient lead time on orders you will be unlikely to meet the target. There will be matters outside your control and matters over which you need complete control. It is the latter that you can do something about and take corrective action should the target not be achieved.

Firstly, you need to estimate the production cycle time during the production trial runs in the product and process validation phase, assess risk areas and build in appropriate contingencies. An assessment of your supplier's previous delivery performance will also enable you to predict their future performance. When new processes become stabilized over long periods and the frequency of improvement reduces as more and more problems are resolved, you will be able to reduce lead time.

The planning and delivery procedures need to record estimated and actual delivery dates, and the data collected and analysed through delivery performance monitoring. When targets are not met you should investigate the cause under the corrective action procedures and formulate corrective action plans. Where the cause is found to be a failure of the customer to supply some vital information or equipment, it would be prudent not to wait for the periodic analysis but react promptly.

Customer notification

A means for notifying customer of pending delivery is often necessary. Your organization might be linked with the customer electronically so that demands are transmitted from the customer to trigger the delivery process. However, the customer may need to change quantities and delivery dates due to variations in production. This does not mean the changes will always be to shorten delivery times but on occasions the delivery times may need to be extended owing to problems on the assembly line or as a result of problems with other suppliers. The customer may not have made provision to store your product so needs to be able to urgently inform you to hold or advance deliveries. If the customer reduced the quantity required from that previously demanded, you could be left with surplus product and consequently need protection through the contract for such eventualities.

Post-delivery processes (7.5.1f)

The standard requires *the implementation of post-delivery activities.*

What does this mean?
Post-delivery activities are those performed after delivery of the consignment to the customer and may include:

- Servicing
- Warranty claims
- Technical support
- Maintenance
- Logistics
- Installation.

Why is this necessary?
This requirement responds to the Customer Focus Principle.

Control of post-delivery activities is just as important as pre-delivery if not more so as the customer may be losing use of the product and want prompt resolution to the problems encountered. Post-delivery performance is often the principle reason why customers remain loyal or choose a competitor. Even if a product does give trouble, a sympathetic, prompt and courteous post-delivery service can restore confidence.

How is this implemented?
The wide range of post-delivery services makes a detailed analysis impractical in this book. However, there are some simple measures that can be taken that would apply to all types of post-delivery activities:

- Define the nature and purpose of the post-delivery service.
- Define post-delivery policies that cover such matters as handling complaints, offering replacement product, and service or installation staff conduct.

- Establish conditions of post-delivery contracts with customers.
- Specify objectives and measures for each feature of the service such as response time and resolution time.
- Communicate the policies and objectives, and ensure their understanding by those involved.
- Define the stages in the process needed to achieve these objectives.
- Identify the information needs and ensure control of this information.
- Identify and provide the resources to deliver the service.
- Install verification stages to verify achievement of stage outputs.
- Provide communication channels for feeding intelligence into production and service design processes.
- Determine methods for measuring process performance.
- Measure process performance against objectives.
- Determine the capability of the process and make changes to improve performance.
- Determine process effectiveness and pursue continual improvement.

Control plan (7.5.1.1)

The standard requires the organization *to develop control plans and to review and update these plans when any changes occurs effecting product, manufacturing process, measurement, logistics, supply sources or FMEA.*

What does this mean?
The actions that are required at each stage of a process to ensure that all process outputs will be in a state of control need to be defined and the vehicle for doing this is a Control Plan: a universal format for capturing process-control data (see Figure 7.15).

The Control Plan does not replace operator instructions that describe how particular operations are to be performed for best results. Control Plans do not specify every conceivable characteristic, i.e. they specify those that need to be controlled and are therefore critical to output quality.

Why is this necessary?
The purpose of the control plan is to ensure that all process outputs will be in a state of control by providing process monitoring and control methods to control product and process characteristics. The control plan is covered in Section 6 of the *APQP Manual* and consists of forms containing data for identifying process characteristics and helps to identify sources of variation in the inputs that cause product characteristics to vary.

How is this implemented?
Control plans are developed during product and process design and development not production. Three types of control plan are required. During the product design and

	Prototype		Pre-launch		Production	Key contact or phone			Date (original)		Date (revised)	
Control plan number												
Part number/latest change level						Core team			Customer engineering approval/date (if required)			
Part name/description						Supplier/plant/approval date			Customer quality approval/date (if required)			
Supplier/plant			Supplier code			Other approval/date (if required)			Other approval/date (if required)			

Part/process number	Process name/operation description	Machine, device, jig, tools for manufacturing	Characteristics			Special characteristics class	Product, process specification or tolerance	Methods				Reaction plan
			No.	Product	Process			Evaluation measurement technique	Sample		Control method	
									Size	Frequency		

Figure 7.15 *Control plan format*

development phase, a prototype control plan is required to be produced. During the process design and development phase a pre-launch or pilot production control plan is required, and during the product and process validation phase the production control plan is to be issued.

Pre-launch occurs after prototype testing and prior to full production. Additional inspections and tests may be needed until the production processes have been validated and process capability assured. The additional checks serve to contain nonconformities until variation has been brought within acceptable limits for production.

Perhaps the most important aspect of Control Plans is that they understood to be living documents. They are not supposed to be produced simply for inclusion in the Part Approval Submission. They are meant to be used and particularly updated following product or process modifications. It is therefore important that no other documents are created that takes the place of the Control Plan. This might be case in organizations that already used similar documents but produce Control Plans to satisfy Part Approval Submission requirements. You can't run both in parallel. One has to go.

A typical weakness with regard to Control Plans is that they follow rather than lead change. Process problems arise, all hands run to the workstation and recovery action, containment action, and eventually corrective action is implemented. Meanwhile the Control Plan, which is supposed to be controlling the process, is out of step; it reflects the state of affairs before the problem arose. So if the customer comes in and observes the process when all this is going on, he or she will get the impression that the process is out of control. There is provision in the control plan for a Reaction Plan and this is where the Containment Plans should be identified. As soon as the preventive action is defined and proven, the Control Plan should be updated.

Verification of job set-ups (7.5.1.3)

The standard requires *job set-ups to be verified whenever performed*.

What does this mean?

Before most production jobs can commence some degree of preparatory work is needed to install, align and adjust equipment and tooling so that it produces the outputs of the required quality. The verification performed on first offs following set-up is used to verify that the set-up is correctly.

Why is this necessary?

Whenever a manufacturing process is stopped either to change tooling, correct errors or make modifications, one cannot be sure that the process will run at the same capability as before because errors may have crept in for all sorts of reasons. Verifying that the settings are the same as needed for continued process capability is a precautionary measure that if not taken might result in nonconforming product being produced.

How is this implemented?
In setting up a job prior to commencing a production run, you need to verify that all the requirements for the part are being met. You will therefore need job-set-up instructions so as to ensure each time the production of a particular part commences the process is set up against the same criteria. In addition, process parameters may change whenever there is material changeover, a job change or if significant time periods lapse between production runs.

Documentation verifying job set-ups should include documentation to perform the set-up and records that demonstrate that the set-up has been performed as required. This requires that you record the parameters set, the sample size and retain the control charts used which indicate performance to be within the central third of the control limits. These records should be retained as indicated in Clause 4.2.4 of the standard.

Preventive and predictive maintenance (7.5.1.4)

The standard requires the organization *to identify key process equipment and provide resources for machine or equipment maintenance and develop an effective planned total preventive maintenance system.*

What does this mean?
A planned total preventive maintenance system is a system that enables the organization to meet the process equipment availability targets needed to satisfy customer production schedules. It is a system in which there is planned, predictive, preventive and corrective maintenance.

Why is this necessary?
As implied in the definition above, an effective maintenance programme is needed for the organization to achieve customer production schedules. These schedules will be based on specific limits to equipment downtime so that if these limits are exceeded, product schedules will be adversely affected.

How is this implemented?
Equipment maintenance is addressed in Chapter 6 under the heading *Infrastructure*.

Management of production tooling (7.5.1.5)

The standard requires the organization to *provide resources for tool and gauge design, fabrication and verification activities.*

What does this mean?

An item is a tool when it comes into contact with a part and produces a change to that part. Many general-purpose tools used in manufacturing industry are designed by tool manufacturers. The purchase of these tools is governed by Clause 7.4 and their use governed by Clause 7.5.1c. If you subcontract the design of tooling, Clause 7.4 again applies but planning, organizing and controlling the production tools so as to deliver the necessary capability to the production process is what this requirement addresses.

Why is this necessary?

Clearly if tooling is not properly used and adequately controlled, product quality will not be maintained. A typical example is with automated machining stations. Operators no longer feed in the material, fix the tool, tighten the chuck, adjust the speed, and feed and control the depth of cut; it is all done automatically and much faster. But tools still wear and lose their cutting edge. The notion of waiting for the nonconforming product to be produced before the tool is replaced is no longer tolerable. Production schedules can only be met if the tool change programmes in place replace tools before the product exhibits signs of tool wear.

How is this implemented?

There are five basic elements to tooling management: tool design, tool manufacture, tool storage, tool use and tool maintenance. All five are the responsibility of the organization but apart from tool use and storage the others can be subcontracted.

Depending on the layout of the production line or cells, provision will need to be made for storing cutting tools close to the operation not miles away in a tool store. Rapid tool change is essential in order to keep maintain throughput.

Most of the shaping, forming, pressing, moulding tools and inspection gages, etc. may need to be especially designed and fabricated. This will probably require a tool design office where the tools, jigs, fixtures and gages are designed, and a toolroom where the tools are manufactured and inspected. Control of tooling is extremely important, as in some cases you will be reliant on the contour of the tool to form the part and you will be unable to check the part economically by other means. In such cases it is simpler to frequently check the tool for wear in order to detect wear before it produces a nonconforming part.

You need to possess either the necessary competence to design and make tools or the ability to control any subcontractors you employ to do this work for you. You need appropriate numbers of staff to do the job, equipped with design and manufacturing resources that enable them to deliver effective tools when needed. Tooling engineers should participate in design reviews during the product design and development phase, and undertake the following activities where appropriate:

- Design review of tooling
- Error-proofing using the results of failure modes analysis

- Tool wear analysis
- Tool accuracy analysis
- Tool maintenance planning
- Preparation of tool set-up instructions.

Certain tools are perishable, i.e. they are consumed during the process. Others are reusable after maintenance and this is where adequate controls need to be in place. The tool control system needs to cover tool selection, set-up, tool change and tool maintenance. You will need procedures for withdrawing maintainable tools from service, performing the maintenance and then putting the tools back into service. You need to build in safeguards that prevent worn tools being used and to replenish tools when their useful life has expired.

If you do subcontract tool maintenance, you need to keep track of assignments so that you are not without vital tools when you need them.

Production scheduling (7.5.1.6)

The standard requires *production to be scheduled in order to meet customer requirements, such as just-in-time supported by an information system that permits access to production information at key stages of the process and is order driven.*

What does this mean?
This requirement makes it very clear that customers in the automotive industry do not expect you to make for stock. You therefore need to have forecasts of orders expected or to have received the actual orders before scheduling production. Throughout production you need a means of tracking progress in order that delivery time from the start can be predicted.

Why is this necessary?
Some vehicle makers gear up to produce cars at a predetermined rate, and this requires materials and components to be supplied as a predetermined rate. However, the end users don't buy at a predetermined rate and Toyota found the answer to this through their *kanban* system that pulls parts when required from the preceding process that of course might be in a supplier's plant.

A problem that might arise is that the customer changes the order quantity or the delivery date after you have scheduled production but by having an information system that permits the customer access to production information at key stages, the schedules can be modified so that the demand for materials are kept in sync with the customer demands.

Another problem that may arise when using traditional ordering regimes, is that more product is produced than is needed by the customer. The customer either pays for unwanted stock or it is left with the supplier for use on future orders that may or may not materialize.

How is this implemented?
The customer will specify the IT interfaces so that your stocks and schedules are accessible from their own computer systems. In sophisticated systems, the scheduling is done by the customer and all you do is react to the demand so you need someone watching the screen all the time and making sure the resources needed to action these schedules are initiated in good time.

With such a system, it is as though the customer controls the complete supply chain but in reality it is often only the first-tier suppliers. Thus further down the chain suppliers have to try and predict the loading and maintain buffer stocks or alternate sources of supply. Clearly in situations like this, the chain is broken when one supplier cannot supply for reasons outside their control such as industrial dispute, scarcity of raw materials or natural disasters. Sometimes the vehicle maker has to lay off operators, reduce the shifts or stop production completely if a vital component cannot be produced to the schedule. The objective is to foresee such events and put in place provisions to minimize their impact. The *kanban* system may well work for the vehicle makers, but there is a consequential impacts on the supply chain. When everything functions in unison it works, but when the unexpected arises the system can fall like a house of cards.

Feedback of information from service (7.5.1.7)

The standard requires a process *to be established and maintained for communicating for information on service concerns to manufacturing, engineering and design activities*.

What does this mean?
Whilst you may not service your products, others may well do so and the standard requires that you collect information generated by the servicing organizations and convey it to those who can use it to improve the product and the manufacturing processes.

Why is this necessary?
One factor that all organizations need to realize is that they often don't use the products they supply to their customers and so might not be aware of any difficulties. When developing new products the designers use information from previous designs from which to learn and conduct FMEA using this and their imagination but sometimes, things get past even the most rigorous of analysis. This is why feedback from service is so valuable source of information.

How is this implemented?

You will need liaison links with servicing organizations and enlist their support in reporting to you any concerns they have about the serviceability or maintainability of the product, the availability of spare parts and the usability of the manuals and other information you have provided.

You should set up a common entry point for such data and put in place an evaluation function to convey appropriate data to the manufacturing, engineering and design activities. A corrective action form or improvement form could be used to convey the data and obtain a written response of the action to be taken. A log of servicing reports would assist in tracking servicing concerns and demonstrate you were making effective use of the data.

Servicing agreement with customer (7.5.1.8)

The standard requires the organization *to verify the effectiveness of service agreements.*

What does this mean?

Where there is a servicing agreement between the organization and its customer, it is necessary to verify that the level of servicing provided actually results in customer satisfaction

Why is this necessary?

Having an agreement in place might have not bearing on the level of service provided. Often service agreements are expressions of intent and not reality. Therefore, it is necessary to verify that the service levels provide actually match those expected.

How is this implemented?

What you need to do is establish the conditions of the servicing agreement and check that they have been conveyed to those affected by them, that these personnel have the competence to achieve such levels of service and that they are in fact delivering the levels of service required. To meet such requirements may require the setting up of a service organization, designing the necessary processes and equipping them with the relevant resources. As with any processes, the objectives and measures of success need to be defined and performance indicators established. Then periodic reviews need to be undertaken to establish whether the output as measured satisfies the conditions stated in the service-level agreement.

Validation of processes (7.5.2 and 7.5.2.1)

The standard requires the organization to *validate all production and service provision processes.*

What does this mean?

The automotive sector does not make any distinction between processes that deliver output that cannot be verified by subsequent measurement and those delivering output that can be verified by subsequent measurement. In both cases the processes are required to be validated; in other words, proven to be capable of delivering the required output.

Why is this necessary?

This requirement responds to the Factual Approach Principle.

Process validation is performed to ensure the manufacturing processes will deliver conforming product at the throughput rate required before production is authorized. If production were to commence without processes being validated, the result output would be unpredictable. This might be acceptable for small batches and craft workshops where minor adjustments can be made without jeopardizing quality. But when the qualities to be produced run into thousands the entire supply chain might be adversely impacted by substandard product reaching the assembly line.

How is this implemented?

Process validation will precede the process approval stipulate in Clause 7.3.6.3. To limit the potential for deficiencies to escape detection before the product is released, measures should be taken that ensure the suitability of all equipment, personnel, facilities and prevent varying conditions, activities or operations. A thorough assessment of the processes should be conducted to determine their capability to maintain or detect the conditions needed to consistently produce conforming product. The limits of capability need to be determined and the processes applied only within these limits.

You should produce and maintain a list of manufacturing processes that have been validated as well as a list of the personnel who are qualified to operate them. In this way you can easily identify an unqualified process, an unauthorized person or an obsolete list if you have neglected to maintain it.

Where process capability relies on the competence of personnel, personnel operating such processes need to be appropriately educated and trained and undergo examination of their competency. If subcontracting manufacturing processes you need to ensure that the supplier only employs qualified personnel and has qualified process equipment and facilities.

Where there is less reliance on personnel but more on the consistency of materials, environment and processing equipment, the particular conditions need to be specified and where necessary restrictions placed on the use of alternative materials, equipment and variations in the environment. Operating instructions should be used that define the set-up, operation and shutdown conditions and the sequence of activities required to produce consistent results. The resultant product needs to be thoroughly tested using such techniques that will enable the performance characteristics to be measured. This may involve destructive tests to measure tensile and compressive strength, purity, porosity, adhesion, electrical properties, etc. In production, samples should be taken at set frequencies and the tests repeated.

In production you need to ensure that only those personnel, equipments, materials and facilities that were qualified are employed in the process otherwise you will invalidate the qualification and inject uncertainty into the results.

The records of qualified personnel using special processes should be governed by the training requirements. Regarding the equipment, you will need to identify the equipment and facilities required within the process specifications and maintain records of the equipment. This data may be needed to trace the source of any problems with product that was produced using this equipment. To take corrective action you will also need to know the configuration of the process plant at the time of processing the product. If only one piece of equipment is involved, the above records will give you this information but if the process plant consists of many items of equipment which are periodically changed during maintenance, you will need to know which equipment was in use when the fault was likely to have been generated.

Identification and traceability (7.5.3)

Identifying product (7.5.3.1)

The standard requires the organization to *identify the product by suitable means throughout product realization*.

What does this mean?
The requirements for product identification are intended to enable products and services with one set of characteristics to be distinguishable from products or services with another set of characteristics.

The option of applying this requirement "where appropriate" has been removed in ISO/TS 16949 implies that there are no situations where product identity is unnecessary. There are of course situations where attaching an identity to a product would be impractical such as for liquids or items too small but the product nevertheless has an identity that is conveyed through the packaging and associated information. Thus identifying a product by suitable means might require the product to be labelled or might require the container to be labelled.

Why is this necessary?
This requirement responds to the Process Approach Principle.

Product identity is vital in many situations to prevent inadvertent mixing, to enable reordering, to match products with documents that describe them and to do that basic of all human activities (to communicate). Without codes, numbers, labels, names and other forms of identification we cannot adequately describe the product or service to anyone else. The product must be identified in one way or another otherwise it cannot be matched to its specification.

How is this implemented?

Separate product identity is necessary where it is not inherently obvious. If products are so dissimilar that inadvertent mixing would be unlikely to occur, a means of physically identifying the products is probably unnecessary. "Inherently obvious" in this context means that the physical differences are large enough to be visible to the untrained eye. Functional differences, therefore, no matter how significant as well as slight differences in physical characteristics such as colour, size, weight and appearance would constitute an appropriate situation for documented identification procedures.

Identifying product should start at the design stage when the product is conceived. The design should be given a unique identity; a name or a number and that should be used on all related information. When the product emerges into production, the product should carry the same number or name but in addition it should carry a serial number or other identification to enable product features to be recorded against specific products. If verification is on a go or no-go basis, product does not need to be serialized. If measurements are recorded some means has to be found of identifying the measurements with the product measured. Serial numbers, batch numbers and date codes are suitable means for achieving this. This identity should be carried on all records related to the product.

Apart from the name or number given to a product you need to identify the version and the modification state so that you can relate the issues of the drawing and specifications to the product they represent. Products should either carry a label or markings with this type of information in an accessible position or bear a unique code number that is traceable to such information.

You may not possess any documents that describe purchased product. The only identity may be marked on the product itself or its container. Where there are no markings, information from the supplier's invoice or other such documents should be transferred to a label and attached to the product or the container. Information needs to be traceable to the products it represents.

The method of identification depends on the type, size, quantity or fragility of the product. You can mark the product directly (provided the surface is not visible to the end user unless of course identity is part of the brand name) tie a label to it or the container in which it is placed. You can also use records remote from the product providing they bear a unique identity that is traceable to the product.

Marking products has its limitations because it may damage the product, be removed or deteriorate during subsequent processing. If applied directly to the product, the location and nature of identification should be specified in the product drawings or referenced process specifications. If applied to labels that are permanently secured to the product, the identification needs to be visible when the product is installed so as to facilitate checks without its removal. If the identity is built into the forging or casting, it is important that it is legible after machining operations. One situation which can be particularly irritating to customers is placing identification data on the back of equipment and then expecting the customer to state this identity when dealing with a service call thus causing delay while the customer dives under the car to locate the serial number and drops the mobile phone in the panic!

Verification status (7.5.3)

The standard requires the organization to *identify the status of the product with respect to measurement and monitoring requirements.*

What does this mean?
Product status with respect to monitoring and measurement means an indication as to whether the product conforms or does not conform to specified requirements. Thus identifying product status enables conforming product to be distinguishable from nonconforming product.

Why is this necessary?
This requirement responds to the Factual Approach Principle.

Measurement does not change a product but does change our knowledge of it. Therefore, it is necessary to identify which products conform and which do not so that inadvertent mixing, processing or delivery is prevented.

How is this implemented?
The most common method of denoting product status is to attach labels either to the product or to containers holding the product. *Green labels* for acceptable good and *Red labels* for reject goods. Labels should remain affixed until the product is either packed or installed. Labels should be attached in a way that prevents their detachment during handling. If labels need to be removed during further processing, the details should be transferred to inspection records so that at a later date the status of the components in an assembly can be checked through the records. At dispatch, product status should be visible. Any product without status identification should be quarantined until re-verified

and found conforming. Once a product has passed through the product realization process and is in use, it requires no product status identity unless it is returned to the production process for repair or other action.

It should be possible when walking through a machine shop, e.g. to identify which products are awaiting verification, which have been verified and found conforming and which have been rejected. If by chance, some product was to become separated from its parent batch, it should still be possible to return the product to the location from whence it came. A machine shop is where this type of identification is essential – it is where mix-ups can occur. In other places, where mix-ups are unlikely, verification status identification does not need to be so explicit.

Identifying product status is not just a matter of tying a label on a product. The status should be denoted by an authorized signature, stamp, mark or other identity which is applied by the person making the accept or reject decision and which is secure from misuse. Signatures are acceptable as a means of denoting verification status on paper records but are not suitable for computerized records. Secure passwords and "write-only" protection has to be provided to specific individuals. Signatures in a workshop environment are susceptible to deterioration and illegibility that is why numbered inspection stamps with unique markings have evolved. The ink used has to survive the environment and if the labels are to be attached to the product for life, it is more usual to apply an imprint stamp on soft metal or bar code.

Small and fragile products should be held in containers and the container sealed and marked with the product status. Large products should either carry a label or have a related inspection record.

In some situations, the location of a product can constitute adequate identification of product status. However, these locations need to be designated as "Awaiting Inspection, Accepted Product or Reject Product", or other such labels as appropriate to avoid the inadvertent placement of items in the wrong location. The location of product in the normal production flow is not a suitable designation unless an automated transfer route is provided.

With software the verification status can be denoted in the software as a comment or on records testifying its conformance with requirements.

With documentation you can either denote verification status by an approval signature on the document or by a reference number, date and issue status that is traceable to records containing the approval signatures.

If you use stamps, you will need a register to allocate stamps to particular individuals and to indicate which stamps have been withdrawn. When a person hands in his stamp it is good practice to avoid using the same number for 12 months or so to prevent mistaken identity in any subsequent investigations.

Traceability

The standard requires the organization to *control and record the unique identification of the product, where traceability is a requirement (see Clause 5.5.7)*.

What does this mean?

Traceability is a process characteristic. It provides the ability to trace something through a process to a point along its course either forwards through the process or backwards through the process and determine as necessary, its origin, its history and the conditions to which it was subjected.

Why is this necessary?

This requirement responds to the Factual Approach Principle.

One needs traceability to find the root cause of problems. If records cannot be found which detail what happened to a product then nothing can be done to prevent its recurrence. Although the standard only requires traceability when required by contract or law, it is key to enabling corrective action.

In situations of safety it is necessary to be able to locate all products of a batch in which a defective product has been found so as to eliminate them before there is a disaster. It is also very important any industry where human life may be at risk due to a defective product being in circulation.

Traceability is also important to control processes. You may need to know which products have been through which processes and on what date if a problem is found some time later. The same is true of test and measuring equipment. If on being calibrated a piece of test equipment is found to be out of calibration then it is important to track down all the equipment which has been validated using that piece of measuring equipment. This in fact is a requirement of ISO 9001 Clause 7.6 but no requirement for traceability is specified.

How is this implemented?

Providing traceability can be an onerous task. Some applications require products to be traced back to the original ingot from which they were produced. Traceability is achieved by coding items and their records such that you can trace an item back to the records at any time in its life. The chain can be easily lost if an item goes outside your control. For example, if you provide an item to a development organization for investigation and it is returned sometime later, without a certified record of what was done to it, you have no confidence that the item is in fact the same one, unless it has some distinguishing features. Traceability is only helpful when the chain remains unbroken. It can also be costly to maintain. The system of traceability that you maintain should be carefully thought out so that it is economic. There is little point in maintaining an elaborate traceability system for the once in a lifetime event when you need it, unless your very survival or society's survival depends on it. However, if there is a field failure; in order to prevent recurrence you will need to trace the component back through the supply chain to establish which operation on which component was not performed correctly simply to rule out any suggestion that other vehicles might be affected.

The conventions you use to identify product and batches need to be specified in the product specifications and the stage at which product is marked specified in the relevant procedures or plans. Often such markings are automatically applied during processing, as is the case with printed circuits, mouldings, ceramics, castings, products, etc. Process setting up procedures should specify how the marking equipment or tools are to be set up.

If you do release a batch of product prior to verification being performed and one out of the batch is subsequently found to be nonconforming, you will need to retrieve all other from the same batch. This may not be as simple as it seems. In order to retrieve a component which has subsequently been assembled into a printed circuit board, which has itself been fitted into a unit along with several other assemblies, not only would you need a good traceability system but also one that is constantly in operation.

It would be considered prudent to prohibit the premature release of product if you did not have an adequate traceability system in place. If nonconformity will be detected by the end-product tests, allowing production to commence without the receipt tests being available may be a risk worth taking. However, if you lose the means of determining conformance by premature release, don't release the product until you have verified it as acceptable.

Customer property (7.5.4 and 7.5.4.1)

Care of customer property (7.5.4 and 7.5.4.1)

The standard requires the organization to *exercise care with customer property including customer-owned tooling and packaging while it is under the organization's control or being used by the organization.*

What does this mean?

Customer property is any property owned or provided by the customer. The product being supplied may have been produced by a competitor, by the customer or even by your own firm under a different contract. Customer property is any property supplied to you by your customer and not only what is to be incorporated into product to be supplied to customers. Customer-owned tooling and returnable packaging also constitutes customer-supplied product. The property being used may be supplied by the customer such as tools, software and equipment, or made available for the organization's use such as test and development facilities on customer premises.

Documentation is not considered customer property because it is normally freely issued and ownership passes from customer to supplier on receipt. However, if the customer requires the documentation to be returned at the end of the contract, it should be treated as customer property.

Why is this necessary?
This requirement responds to the Customer Focus Principle.

Anything you use that does not belong to you should be treated with due care particularly if it has been supplied for your use and is expected to be returned in good condition.

How is this implemented?
For customer property that is used on your own premises you should maintain a register containing the following details:

* Name of product, part numbers, serial numbers and other identifying features.
* Name of customer and source of product if different.
* Delivery note reference, date of delivery.
* Receipt inspection requirements.
* Condition on receipt including reference to any rejection note.
* Storage conditions and place of storage.
* Maintenance specification if maintenance is required.
* Current location and name of custodian.
* Date of return to customer or embodiment into supplies.
* Part number and serial number of product embodying the customer-supplied product.
* Dispatch note reference of assembly containing the product.

These details will help you keep track of the customer-supplied product whether on embodiment loan or contract loan (see Appendix A for definitions) and will be useful during customer audits or in the event of a problem with the item either before or after dispatch of the associated assembly.

There might also be a need for a definition of responsibilities – a table showing which of the three parties (customer, supplier and your organization) is responsible for product acquisition, verification, repair, return to supplier, defect investigation, etc., and what the associated financial liabilities are.

Identification of customer-supplied property (7.5.4 and 7.5.4.1)

The standard requires the organization to *identify customer property provided for use or incorporation into the product. The automotive additions require such product to be permanently marked so that the ownership of each item is visible and can be determined.*

What does this mean?
Identifying customer property means attaching labels or other means of identification that denote its owner.

Why is this necessary?
This requirement responds to the Process Approach Principle.

If customer property carries an identity that distinguishes it from other product, it will prevent inadvertent disposal or unauthorized use.

How is this implemented?
Customer property may carry suitable identification but if not, labels, containers or other markings may be necessary to distinguish it from organization-owned property. As customer property may have been supplied by the organization originally as in the case of a repair service, labels indicating the owner should suffice. In a vehicle service area for instance, a label is attached to the car keys rather than labelling the car itself.

When deciding the type of marking, consideration needs to be given to the conditions of use. Markings may need to be permanent in order to be durable under the anticipated conditions of use. It would be wise to seek guidance from the customer if you are in any doubt as to where to place the marking or how to apply it. Metal identification plates stamped with the customer's identity, date of supply, contract and limitations of use are durable and permanent.

Verification of customer-supplied property (7.5.4)

The standard requires the organization to *verify customer property provided for use or incorporation into the product.*

What does this mean?
Customers may supply product purchased from other suppliers for installation in an assembly purchased from your organization.

Why is this necessary?
This requirement responds to the Factual Approach Principle.

Product needs to be verified before incorporation into the organization's product regardless of its source firstly, to establish the condition of the item on receipt in the event that it is damaged, defective or is incomplete and secondly, to verify that it is fit for the intended purpose before use. If you fail to inspect the product on receipt you may find difficulty in convincing your customer later that the damage was not your fault.

How is this implemented?

When property is received from a customer it should be processed in the same way as purchased product so that it is registered and subject to receipt inspection. The inspection you carry out may be limited if you do not possess the necessary equipment or specification, but you should reach an agreement with the customer as to the extent of any receipt inspection before the product arrives. You also need to match any delivery note with the product because the customer may have inadvertently sent you the wrong product. Unless you know what you are doing it is unwise to energize the product without proper instructions from the customer.

Protection of customer-supplied property (7.5.4)

The standard requires the organization to *protect customer property provided for use or incorporation into the product.*

What does this mean?

Protection means safeguarding against loss, damage, deterioration and misuse.

Why is this necessary?

This requirement responds to the Customer Focus Principle.

As the property will either be returned to the customer on completion of contract or will be incorporated into your products, it is necessary to protect the product from conditions that may adversely affect its quality.

How is this implemented?

Where the customer-supplied property is in the form of products that could be inadvertently degraded, they should be segregated from other products to avoid mixing, inadvertent use, damage or loss. Depending on the size and quantity of the items and the frequency with which your customer supplies such products you may require a special storage areas. Wherever the items are stored you should maintain a register of such items, preferably separate from the store in inventory control or the project office for example. The authorization for releasing customer-supplied property from stores may need to be different for inventory control reasons. You also need to ensure that such products are insured. You will not need a corresponding purchase order and they may not therefore be registered as stock or capital assets. If you receive customer-supplied property very infrequently, you will need a simple system that is only activated when necessary rather than being built into your normal system. Under such circumstances it is easy to lose these products and forget they are someone else's property. You need to alert staff to take extra care especially if they are high-value items which cannot readily be replaced.

Maintenance of customer-supplied property (7.5.4)

The standard requires the organization to *maintain customer property provided for use or incorporation into the product*.

Previously the standard required the supplier to establish and maintain procedures for maintenance of customer-supplied property provided for incorporation into the supplies or for related activities.

There is no change in requirement.

What does this mean?
Maintenance of customer-supplied property means retaining the property in the condition in which it was originally provided.

Why is this necessary?
This requirement responds to the Process Approach Principle.

Customer-supplied property that is issued for incorporation into supplies does not often require maintenance; however, items for use in conjunction with the contract may be retained for such duration that maintenance is necessary.

How is this implemented?
If the property requires any maintenance you should be provided with a maintenance specification and the appropriate equipment to do the job. Maintenance may include both preventive and corrective maintenance, but you should clarify with your customer which it is. You may have the means for preventive maintenance such as lubrication and calibration but not for repairs. Always establish your obligations in the contract regarding customer-supplied property, because you could take on commitments for which you are not contractually covered if something should go wrong. You need to establish who will supply the spares and re-certify the equipment following repair.

Reporting problems to the customer (7.5.4)

The standard requires *occurrence of any customer property that is lost, damaged or otherwise found to be unsuitable for use to be recorded and reported to the customer*.

What does this mean?
While customer property is on your premises, it may be damaged, develop a fault or become lost. Also when using customer property on customer premises, events may occur that result in damage or failure to the property.

Why is this necessary?
This requirement responds to the Customer Focus Principle.

It is necessary to record and report any damage, loss or failure to the customer so that as owners they may decide the action that is required. Normally, the organization does not have responsibility to alter, replace or repair customer property unless authorized to do so under the terms of the contract.

How is this implemented?
The customer is responsible for the product they supply wherever it came from in the first place. It is therefore very important that you establish the condition of the product before you store it or use it. In the event that you detect that the product is damaged, defective or is incomplete, you should place it in a quarantine area and report the condition to the customer. Even if the product is needed urgently and can still be used, you should obtain the agreement of your customer before using inferior product, otherwise you may be held liable for the consequences.

You could use your own reject note or nonconformity report format to notify the customer of a defective product but these are not appropriate if the product is lost. You also need a customer response to the problem and so a form that combined both a statement of the problem and of the solution would be more appropriate.

Preservation of product (7.5.5)

The standard requires the organization to *preserve conformity of product during internal processing and delivery to the intended destination* and goes on to require these measures *to include identification, handling, packaging, storage and protection*.

What does this mean?
These requirements are concerned with conformity control, i.e. ensuring that products remain conforming once they have been certified as conforming. They apply to service operations that involve the supply of product such as maintenance.

Identification in this context means identifying the product through the packaging and can apply at any stage in the product realization process. Handling refers to the manner by which product is moved by hand or by machine. Packaging refers to the materials employed to protect the product during movement and storage refers to the place where product is held pending use, shipment or further processing. Protection applies at all stages in product realization.

Why is this necessary?
This requirement responds to the Process Approach Principle.

As considerable effort will have gone into producing a conforming product, it is necessary to protect it from adverse conditions that could change the physical and functional characteristics. In some cases, preservation is needed immediately the characteristics have been generated (e.g. surface finish). In other cases, preservation is only needed when the product leaves the controlled environment (e.g. chemicals and electronic goods). Preservation processes need to be controlled in order that product remains in its original condition until required for use.

How is this implemented?
Determination of preservation requirements commences during the design phase or the manufacturing-planning phase by assessing the risks to product quality during its manufacture, storage, movement, transportation and installation. Packaging design should be governed by the requirements of Clause 7.3 although if you only select existing designs of packaging these requirements would not be applicable.

The preservation processes should be designed to prolong the life of the product by inhibiting the effect of natural elements. While the conditions in the factory can be measured, those outside the factory can only be predicted.

Having identified there is a risk to product quality you may need to prepare instructions for the handling, storage, packing, preservation and delivery of particular items. In addition to issuing the instructions you will need to reference them in the appropriate work instructions in order that they are implemented when necessary. Whatever the method, you will need traceability from the identification of need to implementation of the provisions and from there to the records of achievement.

Identification
Packages for export may require different markings than those for the home market. Those for certain countries may need to comply with particular laws. Unless your customer has specified labelling requirements, markings should be applied both to primary and secondary packaging as well as to the product itself. Markings should also be made with materials that will survive the conditions of storage and transportation. Protection can be given to the markings while in storage and in transit but this cannot be guaranteed while products are in use. Markings applied to the product therefore need to be resistant to cleaning processes both in the factory and in use. Markings on packaging

are therefore essential to warn handlers of any dangers or precautions they must observe. Limited Life Items should be identified so as to indicate their shelf life. The expiry date should be visible on the container and provisions should be made for such items to be removed from stock when their indicated life has expired.

While a well-equipped laboratory can determine the difference between products and materials the consumer needs a simple practical method of identification and labelled packets is often a reliable and economic alternative.

Handling

Handling provisions serve two purposes both related to safety. Protection of the product from the individual and protection of the individual handling the product. This latter condition is concerned with safety and addressed through other provisions; however, the two cannot and should not be separated, and handling procedures should address both aspects.

Handling product can take various forms depending on the hazard you are trying to prevent from happening. In some cases notices on the product will suffice, such as "LIFT HERE" or "THIS WAY UP" or the notices on batteries warning of acid. In other cases you will need to provide special containers, or equipment. There follows a short list of handling provisions that your procedures may need to address:

- Lifting equipment
- Pallets and containers
- Conveyors and stackers
- Design features for enabling handling of product
- Handling of electrostatic-sensitive devices
- Handling hazardous materials
- Handling fragile materials.

Packaging

Packing processes should be designed to protect the product from damage and deterioration under the conditions that can be expected during storage and transportation. You will need a means of identifying the packaging and marking requirements for particular products and of identifying processes for the design of suitable packaging including the preservation and marking requirements. Depending on your business you may need to devise packages for various storage and transportation conditions, preservation methods for various types of product and marking requirements for types of product associated with their destination.

Unless your customer has specified packaging requirements, there are several national standards that can be used to select the appropriate packaging, marking and preservation requirements for your products. Your procedures should make provision for the selection to be made by competent personnel at the planning stage and for the requirements thus selected to be specified in the packing instructions to ensure their implementation.

Packing instructions should not only provide for protecting the product but also for including any accompanying documentation such as:

- assembly and installation instructions,
- licence and copyright notices,
- certificates of conformity,
- packing list identifying the contents of the container,
- export documents,
- warranty cards.

The packing instructions are likely to be one of the last instructions you provide and probably the last operation you will perform for a particular consignment. This also presents the last opportunity for you to make mistakes! They may be your *last* mistakes but they will be the *first* your customer sees. The error you made on component assembly probably won't be found, but the slightest error in the packaging, the marking, the enclosures will almost certainly be found therefore this process needs careful control. It may not be considered so skilled a process but all the same it is vital to your image.

Storage

In order to preserve the quality of items that have passed receipt verification they should be transferred to stockrooms in which they are secure from damage and deterioration. You need secure storage areas for several reasons:

- For preventing personnel from entering the stock rooms and removing items without authorization.
- For preventing items from losing their identity. (Once the identity is lost it is often difficult, if not impossible to restore complete identification without testing material or other properties.)
- For preventing vermin damaging the stock.
- For preventing climatic elements causing stock to deteriorate.

While loss of product may not be considered a quality matter, it is if the product is customer property or if it prevents you from meeting your customer requirements. Delivery on time is a quality characteristic of the service you provide to your customer and therefore secure storage is essential.

To address these requirements you will need to identify and specify the storage areas that have been established to protect product pending use or delivery. Although it need be only a brief specification, the requirements to be maintained by each storage area should be specified based on the type of product, the conditions required preserving its quality, its location and environment. Products that require storage at certain temperatures should be stored in areas that maintain such temperatures. If the environment in the area in which the room is located is either uncontrolled or at a significantly higher or lower temperature, an environmentally controlled storage area will be required.

All items have a limit beyond which deterioration may occur and therefore temperature, humidity, pressure, air quality, radiation, vibration, etc. may need to be controlled. At some stage, usually during design or manufacturing or service planning, the storage conditions need to be defined and displayed. In many cases, dry conditions at room temperature are all that is necessary but problems may occur when items requiring non-standard conditions are acquired. You will need a means of ensuring that such items are afforded the necessary protection and your storage procedures need to address this aspect. It is for this reason that it is wiser to store items in their original packaging until required for use. If packets need to be opened to verify product identity, etc., the packaging design is already non-compliant.

Any area where product is stored should have been designated for that purpose in order that the necessary controls can be employed. If you store product in undesignated areas then there is a chance that the necessary controls will not be applied. Designation can be accomplished by placing notices and markers around the area to indicate the boundaries where the controls apply.

Each time the storage controller retrieves an item for issue, there is an opportunity to check the condition of stock. However, some items may have a slow turnover in certain storage areas, e.g. where spares are held pending use. It is also necessary to plan and carry out regular checks of the overall condition of the stockroom for damage to the fabric of the building. Rainwater may be leaking on to packaging and go undetected until that item is removed for use.

Some items such as electrolytic capacitors and two part adhesives may deteriorate when dormant. Others such as rubber materials, adhesive tape and chemicals deteriorate with the passage of time regardless of use. These are often referred to as "Shelf Life Items" or "Limited Life Items". Dormant electronic assemblies can deteriorate in storage and in the unlikely event that product would remain in storage for more than 1 year; provision should be made to retest equipment periodically or prior to release.

The assessment interval should depend on the type of building, the stock turnover, the environment in which the stock is located and the number of people allowed access. The interval may vary from storage area to storage area, and should be reviewed and adjusted as appropriate following the results of the assessment.

Segregation

Segregation is vital where products can only be positively identified by their containers. It is also important to prevent possible mixing or exposure to adverse conditions or cross-contamination. Examples where segregation makes sense are:

- toxic materials,
- flammable materials,
- limited life items.

Segregation is not only limited to the product but also to the containers and tools used with the product. Particles left in containers and on tools – no matter how small, can cause blemishes in paint and other finishes as well as violate health and safety regulations. If there are such risks in your manufacturing process then procedures need to be put in place that will prevent product mixing.

Segregation may also be necessary in the packaging of products not only to prevent visible damage but also to prevent electrical damage as with electrostatic-sensitive devices. Segregation may be the only way of providing adequate product identity as is the case with fasteners.

Cleaning

Where applicable, preservation processes should require that the product be cleaned before being packed and preservative applied. In other cases the product may need to be stored in sealed containers in order to retard decay, corrosion and/or contamination.

Inventory (7.5.5.1)

The requirements on inventory are addressed under the heading *Provision of resources* in Chapter 6.

Control of monitoring and measuring devices (7.6)

Determining monitoring and measurements to be undertaken (7.6)

The standard requires the organization to *identify the monitoring and measurement to be undertaken to provide evidence of conformity of product to determined requirements.*

What does this mean?

This requirement should strictly be located in Section 8 but has been included in Section 7 because it is believed that there are some applications where it does not apply. Clearly there are no applications where monitoring and measurement do not apply but there may be applications where physical calibration of measuring devices may not be applicable in the traditional sense of the word *devices*. The monitoring and measurements referred to are those required to carry out product verification rather than the measurements to calibrate a measuring device. The requirement is under product realization and not a subsection, and therefore applies equally to product or service design, purchasing and production or service delivery. It should not be interpreted as only applying to the characteristics of a product that can be measured through examination. It applies equally to performance characteristics that are inherent in the product design such as durability, safety and security, and to intangible characteristics such as courtesy, respect and integrity in service design.

Measurement is the process of associating numbers with physical quantities and phenomena. Measurements may be made by unaided human senses in which case they are often called estimates or, more usually, by the use of instruments, which may range in complexity from simple rules for measuring lengths to highly sophisticated systems designed to detect and measure quantities entirely beyond the capabilities of human senses. A unit is the name of a quantity, such as kilogram or metre; a standard is the physical embodiment of a unit.

Determined requirements are those of the customer, the regulators and the organization as addressed previously in Clause 7.2.1 of the standard. It is therefore not only a question of measuring the characteristics of the product but relating these characteristics to the defined requirements. For tangible product, the requirements may be expressed in performance terms (performance requirements) or in terms of form, fit and function (conformity requirements). Performance requirements may not be directly measurable. For instance safety requirements have to be translated into physical characteristics that are deemed to satisfy the requirement. You do not measure safety but the absence of hazard and therefore on a metal enclosure you would measure the absence of sharp edges and with liquids you would measure the integrity of containers. You can only search for what is known. Only when we know that a certain quantity of a substance is toxic can we search for it.

Where the product of a process is intangible the range of measurements varies widely. Knowledge is measured as the output of a course of study, skill or competence is measured as an output of training. Perceptions and behaviours are measured where they are vital to service quality. Comfort in a vehicle may be function of ride quality measured by using a laser profilometer on a road surface of known characteristics.

Why is this necessary?
This requirement responds to the Factual Approach Principle.

The reason for requiring monitoring and measurements to be identified is such that you have a means of relating the requirements to the characteristics to be measured so as to verify their achievement. If you did not identify the monitoring and measurements to be undertaken you would have no knowledge of whether or not requirements had been met before you delivered the product or service.

How is this implemented?

The measurements to be made should be derived from the characteristics defined in the product requirement. In some cases these may be directly measurable, such as size and weight, but often the characteristics need to be translated into measurable parameters. One way to identify the measurements is to produce a verification matrix showing the product characteristic, the measures, the units of measure, the target values and the level at which the measurements will be taken.

The product characteristic may be a specific dimension or an attribute. Length, mass and time are specific dimensions; response, safety, reliability, on-time delivery are attributes.

There are many different ways of expressing the units of measure (Table 7.4).

With tangible products the level of measurement may be at vehicle, equipment, part or material level. With intangible products the measurement may be at the relationship, encounter or transaction level. The relationship is the long-term interaction between customer and supplier, the encounter tends to be the short-term interaction involving

Table 7.4 *Units of measure*

Measure	Units
Length	Metres
Mass	Kilograms
Time	Seconds
Response	Time between receipt of call and engineer on case
Safety	Incidents per passenger mile
Reliability	Mean time between failure (MTBF)
Maintainability	Mean time to repair
On-time delivery	Per cent delivered by agreed date
Conformity	Per cent defective
Courtesy	Ratio of complaints to transactions

a single purchase and the transaction is a specific activity between representatives of the customer and supplier. A single encounter may involve several transactions and a relationship may involve years of encounters.

The measurements to be made should be identified in test specifications, process specifications and drawings, etc., but often these documents will not define how to take the measurements. The method of measurement should be defined in verification procedures that take into account the measurement uncertainty, the devices used to perform the measurements, the competency of the personnel and the physical environment in which measurements are taken. There may be a tolerance on variable parameters so as to determine the accuracy required. You may use general tolerances to cover most dimensions and only apply specific tolerances where it is warranted by the application.

Determining monitoring and measuring devices needed (7.6)

The standard requires the organization to *identify the monitoring and measuring devices needed to provide evidence of conformity of product to determined requirements.*

What does this mean?
The integrity of products depends on the quality of the devices used to create and measure their characteristics. This part of the standard specifies requirements for ensuring the quality of such devices.

A device is something constructed for a particular purpose and need not be a piece of hardware. It could be software or a method or sensor that captures information. It includes devices used during design and development for determining product characteristics and for design verification. Some characteristics cannot be determined by calculation and need to be derived by experiment. In such cases the accuracy of devices you use must be controlled, otherwise the parameters stated in the resultant product specification may not be achievable when the product reaches production.

Wherever there is a measurement to be made, a parameter or phenomena to be monitored, there is always a device or sensor employed to sense the measurements. The device may be human senses (hearing, sight, touch, smell and taste) a physical instrument measuring mass, length, time, electric current, temperature, etc.; a perception obtained from surveys, interviews, questionnaires or behaviour obtained from psychometric tests or knowledge obtained from written examination. There may be others, but clearly measurement is a wide subject and beyond the scope of this book to explore in any detail.

The sensor senses what is taking place and registers it in a form suitable for transmission to a receiver. The measuring and monitoring device should encompass the sensor, the transmitter and the receiver because the purpose of measurement is to take decisions and without receipt of the information no decisions can be taken. Also, you need to be aware that the transmitter and receiver may degrade the accuracy and precision of the measurement.

Why is this necessary?

This requirement responds to the Process Approach Principle.

It is necessary to identify the measuring and monitoring device so that a device capable of the appropriate accuracy and precision is used to take the measurements or monitor the parameters. If the devices you use to create and measure characteristics are inaccurate, unstable, damaged or in any way defective the product will not possess the required characteristics and furthermore you will not know it. You know nothing about an object until you can measure it, but you must measure it accurately and precisely. The devices you use therefore need to be controlled.

How is this implemented?

When identifying measuring and monitoring devices you need to identify the characteristic, the unit of measure, the target value and then choose an appropriate measuring or monitoring device. As the type of product may vary considerably, the range of measuring devices also varies widely. Considering that the term product in the ISO 9000 includes service, hardware, software and processed material, and that documents are considered to be software rather than information products, the range of characteristics for which measuring devices are required is enormous. It is relatively easy to identify the measuring and monitoring devices for hardware product and processes material, but less easy for services, software and information.

In many cases there will be a device available to do the measuring, but if you propose a new unit of measure a new sensor may be required.

Physical measurement

There are two categories of equipment that determine the selection of physical devices: *general-purpose* and *special to type equipment*. It should not be necessary to specify all the general-purpose equipment needed to perform basic measurements that should be known by competent personnel. You should not need to tell an inspector or tester which micrometer, vernier caliper, voltmeter or oscilloscope to use. These are the tools of the trade and they should select the tool that is capable of measuring the particular parameters with the accuracy and precision required. However, you will need to tell them which device to use if the measurement requires unusual equipment or the prevailing environmental conditions require that only equipment be selected that will operate in such an environment. In such cases, the particular devices to be used should be specified in the verification procedures. In order to demonstrate that you selected the appropriate device at some later date, you should consider recording the actual device used in the record of results. With mechanical devices this may not be necessary because wear will be normally detected by periodic calibration well in advance of a problem with the operation of the device.

With electronic devices subject to drift with time or handling, a record of the device used will enable you to identify suspect results in the event of the device being found to be outside the limits at the next calibration. A way of reducing the effect is to select devices that are several orders of magnitude more accurate that needed.

Service measurements

In the service sector there are many measurements that cannot be made using physical devices. The most common device is the customer survey either used directly by an interview with customers or by mail. Many service organizations develop metrics for monitoring service quality relative to the type of service they provide. Some examples are provided in Table 7.5.

Defining the monitoring and measuring processes (7.6)

The standard requires *processes to be established to ensure that monitoring and measuring can be carried out and are carried out in a manner consistent with the monitoring and measuring requirements.*

Table 7.5 *Service quality measures*

Service provided	Measures
Laboratory	Turn round time Conformity with requirements Calibration accuracy Time to respond to complaints
Distribution	Time to respond to complaints Supply delivery time Received condition
Data analysis	Report accuracy Conformity with requirements Time to respond to complaints

What does this mean?

A measurement process consists of the *operations* (i.e. the measurement tasks and the environment in which they are carried out), *procedures* (i.e. how the tasks are performed), *devices* (i.e. gauges, instruments, software, etc. used to make the measurements) and the *personnel* used to assign a quantity to the characteristics being measured and the measurement system (i.e. the units of measure and the process by which standards are developed and maintained).

Why is this necessary?

This requirement responds to the Systems Approach Principle.

If the measurements of product and service are to have any meaning, they have to be performed in a manner that provides results of integrity – results that others inside and outside the organization can respect and rely on as being accurate and precise. If the integrity of measurement is challenged and the organization cannot demonstrate the validity of the measurements, the quality of the product remains suspect.

How is this implemented?

Measurement process

Measurement begins with a definition of the measure, the quantity that is to be measured, and it always involves a comparison of the measure with some known quantity of the same kind. If the measure is not accessible for direct comparison, it is converted or "transduced" into an analogous measurement signal. As measurement always involve some interaction between the measure and the observer or observing instrument, there is always an exchange of energy, which, although in everyday applications is negligible, can become considerable in some types of measurement and thereby limit accuracy.[4]

Any measuring requirement for a quantity requires the measurement process to be capable of accurately measuring the quantity with consistency. For this to happen, the factors that affect the result need to be identified and a process designed that take into account the variations in these factors and delivers a result that can be relied on as being accurate within defined limits.

Controlling measurements

One necessary measuring and monitoring requirement is that the measurements carried out are controlled, i.e. regulated in a manner that will ensure consistent results. Control in this instance means several things:

- Knowing what devices are used for product verification purposes so that you can distinguish between controlled and uncontrolled devices – you will need to maintain a list of devices for this purpose.

- Knowing where the equipment is located so that you can recall it for calibration and maintenance – you will need a recall notice for this purpose.
- Knowing who the current custodian is so that you have a name to contact.
- Knowing what condition it is in so that you can prohibit its use if the condition is unsatisfactory – you will need a defect report for this purpose.
- Knowing when its accuracy was last checked so that you can have confidence on its results – calibration records and labels fulfil this need.
- Knowing what checks have been made using the instrument since it was last checked so that you can repeat them should the instrument be subsequently found out of calibration – this is only necessary for instruments whose accuracy drifts over time, i.e. electronic equipment. (It is not normally necessary for mechanical devices. You will need a traceability system for this purpose.)
- Knowing that the measurements made using it are accurate so that you rely on the results – a valid calibration status label will fulfil this purpose.
- Knowing that it is only being used for measuring the parameters for which it was designed so that results are reliable and equipment is not abused. The abuse of measuring devices needs to be regulated primarily to protect the device but also if high pressures and voltages are involved, to protect the user. Specifying the devices to be used for making measurements in your work instructions will serve to prevent the abuse of devices.

You may not need to know all these things about every device used for product verification but you should know most of them. This knowledge can be gained by controlling:

- the selection of measuring devices,
- the use of measuring devices,
- the calibration of measuring devices.

You may know where each device is supposed to be, but what do you do if a device is not returned for calibration when due? Your procedures should track returns and make provision for tracking down any maverick devices, because they could be used on product acceptance. A model process flow for control of measuring and monitoring devices is illustrated in Figure 7.16.

Measurement uncertainty
There is uncertainty in all measurement processes. There are uncertainties attributable to the measuring device being used, the person carrying out the measurements and the environment in which the measurements are carried out. When repeated measurements are taken with the same device on the same dimension of the same product and the results vary, this is measurement uncertainty. When you make a measurement with a calibrated instrument you need to know the specified limits of permissible error (how

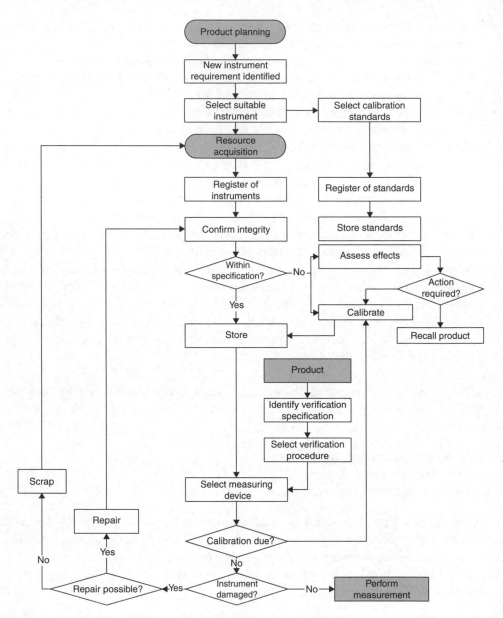

Figure 7.16 *Measuring device control process flow*

close to the true value the measurement is). If you are operating under stable environmental conditions, you can assume that any calibrated device will not exceed the limit
of permissible error. Stable conditions exist when all variation is under statistical control.
This means that all variation is due to common causes only and none due to special
causes. In other cases, you will need to estimate the amount of error and take this into

account when making your measurements. Test specifications and drawings, etc. should specify characteristics in true values, i.e. values that do not take into account any inherent errors. Your test and inspection procedures, however, should specify the characteristics to be measured taking into account all the errors and uncertainties that are attributable to the equipment, the personnel and the environment when the measurement system is in statistical control. This can be achieved by tightening the tolerances in order to be confident that the actual dimensions are within the specified limits.

Measurement laboratories

In order to maintain the integrity of measurement, physical measurements need to be undertaken in a controlled environment often referred to as a "laboratory". The controlled environment consists of a workspace in which the temperature, humidity, pressure, cleanliness, access and the integrity of the measuring devices and supporting equipment are controlled and the personnel qualified. Such areas can be assessed separately to other areas of the organization against international metrological standards. Other organizations are established solely for the purpose of providing measurement services to industry and therefore undertake certification of measurement devices – testifying the devices have been calibrated against standards traceable to national or international standards. Here we enter the world of accreditation as opposed to certification.

Accreditation and certification   Laboratory accreditation is defined by ISO as formal recognition that a laboratory is competent to carry out specific tests or specific types of tests. The key words in this definition are "competent" and "specific tests". Each accreditation recognizes a laboratory's technical capability (or competence) to do specific tests, measurements or calibrations. In that sense, it should be recognized as a standalone form of very specialized technical certification, as distinct from a purely quality system certification as provided by ISO 9001. An accredited organization is authorized to issue certificates of conformity to national or international standards. ISO 9001 certification does not authorize organizations to issue such certificates. Accreditation is awarded for a specific scope of service or range of products as is certification, except that for laboratory accreditation they are accredited for very specific tests or measurements (usually within specified ranges of measurement) with associated information on uncertainty of measurement and for particular product and test specifications. An ISO 9001 certificate for a laboratory does not accurately specify the performance characteristics of the product that the certificated organization is capable of supplying.

Use of laboratories   Wherever physical measurement is performed you need to be confident in the results and therefore whether the measurements are performed in-house or by external laboratories, both areas should meet the same standards in order that results are consistently accurate and precise. Wherever the calibration is performed, the same standards should therefore apply. Calibrating equipment in-house should not absolve you from complying with the same requirements that you would need to impose on an external test house.

The standard that applies to measurement laboratories is now ISO/IEC 17025. Previously ISO Guide 25 applied but this is now being replaced by ISO/IEC 17025. Laboratories meeting the requirements of ISO/IEC 17025, for calibration and testing activities, will comply with the relevant requirements of ISO 9001 when they are acting as suppliers producing calibration and test results. However, laboratories meeting the requirements of ISO 9001 would not meet the requirements of ISO/IEC 17025. Therefore, in-house laboratories should be assessed against ISO/IEC 17025 to obtain the same level of confidence as obtained from external laboratories.

Calibrating and verifying measurements (7.6a)

The standard requires measuring and monitoring devices to *be calibrated or verified at specified intervals or prior to use, against measurement standards traceable to international or national standards*.

What does this mean?

In a measurement system the physical signal is compared with a reference signal of known quantity. The reference signal is derived from measures of known quantity by a process called *calibration*. The known quantities are based on standards that in the majority of cases are agreed internationally.

The International System of Units was adopted by the *11th General Conference on Weights and Measures* in 1960, it is abbreviated SI in all languages. Several base measuring units are defined, the three primary one being mass, length and time. Since 1983 the standard for the unit of *length*, the metre has been defined as the distance travelled by light in a vacuum in $1/299,792,458$ second. The standard for the unit of *mass*, the kilogram, is a cylinder of platinum–iridium alloy kept by the International Bureau of Weights and Measures, located in Sèvres, near Paris. The standard for the unit of *time*, the second, is defined as the duration of $9,192,631,770$ cycles of the radiation associated with a specified transition, or change in energy level, of the cesium-133 atom.[5] All physical measurements should be traceable to these standards and their derivatives. This means that an instrument will give the same reading when measuring a quantity under the same environmental conditions, wherever the measurement is taken.

With non-physical measurement systems there is still a need for calibration but we tend to use the term integrity. The reference signal of known quantity is the standard and you can derive standards for anything – the only proviso is that those who benefit from the measurements agree them. Therefore, if you set out to measure customer satisfaction the standard used needs to be agreed with the customer. Standards that are not agreed by those they affect lack integrity.

Why is this necessary?

This requirement responds to the Factual Approach Principle.

Variations can arise in measurements taken in different locations due to the measuring device not being calibrated to the same standards as other devices. With the introduction of the SI system of units, this variation could be eliminated providing the quantity used to calibrate the measuring device was traceable to national or international standards.

How is this implemented?

Calibration of monitoring and measuring devices

Calibration is concerned with determining the values of the errors of a measuring instrument and often involves its adjustment or scale graduation to the required accuracy. You should not assume that just because a device was once accurate it would remain so forever. Some devices if well treated and retained in a controlled environment will retain their accuracy for very long periods. Others if poorly treated and subjected to environmental extremes will lose their accuracy very quickly. Ideally you should calibrate measuring devices before and thereafter in order to prevent an inaccurate device being used in the first place and afterwards to confirm that no changes have occurred during use. However, this is often not practical and so intervals of calibration are established which are set at such periods as will detect any adverse deterioration. These intervals should be varied with the nature of the device, the conditions of use and the seriousness of the consequences should it produce incorrect results.

It is not necessary to calibrate all measuring and monitoring devices. Some devices may be used solely as an indicator such as a thermometer, a clock or a tachometer – other equipment may be used for diagnostic purposes, to indicate if a fault exists. If such devices are not used for determining the acceptability of products and services or process parameters, their calibration is not essential. However, you should identify such devices as for "Indication-Purposes Only" if their use for measurement is possible. You don't need to identify all clocks and thermometers fixed to walls unless they are used for measurement.

There are two systems used for maintaining the accuracy and integrity of measuring devices: a calibration system and a verification system. The calibration system determines the accuracy of measurement and the verification system determines the integrity of the device. If accuracy is important then the device should be included in the calibration system. If accuracy is not an issue but the device's form, properties or function is important then it should be included in the verification system. You need to decide the system in which your devices are to be placed under control and identify them accordingly.

There are two types of devices subject to calibration: those that are adjustable and those that are not. An adjustable device is one where the scale or the mechanism is capable of adjustment (e.g. micrometer, voltmeter and load cell). For non-adjustable devices, a record of the errors observed against a known standard can be produced which can be taken into account when using the device (e.g. slip gage, plug gage, surface table and thermometer).

Comparative references are not subject to calibration. They are, however, subject to verification. Such devices are those which have form or function where the criteria is either pass or fail, i.e. there is no room for error or where the magnitude of the errors does not need to be taken into account during usage. Such devices include software, steel rules or tapes, templates, forming and moulding tools. Devices in this category need carry no indication of calibration due date. The devices should carry a reference number and verification records should be maintained showing when the device was last checked. Verification of such devices include checks for damage, loss of components, function, etc.

Some electronic equipment has self-calibration routines built into the start-up sequence. This should be taken as an indication of serviceability and not of absolute calibration. The device should still be subject to independent calibration at a defined frequency. (*Note*: **Use, not function, determines need for calibration.**)

Traceability

If you calibrate your own devices you will need (in addition to the "working standards" which you use for measurement) calibration standards for checking the calibration of the working standards. The calibration standards should also be calibrated periodically against national standards held by your national measurement laboratory. This unbroken chain ensures that there is compatibility between measurements made in different locations using different measuring devices. By maintaining traceability you can rely on obtaining the same result (within the stated limits of accuracy), wherever and whenever you perform the measurement providing the dimensions you are measuring remain stable. The relationship between the various standards is illustrated in Figure 7.17.

Determining calibration frequency

ISO 10012 requires that measuring equipment be confirmed at appropriate intervals established on the basis of stability, purpose and usage. With new equipment it is customary to set the frequency at 12-month intervals unless recommended otherwise by the manufacturer. Often this frequency remains despite evidence during calibration that accuracy and precision is no longer stable. Such action indicates that the calibration staff have not been properly trained or that cost rather than quality is driving calibration services. Calibrations should be performed prior to any significant change in accuracy that can be anticipated. The results of previous calibrations will indicate the amount of drift – if drift is detected, the intervals of confirmation should be shortened. Conversely,

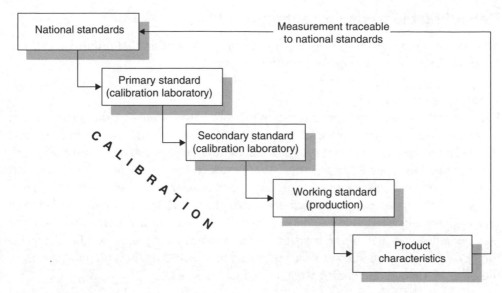

Figure 7.17 *Traceability of standards*

if drift is not detected, the intervals may be lengthened if two previous confirmations indicate such action would adversely affect confidence in the accuracy of the device. Environment, handling, frequency of use and wear are factors that can affect the stability of devices, therefore regardless of the calibration results, both previous and future conditions need to be taken into account. In order to demonstrate you have reviewed the results and determined the appropriate calibration frequency, provision should be made on the calibration records for the frequency to be decided at each calibration. Specifying a date is insufficient if the calibration instructions specify a frequency because it is unreasonable to expect the person subsequently performing calibration to detect whether any change has been made.

Reference materials
Comparative references are devices that are used to verify that an item has the same properties as the reference. They may take the form of colour charts or materials such as chemicals which are used in spectrographic analysers or those used in tests for the presence of certain compounds in a mixture or they could be materials with certain finishes, textures, etc. Certificates should be produced and retained for such reference material so that those who will use them know their validity. Materials that degrade over time should be dated and given a use by date. Care should be taken to avoid cross-contamination and any degradation due to sunlight (as can happen with colour charts). A specification for each reference material should be prepared so that its properties can be verified.

Recording the basis for calibration (7.6a)

The standard requires that *where no international or national standards exist, the basis used for calibration or verification be recorded.*

What does this mean?

For physical and chemical measurements that are based on the fundamental units of measure (metre, kilogram, second, ampere, etc.), there are national or international standards but for other measures no national or international standard may exist. Each industry has developed a series of measures by which the quality of its goods and services are measured and has accordingly developed standards that represent agreed definitions of the measures. In the service sector involving interrelationships between people, standards become more difficult to define in quantitative terms and therefore may be defined qualitatively. For instance, the performance of a person handling customer calls may be defined by a series of results to be achieved. In setting the standard, the effort is focused on defining what a good job looks like.

Why is this necessary?

This requirement responds to the Factual Approach Principle.

Without a sound basis for comparison, the effort of measurement is wasted.

How is this implemented?

In some situations there may be no national standard against which to calibrate your devices. If you face this situation, you should gather together a group of experts within your company or trade association and establish by investigation, experimentation and debate what constitutes the standard. Having done this you should document the basis of your decisions and produce a device or number of devices that can be used to compare the product or result with the standard using visual, quantitative or other means. The device may be a physical instrument but could be information such as a set of agreed criteria.

Where you devise original solutions to the measurement of characteristics, the theory and development of the method should be documented and retained as evidence of the validity of the measurement method. Any new measurement methods should be proven by rigorous experiment to detect the measurement uncertainty and cumulative effect of the errors in each measurement process. The samples used for proving the method should also be retained so as to provide a means of repeating the measurements should it prove necessary.

Adjustment of devices (7.6b)

The standard requires measuring equipment to *be adjusted or re-adjusted as necessary*.

What does this mean?
Adjustment is only possible with devices that have been designed to be adjustable.

Why is this necessary?
This requirement responds to the Process Approach Principle.

When a measuring device is verified, it may be found within specification and adjusted if the parameters have drifted towards the upper or lower limits. If the device is found outside specification it can be adjusted or re-adjusted (on subsequent occasions) within the specified limits.

How is this implemented?
Mechanical devices are normally adjusted to the null position on calibration. Electronic devices should only be adjusted if found to be outside the limits. If you adjust the device at each calibration you will not be able to observe drift. Adjustments, if made very frequently, may also degrade the instrument. If the observed drift is such that the device may well be outside the specified limits by the next calibration, adjustment will be necessary.

In addition to calibrating the devices, you will need to carry out preventive and corrective maintenance in order to keep them in good condition. Preventive maintenance is maintenance to reduce the probability of failure, such as cleaning, testing, inspecting, replenishment of consumables, etc. Corrective maintenance is concerned with restoring a device (after a failure has occurred) to a condition in which it can perform its required function. These activities may cover a wide range of skills and disciplines depending on the nature of the measuring devices you use. It will include software development skills if you use test software, for instance, or electronic engineering if you use electronic equipment. You can of course subcontract the complete task to a specialist who will not only maintain the equipment but, on request, carry out calibration. Take care to confirm that the supplier is qualified to perform the calibrations to national standards and to provide a valid certificate of calibration.

Indicating calibration status (7.6c)

The standard requires measuring devices to *be identified to enable the calibration status to be determined*.

What does this mean?
Calibration status is the position of a measuring device relative to the time period between calibrations. If the date when calibration is due is in the future, the device can be considered calibrated; i.e. if the current date is beyond the date when calibration is due, the device is not necessarily inaccurate but remains suspect until verified. However, devices can also be suspected of being damaged if dropped or damaged even when

the date of calibration is due is in the future. The requirement only applies to physical devices subject to wear, drift or variation with use or time.

Why is this necessary?
This requirement responds to the Factual Approach Principle.

While a robust calibration system should ensure no invalid devices are in use, system failures are a possibility. As the consequences of failure are greater than the effort involved in checking the validity of devices before use, it is prudent to provide a means for checking calibration status.

How is this implemented?
All devices subject to calibration should display an identification label that either directly or through traceable records, indicates the authority responsible for calibrating the device and the date when the next calibration is due. Don't state the actual calibration date because this would be no use without the user knowing the calibration frequency. Measuring equipment should indicate its calibration status to any potential user. Measuring instruments too small for calibration status labels showing the due date may be given other types of approved identification. It is not mandatory that users identify the due date solely from the instrument itself but they must be able to determine that the instrument has been calibrated. Serial numbers alone do not do this unless placed within a specially designed label that indicates that the item has been calibrated or you can fix special labels that show a circular calendar marked to show the due date. If you do use serial numbers or special labels, they need to be traceable to calibration records that indicate the calibration due date.

Devices used only for indication purposes or for diagnostic purposes should also display an identity that clearly distinguished them as not being subject to calibration. If devices are taken out of use for prolonged periods, it may be more practical to cease calibration and provide a means of preventing inadvertent use with labels indicating that the calibration is not being maintained. You may wish to use devices that do not fulfil their specification either because part of the device is unserviceable or because you were unable to perform a full calibration. In such cases, you should provide clear indication to the user of the limitation of such devices.

Safeguarding monitoring and measuring devices (7.6d)

The standard requires monitoring and measuring devices to *be safeguarded from adjustments that would invalidate the measurement result.*

What does this mean?
Once a device has been calibrated or verified there need to be safeguards in place to prevent unauthorized or inadvertent adjustment.

Why is this necessary?
This requirement responds to the Factual Approach Principle.

The purpose of this requirement is to ensure that the integrity of the measurements is maintained by precluding errors that can occur if measuring equipment is tampered with.

How is this implemented?
To safeguard against any deliberate or inadvertent adjustment to measuring devices, seals should be applied to the adjustable parts or where appropriate to the fixings securing the container. The seals should be designed so that tampering will destroy them. Such safeguards may not be necessary for all devices. Certain devices are designed to be adjusted by the user without needing external reference standards, e.g. zero adjustments on micrometers. If the container can be sealed then you don't need to protect all the adjustable parts inside.

Your procedures need to specify:

*   those verification areas that have restricted access and how you control access;
*   the methods used for applying integrity seals to equipment;
*   the authority permitted to apply and break the seals;
*   the action to be taken if the seals are found to be broken either during use or during calibration.

Protection of monitoring and measuring devices (7.6e)

The standard requires monitoring and measuring devices to *be protected from damage and deterioration during handling, maintenance and storage*.

What does this mean?
Each monitoring and measuring device has a range within which accuracy and precision remains stable – use the device outside this range and the readings are suspect.

Why is this necessary?
This requirement responds to the Process Approach Principle.

Physical monitoring and measuring devices can be affected by inappropriate handling, maintenance and storage, and thus jeopardize their integrity. Often measuring devices are very sensitive to vibration, dirt, shock and tampering and thus it is necessary to protect them so as to preserve their integrity.

How is this implemented?
When not in use, measuring devices should always be stored in the special containers provided by the manufacturer. Handling instructions should be provided with the storage

case where instruments may be fragile or prone to inadvertent damage by careless handling. Instruments prone to surface deterioration during use and exposure to the atmosphere should be protected and moisture absorbent or resistant materials used. When transporting measuring devices you should provide adequate protection. Should you employ itinerant service engineers, ensure that the instruments they carry are adequately protected as well as calibrated.

Action on equipment found out of calibration (7.6)

The standard requires the validity of previous measuring results to *be assessed and recorded when equipment is found not to conform to requirements and for appropriate action to be taken on the equipment and any product affected.*

What does this mean?

This is perhaps the most difficult of requirements to meet for some organizations. It is not always possible or practical to be able to trace product to the particular devices used to determine its acceptability. The requirements apply not only to your working standards but also to your calibration standards. When you send calibration standards away for calibration and they are subsequently found to be inaccurate, you will need a method of tracing the devices they were used to calibrate. If you have a small number of measuring devices and only one or two of each type, it may not be too difficult to determine which products were accepted using a particular device. In large organizations that own many pieces of equipment that is constantly being used in a variety of situations, meeting the requirements can be very difficult.

Why is this necessary?

This requirement responds to the Process Approach Principle.

If a measurement has been taken with a device that is subsequently found inaccurate, the validity of the measurement is suspect and therefore an assessment is needed to establish the consequences. In most cases, the devise used is accurate to an order of magnitude greater than that required; therefore, if found outside tolerance, it may not mean that the product measured is nonconforming. However, if measurements are taken at the extreme of device accuracy, the product may well be nonconforming if the device is found to be inaccurate.

How is this implemented?

One way of meeting this requirement is to record the type and serial number of the devices used to conduct measurements, but you will also need to record the actual measurements made. Some results may be made in the form of ticks or pass or fail and not by recording actual readings. In these cases, you will have a problem in determining whether the amount by which the equipment is out of specification would be sufficient to reject the product. In extreme circumstances, if the product is no longer in the factory, this situation could result in your having to recall the product from your customer or distributor.

In order to reduce the effect, you can select measuring devices that are several orders of magnitude more accurate than your needs so that when the devices drift outside the tolerances, they are still well within the accuracy you require. There still remains a risk that the device may be wildly inaccurate due to damage or malfunction. In such cases, you need to adopt the discipline of re-calibrating devices that have been dropped or are otherwise suspect before further use.

You need to carefully determine your policy in this area paying particular attention to what you are claiming to achieve. You will need a procedure for informing the custodians of unserviceable measuring devices and one for enabling the custodians to track down the products verified using the unserviceable device and assess the magnitude of the problem. You will need a means of ranking problems in order of severity so that you can resolve the minor problems at the working level and ensure that significant problems are brought to the attention of the management for resolution. It would be irresponsible for a junior technician to recall 6-month production from customers and distributors based on a report from the calibration laboratory. You need to assess what would have happened if you had used serviceable equipment to carry out the measurements. Would the product have been reworked, repaired, scrapped or the requirement merely waived. If you suspect previously shipped product to be nonconforming and now you have discovered that the measurements on which their acceptance was based were inaccurate, you certainly need to notify your customer. In your report to your customer, state the precise amount by which the product is outside specification so that the customer can decide whether to return the product – remember the product specification is but an interpretation of what constitutes fitness for use. Out of "specification" doesn't mean unsafe, unusable, un-saleable, etc.

Calibration and verification records (7.6 and 7.6.2)

The standard requires *records of the results of calibration and verification to be maintained*. The additional automotive requirements require *records of calibration or verification for all gauges, measuring and test equipment that are needed to provide evidence of product conformity are to comply with specified requirements*.

What does this mean?
Records of the results of calibration and verification are those records indicating the accuracy of the device prior to any adjustment and records after adjustment. These records apply not only to equipment design and produced by the organization, but also

those owned by the organization and those owned by employees and customers when being used for product acceptance.

Why is this necessary?
This requirement responds to the Factual Approach Principle.

It is important to record calibration and verification in order to determine whether the device was inside the prescribed limits when last used. It also permits trends to be monitored and the degree of drift to be predicted.

Calibration records are also required in order to notify the customer if suspect product or material has been shipped.

How is this implemented?
Calibration and verification records are records of activities that have taken place. Records should be maintained not only for proprietary devices but also for devices you have produced and devices owned by customers and employees.

These records should include where appropriate:

- The precise identity of the device being calibrated or verified (type, name, serial number, configuration if it provides for various optional features).
- The modification status if relevant (applies to specially designed test equipment and gauges).
- The name and location of the owner or custodian.
- The date calibration was performed.
- Reference to the calibration or verification procedure, its number and issue status.
- The condition of the device on receipt.
- The results of the calibration or verification in terms of readings before adjustment and readings after adjustment for each designated parameter, e.g. any out-of-specification readings.
- An impact assessment of any out-of-specification conditions.
- The date fixed for the next calibration or verification.
- The permissible limits of error.
- The serial numbers of the standards used to calibrate the device.
- The environmental conditions prevailing at the time of calibration.
- A statement of measurement uncertainty (accuracy and precision).
- Details of any adjustments, servicing, repairs and modifications carried out.
- The name of the person performing the calibration or verification.
- Details of any limitation on its use.
- Notification to the customer if suspect product has been shipped.

Clearly not all this information would be presented on one record but the records should be indexed so that all this information is traceable both forwards and backwards.

For example, the record containing the results of an assessment of out-of-specification conditions should carry a reference to the related calibration record and vice versa.

The records required are only for formal calibrations and verification and not for instances of self-calibration or zeroing using null adjustment mechanisms. Whilst calibration usually involves some adjustment to the device, non-adjustable devices are often verified rather than calibrated. However, it is not strictly correct to regard all calibration as involving some adjustment. Slip gauges and surface tables are calibrated but not adjusted. An error-record is produced to enable users to determine the uncertainty of measurement in a particular range or location and compensate for the inaccuracies when recording the results.

Software validation (7.6)

The standard requires *confirmation of the ability of software used for measuring and monitoring of specified requirements to satisfy intended application to be undertaken prior to initial use and reconfirmed as necessary.*

What does this mean?
These requirements apply not only to production, installation and servicing but also to design, development and operations.

Why is this necessary?
This requirement responds to the Factual Approach Principle.

Software is used increasingly to drive equipment used for measurement or to interpret results. The integrity of the software therefore has a bearing on the integrity of the measurement and therefore needs to be verified prior to use. Although software does not degrade or wear out, it can be corrupted such that it no longer does the job it was intended to. In many cases software malfunction will be apparent by the absence of any result at all, but in some cases, a spurious result may be generated that appears to the observer as correct. Re-confirmation is necessary therefore after a period where the equipment may have been used in situations where intended or unintended changes to the configuration could have been made.

How is this implemented?
The integrity of software is critical to the resultant product whether it be a deliverable product to a customer or a product being developed or in use. The design of software should be governed by the requirements of Clause 7.3 although this is not mandated by the standard, indeed, if these requirements are applied, the design verification requirements should adequately prove that the software is capable of verifying the acceptability of product. However, the design control requirements may be impractical for many minor verification devices. The hardware that provides the platform for the software should also be controlled, and whilst it may not measure any parameters, its

malfunction could result in nonconforming product being accepted. Complex hardware of this nature should be governed by the design controls of Clause 7.3 if designed in-house because it ensures product quality. If bought out, you should obtain all the necessary manuals for its operation and maintenance, and it should be periodically checked to verify it is fully operational.

To control software you need to consider what it is that you need to control. As a minimum you should control its use, modification, location (in terms of where it is installed), replication and disposal.

Use is controlled by specifying the software by type designation and version in the development and production test procedures or a register that relates products to the software that has to be used to verify its acceptability. You should also provide procedures for running the software on the host computer or automatic test equipment. They may of course be menu driven from a display screen and keyboard rather than paper procedures.

Modifications should be controlled in a manner that complies with the requirements of Clauses 7.3.7 of the standard.

The location could be controlled by index, register, inventory or other such means which enables you to identify on what machines particular versions of the software are installed, where copies and the master tapes or disks are stored.

Replication and disposal could be controlled by secure storage and prior authorization routines where only authorized personnel or organizations carry out replication and disposal.

Measurement system analysis (7.6.1)

The standard requires statistical studies *to be conducted to analyse the variation present in the results of each type of measuring system and that the analytical methods used are to conform to those in customer reference manuals on measurement systems analysis.*

What does this mean?
Measurement systems must be in statistical control so that all variation is due to common cause and not special cause. ISO/TS 16949 therefore requires that you devise a measurement system for all measurements specified in the Control Plan in which all variation is in statistical control.

Why is this necessary?
It is often assumed that the measurements taken with a calibrated device are accurate and indeed they are if we take account of the variation that is present in every measuring

system and bring the system under statistical control. Variation in measurement systems arises due to bias, repeatability, reproducibility, stability and linearity that are summarized as follows:

- *Bias* is the difference between the observed average of the measurements and the reference value.
- *Repeatability* is the variation in measurements obtained by one appraiser using one measuring device to measure an identical characteristic on the same part.
- *Reproducibility* is the variation in the average of the measurements made by different appraisers using the same measuring instrument when measuring an identical characteristic on the same part.
- *Stability* is the total variation in the measurements obtained with a measurement system on the same part when measuring a single characteristic over a period of time.
- *Linearity* is the difference in the bias values through the expected operating range of the measuring device.

It is only possible to supply parts with identical characteristics if the measurement systems as well as the production processes are under statistical control. In an environment in which daily production quantities are in the range of 1000–10,000 units, inaccuracies in the measurement system that go undetected can have a disastrous impact on customer satisfaction and hence profits.

How is this implemented?
Gauge and test equipment requirements are required to be formulated during product design and development, and this forms the input data to the process design and development phase. During this phase a measurement system analysis plan is required to accomplish the required analysis. During the product and process validation phase, measurement system evaluation is required to be carried out during or prior to the production trial run and during full production continuous improvement is required to reduce measurement system variation.

The *Measurement Systems Analysis Manual* provides excellent guidelines for selecting procedures to assess the quality of a measurement system. It includes an introduction to measurement systems, explains the factors that cause variation in a measurement system, has guidance for preparing for a measurement system study and includes step-by-step procedures for determining the degree of each type of variation present in a measurement system.

For those suppliers not conforming to Ford, General Motors and Daimler-Chrysler requirements, other analytical methods and acceptance criteria may be used if approved by the customer. If you lack any documented methods, the *MSA Reference Manual* is recommended.

Laboratory requirements (7.6.3)

Internal laboratory

The standard requires the organization's measurement laboratories *to have a defined scope that includes its capability to perform the required inspections, tests or calibration services and that it shall meet certain technical requirements.*

What does this mean?
Every measurement laboratory will be limited in what it can measure with any degree of accuracy and precision therefore this requirement implies that there needs to be a definition of the type, accuracy and precision of measurement the organization's internal measurement laboratory is capable of performing.

Why is this necessary?
The manager probably knows the capability of his or her laboratory but this capability might not be defined so that others also know. It will also change continually as more and more equipment is brought in to meet new demands from product development. Having an up-to-date definition of measurement capability will therefore enable those charged with reviewing contracts and specifications to determine whether the organization has the capability to meet customer requirements (Clause 7.2.2), whether manufacture of the product is feasible (Clause 7.2.2.2) or whether new capability needs to be developed (Clause 7.6 first requirement).

How is this implemented?
The solution to this requirement is an exposition, defining the capability of the measurement laboratory in terms of:

- the types of measurement undertaken (mass, length and time including derivatives of these);
- the range of accuracy to which measurement are made (the quantities and number of decimal places);
- the list of equipment categorized as primary, secondary and working standards;
- the national approvals granted such as ASTM and NAMAS.

The procedures employed by the laboratory would form part of the organization's quality management system and would therefore cover the competency of personnel, maintenance of records and documentation controls.

External laboratory

The standard requires external, commercial or independent laboratory facilities used for inspection, test or calibration services by the organization *to have a defined laboratory*

*scope that includes the capability to perform the required inspections, test or cali-
bration and is to be acceptable to the customer or accredited to ISO/IEC 17025 or
national equivalent.*

What does this mean?

Basically this means that the organizations you use to supply measurement services
need to have the capability of carrying out the measurements with the accuracy and
precision required. Some laboratories might be approved to carry out some measurements but not others, simply because they don't have the necessary equipment, methods,
environment or competent personnel. ISO/IEC 17025 is an international standard
that measurement laboratories need to meet in order provide assurance of competence
to perform the calibrations or tests specified on its accreditation certificate. ISO/IEC
17025 supersedes ISO/IEC Guide 25:1990.

Laboratory accreditation is defined by ISO as formal recognition that a laboratory is competent to carry out specific tests or specific types of tests. The key words in this definition
are "competent" and "specific tests". Each accreditation recognizes a laboratory's technical capability (or competence) to do specific tests, measurements or calibrations. In that
sense, it should be recognized as a stand-alone form of very specialized technical certification, as distinct from a purely quality system certification as provided by ISO 9000. An
accredited organization is authorized to issue certificates of conformity to national or
international standards. ISO 9000 certification does not authorize organization to issue
such certificates. Accreditation is awarded for a specific scope of service or range of products as is certification, except that for laboratory accreditation they are accredited for very
specific tests or measurements (usually within specified ranges of measurement) with
associated information on uncertainty of measurement and for particular product and test
specifications. An ISO 9000 certificate for a laboratory does not accurately specify the
performance characteristics of the product that the organization is capable of supplying.

Why is this necessary?

The standards required of the measurement facilities in-house should equally apply to
any measurement services provided by external sources and vice versa. Who performs
the measurement is irrelevant. What is important is that the measurements are
carried out within a measuring system that is capable. If the equipment, environment,
methods and personnel are deficient in any way the results cannot be relied on to be
accurate.

How is this implemented?

The simplest method is to select only those measurement laboratories that are
accredited by the national accreditation service to carry of the range of measurements
you require to the stated accuracy. If there is no suitable accredited measurement laboratory for particular calibrations, arrangements need to be made to qualify the laboratory

with the agreement of the customer. Such qualification would require an assessment of the laboratory to ISO/IEC 17025 by an approved second party, i.e. it could be another laboratory already accredited to ISO/IEC 17025.

Summary

In this chapter, we have examined the requirements contained in Section 7 of ISO/TS 16949. This is the largest section of the standard and covers the result-producing process of an organization. Most of the requirements contained in QS-9000 and ISO/TS 16949:1999 are included in this section. We have examined the product realization processes both in general and specific terms through a series of processes that took us from receipt of a customer enquiry to post-delivery activities. However, we have learnt that these processes are not necessarily triggered by customer orders and can also be triggered by a marketing process feeding requirements that have been derived from an understanding of market needs into a design process. This changes contract review into product requirement review and brings in all those functions that might have been excluded from the management system such as marketing, sales and design. We have learnt that we cannot ignore the first two requirements of Clause 7.6 as they apply to all organizations and that the integrity of every device used to determine whether or not a requirement has been met does need to be managed. A recurring theme through this section of the standard is the absence of requirements for documented procedures. Taking one example to illustrate this trend; instead of requiring organizations to establish and maintain documented procedures for customer communication, the standard requires the organization to determine and implement effective arrangements for communicating with customers, thereby focusing on results rather than on the documentation. We have learnt that every one of the requirements is important and examined several solutions to meeting them. We have also seen how many of the product realization clauses can be linked together to form a cycle that covers the core business process of Demand Fulfilment – a process that needs to be managed effectively in order to achieve and sustain customer satisfaction.

Product realization requirements checklist

These are the topics that the requirements of ISO/TS 16949 address consecutively. Topics 85–111 appeared in Chapter 6.

Product realization
Planning of product realization
112   Planning product realization processes
113   Consistency of process planning
114   Determining product quality objectives and requirements
115   Establishing product-specific processes
116   Providing product-specific documents
117   Providing product-specific resources
118   Determining product-specific verification and validation activities and methods
119   Determining product-specific acceptance criteria
120   Determining product-specific records
121   Determining process-specific records
122   Ensuring suitability of planning output

Planning of product realization – Supplemental
123   Production of quality plan

Acceptance criteria
124   Defining acceptance criteria
125   Sampling acceptance criteria

Confidentiality
126   Confidentiality of customer products, projects and information

Change control
127   Change control process
128   Assessment the effects of change
129   Verifying and validating changes
130   Customer review of major changes
131   Customer requirements for change verification and identification

Customer-related processes
Determination of requirements related to the product
132   Determining customer specified requirements
133   Determining requirements for intended use
134   Determining statutory and regulatory requirements
135   Determining organizational requirements

Customer-designated special characteristics
136   Conformity with customer-designated special characteristics

Review of requirements related to the product

137   Reviewing product requirements
138   Timing of product requirements review
139   Ensuring product requirements are defined
140   Resolving differing requirements
141   Ensuring the organization has the ability to meet requirements
142   Maintaining records of the results of product reviews
143   Confirming undocumented requirements
144   Amending documents affected by changed product requirements
145   Informing personnel of changed requirements

Review of requirements related to the product – Supplemental

146   Waiving requirement for formal review

Organization manufacturing feasibility

147   Investigation of manufacturing feasibility

Customer communication

148   Arrangements for communicating product information to customers
149   Arrangements for dealing with customer enquiries
150   Arrangements for dealing with contracts and orders
151   Arrangements for dealing with amendments to contracts and orders
152   Arrangements for dealing with customer feedback

Customer communication – Supplemental

153   Language and format of customer communication

Design and development

Design and development planning

154   Planning product design and development
155   Controlling product design and development
156   Determining design and development stages
157   Determining review, verification and validation for each stage
158   Determining design and development responsibilities and authority
159   Managing design and development interfaces
160   Updating design and development planning

Multidisciplinary approach

161   Using a multidisciplinary approach

Design and development inputs

162   Determining design and development inputs
163   Determining functional and performance requirements
164   Determining statutory and regulatory requirements
165   Determining information from previous design

Product realization – Food for thought

1 Do your marketing, sales and design processes include provisions for measuring the extent to which the process objectives are being achieved?

2 Are you confident that your sales personnel will not commit the organization beyond the capability of its processes?

3 When preparing plans for product realization how do you know you have taken account of all the factors that will affect successful implementation?

4 How do you know that the plans made for fulfilling product requirements will reach those who will create the product and process features necessary for successful implementation?

5 Do you assume customers will define the characteristics necessary to give satisfaction or do you recognize they are not experts and endeavour to find out what their expectations really are?

6 How do you know you have identified all the relevant regulations that apply to the customer transaction?

7 Would your customer expect you to proceed knowing there are issues to be resolved or to wait while you sought resolution?

8 However simple the order, do you always confirm understanding before proceeding?

9 How confident are you that the sales literature does not lead potential customers to expect more than you are prepared to provide?

10 Is customer feedback collected from all the points of contact with customers or only through the mailbox?

11 At what point do you bring design under control, before the design is released or before you spend money?

12 Are you confident that you won't make the same mistakes on the next new design as you did on the last design?

13 How do you stop your designers reinventing solutions to problems solved previously?

14 How do you ensure that design weaknesses revealed through risk analysis techniques are eliminated, reduced or at least controlled before the design is released?

15 What research is performed to discover the probability of success with new designs?

16 Is your purchasing process sufficiently flexible to prevent inappropriate conditions being placed on your suppliers?

17 Does your supplier selection procedure permit value-based decisions to be made or is it one size that fits all?

18 How do you know that the products you purchase for incorporation into supplies will satisfy the regulatory conditions that apply to the item you deliver to your customer?

19 How do your receipt inspectors know of the decisions your supplier verification personnel have made on the shipments received?

20 How do you know the processes are capable of achieving the required product or service features before commencing production or service delivery?

21 Does your system include all the distribution channels that have been established for delivering product to customers?

22 If a product was returned because of a failure, are you confident that you could find out what work has been done on it as it passed through the production process?

23 If an item of customer property was recalled, could you find it and return it in its original condition if requested to do so?

24 Have you put in place methods that will ensure the integrity of all of the devices used for monitoring and measurement?

Chapter 8

Measurement, analysis and improvement

There can be no improvement where there are no standards ... standards exist only
to be superseded by better standards. Every standard, every specification and every
measurement cries out for constant revision and upgrading.

Masaaki Imai

Summary of requirements

Measurement, analysis and improvement processes are vital to the achievement of quality. Until we measure using devices of known integrity, we know little about a process or its outcomes. But if we measure using instruments that are unfit for purpose, we will be misled by the results. With the results of valid measurement we can make a judgement on the basis of facts. The facts will tell us whether we have met the target. Analysis of the facts will tell us whether the target can be met using the same methods or better methods or whether the target is the right target to aim for. Measurements without a target value to compare results of measurement are measurements without a purpose. The target value is therefore vital but arbitrary values de-motivate personnel. Targets should always be focused on purpose so that through the chain of measures from corporate objectives to component dimensions there is a soundly based relationship between targets, measures, objectives and the purpose of the organization, process or product.

> **Measurement Axioms**
>
> What gets measured gets managed.
> What gets reported gets attention.

Measurement tells us whether there has been a change in performance. Change is a constant. It exists in everything and is caused by physical, social or economic forces. When we measure the same parameter on different items we expect slight variation. However, if we measure the same parameter using the same device we might not expect there to be a change, but the inaccuracies inherent in the measuring system will lead to a variation in readings. To understand change we need to understand its cause.

Some change is represented by variation about a norm and is predictable – it is a natural phenomenon of a process and when it is within acceptable limits it is tolerable. Other change is represented by erratic behaviour and is not predictable but its cause can be determined and eliminated through measurement, analysis and improvement.

Measurement, analysis and improvement are strictly subprocesses within each business process. However, parent processes will often capture data from monitoring and measurements within subprocesses. This may happen when assessing a variety of data from individual processes to determine customer satisfaction or for discovering common cause problems and subsequently devising company wide improvement programmes.

There is a sort of logic in the structure of the requirements in this section but there are some gaps. It would have assisted understanding if the same terms as used in Clause 8.1 had been used in the headings of Clauses 8.2 to 8.5. In that way the relationships would have been more obvious. The general requirements of Clause 8.1 are amplified by Clauses 8.2 to 8.5, so the requirements in Clause 8.1 are not separate to those in Clauses 8.2 to 8.5 with the exception of Clause 8.3 on the control of nonconforming product and those on statistical techniques. This later requirement is stated once because it applies to all monitoring, measurement and analysis processes. Clause 8.3 on nonconforming product appears in Section 8 not because it has anything to do with measurement, analysis and improvement but because its inclusion in Section 7 would imply that it could be excluded from the management system (see ISO 9001 Clause 1.2).

It should not be assumed that Section 8 includes all requirements on measurement, analysis and improvement.

Measurement and monitoring is also addressed by:

- *Management representative (5.5.2)* in the context of reporting on system performance.
- *Management review (5.6)* in the context of reviewing system adequacy.
- *Control of monitoring and measurement devices (7.6).*
- *Design and development verification (7.3.5).*
- *Design and development validation (7.3.6).*
- *Verification of purchased product (7.4.3).*

Analysis is also addressed by:

- *Management review (5.6)* in the context of changes that could affect the management system.
- *Control of design and development changes (7.3.7)* in the context of evaluation of the effects of change.
- *Control of monitoring and measuring devices (7.6)* in the context of measurement systems.

Improvement is also addressed by:

- *Management review (5.6)* in the context of changes to quality policy.
- *Control of design and development changes (7.3.7)*.
- *Internal communication (5.5.3)* in the context of communicating the effectiveness of the management system.
- *Provision of resources (6.1)* in the context of resources needed for continual improvement.

We can link the requirements of Section 8 together in a number of separate cycles (indicating the headings from ISO/TS 16949 in bold italics type).

During the design and development of the business and work processes we would undertake a review of the established practices to identify potential problems and undertake **preventive action** to prevent occurrence of such problems. Before implementing the management system processes or any changes thereto, we would perform **internal audits** (or undertake **process validation**) to determine whether these processes met the relevant requirements of the standard, enabled the organization to fulfil its **policies** and **objectives**, and produce the required products and services. Any potential problems discovered would be subject to **preventive action** to prevent occurrence of such problems. At appropriate stages in the production process we would **monitor and measure product** for compliance with specified requirements, periodically undertake **product audits** to establish the effectiveness of the process controls, initiate **control of nonconforming product** on detecting nonconformity and undertake **corrective action** when process targets had not been met. After introducing new or changed practices and periodically thereafter, we would perform **internal audits** to determine whether the planned arrangements were being implemented as intended and undertake **corrective action** to bring about **improvement** by better control. Periodically we would **monitor and measure processes** for their ability to achieve planned results and undertake **process audits** to establish whether the achieved results arose from implementing the planned arrangements and if necessary, undertake corrective action to reduce variation and bring about **improvement** by better control. We would also **analyse data** resulting from these reviews and bring about **improvements** by better utilization of resources. Sometime after establishing the **organization's purpose**, setting **policies** and **objectives** that were **customer focused** and installed the enabling **processes** we would collect and **analyse data** in order to monitor **customer satisfaction** and undertake **corrective action** to bring about **improvement** by better control.

One observes from this consolidation that the order in which the clauses are mentioned is not remotely the same as the order they are addressed in the standard, that some clauses in the standard appear several times and others are drawn from other sections thus demonstrating that you cannot treat the clauses in isolation or in any particular sequence.

Monitoring, measurement, analysis and improvement processes (8.1)

Processes to demonstrate conformity (8.1)

The standard requires the organization *to plan and implement the monitoring, measurement, analysis and improvement processes needed to demonstrate conformity of the product including determination of applicable methods such as statistical techniques and the extent of their use.*

What does this mean?

The requirement for monitoring, measurement, analysis and improvement processes to be planned means that these activities should not be left to chance. It is odd that the requirements are limited to the processes needed to demonstrate conformity of the product and not all processes. However, this should not be taken too literally, because almost every process in an organization will impact the product either directly or indirectly. For example, if training processes are not monitored, untrained staff could produce nonconforming product. If management processes are not monitored result-producing processes may be starved of resources and consequently impact the product.

The monitoring processes that are needed to demonstrate conformity of product are addressed in the standard by:

- The monitoring of customer satisfaction (8.2.1).
- Monitoring of product (8.2.4) including design and development review (7.3.4).

These are the processes that keep operations and operating conditions under periodic or continual observation in order to be alerted to events before they occur. This is so that action can be taken to prevent nonconformity. Typical monitoring processes are those employed to check that machines are functioning correctly, that processes are under statistical control, that there are no serious bottlenecks or other conditions that may cause abnormal performance.

The measurement processes that are needed to demonstrate conformity of product are addressed in the standard by:

- Measurement of product (8.2.4) including
  - Design and development verification (7.3.5).
  - Design and development validation (7.3.6).
  - Verification of purchased product (7.4.3).
  - Control of monitoring and measuring devices (7.6).

These are the processes that determine the characteristics of product in order to establish whether the product meets defined requirements. Typical measurement processes involve physical measurement of a product using inspection, test or demonstration techniques, or non-physical measurement of services using observation or examination techniques. These processes are found in design, production and service delivery operations.

The analysis processes to demonstrate conformity of product are addressed in the standard by:

- Design and development review (7.3.4). (This is both a monitoring and an analysis process.)
- Corrective action (8.5.2).
- Preventive action (8.5.3).

These are those processes that convert product data into knowledge from which decisions of conformity against prescribed standards can be made. Typical analysis processes are chemical and metallurgical analysis to ascertain the properties or composition of a material, stress analysis to ascertain the load bearing capacity of a structure. Other types of analysis may be used to compare current designs with previously validated designs, determine time-dependent characteristics, assess vulnerability, susceptibility and durability.

The improvement processes needed to demonstrate conformity of the product are those processes that eliminate the cause of nonconformities and prevent their occurrence. These include the corrective action processes and preventive action processes. (These are dealt with under the relevant headings.)

There are various ways of monitoring, measuring and analysing things, and some methods are better than others. The standard requires that applicable methods be determined and one field of data collection and analysis that can be used to demonstrate conformity of the product is to use statistical techniques. Any technique that uses statistical theory to reveal information is a statistical technique. However, there is a difference between statistical theory and statistics. Any set of figures that are intended to describe an entity or phenomena can be regarded as statistics. When one makes a prediction from these figures that attributes a particular quality to an entity or phenomena one needs to apply statistical theory for the prediction to be valid. Techniques, such as Pareto analysis, histograms, correlation diagrams and matrix analysis, are regarded as statistical techniques but although numerical data is used, there is no probability theory involved. These techniques are used for problem-solving not for making product acceptance decisions. Other techniques such as statistical process control (SPC), reliability prediction and maintainability prediction use probability theory to provide a result which may not be absolute fact but which is the most probable result that can be deduced from the facts about a product or a number of products.

Why is this necessary?
This requirement responds to the Factual Approach Principle.

Monitoring, measurement, analysis and improvement processes are necessary to control the quality, cost and delivery of output. Where monitoring and measurement processes needed to demonstrate conformity of the product are performed, they should form part of a plan the intention of which is to discover whether the product conforms to requirements. Having discovered that information, it should also be planned that the information is passed through analysis processes on to the decision-makers so that decisions on product acceptability are made on the basis of fact and not opinion. Also as part of the plan, it is necessary to make provision for dealing with unacceptable product, either making it conform or preventing its unintended use or further processing.

How is this implemented?
Planning the monitoring, measurement and analysis processes
Whether monitoring or measuring, the processes are very similar and consist of a uniform series of elements:[1]

* The characteristic to be measured or monitored.
* The units of measure.
* The standard or specification to be achieved.
* The sensor for detecting variance.
* The human and physical resources for collecting, analysing, transmitting and presenting the data.
* Interpretation and verification of results.
* Decision on the action needed.
* Taking action.

Some of these elements were addressed in Chapter 7 under the heading *Control of monitoring and measuring devices.*

In planning the monitoring and measurement processes needed to demonstrate conformity of the product the first thing to do is to identify the characteristics that need to be achieved and then determine how and where they are going to be verified. It is also necessary to determine the conditions that affect the achievement of these characteristics such as key process parameters and establish how these will be monitored.

A useful approach is to develop for each product a Verification Matrix that identifies the requirement to be achieved, the level at which the requirement is achieved and the method to be employed. Some requirements may be verified during design verification and not require confirmation during production because they are inherent features of the design. Other requirements may need to be reconfirmed on each product due to variations in materials or processes. Some characteristics may be only accessible for verification at the component or subassembly levels whereas others can be verified at end product level.

Another useful approach is a control plan. The aim of the control plan is to ensure that all process outputs will be in a state of control by providing process monitoring and control methods that control product and process characteristics. Apart from product identification data, the type of information contained in a control plan for each manufacturing operation is as follows:

- Operation description
- Characteristic
- Specification
- Measurement technique
- Sample size and frequency
- Control method (control chart, 100% inspection, functional test, check sheet, etc.)
- Reaction plan (what to do in the event of a nonconformity).

For each production process, there should be a process specification that defines the standard operating conditions that need to be maintained and the means by which variation in these conditions is to be detected. Various instruments may be needed to provide a visual indication of operating conditions and the specification should also indicate the frequency of checks if they are located remote from the operator.

Determining measurement methods
Within your procedures you will need a means of determining when statistical techniques will be needed to determine product characteristics and process capability. One way of doing this is to use checklists when preparing customer specifications, design specifications, and verification specifications and procedures. These checklists need to prompt the user to state whether the product characteristics or process capability will be determined using statistical techniques and if so which techniques are to be used.

Techniques for establishing and controlling process capability are essentially the same – the difference lies in what you do with the results. Firstly, you need to know if you can make the product or deliver the service in compliance with the agreed specification. For this you need to know if the process is capable of yielding conforming product. Statistical Process Control (SPC) techniques will give you this information. Secondly, you need to know if the product or service produced by the process actually meets the requirements. SPC will also provide this information. However, having obtained the results you need the ability to change the process in order that all product or service remains within specified limits and this requires either real-time or off-line process monitoring to detect and correct variance. To verify process capability you periodically rerun the analysis by measuring output product characteristics and establishing that the results demonstrate that the process remains capable.

When carrying out quality planning you will be examining intended product characteristics and it is at this stage that you will need to consider how achievement is to be measured and what tool or technique is to be used to perform the measurement. When you have chosen the tool you need to describe its use in the control plan. If using APQP, you will generate Gauges/Testing Equipment Requirements as an output of product design and input to process design that produces the outputs of a Measurement Systems Analysis Plan and Preliminary Process Capability Study Plan.

Where statistical techniques are used for establishing, controlling and verifying process capability and product characteristics, procedures need to be produced for each application. You might for instance need a Process Control Procedure, Process Capability Analysis Procedure, Receipt Inspection Procedure, Reliability Prediction Procedure, etc. The procedures need to specify when and under what circumstances the techniques should be used and provide detail instruction on the sample size, collection, sorting and validation of input data, the plotting of results and application of limits. Guidance will also need to be provided to enable staff to analyse and interpret data, convert data and plot the relevant charts as well as make the correct decisions from the evidence they have acquired. Where computer programs are employed, they will need to be validated to demonstrate that the results being plotted are accurate. You may be relying on what the computer tells you rather than on any direct measurement of the product.

Processes to ensure conformity of management system (8.1)

The standard requires the organization *to plan and implement the monitoring, measurement, analysis and improvement processes needed to ensure conformity of the quality management system including determination of applicable methods such as statistical techniques and the extent of their use.*

What does this mean?

Conformity of the management system means that the system has been designed with the capability of implementing the defined policies and fulfilling the established objectives and targets, and is being operated in a manner consistent with these policies, objectives and targets.

The monitoring processes to ensure conformity of the management system are addressed in the standard by:

- Reporting on management system performance (5.5.2).
- Review of system adequacy (5.6).
- Internal audit (8.2.2).
- Monitoring of processes (8.2.3).

Note: System adequacy is interpreted as a state where the system delivers the required outputs. They may not be produced efficiently or be the right outputs but they are the ones required.

More generally in this context the monitoring processes are those that keep operations and operating conditions under periodic or continual observation. The reason for doing this is to be alerted to events before, during or immediately after they occur so that action can be taken to prevent or minimize their effects. Typical monitoring processes are those employed in the utility industries where advanced warning of potential or actual problems is needed to maintain the supply of electricity, water and gas to consumers. Management monitor expenditure or rate of expenditure in order to forecast whether their current and future commitments can be met. Operators monitor machines to check that they are functioning correctly, that processes are under statistical control. Distribution staff may monitor on-time delivery and ring the alarm bells when this drops below the norm.

The measurement processes to ensure conformity of the management system are addressed in the standard by:

- Validation of processes (7.5.2).
- Quality management system audit (8.2.2.1).
- Manufacturing process audit (8.2.2.2).
- Product audit (8.2.2.3).
- Measurement of processes (8.2.3).
- Monitoring and measurement of manufacturing processes (8.2.3.1).

The measurement processes needed to ensure conformity of the management system are those processes that measure the performance of the business processes. Every process should not only contain provision for measuring output but for measuring whether the process objectives are being achieved. Audits are one means of verifying that the management system conforms in both its design and implementation. Such audits would not only determine if procedures were being followed but whether the desired results were being achieved and any regulations complied with. (Audits are dealt with separately under the relevant heading.)

The analysis processes to ensure conformity of the management system are addressed in the standard by:

- Internal audit (8.2.2) this is in the context of audits of implementation.
- Quality management system audit (8.2.2.1).
- Manufacturing process audit (8.2.2.2).
- Product audit (8.2.2.3).
- Analysis of data (8.4).

More generally the analysis processes needed to ensure conformity of the management system are those processes that covert process data into knowledge from which decisions of conformity against prescribed standards can be made. Examples include the analysis of variation in processes, process capability studies, behavioural analysis, organizational analysis, workflow analysis, analysis of constraints, risk analysis, hazard analysis and defect analysis.

The improvement processes needed to ensure conformity of the management system are the corrective action processes that reduce variation in process performance and restore the *status quo*. (Corrective action is dealt with separately under the relevant heading.)

The statistical techniques in this context are those used for measuring process capability.

Why is this necessary?
This requirement responds to the Factual Approach Principle.

As the management system is the means by which the organization achieves its object-ives, it is necessary to ensure that the processes are well designed and are operating properly. Much of the focus in QS-9000 was on measuring product and auditing pro-cedures. It has been widely accepted that both these techniques are not only ineffective in ensuring customer satisfaction but are also uneconomic. A more effective technique is to monitor and measure processes, and strive to reduce variation in these processes. By setting up processes that are capable of delivering conforming output and then monitoring, measuring and improving these processes so that process capability is assured, less dependence needs to be put on measuring product and auditing proced-ures and as a result less resources are utilized.

How is this implemented?
These requirements are expanded in the other sections of Clause 8 of the standard and therefore the methods for implementing them are addressed under the appropriate heading.

Processes to continually improve the effectiveness of the management system (8.1)

The standard requires the organization *to plan and implement the monitoring, measurement, analysis and improvement processes needed to continually improve the effectiveness of the quality management system including determination of applicable methods such as statistical techniques and the extent of their use.*

What does this mean?
The Monitoring and measurement processes to continually improve the effectiveness of the management system are addressed in the standard by:

*   Management review (5.6).
*   Review of requirements related to product (7.2.2).
*   Monitoring and measurement of processes (8.2.3).

Continual improvement of the effectiveness of the management system is addressed in more detail against Clause 8.5.1 of the standard. However, it is important to distinguish between improving the management system and improving the effectiveness of the management system. The management system can be improved by correcting deficiencies in process design and implementation but this is improving conformity – making the process perform as it should do. Effectiveness is about doing the right things therefore improving the effectiveness of the management system means making the objectives, standards or targets established for activities, tasks or processes meet the needs of the organization to accomplish its purpose or mission. The objectives, targets and standards focus action therefore if they are not the right objectives, targets and standards the action, no matter how well it is performed, will be the wrong action for the organization. It may be the right action in another organization or in different circumstances.

The only analysis process that serves continual improvement in system effectiveness is the process for scanning the environment for changes that impact the objectives and targets that have been established. This is addressed by Clause 5.6.2f.

The improvement process for improving the effectiveness of the management system is the *change management process*. This process would be triggered by the results of scanning the environment for changes as stated above and would manage the change through the organization.

Why is this necessary?
This requirement responds to the Continual Improvement Principle.

Many organizations are managed as a series of functions. A functional hierarchy is a common approach to organizational design. However, the structure tends to dictate the way the objectives are set, and therefore each function sets its objectives based on what it believes is necessary to meet the organization's purpose and mission. This often leads to conflicting objectives where one function pursues objectives that compromise other functions. When functional objectives are based on quotas, such as number of orders processed, number of designs produced, number of products produced, the challenge is to do more rather than do better. Effectiveness is not a measure of meeting quotas but fulfilling purpose. It is therefore necessary to monitor and measure the effectiveness of the system as a whole – not the functions. A typical example is where the purchasing function is measured on the costs of purchased product and naturally strives to drive down suppliers on price. The net result is that either the supplier goes out of business or uses substandard materials and takes risks. The results submitted by the purchasing function to the Executive look good; it has halved the purchasing costs but production can't make the product and warranty costs go up. This is why it is necessary to manage the system as a series of processes not as a series of functions.

How is this implemented?

The methods for improving the effectiveness of the management system are dealt with under the heading *Improvement*.

Identification of statistical tools (8.1.1)

The standard requires appropriate statistical tools *to be determined for each process during advanced quality planning and included in the control plan.*

What does this mean?

Firstly, the requirement applies to product development and therefore not to the use of statistical tools in marketing, business development or even service development. When planning production the characteristics susceptible to variation will be identified. The techniques need to detect and control this variation also needs to be identified. The techniques chosen will depend on what is to be measured, the quantities, the accuracy and precision of measurement and probability of error. The chosen controls will be indicated in the control plan against the relevant operation. (Normally Column 25 in the Control Plan.)

Why is this necessary?

There is a right time and a wrong time to determine the methods of measurement. Doing it when you have a thousand units to measure is definitely the wrong time. Therefore, when planning production, it is prudent to establish measurement methods and specifically the methods by which variation in product characteristics and process capability is to be detected.

How is this implemented?

The Control Plan will identify the process stages and from each stage there will be a process output, i.e. something that is measurable or countable. This is normally entered in the column headed "Characteristics" in the Control Plan. The tools used to measure these characteristics with the accuracy and precision required (defined under Specification/Tolerance in the Control Plan) are identified in the column headed Evaluation and Measurement Technique. The statistical tools, if any, are defined in the column headed Control Method. In a typical Control Plan you will see reference to check sheets, x bar-R charts, x-MR chart, etc.

Here is not the place to start a treatise on statistical concepts, there are many books one can acquire on this subject and perhaps the most appropriate in this context is the *Fundamental Statistical Process Control Reference Manual*, i.e. one of the five core quality tools required by the US automotive industry. In the UK the SMMT has published two part guidelines to Statistical Process Control. These manuals identify the tools, their purpose and applicability and how the tools can be applied.

Knowledge of basic statistical concepts (8.1.2)

The standard requires basic statistical concepts *to be understood throughout the organization*.

What does this mean?
Statistical concepts are ways of collecting, analysing, presenting and interpreting facts and figures. These might include probabilities, distributions and relationships, etc. that are used to draw conclusions, make comparisons or make decisions. Knowledge of these concepts means that people have sufficient understanding so as not to feel helpless or frightened when exposed to statistical information.

Why is this necessary?
A basic understanding of statistical concepts that is well founded and frequently refreshed will pay dividends because it will heighten the sensitivity of managers and staff to variation. Statistically aware personnel are more likely to question results, not only when presented with a graph or chart but when observing operations, listening to conversations or reading reports.

With modern tools and techniques, the technical expertise has been taken out of statistics, i.e. staff simply put in the figures and out come the charts often leading to complacency. A wider appreciation of the concepts and heightened awareness should enable observers to question why the dots on the chart vary or remain the same and to reveal significant process problems. This is what we mean by empowerment. When staff don't feel helpless they are empowered to act. All managers need a basic appreciation but those in production ought to be able to apply the techniques their staff use so that they can detect when they are not being applied correctly. A manager who can look at a set of figures or perform a simple calculation and immediately detect that there is something amiss is a valuable asset in any organization. Simply plotting measurements will reveal whether flinching has been used. (Flinching is the tendency of inspectors to falsify the results of measurements that are borderline.)

Auditors need to be able to determine whether the right techniques are being applied and whether the techniques are being applied as directed. Remember that the auditor's task is to determine whether the system is effective, so the ability to detect differentiate between the use of inappropriate techniques is essential. Putting a tick in a box opposite a requirement for SPC charts should imply that the charts have been produced correctly not simply that the charts exist.

The staff assigned to quality planning need an even wider appreciation of statistical concepts and it is probably useful to have an expert in your company on whom staff can call on from time to time. If the primary technique is SPC then you should appoint an SPC Co-ordinator who can act as mentor and coach to the other operators of SPC techniques.

How is this implemented?

Equipping personnel with the awareness necessary can be accomplished firstly establishing one or more teams of those who need knowledge of statistical concepts to perform their jobs as well as possible. Tutors, mentors or coaches are appointed to train and lead the teams. By exposure to case studies and the techniques in action followed by classroom instruction and then mentoring coupled with recommended reading, the teams gradually become aware of statistical concepts and their fear or helplessness subsides so they emerge feeling empowered. It is essential that the mentor observes behaviour and has the respect of those they observe. No one is too high in an organization to escape this awareness. Perhaps at the upper levels, the examples and case studies need to be tailored to situations in which the executives might find themselves. But it is also important for them to realize the consequences of failing to observe variation and taking action even with the most mundane of jobs.

Customer satisfaction (8.2.1 and 8.2.1.1)

The standard requires the organization *to monitor information relating to customer perception as to whether the organization has met customer requirements through continual evaluation of performance of the realization processes*. In addition the standard requires *the organization to monitor the performance of manufacturing processes to demonstrate compliance with customer requirements for product quality and efficiency of the process*. It also requires *the methods of obtaining and using this information to be determined and for performance indicators to be based on objective data*.

What does this mean?

By combining definitions of the terms 'customer satisfaction' and 'requirement' ISO 9000 defines customer satisfaction as the customers perception of the degree to which the customer's stated or implied needs or expectations have been fulfilled.

In order to satisfy customers you therefore have to go beyond the stated requirements. Customers will differ in their perceptions as to whether a transaction has been satisfactory. The term perception is used because satisfaction is a subjective and human condition unlike acceptance that is based on objective evidence. Customers may accept a product but not be wholly satisfied with it or the service they have received. Whether or not you have done your utmost to please the customer, if the customers perception is that you have not met their expectations, they will not be satisfied. You could do exactly the same for two customers and find that one is ecstatic about the products and the services you provide and the other is dissatisfied.

Information relating to customer perception is any meaningful data from which a judgement can be made about customer satisfaction and would include compliments, complaints, sales statistics, survey results, etc. The requirement refers to the monitoring of customer perception rather than the measurement of customer satisfaction. One difference is that monitoring involves systematic checks on a periodic or continuous basis, whereas measurement may be a one-off event.

The supplemental automotive requirements contain some duplication. It requires manufacturing processes to be monitored to demonstrate compliance with *customer requirements for product quality and efficiency of the process*. This requirement appears to be already addressed by the more general requirement to monitor customer satisfaction through *continual evaluation of performance of the realization processes*. The two requirements may differ in detail but by meeting the more general requirement one must meet the more detailed requirement.

Why is this necessary?
This requirement responds to the Factual Approach Principle.

The primary purpose of the management system is to enable the organization to achieve its objectives one of which will be the creation and retention of satisfied customers. It therefore becomes axiomatic that customer satisfaction needs to be monitored.

How is this implemented?
There are several ways of monitoring information relating to customer perceptions.

Repeat orders
The number of repeat orders (e.g. 75% of orders are from existing customers) is one measure of whether customers are loyal but this is not possible for all organizations particular those that deal with consumers and do not capture their names. Another measure is the period over which customers remain loyal (e.g. 20% of our customers have been with us for more than 10 years). A marked change in this ratio could indicate success or pending disaster.

Competition
Monitoring what the competition is up to is an indicator of your success or failure. Do they follow your lead or are you always trying to catch up? Monitoring the movement of customers to and from your competitors is an indicator of whether your customers are being satisfied.

Referrals
When you win new customers find out why they chose your organization in preference to others. Find out how they discovered your products and services. It may be from advertising or maybe, your existing customers referred them to you.

Demand
Monitoring the demand for your products and services relative to the predicted demand is also an indicator of success or failure to satisfy customers. It could also be an indicator of the effectiveness of your sales promotion programme; therefore analysis is needed to establish which it is.

Effects of product transition
When you launch a new product or service, do you retain your existing customers or do they take the opportunity to go elsewhere?

Surveys
There are several types of survey than can be used. There is the impersonal form and the personal form. The impersonal form relies on responses to questionnaires and seeks to establish customer opinion on a number of topics ranging from specific products and services to general perceptions about the organization. The questionnaires can be sent to customers in a mail shot, included with a shipment or filled in before a customer departs (as with hotels and training courses). These questionnaires are somewhat biased because they only gather information on the topics perceived as important to the organization.

It should be noted that questionnaires by themselves are not an effective means of gathering customer opinion. Customers don't like them and are not likely to take them seriously unless they have a particular issue they want to bring to your attention. It is much better to talk face to face with your customer using an interview checklist. Think for a moment how a big customer like Ford and General Motors (GM) would react to thousands of questionnaires from their suppliers. They would either set up a special department just to deal with the questionnaires or set a policy that directs staff not to respond to supplier questionnaires. Economics alone will dictate the course of action that customers will take.

The personal form of survey is conducted through interview such as a customer service person approaching a customer with a questionnaire while the customer is on the organization's premises. This may apply to hotels, airports, entertainment venues and large restaurants. With this method there is the opportunity for dialogue and capturing impromptu remarks that hide deep-rooted feelings about the organization.

Focus meetings

A personal form of obtaining information on customer satisfaction is to arrange to meet with your customer. Seek opinions from the people within the customers organization such as from Marketing, Design, Purchasing, Quality Assurance and Manufacturing departments, etc. Target key product features as well as delivery or availability, price and relationships. This form is probably only suitable where you deal with other organizations.

Complaints

The process for handling customer complaints is addressed in Chapter 7 under the heading *Customer feedback*. Here the topic is *monitoring* and therefore you should be looking at the overall number of complaints, the upward or downward trends and the distribution of complaints by type of customer, location and nature of complaint. Coding conventions could be used to assign complaints to various categories covering the product (or parts thereof) packaging, labelling, advertising, warranty, support, etc.

Any complaint, no matter how trivial, is indicative of a dissatisfied customer. The monitoring methods need to take account of formal complaints submitted in writing by the customer and verbal complaints given in conversation via telephone or meeting. Everyone who comes into contact with customers should have a method of capturing customer feedback and communicating it reliably to a place for analysis.

Compliments

Compliments are harder to monitor because they can vary from a passing remark during a sales transaction to a formal letter. Again, all personnel who come into contact with customers should have a non-intrusive method for conveying to the customer that the compliment is appreciated and will be passed on to the staff involved.

Performance indicators

In general, the foregoing indicators provide data about the organization as a whole rather than specific product. In addition, it is necessary and probably much easier to gather data direct from customers about specific product. ISO/TS 16949 Clause 8.2.1.1 suggests such data include:

- performance of delivered parts,
- customer disruptions including the end user,
- delivery performance against schedule including incidents of premium freight,
- customer notifications related to quality or delivery issues.

The vehicle makers use different terms when notifying suppliers of their dissatisfaction with product quality or delivery but all come down to one thing. When an automotive customer notifies an organization that its certification is suspended owing to a major nonconformity this is a notification that comes under Clause 8.2.1.1. The ISO/TS

16949 certificate is suspended, the date of suspension is the date that the status was invoked and the organization has in most cases 90 days to close the nonconformity to the customers satisfaction. The certification is notified and brought in to perform a re-assessment which if unsatisfactory the certificate will be withdrawn.

Internal audit (8.2.2)

Auditing for conformance with planned arrangements (8.2.2a)

The standard requires the organization *to conduct internal audits at planned intervals to determine whether the quality management system conforms to the planned arrangements (see Clause 7.1).*

What does this mean?
ISO 9000 defines an audit as a systematic, independent and documented process for obtaining audit evidence and evaluating it objectively to determine the extent to which agreed criteria are fulfilled. In other words an audit is an examination of performance to determine whether defined performance standards have been met. Audits are a means of verification and as such involve monitoring and measurement. Both these words are used in the standard and it seems as though they are intended to convey different concepts but are actually part of the same concept (see Chapter 1).

The Planned arrangements in this context are the agreed criteria and these are the intentions of management as expressed by the quality policy, quality objectives, particular requirements for products as expressed by customers and particular regulations as expressed by laws and statutes. The purpose of this audit is therefore to establish that the management system has been designed to implement the agreed quality policy, quality objectives and product requirements, i.e. does it possess all the characteristics necessary for it to fulfil its purpose? Principal among these characteristics are the organization's processes.

Clause 7.1 refers to Clause 4.1 and therefore, the audit should verify that in designing the management system, the organization has identified all the necessary processes, determined their sequence and interaction, the criteria and methods for effective operation and control, provided the necessary information and resources and installed the necessary monitoring, measurement analysis and improvement processes.

Planned intervals can be any time interval and as the quality policy is unlikely to change very much in a 5-year period, annual audits may not detect any significant variations. Quality objectives also may change once a year and so quarterly audits may not detect any significant change. However, product requirements vary with every new product and therefore the intervals of audit need to be set relative to changes in policies, objectives and products rather than a set frequency.

Why is this necessary?
This requirement responds to the Factual Approach Principle.

The purpose of quality audits is to establish, by an unbiased means, factual information on quality performance. Quality audits are the measurement component of the quality system. Having established a quality system it is necessary to install measures that will inform management whether the system is being effective. Installing any system without some means of being able to verify whether it is doing its intended job is a waste of time and effort. Audits gather facts, they should not change the performance of what is being measured and should always be performed by someone who has no influence over what is being measured. Audits should not be performed to find faults, to apportion blame or to investigate problems; other techniques should be used for these purposes. One of the most common deficiencies as Deming points out[2] is that a goal without a method for reaching it is useless, but it is a common practice for management to set goals without describing how they are going to be accomplished. A management system audit is needed to reveal such deficiencies.

How is this implemented?
As indicated above, this audit should be triggered by a change in policy, objectives or product requirements. A practical approach would therefore be to design three types of Management System Audits or System Audits. The audits should be designed to verify that the system is capable of:

- implementing the agreed policies (a policy audit);
- enabling the organization to achieve the agreed objectives (a strategic audit);
- enabling the organization to meet specific project requirements (a project audit);
- enabling the organization to meet specific product requirements (a product audit).

Policy audit
The policy audit should establish that:

- the corporate policies are derived from an analysis of the factors critical to accomplishment of the organization's purpose (Are they soundly based?);
- the corporate policies are communicated throughout the organization (Does everyone understand them?);
- measures have been established for determining whether the corporate policies are being implemented (How will we know if they are not being implemented?);
- the operational policies do not conflict with the corporate policies (How do departmental policies relate to corporate policies?).

Strategic audit

The strategic audit should establish that:

- there is a defined process for establishing the organizations goals and objectives;
- an analysis of current and future needs of customers and other interested parties has been carried out;
- the requirements which the organization needs to meet to fulfil its mission have been determined;
- objectives have been established for achieving these requirements;
- priorities for action have been set;
- the products, services and projects that need to be developed or abandoned to achieve these objectives have been identified;
- the risks to success have been quantified;
- the processes for achieving the objectives have been designed and constructed;
- the information, resources, criteria and methods for effective operation of these processes have been identified, developed and provided;
- the necessary monitoring, measurement, analysis and improvement processes have been designed and installed.

Project audit

The project audit should be conducted for each new project (i.e. an undertaking that requires the development of new or modified products, processes and services). The project audit should establish that:

- any changes to the management system processes that are needed to achieve specific product requirements have been identified;
- the processes have been modified or new processes designed;
- the information, resources, criteria and methods for effective operation and control of these processes have been identified, developed and provided;
- the necessary monitoring, measurement, analysis and improvement processes have been designed and installed.

By judicious implementation of these system audits, the integrity of the management system will be confirmed.

Auditing for conformity (8.2.2a and 8.2.2.1)

The standard requires the organization *to conduct internal audits at planned intervals to determine whether the quality management system conforms to the requirements of this Technical Specification and any additional quality management system requirements.*

What does this mean?
ISO/TS 16949 contains a series of requirements for which there are numerous solutions depending on the nature of the organization and its markets, products and

services. As the organization, its processes and products change, periodic confirmation that the system conforms to the *Technical Specification* is necessary to verify it remains in conformity.

Why is this necessary?

This requirement responds to the Factual Approach Principle.

It is debatable whether this requirement is really necessary. If the organization has conducted system audits of the type described above, has conducted implementation audits of the type described below and is conducting management reviews, it will know whether the management system is effective. There is therefore no justification for verifying that the system conforms to the requirements of ISO/TS 16949 because it puts the emphasis on conformity rather than performance. It suggests that if the management system was deemed effective by the management review but an ISO/TS 16949 conformity audit revealed that certain requirements were not being met, the system should be changed simply to bring it into conformity with the standard. Why would anyone want to do this? The system could be focused on the wrong objectives, but it does not require a complete ISO/TS 16949 audit for this to be discovered.

The argument for this requirement is that all organizations should have control over their own operations and not rely on external Certification Bodies to detect system nonconformities. It would therefore be logical for organizations to perform internal audits against the same standards as the external bodies and only use the external bodies as confirmation that they meet the requirements of the standard. Having performed an initial assessment against the requirements of ISO/TS 16949, one should only need to repeat the assessment when the system changes.

How is this implemented?

As the requirement is placed under the heading of internal audit, the Certification Body audit cannot be a substitute even though it is performed with the same purpose in mind. There is a school of thought that believes a Certification Body audit will ultimately reduce to a confirmation that the internal conformity audit has been carried out by competent personnel and has demonstrated that the system is in conformity with the requirement of ISO/TS 16949. This is not beyond the bounds of possibility. All it needs is for the Certification Body to recognize the competence and impartiality of the internal auditors, reduce their certification costs and a new auditing regime will emerge.

Conformity with the requirements of ISO/TS 16949 can be confirmed in one of two ways:

(a) by performing a full internal system audit against ISO/TS 16949;
(b) by analysing the results of policy, strategic, project and process audits, and determining conformity by correlation.

Table 8.1 *Sample conformity audit*

No.	Requirement	Response
1	Is there a quality policy	Yes
2	It is appropriate to the purpose of the organization	No because there is no published statement of the organization's purpose or mission
3	Does it include a commitment to comply with requirements	Yes
4	Does it include a commitment to improve the effectiveness of the quality management system	Yes
5	Does it provide a framework for establishing and reviewing quality objectives	No because the statements are motherhood statements not definable values or principles
6	Is it communicated within the organization	Yes
7	Is it understood within the organization	Only 5 out of 20 understood how it applied to what they do
8	Is it reviewed	Yes
9	Is it reviewed for continuing suitability	No because there is no criteria for determining its suitability
10	Is all this directed by top management	No because top management are simply presented with a statement to approve

In order to meet this requirement by analysis of data, rather than audit you would first need a matrix showing all the requirements of ISO/TS 16949 against the processes in which the requirements were implemented. One would also need to add in any additional customer-specific quality management requirements. For each process a list of applicable requirements can then be identified and cross-referenced to the process stage where it is implemented. The audit results from the policy, strategic, project and process audits can then be cross-referenced and through data synthesis, conformity with the requirements of ISO/TS 16949 can be confirmed. A report can be printed and the assumptions, conclusions and recommendations added. It won't have escaped your notice that the requirements are expressed in such a way that a single response is almost impossible. For example, with the requirement for a quality policy there are 10 separate requirements that could elicit a different response if you asked each of the questions in Table 8.1.

Auditing for effective implementation and maintenance (8.2.2b and 8.2.2.2)

The standard requires the organization *to conduct internal audits at planned intervals to determine whether the quality management system is effectively implemented and maintained.* The automotive additions require audits of the manufacturing process *to determine its effectiveness and audit of products at defined frequency and at appropriate stages of production and delivery to verify conformity to all specified requirements, such as product dimensions, functionality, packaging and libelling.*

What does this mean?

Effective implementation should not be confused with system effectiveness. The evaluation of system effectiveness serves to explore better ways of doing things, whereas, an evaluation of effective implementation serves to explore whether the processes are being run as intended and delivering the required outputs, i.e. people are doing what they are required to do and the results are having the desired effect.

Effectively maintained means that the processes continue to remain capable despite changes in the quantity, condition or nature of the human, physical and financial resources.

In the past it has been assumed that if people were found to be following the procedures as documented, the system was effective. Conversely, if the people were not found to be following the procedures, the system was somehow ineffective. But this was not the reason for the system, i.e. it was not the purpose of the system to force people to follow procedures. The purpose was to ensure results, therefore a system is effective only if it can be demonstrated that the desired results are being achieved.

Why is this necessary?

This requirement responds to the Factual Approach Principle.

In order to manage the organization effectively, it is necessary to know whether the system for achieving the organization's objectives does the job for which it has been designed. Management also needs a system that does not collapse every time something changes – the system has to be robust. It has to cope with changes in personnel, changes in customer requirements, changes in the environment and in resources. Consequently to remain robust the system has to be effectively maintained.

How is this implemented?

The management system comprises a series of interconnected processes and each contributes to its overall effectiveness. ISO/TS 16949 requires audits to determine whether the system to be effectively implemented and maintained but also requires manufacturing process audits and product audits. As the whole system is required to be audited it would appear logical to perform product audits to establish whether the end product conforms to specification and process audits to establish whether the product resulted from implementing the planned arrangements. The various activities specified by the planned arrangements should result in an output that conforms in full with the specified requirements. However, there is variation in all processes and although the processes may be deemed capable, incidents can occur that escape detection. The product audit is performed to verify that output emerging from the process meets the specified requirements and therefore determines whether the controls in place are effective. Errors that are detected indicate that the controls are not effective as they should have been removed before output was released.

Process audit (8.2.2.2)

There are different views on what constitutes a process audit. Is it clear from the IATF bulletin S1-04-04 that ISO/TS 16949 audits must not be driven by a clause or section-driven checklist. Such an audit would be a compliance audit, i.e. verifying that specific requirements have been complied with but not necessarily verifying that it resulted in a system that was capable of delivering product that met customer requirements. Sometimes referred to as a "tick-in-a-box" audit because on observing evidence of compliance the auditor places a tick in a box on a clause-based checklist. A process audit is much different. It focuses on results and the provisions in place that deliver these results. The old *Quality System Assessment Booklet* containing all the clauses of ISO/TS 16949 rephrased

---

IATF process audits

The IATF process audit requires auditors to:

1. identify the organizations processes based on the documented QMS and other information provided;
2. analyse the identified processes for relevance to products delivered to automotive customers;
3. analyse the identified customer-oriented processes for risks, interfaces and contribution;
4. prioritize the audit activities on status and importance;
5. finalize and audit plan that takes account of process steps and sequence;
6. give consideration to the organizations definition of its processes;
7. examine processes where they occur when practical;
8. consider the octopus and turtle models used during IATF auditor qualification.

as questions has now been withdrawn and IATF have forbidden its use by certification body auditors.

Instead of the audit results being presented as a list of clauses against the site where compliance was verified, the records are required to identify the processes verified against the clauses that where applicable to them. Auditors now use the Octopus to identify the Customer Oriented Processes (COPs) and apply the "Turtle Diagram" in Figure 8.1 to address the relevant question when conducting a process audit.

The methodology is as follows assuming the information is gathered by asking questions rather than by studying documentation:

1 Establish a particular customer requirement that has to be satisfied.
2 Establish what outputs are intended to satisfy this requirement.
3 Identify the customer oriented processes (COPs) that deliver these outputs.
4 Identify the process stages to convert the customer requirement into a satisfied customer.
5 Identify the provisions that have been made in the process for preventing failure such as error-proofing.
6 Establish for each stage that:
   (a) the outputs relate to the process outputs and there is a clear linkage between the interfacing stages;
   (b) the materials and equipment being used are those that have been specified and are being used under controlled conditions;
   (c) the competences required to produce the stage output have been specified and that the competency of those producing these outputs has been assessed under controlled conditions and found acceptable;

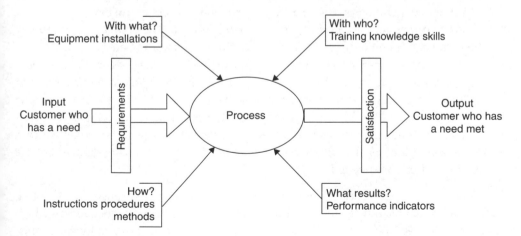

Figure 8.1 *Turtle diagram*

(d) the information being used to produce the stage outputs is that which has been defined and that it is being used under controlled conditions.

7 Identify the performance indicators that are being used to determine whether the process outputs satisfy customer requirements.

8 Verify that these performance indicators either have customer approval or where derived from customer requirements.

9 Establish that performance is being measured using these indicators, that results are being recorded and reported and that appropriate action is being taken when targets are not met.

10 Establish that the methods of measuring performance are soundly based and consistent with the accuracy and precision required for the process outputs.

The turtle diagram is a variation on a diagram that appeared in a 1993 *GM Booklet* see Figure 8.2 which shows that over 10 years ago GM's strategy incorporated key attributes of the Toyota Production System such as Pull System, Small Lot Strategy and Error-Proofing.

Product audit (8.2.2.3)

As product audits are performed to verify that the planned controls have been effective it follows that some policy decisions need to be taken regarding:

- *When product audits are to be carried out*: often this is after product release but before shipment. However, product audits are also appropriate on product returned from customers.
- *Who is to perform product audits*: often this is a group independent of production so as to be unbiased or having freedom of action and decision.
- *What the sampling criteria are to be*: often this might be 1 in 10, 1 in 100, 1 in 1000, etc. depending on the production rate and product complexity.
- *What happens to the batch from which the sample is taken while the audit takes place*: often the batch is held pending the results of the audit.
- *What the acceptance criteria are to be*: these must address the same characteristics as specified by the customer with the agreed limits. It might be necessary to operate the product under actual or simulated conditions in order to measure some of the characteristics. However, some characteristics might not have been specified for manufacture such as the degree of surface abrasions to be expected from a crankshaft that has run 24 hours under a specific load. Some additional criteria might be therefore needed.

Clearly it might be necessary to develop strip down instructions because it may not be appropriate to simply reverse the assembly instructions. Forms will be needed on which to record the measurements and any observations, such as unexpected wear marks, contamination, parts incorrectly oriented, parts missing, low torque, etc.

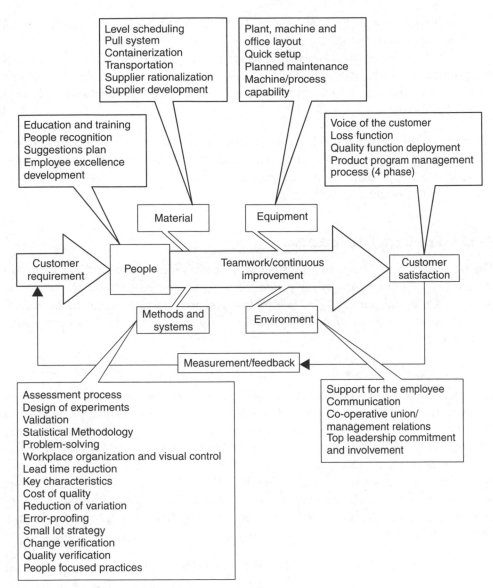

Figure 8.2 *GM process model (1993)[3]*

Product auditors need to be extremely vigilant. It's a job for an analytical person, i.e. someone who is not daunted by tedium, repetition and attention to detail and it needs to be a person who is cautious, intellectual, works slowly and is good at problem-solving. Never put the product auditor under pressure: if you do, you will encourage him or her to miss something important. There is no point in employing product auditors and ignoring the results after all, you will probably put some of your best people onto this job.

Documenting the evidence is vital simply because someone or something created it and that someone or something is still around and may be creating more of it as the report is written. In the modern world of digital photography, here is a tool that seems designed for product auditors. It costs virtually nothing to capture exactly what was observed and even date the picture so there is no doubt. But because its easily done, it can be over done.

Traceability is important also. The report should provide sufficient information to identify all products produced in the same batch and likely to possess the same characteristics. If you can't isolate all related product you run the risk of releasing suspect nonconforming product to customers.

Planning audits (8.2.2 and 8.2.2.4)

The standard requires the audit programme *to be planned taking into consideration the status and importance of the processes and areas to be audited as well as the results of previous audits.* The automotive additions require audits *to be scheduled to an annual plan and to cover all quality management-related processes, activities.*

What does this mean?
The audit programme is defined in ISO 9000 as a set of one or more audits planned for a specific time frame and directed towards a specific purpose. The *programme* will therefore have dates on which the audits are to be conducted. As the programme should be directed at the purpose of the audits, there may be a need for different types of audit programmes depending on whether the audits are of the quality system, contracts, projects, processes, products or services. It would therefore be expected that all audits in a particular audit programme would serve the same purpose. The audit programme would also be presented as a calendar chart showing where and when the audits will take place.

Status has three meanings in this context: the first to do with the relative position of the process or area in the scheme of things; the second to do with the maturity of the process and the third to do with the performance of process.

On the importance of the process, you need to establish to whom is it important: to the customer, the managing director, the public or your immediate superior? You also need to establish the importance of the activity on the effect of non-compliance with the planned arrangements. Importance also applies to what may appear minor decisions in the planning or design phase but if the decisions are incorrect it could result in major problems downstream. If not detected, getting the units of measure wrong can have severe consequences particularly if the customer specified dimensions in metric units and the purchase order has them in imperial units. Rather than check the figures or the units of measure, audits should verify that the appropriate controls are in place to detect such errors before it is too late.

The status and importance of the activities will determine whether the audit is scheduled once a month, once a year, or left for 3 years – any longer and the activity might be considered to have no value in the organization.

Why is this necessary?

This requirement responds to the Process Approach Principle.

There is little point in conducting in-depth audits on processes that add least value. There is also little point auditing processes that have only just commenced operation. You need objective evidence of compliance and that may take some time to be collected. Where the results of previous audits have revealed a higher than average performance in an area (such as zero nonconformities on more than two occasions) the frequency of audits may be reduced. However, where the results indicate a lower than average performance (such as a much higher than average number of nonconformities) the frequency of audits should be increased.

How is this implemented?

An audit programme should be developed for each type of audit as indicated above. Therefore one might prepare audit programmes for the following:

- Management audits (a combination of policy audits and strategic audits).
- Project audits (for new developments).
- Product audits (for existing products and services).
- Process audits (for all the organization's processes).
- Conformity audits (for all regulations and standards that apply).

The conformity audits could focus on *all* regulatory requirements, not simply those of ISO/TS 16949, because they serve the same purpose – to verify conformity with standards.

This requirement focuses on the criteria for choosing the areas, activities, processes, etc. to audit and therefore following initial audits to verify that the system is in place and functioning as planned, subsequent audits should be scheduled depending on status and importance.

An audit of one requirement of a policy, standard, process, procedure, contract, etc. in one area only will not be conclusive evidence of compliance if the same requirements are also applicable to other areas. Where operations are under different managers but performing similar functions you cannot rely on the evidence from only one area – management style, commitment and priorities will differ. In order to ensure that a particular audit programme is comprehensive you will need to draw up a matrix showing the areas/processes/products, etc. to be audited and the dates when the audits are to be carried out. Supporting each audit programme an analysis of the status and importance should be performed and the key aspects to be audited identified. The programme also has to include shift working so that auditors need to be very flexible. One audit per year covering 10% of the quality system in 10% of the organization is hardly comprehensive. However, there are cases where such an approach is valid. If sufficient confidence has been acquired after conducting a comprehensive series of audits over sometime, the audit programme can be adjusted so that it targets only those areas where change is most likely, thus auditing more stable areas less frequently.

The management system will contain many provisions, not all of which may be verified on each audit. This may either be due to time constraints or work for which the provisions apply not being scheduled. It is therefore necessary to record those aspects that have or have not been audited and devise the programme so that over a 1- to 3-year cycle all provisions are audited in all areas at least once.

Defining audit criteria, scope, frequency and methods (8.2.2)

The standard requires *the audit criteria, scope, frequency and methods to be defined.*

What does this mean?
The audit criteria are the standards for the performance being audited. They may include policies, procedures, regulations or requirements. Examinations without such a standard are surveys – not audits.

The scope of the audit is a definition of what the audit is to cover – the boundary conditions including the areas, locations, shifts, processes, departments, etc.

The frequency is the interval over which the audit is to be repeated and can be daily, weekly, monthly, quarterly, annually or longer.

The methods are the manner by which the audit is to be planned, conducted, reported and completed.

In defining these aspects, they may be described through procedures, standards, forms and guides.

Why is this necessary?
This requirement responds to the Process Approach Principle.

Without defined methods of auditing it is likely that each auditor will choose a different way of performing the audit – some will be good and some not so good. In order to run an effective management system, auditing should aim for best practice and therefore defining auditing methods enables best practice to be defined for the benefit of the organization.

How is this implemented?

For each audit the auditor should as a matter of routine always define the standard against which the audit is to be carried out and the scope of the audit. The frequency of the audit should be defined in the audit programme and the method within the auditing procedures.

Selection of auditors (8.2.2 and 8.2.2.5)

The standard requires the selection of auditors and the conduct of audits *to ensure objectivity and impartiality of the audit process and for auditors not to audit their own work*. The automotive additions require internal auditors *to be qualified to audit the requirements of ISO/TS 16949*.

What does this mean?

This requirement means that auditors should be selected on the basis of their objectivity and impartiality, such that their association with the work being audited should not influence their judgement. The requirement suggests that anyone auditing their own work may be influenced to overlook, hide or ignore facts pertinent to the audit. Being qualified means having the appropriate qualifications for a particular position and the qualifications for an auditor largely depends on what is required to be audited. Auditors qualified to audit the requirements of ISO/TS 16949 means auditors who have the ability to verify that all or some of the requirements of the technical specification will be, are being and have been met by the organization. For example, a product auditor needs to be able to perform product audits meeting Clause 8.2.2.3. A process auditor needs to be able to perform process audits as defined by the IATF (see above). He or she does not need to be qualified to audit design, purchasing, etc. Likewise a person auditing design needs to be qualified to audit Clause 7.3 but not necessarily any other clause.

Why is this necessary?

This requirement responds to the Leadership Principle.

If personnel have personally produced a product, they are more likely to be biased and oblivious to any deficiencies than someone totally unconnected with the product. They may be so familiar with the product that they are blind to its full strengths and weaknesses. A second pair of eyes often catches the errors overlooked by the first pair of eyes. However, auditors are human and if there is a personal relationship between the auditor and the auditee, the judgement of the auditor may be prejudiced. Depending on the nature of any problems found, the auditor being a friend, relation or confidant of the auditee, may be reluctant to or may be persuaded not to disclose the full facts if the findings indicate a serious deficiency. Even a customer may fail to exercise objectivity when it is found that the cause of problems is the inadequacy of the customer requirement!

Regarding auditor qualification, it is necessary for those auditing automotive quality management systems to have demonstrated their knowledge and skill in this arena. Auditors with no knowledge of the automotive industry would be unlikely to know "what to look at" and "what to look for"; these are the two essential attributes of an auditor.

How is this implemented?
Apart from the requirement for auditors not to audit their own work, any other competent person with experience in the industry could be selected as an auditor. The requirement for objectivity and impartiality does not mean that one must rule out supervisors, managers, friends, relations or internal customers as auditors. These conditions do not necessarily mean such a person cannot be objective and impartial – there is simply an inherent risk. This risk is overcome by the selection being made on a person's character and track record. It would be foolish to limit the selection of auditors to those who are totally independent because in some small organizations there may be no one who fits this criterion. The difficulty arises in demonstrating subsequent to the audit, that the selected auditor exercised objectivity and impartiality. In organizations that observe a set of shared values, where honesty and trust are prevalent and frequently reinforced, it should not be necessary to demonstrate the selected auditors meet this criterion. For other organizations a solution is for the auditors to be selected on the basis of having no responsibility for the work audited and no personal relationship with any of the auditees concerned.

By being divorced from the audited activities, the auditor is unaware of the pressures, the excuses, the informal instructions handed down and can examine operations objectively without bias and without fear of reprisals. To ensure their objectivity and impartiality, auditors need not be placed in separate organizations. Although it is quite common for quality auditors to reside in a quality department, it is by no means essential. There are several solutions to retaining impartiality:

- Auditors can be from the same department as the activities being audited, provided they do not perform the activities being audited.
- Separate independent quality audit departments could be set up staffed with trained auditors.
- Competent personnel at any level could carry out audits.

A competent person in this regard is one that can satisfy certain acceptance criteria. A common trap that many fall into is to put people through an ISO/TS 16949 training course and imagine that by magic, they become qualified auditors. It is true that such people have been exposed to the requirements of ISO/TS 16949 but they could have had equal exposure simply by reading the document and many of these 1–2 days courses are no more than tutored reading. An in-depth understanding is what is required and this comes from application followed by examination. Training courses that offer audit simulation or "live audits" provide a more effective learning environment. Slide shows teach little if anything. n. Examinations that test memory are not effective. An examination that explains a situation or presents objective evidence and asks the auditor to study the evidence and write observations or nonconformities as appropriate explaining the rationale for his or her conclusions is more conducive to yielding competent ISO/TS 16949 auditors but not as effective as a witnessed audit where the auditor is observed asking questions of people at their place of work, assessing the answers and evidence presented, drawing conclusions and writing a report of their findings.

Audit procedures (8.2.2)

The standard requires *the responsibilities and requirements for planning and conducting audits and for reporting results and maintaining records to be defined in a documented procedure.*

What does this mean?
This means that the "who, when, where, what, why and the how" of planning, conducting and reporting audits should be defined and documented. It also means that the methods of maintaining audit records following the audit should also be defined and documented.

Although the requirement calls for *a* documented procedure, it is not essential that your documentation be limited to a single document, because you may need policies, forms, guides and standards to describe the audit process.

Why is this necessary?
This requirement responds to the Process Approach Principle.

Auditing is a process and as such should be documented in order to achieve consistency, to record best practice and to provide the basis for improvement.

How is this implemented?
The audit procedures should cover the following:

* Preparing the annual audit programme
* The selection of auditors and team leader if necessary

- Planning particular type of audits
- Conducting the audit
- Recording observations
- Determining corrective actions
- Reporting audit findings
- Implementing corrective actions
- Confirming the effectiveness of corrective actions
- The forms on which you plan the audit
- The forms on which you record the observations and corrective actions
- Any warning notices you send out of impending audits, overdue corrective actions and escalation actions.

The detail plan for each audit may include dates if it is to cover several days but the main substance of the plan will be what is to be audited, against what requirements and by whom. At the detail level, the specific requirements to be checked should be identified based on risks, past performance and when it was last checked. *Overall plans* are best presented as programme charts and *detail plans* as checklists. Audit planning should not be taken lightly. Audits require effort from auditees as well as the auditor so a well-planned audit designed to quickly discover pertinent facts is far better than a rambling audit that jumps from area to area looking at this or that without any obvious direction.

Although checklists may be considered a plan, in the context of an audit they should be considered only as an aid in preparing the auditor to follow trails that may lead to the discovery of pertinent facts. However, there is little point in drawing up a checklist then putting it aside. Its rightful place is after the audit to verify that there is evidence indicating:

- those activities that were compliant;
- those activities that did not comply;
- those activities that were not checked;
- those activities where there were opportunities for improvement.

Audits of practice against procedure or policy should be recorded as they are observed and you can either do this in note form to be written up later or directly on to observation forms especially designed for the purpose. Some auditors prefer to fill in the forms after the audit and others during the audit. The weakness with the former approach is that there may be some dispute as to the facts if presented some time later. It is therefore safer to get the auditee's endorsement to the facts at the time they are observed. In other types of audits there may not be an auditee present. Audits of process documentation against policy can be carried out at a desk. One can check whether the documents of the quality system address the relevant clauses of the standard at a desk without walking around the site, but you can't check whether the system is documented unless you examine the operations in practice as there may be activities that make the system work that are not documented. An Audit Findings Report is illustrated in Figure 8.3.

AUDIT FINDINGS REPORT			
Process audited		**SAR/**	
Auditor	Location		Date

AUDIT OBJECTIVES	X	AUDIT CRITERIA
To find evidence of conformity		
To verify the integrity of performance measurement		
To find evidence of applied values and policies		
To find opportunities for improvement		
To verify that the system description is up to date		

AUDIT FINDING

As relevant define:

1 What was found
2 Where was it found
3 What was wrong with it
4 Why this needs action

PROVISIONAL IMPACT ASSESSMENT

Impact on Customer and Business

PROPOSED ACTION

Remedial action

As relevant define:

1 Action to remedy the problem
2 Action to seek out and correct others
3 Action to contain the problem

Target date			
Root Cause Analysis and Corrective Action Required	YES	NO	If YES, CAR required by:

Related NCRs/CARs/SARs

Findings agreed by:	Actions agreed by:	Date

EFFECTIVENESS OF ACTIONS

All actions taken and effective?	Yes	No	Reviewed by:	Date closed

Figure 8.3 *Audit findings report*

You could use a Nonconformity Report but this is usually designed to collect product data so it may not be suitable, apart from the title being a little provocative. The Audit Findings Report simply conveys findings not the conclusions of the audit, i.e. that would be addressed in a summary report. The Form illustrated in Figure 8.3 does not have provision for Corrective Action unlike so many Audit Reports simply because, there is another form for this (see Figure 8.5). What is stated on so many Audit Reports as Corrective Action is the Remedial Action because it invariably does not get to the root cause. Many such forms contain actions to correct documentation that should have been correct. The authors of such reports obviously believed that by correcting the documents they were preventing recurrence but all they were doing was making good. Most auditors do not explore the reason why the documents were incorrect. Also Audit Findings Report in Figure 8.3 has provision for an impact assessment. Very few audit reports do this, but Clause 6.2.2.3 and 6.2.2.4 clearly requires personnel to be aware of the consequences of nonconformity and auditors are no exception.

Taking action (8.2.2)

The standard requires management responsible for the area audited *to ensure that actions are taken without undue delay to eliminate the detected nonconformities and their causes.*

What does this mean?

Managements responsible for the area audited are those who have the authority to cause change. An auditor may have interviewed a supervisor and found opportunities for improvement but the supervisor may not have the authority to agree to any changes. The auditor would therefore need to report the findings to the person who is authorized to take action.

Action without undue delay means that management are expected to act before the detected problem impacts subsequent results.

Eliminating nonconformities and their causes means that managers should:

- take remedial action to correct the particular nonconformity;
- search for other examples of nonconformity and establish how widespread the problem is;
- establish the root cause of the nonconformity and prevent its recurrence.

Why is this necessary?

This requirement responds to the Leadership Principle.

There is simply no point in conducting audits and finding problems if management does not intend to take action to prevent such problems impacting results. It often arises that problems detected by internal audits are perceived by management to have no impact on results, so it delays taking any action.

How is this implemented?

To ensure actions are implemented without undue delay, the auditor needs to be sure that a failure to act will in fact impact performance of the process or the system. Management will not implement actions that have no effect on performance even if the action required restores conformity with ISO/TS 16949. It is therefore sensible for the auditor to explain the impact of the detected nonconformity within the audit report – possibly by using a classification convention from critical to minor. But it would be more effective if the potential impact was actually stated and this requires the auditor to have a greater knowledge of the requirements be they in a contract, standard, policy, procedure or work instruction and the consequences of failing to implement them.

Unless the auditee is someone with responsibility for taking the corrective action, the auditee's manager should determine the actions required. If the action required is outside that manager's responsibility, the manager and not the auditor should seek out the appropriate authority and secure a proposal. Your policy manual should stipulate management's responsibility for taking action without undue delay.

A proposed action may not remove the non-compliance; it may be palliative leaving the problem to recur again at some future time. Target dates should be agreed for all actions and the dates should be met as evidence of commitment. Third-party auditors will search your records for this evidence so you will need to impress on your managers the importance of honouring their commitments. The target dates also have to match the magnitude of the deficiencies. Small deficiencies, which can be corrected in minutes, should be dealt with at the time of the audit otherwise they will linger on as sores and show a lack of discipline. Others, which may take 10–15 minutes, should be dealt with within a day or so. Big problems may need months to resolve and require an orchestrated programme to be implemented. The actions in all cases when implemented should remove the problem, i.e. restore compliance. An action should not be limited to generating another form or procedure because it can be rejected by another manager thereby leaving the deficiency unresolved.

Follow-up audits (8.2.2)

The standard requires follow-up activities *to include the verification of the actions taken and the reporting of verification results.*

What does this mean?

A follow-up audit is an action taken after the audit to verify that agreed actions from the audit have been completed as planned.

Why is this necessary?

This requirement responds to the Factual Approach Principle.

Verification of actions completed is a normal activity of any managed process. The audit remains open until completion of actions has been confirmed.

How is this implemented?

Follow-up action is necessary to verify that the agreed action has been taken and verify that the original nonconformity has been eliminated. Follow-up audits may be carried out immediately after the planned completion date for the actions or at some other agreed time. However, unless the audit is carried out relatively close to the agreed completion date, it will not be possible to ascertain if the action was carried out without undue delay.

The auditor who carries out the follow-up audit need not be the same that carried out the initial audit. In fact, there is some merit in using different auditors in order to calibrate the auditors.

When all the agreed nonconformities have been eliminated the audit report can be closed. The audit remains incomplete until all actions have been verified as being completed. Should any action not be carried out by the agreed date, the auditor needs to make a judgement as to whether it is reasonable to set a new date or to escalate the slippage to higher management. For minor problems, when there are more urgent priorities facing the managers, setting a new date may be prudent. However, you should not do this more than once. Not meeting the agreed completion date is indicative either of a lack of commitment or poor estimation of time and both indicate that there may well be a more deep routed problem to be resolved.

Monitoring and measurement of processes (8.2.3)

Monitoring and measuring methods (8.2.3)

The standard requires the organization *to apply suitable methods for monitoring and where applicable, measurement of the quality management system processes.*

What does this mean?

In all managed processes there will be stages where outputs are verified against inputs – these are product controls. There also need to be stages where the performance of the process itself is measured: these are process controls.

Monitoring is an ongoing activity and, as stated at the beginning of this chapter, is the periodic or continual observation of operations to detect events before they occur so that action can be taken to prevent nonconformity. Measurement on the other hand implies that standards have been set and performance against those standards is being verified. For processes this may involve a range of parameters that are defined as being critical for the process to consistently deliver the correct results.

Why is this necessary?

This requirement responds to the Factual Approach Principle.

It is processes that produce products and therefore measuring products tells us whether the products are correct but does not tell us whether the process is achieving its objectives. Parameters such as the rate at which products are produced, the variability between product characteristics, the resources used by the process and the effect of the process on its environment are parameters that require process measurement.

How is this implemented?

There are several methods available for monitoring and measuring processes. The simplest monitoring method is visual observation by a person trained to detect variations that signal something is not quite right with a process. With industrial processes instruments may be installed to give the observer a visual indicator of performance. Data may be recorded on control charts so that the observer can tell when the performance is deteriorating. With management and administration processes, the observer (often the manager or supervisor) has formal or informal standards against which processes are monitored. Staff sometimes work in an environment in which they can't predict how their manager will react – an environment where the manager has not communicated the standards of performance expected. In other cases the staff work in a predictable environment – where the manager monitors performance against a standard known to the staff. In the former case the processes are not managed – they are reactive. In a managed process, monitoring would only look for the unpredictable where the immediate reaction is "Why did that happen?" or "That shouldn't have happened." Management and administrative processes also lend themselves to management by data rather than observation. Provided that targets have been set, data can be collected, analysed and results produced to show whether performance is on track.

Whatever is being monitored or measured a soundly based method needs to be used to sense the variance from target, transmit the data, analyse it and compute factual results. The integrity of the measurements need to be sound, i.e. there should be no filtering, screening or reduction of the raw data that is not planned and the personnel performing the measurement should be competent to do so.

Process capability

Process capability studies (8.2.3 and 8.2.3.1)

The standard requires the monitoring and measurement methods *to demonstrate the ability of the processes to achieve planned results and* to perform and document process studies on all new manufacturing processes to verify process capability. *The documents are required to include objectives for manufacturing process capability, reliability, maintainability and availability as well as acceptance criteria.*

What does this mean?
The planned results are the performance requirements at corporate and system level that relate to a particular process. There are whatever the process has been designed to achieve but should include needs, expectations and obligations from outside the process that impinge on the process.

To demonstrate ability means that either by observation or through validated records, evidence should be available which shows that the process is capable of achieving the planned results. A process is in control when the average spread of variation coincides with the nominal specification for a parameter. The range of variation may extend outside the upper and lower limits but the proportion of output within the limits can be predicted. This situation will remain as long as the process remains in statistical control. A process is in statistical control when the source of inherent variation is from common causes only, i.e. a source of variation that affects all the individual values of the process output and appears random. Common cause variation results in a stable and repeatable distribution of results over time. When the source of variation causes the location, spread and shape of the distribution to change, the process is not in statistical control. These sources of variation are due to special or assignable causes and must be eliminated before commencing with process capability studies. It is only when the performance of a process is predictable that its capability to meet customer expectations can be assessed.

Process capability studies are studies conducted to obtain information about the inherent variation present in processes that are under statistical control in order to reduce the spread of variation to less than the tolerances specified in the product specification.

Objectives for manufacturing process capability would be expressed as indices such as C_p, C_{pk} and P_{pk}. These terms are defined in Appendix A.

Objectives for process reliability would be expressed as the mean time between process failures (MTBF) over a defined operating period either as hours or as a probability of success. Objectives for process maintainability would be expressed as the mean time to repair/restore (MTTR), a process to operational status following a failure. MTTR would include access time, diagnosis time, repair time, retest time and capability verification time. Availability is not simply MTBF/Total time because there will be planned maintenance time (PMT) and periods when the process is not required. Process availability therefore has to be a ratio of the Uptime to Total time which will equate to MTBF/(MTBF + MTTR + PMT).

Why is this necessary?

This requirement responds to the Factual Approach Principle.

Monitoring and measurement should not be performed merely to acquire information; it should either be performed to establish that processes are achieving the results for which they have been designed or for establishing that the results for which the processes have been designed are not adequate to meet the organization's objectives. Without objectives and targets information from monitoring and measurement is merely interesting but not justification for action. Action should only be taken to bring performance in line with requirements or to change the requirements.

How is this implemented?

In monitoring and measuring processes there are three questions that the process manager needs to be able to answer:

1 How do you know the process is performing as planned?
2 How do you know the process is achieving the results in the best way?
3 How do you know that the results being achieved are those necessary to fulfil organizational goals?

Achieving objectives

Provisions need to be put in place for each process that enable operations to be monitored and its performance to be measured against process objectives and targets. Objectives and targets need to be established for each process and these should be derived from an analysis of the factors that affect the ability of the process to deliver the desired results.

For example, the purpose of a purchasing process might be to acquire physical resources needed to fulfil the organization's objectives. An analysis of these objectives may reveal a need for a reliable supply of raw materials of consistent quality to be delivered in accordance with schedules that change monthly. A purchasing objective might be to secure a supply of the specific raw materials from a supplier who is capable of meeting the delivery schedules together with raw material quality, service quality and cost targets. These may include targets for response to changes in delivery schedules and raw material specification. Clearly a critical factor is the ability of the purchasing process to respond to change. Not only would the supplier have to respond to change but the internal stages of the purchasing process would also have to respond to change. Monitoring this process would involve observing how changes were received and transmitted through all the stages in the process and alerting staff to bottlenecks or other factors that could jeopardize meeting the targets. Measuring this process would involve checking that deliveries were being made on time and that the variation in cost and quality was also on target. Product measures would look at specific deliveries, whereas process measures would look at all deliveries for evidence that the purchasing objective was being met.

A process may be designed to deliver output that meets specification therefore a measure of performance is the ratio of conforming output to total output. If the ratio is less than 1 the process is not capable. Most processes fall into this category because some defective output is often produced, but it is possible to design processes so that they only produce conforming output. This does not mean perfect output, but output that is within the limits defined for the process. The target yield for a process may be 97% implying that 100% is not feasible, therefore a yield of 98% is good and a yield of 96% bad, i.e. it depends what standards have been set.

This type of measurement requires effective data collection, transmission and analysis points so that information is routed to analysts to determine performance and for results to be routed to decision-makers for action. With a process such as order processing, in addition to each order being checked (product verification) the process should be monitored and measured to establish it is meeting the defined objectives for processing time, customer communication, etc. and that there is no situation developing that may jeopardize achievement of the order-processing objectives. Therefore, every process will have at least two verification stages: one in-line for verifying output quality and another for verifying process performance against objectives.

The object of process capability studies is to compute the indices and then take action to reduce common cause variation by preventive maintenance, error-proofing, operator training, revision to procedures and instructions, etc. The inherent limitations of attribute data prevent their use for preliminary statistical studies since specification values are not measured. Attribute data have only two values (conforming/nonconforming, pass/fail, go/no-go, present/absent) but they can be counted, analysed and the results plotted to show variation. Measurement can be based on the fraction defective such as parts per million (PPM). Whilst variables data follows a distribution curve, attribute data varies in steps since you can't count a fraction. There will either be zero errors or a finite number of errors. Figure 2.9 illustrated the difference between a process that is under control and one that is capable.

Achieving best practice
Each result requires resources and even when the required results are being achieved there may be better ways or more efficient ways of achieving them. Therefore targets may be set for efficiency and these too need to be monitored and measured. One method is to monitor resource utilization. Another is to conduct benchmarking against other processes to find the best practice.

Fulfilling purpose
The planned results include organizational objectives and therefore the process objectives should be reviewed periodically to verify they are the right objectives. A 99.73% (three sigma) yield may have been a real challenge 5 years ago but today your competitors

achieve 99.999998% (six sigma) and therefore the objective needs to change for the organization to maintain its position in the market.

Maintaining process capability (8.2.3.1)

The standard requires the organization *to maintain manufacturing process capability as specified by the customer part approval process requirements and ensure adherence to control plans and process flows.*

What does this mean?
When processes have been developed so as to produce conforming output every time, the quality of output needs to be maintained.

Why is this necessary?
The quality of outputs will have been achieved by managing the processes according to the specifications, plans and procedures defined in the control plan. It is therefore necessary that the provisions of the Control Plan are strictly adhered to otherwise performance is likely to deteriorate.

How is this implemented?
It is perhaps a little trite simply to suggest that the control plans are adhered to and that is all that you need to do. But there is more to it than that. The control plans are simply the front end. Behind the control plans are a number of processes that deliver correct materials, correct equipment, competent personnel and the correct environment. Simply following a control plan will not ensure these other processes deliver quality output. So one needs to be conversant with the "Fishbone Diagram" or cause-and-effect diagram for the process, i.e. on what does conforming output depend and what could cause these dependencies to fail in their objective?

Reaction plans (8.2.3.1)

The standard requires the organization *to initiate reaction plans for characteristics that are either not statistically capable or are unstable.*

What does this mean?
A reaction plan is a predetermined response to a predictable failure. The plan is implemented when such a failure occurs and continues to be implemented until the required capability of the process concerned is restored.

Why is this necessary?
Reaction plans are usually produced for parameters that are not statistically capable or are unstable over long periods. It would not be unreasonable to also provide reaction plans

where the severity of failure is above six in the process FMEA. For example, whether or not a cylinder block cleaning process was likely to leave swarf in the oil or water channels, a reaction plan would need to be implemented in the event of unacceptable levels of swarf being discovered during a product audit.

How is this implemented?

It is often the case that until a process failure occurs, reaction plans don't exist except in the imagination of production personnel, i.e. what they might do if failure occurs. When the failure has occurred, the staff can develop effective reaction plans with hindsight. However, this is not effective process management. The planners should have be developed a process FMEA and in doing so will have identified potential failure modes, i.e. everyone of which could have a reaction plan. But, its not cost effective to create plans for events that are unlikely and even if they do happen, are not likely to cause major problems. There will be some events, however, that if they were to happen and escape detection would affect vehicle safety and these require a reaction plan.

Reaction plans should deal with:

*   *The recovery plans* which should detail the actions to be taken to inspect all products produced since the last time the parameter was verified as conforming; this is product which will be subject to a recovery plan and may involve product recall.
*   *All product currently in production*; this is product that will be subject to a containment plan and it might involve some reprocessing.
*   *All product planned for production up to a date* when the corrective action is expected to have deemed effective; this is product that will be produced to a process modified by the containment plan/test product for the presence of the nonconformity and rework, strip and rebuild product. The containment plans may include 100% inspection/test, revised instructions, modified tooling, diagrams of what to look out for, how to avoid the errors, retraining, etc.

Monitoring and measurement of product (8.2.4)

Verifying conformity with product requirements (8.2.4)

The standard requires the organization *to monitor and measure the characteristics of the product to verify that product requirements have been met* and that these activities *be conducted at appropriate stages of the product realization process in accordance with the planned arrangements.*

What does this mean?

Activities that monitor and measure product are often referred to as inspection, test or verification activities. Appropriate stages of the product realization process means the stages at which:

- the achieved characteristics are accessible for measurement;
- an economic means of measurement can be performed;
- the correction of error is less costly than if the error is detected at later stages.

It may be possible to verify some characteristics on the final product just prior to shipment but it is costly to correct errors at this late stage resulting in delayed shipment. It is always more economic to verify product at the earliest opportunity.

The planned arrangements in this case are the plans made for verifying product in terms of what is to be verified, who is to verify it, when is it to be verified, how is it to be verified, where is it to be verified and what criteria is to be used to judge conformity.

Product requirements are all the requirements for the product including customer, regulatory and the organization's requirements. Some of these may be met by inherent design features; others will be met in production, installation or service.

The forms of verification that are used in product and service development, should also be governed by these requirements as a means of ensuring that the product on which design verification is carried out conforms with the prescribed requirements. If the product is non-compliant it may invalidate the results of design verification. Product verification also applies to any measuring and monitoring devices that you design and manufacture to ensure that they are capable of verifying the acceptability of product as required. Product verification is part of product realization and not something separate from it, although the way the requirements are structured may imply otherwise. Whenever a product is supplied, produced or repaired, rebuilt, modified or otherwise changed, it should be subject to verification that it conforms to the prescribed requirements and any deficiencies corrected before being released for use.

Why is this necessary?

This requirement responds to the Factual Approach Principle.

One verifies product to establish that it meets requirements. If one could be certain that a product would be correct without it being verified, product verification would be unnecessary. However, most processes possess inherent variation due to common causes – variations that affect all values of process output and appear random. Although a process may be under statistical control, a special event could disturb performance and without checks on the output, its detection may go unnoticed. One can only check for those events we think might happen which is why our confidence in the "system" is shaken when we discover a condition with a cause we had not predicted.

How is this implemented?

As product requirements may include characteristics that are achieved by design, production, installation or service delivery, a high-level verification matrix is needed to provide traceability from requirement to the means of verification. This will undoubtedly lead to there being a few characteristics that need to be verified only once by design verification with many of the others being verified in production or service delivery. Characteristics that do not vary only need to be checked once. For example, a car designed with four wheels could not possibly be made with two, three or five wheels when put into production, but a body panel with screw inserts could emerge from the process without the inserts!

Having established that characteristics vary, the stage at which they need to be verified should be determined. This leaves three possibilities; on receipt, in-process or on completion. Receipt verification was addressed under the section Purchasing in Chapter 7. Here we address in-process and finished product verification.

In-process verification

In-process verification is carried out in order to verify those features and characteristics that would not be accessible to verification by further processing or assembly. When producing a product that consists of several parts, subassemblies, assemblies, units, equipments and subsystems, each part, subassembly, etc. needs to be subject to final verification but may also require in-process verification for the reasons given above. Your control plans should define all the in-process verification stages that are required for each part, subassembly, assembly, etc. In establishing where to carry out the verification, a flow diagram may help. The verification needs to occur after a specified feature has been produced and before it becomes inaccessible for measurement. This doesn't mean that you should check features as soon as they are achieved. There may be natural breaks in the process where the product passes from one stage to another, or stages at which several features can be verified at once. If product passes from the responsibility of one person to another, there should be a stage verification at the interface to protect the producer even if the features achieved are accessible later. In addition to the control plans addressed previously, your verification arrangements should:

- define the environment for the measurements to be made if critical to the measurements to be made;
- identify the organization that is to perform the verification;
- make provision for the results of the verification to be recorded.

Finished product verification

Finished product verification is in fact the last verification of the product that you will perform before dispatch but it may not be the last verification before delivery if your contract includes installation. There are three definitions of finished product verification:

1   The verification carried out on completion of the product; afterwards the product may be routed to stores rather than to a customer.
2   The last verification carried out before dispatch; afterwards you may install the product and carry out further work.
3   The last verification that you as a supplier carry out on the product before ownership passes to your customer; this is the stage when the product is accepted and consequently the term *product acceptance* is more appropriate and tends to convey the purpose of the verification rather than the stage at which it is performed.

There are two aspects to finished product verification. One is checking what has gone before and the other is accepting the product.

Final verification and test checks should detect whether:

*   All previous verification activities have been performed.
*   The product bears the correct identification, part numbers, serial numbers, modification status, etc.
*   The as-built configuration is the same as the issue status of all the parts, subassemblies, assemblies, etc. specified by the design standard. A Configuration Record containing this data would avoid argument later as to whether or not certain specification changes were embodied in the product.
*   All recorded nonconformities have been resolved and remedial action taken and verified.
*   All concession applications have been approved.
*   All verification results have been collected.
*   Any result outside the stated limits is either subject to an approved concession, an approved specification change or a retest that shows conformance with the requirements.
*   All documentation to be delivered with the product has been produced and conforms to the prescribed standards.

Evidence of conformity (8.2.4)

The standard requires *evidence of conformity with the acceptance criteria to be maintained*.

What does this mean?

Evidence of conformity is the information recorded during product verification that shows the product to have exhibited the characteristics required.

Why is this necessary?

This requirement responds to the Factual Approach Principle.

At a point in the process, product will be presented for delivery to the next stage in the process or to a customer. At such stages a decision is made whether or not to release product and this decision needs to be made on the basis of facts substantiated by objective evidence.

How is this implemented?

This requires that you produce something like an Acceptance Test Plan which contains, as appropriate, some or all of the following:

* Identity of the product to be verified.
* Definition of the specification and acceptance criteria to be used and the issue status that applies.
* Definition of the verification aids and measuring devices to be used.
* Definition of the environment for the measurements to be made.
* Provision for the results of verification to be recorded – these need to be presented in a form that correlates with the specified requirements.

Having carried out these verification activities, it should be possible for you to declare that the product has been verified and objective evidence produced that will demonstrate that it meets the specified requirements. Any concessions given against requirements should also be identified. If you can't make such a declaration, you haven't done enough verification. Whether or not your customer requires a certificate from you testifying that you have met the requirements, you should be in a position to produce one. The requirement for a certificate of conformance should not alter your processes, your quality controls or your procedures. Your management system should give you the kind of evidence you need to assure your customers that your product meets their requirements without having to do anything special.

Your verification records should be of two forms: one which indicates what verification activities have been carried out, and the other which indicates the results of such verification. They may be merged into one record but when parameters need to be recorded it is often cleaner to separate the progress record from the technical record. Your procedures, quality plan or product specifications should also indicate what measurements have to be recorded.

Don't assume that because a parameter is shown in a specification that an inspector or tester will record the result. A result can be a figure, a pass/fail or just a tick. Be specific in what you want recorded because you may get a surprise when gathering the data for analysis. If you use computers, you shouldn't have the same problems but beware, too much data is probably worse than too little! In choosing the method of recording measurements, you also need to consider whether you will have sufficient data to minimize recovery action in the event of the measuring device subsequently being found to be out of calibration. As a general rule, only gather that data you need to determine whether the product meets the requirements or whether the process is capable of producing a product that meets the requirements. You need to be selective so that you can spot the out of tolerance condition or variation in the measurement system. Sometimes, plotting the results as a histogram might indicate abnormalities in the results that are symptomatic of measurement errors. The acceptance criterion is therefore not simply a specified upper and lower limits, but evidence that results are located in a normal distribution that is centred on the nominal condition.

All verification records should define the acceptance criteria, the limits between which the product is acceptable and beyond which the product is unacceptable and therefore nonconforming.

Identifying the person(s) authorizing release (8.2.4)

The standard requires *records to indicate the person(s) authorizing release of product.*

What does this mean?
The person authorizing release of the product is the one whose permission is needed before product can pass beyond a defined stage. Release conditions may include conformity to specification and quantity. Product may be held until the required quantity has been produced. When a product is "in-process" the operator is working on it. Product may be piling up in the output basket, but until the operator indicates it can be released, it remains under his or her control. This requirement means that such decisions are to be recorded in a manner that is traceable to the person who made them. With hardware, software and processed material this often means a signature or stamp on verification records or labels attached to the product. With documentation, it is an approval signature on a document or accompanying forms.

This requirement implies that there will always be a person in the process who decides when product should or should not be released. This may not always be so. With automated processes equipped with product verification instrumentation, the product may well pass straight into dispatch and onto the customer without any human intervention. The person releasing product in this case is the person controlling the process.

Why is this necessary?

This requirement responds to the Factual Approach Principle.

Within an organization you may wish to identify the individual responsible so that you can go back to ask questions. This is more likely in the case of a reject decision as opposed to an acceptance decision. It is also necessary for products with special characteristics to be able to demonstrate that product was released by someone who was aware of the consequences of their actions and were acting responsibly. An organization's reputation could be easily damaged if it emerged that the person releasing product into the supply chain was unqualified.

How is this implemented?

There are two parts to this requirement. The first is that the person who releases product is identified and the second that this person is authorized to release product. It is not enough for a document, record or label to bear a signature; the signature has to be of a person who has the right to release the product. The signature therefore needs to be legible or at least traceable to the individual.

Some organizations maintain a list of authorized signatures as a means of being able to trace signatures to names of people who carry certain authority. If you have a large number of people signing documents and records and there is a possibility that the wrong person may sign a document, the list is a good tool for checking that there has been no abuse of authority. Otherwise, the name of the individual and his or her position below or alongside the signature should be adequate. The management system should have in-built provision for preventing the wrong people releasing product. When such provisions have been made, the authorized person could, under certain circumstances, be influenced to release bad products; it is only the strength of the shared values that would prevent such transgression.

Product release approval (8.2.4)

The standard *requires that unless otherwise approved by a relevant authority and where applicable, by the customer, product release and service delivery is not to proceed until the planned arrangements have been satisfactorily completed.*

What does this mean?

This requirement can impose unnecessary constraints if taken literally. Many activities in planned arrangements are performed to give early warning of nonconformities. This is in order to avoid the losses that can be incurred if failure occurs in later tests and inspections. The earlier you confirm conformance the less costly any rework will be. One should therefore not hold shipment if later activities have verified the parameters, whether or not earlier activities have been performed. It is uneconomic for you to omit the earlier activities, but if you do, and the later activities can demonstrate that the end product meets the requirements, it is also uneconomic to go back and perform those activities that have not been completed. Your planned arrangements could cover installation and maintenance activities which are carried out after dispatch and so it would be unreasonable to insist that these activities were completed before dispatch or to insist on separate plans just to sanitize a point. A less ambiguous way of saying the same thing is to require *no product to be dispatched until objective evidence has been produced to demonstrate that it meets the product requirements and that authorization for its release has been given.*

If planned arrangements cannot be achieved, a concession might be obtained from the recipient to permit release of product that did not fully meet the requirements. The recipient could be the owner of the process receiving product for processing or the external customer receiving product in response to an order.

Why is this necessary?
This requirement responds to the Factual Approach Principle.

Having decided on the provisions needed to produce product that meets the needs and expectations of customers, regulators and the organization itself, it would be pretty foolish to permit release of product before confirming that all that was agreed to be done has been done. However, circumstances may arise where nonconforming product has been produced and instead of shipping such product without informing the intended recipient, an organization committed to quality would seek permission to do so.

How is this implemented?
You need four things before you can release product whether it be to the stores, to the customer, to the site for installation or anywhere else:

1   Sight of the product.
2   Sight of the requirement with which the product is to conform including its packaging, labelling and other product-related requirements.
3   Sight of the objective evidence that purports to demonstrate that the particular product meets the requirement.
4   Sight of an authorized signatory or the stamp of an approved stamp-holder who has checked that the particular product, the evidence and the requirement are in complete accord.

Once the evidence has been verified, the authorized person can make the release decision and endorse the appropriate record indicating readiness for release. Should there be any discrepancies, they should be validated and if proven valid, the nonconforming product process should be initiated.

Layout inspection and functional testing (8.2.4.1)

The standard requires layout inspection and functional verification *to be performed as specified in the control plans and results made available for customer review.*

What does this mean?
A layout inspection is the complete measurement of all part dimensions shown on the design record and a functional verification is testing to ensure that the part conforms to all customer engineering material performance standards and hence fully satisfies the approved design requirements.

Why is this necessary?
When a product undergoes design verification and validation, the tests are conducted on a small sample of product that is representative of the production standard. The variation in materials, environment and characteristics that is possible over long-production runs cannot be fully predicted and therefore, periodic tests are necessary to verify that the product in current production is of the same standard as the product that gained production approval.

How is this implemented?
The frequency of such checks and the sample size will be specified by the customer and could be annually or more or less often depending on quantities produced and other considerations.

The tests and inspections carried out need to be to the same specifications and procedures as those used for the original production part approval and as amended by subsequent approved engineering changes. The results of the tests should be recorded in the same format as the original tests unless otherwise required by the customer.

Appearance items (8.2.4.2)

The standard requires organizations producing parts designated as "appearance items" *to provide appropriate resources, competent personnel, masters and control of such masters.*

What does this mean?
Appearance items are those with surface finish characteristics that are visible to the end user. Your customer will designate these items so you don't need to guess which items they are.

Why is this necessary?
Appearance is a subjective characteristic so means need to be provided to reduce the subjectivity and make judgement more objective.

How is this implemented?

Samples indicating the acceptable range of colour, gloss, metallic brilliance, grain and texture and distinctness of image may be needed and if not provided by your customer, those that you provide will need customer approval.

It is also important when selecting personnel for making appearance decisions, to ensure that they have the requisite physical attributes. Eyesight and colour blindness tests should be conducted when appropriate. Lighting conditions should be appropriate for the evaluations performed, avoiding shadows, glare and other adverse factors. The tests need to be conducted periodically as a safeguard against deterioration in the relevant physical attributes.

Control of nonconforming product (8.3)

Preventing unintended use (8.3 and 8.3.1)

The standard requires the organization to ensure that *product which does not conform to product requirements is identified and controlled to prevent unintended use or delivery and goes on to require the controls and related responsibilities for dealing with nonconforming product to be defined in a documented procedure.* The automotive additions designate product with unidentified or suspect status as nonconforming.

What does this mean?

Nonconforming product is product that does not conform to agreed product requirements when subject to either planned or unplanned verification. Product requirements are not limited to customer requirements (see also Chapter 7) therefore a nonconforming product is one that fails to meet the:

- specified customers requirements,
- intended usage requirements,
- stated or implied needs,
- organization's own requirements,
- customer expectations.

A product is judged either conforming or nonconforming at a verification stage. A product could also be judged nonconforming if it becomes damaged or fails at any other stage but the product is normally considered to be unserviceable in such cases. Unserviceable products, however, are not necessarily nonconforming – they may simply lack lubrication or calibration. A piece of test equipment, the calibration date of which has expired, is not nonconforming – it is merely unserviceable. When checked against a standard it may be found to be out of calibration and then it is nonconforming, but it could be found to be within the specified calibration *limits*.

A product may be suspected of being nonconforming as might be the case with a batch of product that has failed the sampling inspection. Only the samples checked are definitely nonconforming – the others in the batch are only suspected as being nonconforming. We should therefore look further than the product that has been found to be nonconforming and seek out other products that may possess the same characteristics as those found to be nonconforming. These other products may have already been released to customers. This latter situation can arise if you discover the measuring or processing equipment to be inaccurate or malfunctioning. Any product that has passed through that process since it was last confirmed as serviceable is now suspect. Another example of suspect product is when the product is mishandled but shows no obvious signs of damage. This may arise when product is dropped or not handled in stipulated clean conditions or in accordance with electrostatic safe handling procedures. A further example of suspect product is where a product carries no indication of verification status either by its location, label or marking. Suspect product should be treated in the same manner as nonconforming product and quarantined until dispositioned (i.e. dealt with). However, until nonconformity can be proven, the documentation of the nonconformity merely reveals the reason for the product being suspect.

This requirement relates to the controls exercised over the product itself whereas Clause 8.5.2 on corrective action addresses the measures needed to prevent recurrence of the nonconformity. The scope of procedures for control of nonconforming product should therefore focus on the product and it correction not the cause.

Why is this necessary?
This requirement responds to the Process Approach Principle.

Nonconforming product needs to be prevented from use or delivery, simply because the organization should not knowingly supply nonconforming product.

How is this implemented?
The only sure way of preventing inadvertent use of nonconforming product is to destroy it, but that may be a little drastic in some cases. It may be possible to eliminate the nonconformity by repair, completion of processing or rework. A more practical way of preventing the inadvertent use or installation of nonconforming or unserviceable products is to identify the product as *nonconforming* or *unserviceable* and then place it in an area where access to it is controlled. These two aspects are covered further below.

Identifying nonconforming product

The most common method is to apply labels to the product that are distinguishable from other labels. It is preferable to use red labels for nonconforming and unserviceable items and green labels for conforming and serviceable items. In this way you can determine product status at a distance and reduce the chance of confusion. You can use segregation as a means of identifying nonconforming product but if there is the possibility of mixing or confusion then this means alone should not be used.

On the labels themselves you should identify the product by name and reference number, specification and issue status if necessary and either a statement of the nonconformity or a reference to the service or nonconformity report containing full details of its condition. Finally the person or organization testifying the nonconformity should be identified either by name or inspection stamp.

Controlling nonconforming product

To control nonconforming product you need to:

(a) know when it became nonconforming;
(b) know who decided it was nonconforming;
(c) know of its condition;
(d) know where it is located;
(e) know that it is unable to be used.

On detection of a nonconformity, details of the product and the nonconformity should be recorded so as to address (a), (b) and (c) above.

Segregating a nonconforming product (or separating good from bad) places it in an area with restricted access and addresses (d) and (c) above. Such areas are called quarantine areas or quarantine stores. Products should remain in quarantine until disposal instructions have been issued. The store should be clearly marked and a register maintained of all items that enter and exit the store. Without a register you won't be able to account for the items in store, check whether any are missing, or track their movements. The quarantine store may be contained within another store providing there is adequate separation that prevents mixing of conforming and nonconforming articles. Where items are too large to be moved into a quarantine store or area, measures should be taken to signal to others that the item is not available for use and cordons or floor markings can achieve this.

Documented procedures

Documented procedures should specify the authorities who make the disposition, where it is to be recorded and what information should be provided in order that it can be implemented and verified as having been implemented.

In order to implement these requirements your nonconformity control procedures should include the following actions:

- Specify how product should be scrapped, or recycled, the forms to be used, the authorizations to be obtained.
- Specify the various repair procedures, how they should be produced, selected and implemented.
- Specify how modifications should be defined, identified and implemented.
- Specify how production permits (deviations) and concessions (waivers) should be requested, evaluated and approved or rejected.
- Specify how product should be returned to its supplier, the forms to be completed and any identification requirements in order that you can detect product on its return.
- Specify how regrading product is to be carried out, the product markings, prior authorization and acceptance criteria.

When making the disposition your remedial action needs to address:

- action on the nonconforming item to remove the nonconformity;
- a search for other similar items which may be nonconforming (i.e. suspect product);
- action to recall product containing suspect nonconforming product.

If you need to recall product that is suspected as being defective you will need to devise a *Recall Plan*, specify responsibilities and timescales and put the plan into effect. Product recall is a *Remedial Action* not a *Corrective Action* because it does not prevent a recurrence of the initial problem.

Defining responsibility
The decision on product acceptance is a relatively simple one because there is a specification against which to judge conformance. When product is found to be nonconforming there are three decisions you need to make based on the following questions:

1 Can the product be made to conform?
2 If the product cannot be made to conform, is it fit for use?
3 If the product is not fit for use, can it be made fit for use?

The authority for making these decisions will vary depending on the answer to the first question. If, regardless of the severity of the nonconformity, the product can be made to conform simply by rework or completing operations, these decisions can be taken by operators or inspectors, providing rework is economical. Decisions on scrap, rework and completion would be made by the fund-providing authority rather than the design authority. If the product cannot be made to conform by using existing specifications, decisions requiring a change or a waiver of a specification should be made by the authority responsible for drawing up or selecting the specification.

It may be sensible to engage investigators to review the options to be considered and propose remedial actions for the authorities to consider. In your procedures you should identify the various bodies that need to be consulted for each type of specification. Departures from customer requirements will require customer approval; departures from design requirements will require design approval; departures from process requirements will require process-engineering approval; etc. The key lies in identifying who devised or selected the requirement in the first place. All specifications are merely a substitute for knowledge of fitness for use, i.e. any departure from such specification must be referred back to the specification authors for a judgement.

Correction of nonconforming product (8.3a and 8.3.2)

The standard requires the organization when appropriate to *deal with nonconforming product by taking action to eliminate the detected nonconformity*. The automotive additions require *instructions for rework to be accessible to and utilized by the appropriate personnel*.

What does this mean?
Action to remove the detected nonconformity is a remedial action and can include the completion of operations, rework, repair or modification. Sometimes a product may be inadvertently submitted for verification before all operations have been completed. Rework means the continuation of processing that will make an item conform to specification. Rework requires only normal operations to complete the item and does not require any additional instructions. Rework when applied to documents means correcting errors without changing the original requirement. Repair is an action that restores an item to an acceptable condition but unlike rework, it may involve changing the product so that it differs from the specification but fulfils the intended use requirement. Modification may involve changing the requirement, as would be the case if the product requirement were found to be incorrect when assembling or operating product.

In some cases it may not be cost effective to attempt to eliminate the nonconformity and therefore such action would be inappropriate and the product should be disposed of.

Why is this necessary?
This requirement responds to the Process Approach Principle.

Remedial action is warranted when there are cost benefits from attempting to eliminate the nonconformity.

How is this implemented?
To implement this requirement you will need a form or other such document in which to record the decision and to assign the responsibility for the remedial action. These

documents also need to stipulate the verification requirements to be implemented following rework. When deciding on repair or rework action, you may need to consider whether the result will be visible to the customer on the exterior of the product. Rework or repairs that may not be visible when a part is fitted into the final assembly might be visible when these same parts are sold as service spares. To prevent on-the-spot decisions being at variance each time, you could:

(a)   identify in the drawings, plans, etc. those products that are supplied for service applications: i.e. for servicing, maintenance and repair;
(b)   provide the means for making rework invisible where there are cost savings over scrapping the item;
(c)   stipulate on the drawings, etc. the approved rework techniques.

Authorizing use of nonconforming product (8.3b, 8.3.3 and 8.3.4)

The standard requires the organization when appropriate to *deal with nonconforming product by authorizing its use, release or acceptance under concession by a relevant authority and where applicable by the customer*. The automotive additions require *the organization to obtain customer concessions or deviation permit prior to further processing whenever the product or manufacturing process is different from that which is currently approved*. They also require *records of expiration dates or quantity authorized to be maintained and for customers to be promptly notified in the event that nonconforming product has been shipped*.

What does this mean?

If you choose to accept a nonconforming item as is without rework, repair, etc., then you are in effect granting a *concession* or waiving the requirement *only* for that particular item. If the requirements cannot be achieved at all then this is not a situation for a concession but a case for a change in requirement. If you know in advance of producing the product or service that it will not conform to the requirements, you can then request a deviation from the requirements. This is often referred to as a *production permit* or *deviation*. Concessions apply *after* the product has been produced. Production permits or deviations apply *before* it has been produced. Both are requests that should be made to the acceptance authority for the product. The relevant authority is the authority that specified the requirement that has not been met. This authority could therefore be the customer or the designer.

Why is this necessary?

This requirement responds to the Process Approach Principle.

Product that does not conform to requirement may be fit for use. All specifications are but a substitute for knowledge of fitness for use. Any departure from such specification should be referred back to the specification authors for a judgement.

How is this implemented?

In order to determine whether a nonconforming product could be used, an analysis of the conditions needs to be made by qualified personnel. There are two ways of doing this. Either you refer all such nonconformities to the relevant authority or the authority appoints representatives who are capable of making these decisions within prescribed limits. A traditional method is to classify nonconformities, assign authority for accepting concessions for each level and define the limits of their authority. These levels could be as follows:

- *Critical Nonconformity*: A departure from the requirements which renders the product or service unfit for use.
- *Major Nonconformity*: A departure from the requirements included in the contract or customer specification.
- *Minor Nonconformity*: A departure from the requirements not included in the contract or customer specification.

The only cases where you need to request concessions from your customer are when you have deviated from one of the customer requirements and cannot make the product conform. Even when you repair a product, providing it meets all of the customer requirements, there is generally no need to seek a concession from your customer. While it is generally believed that nonconformities indicate an out of control situation, provided that you detect and rectify them before release of the product, you have quality under control and have no need to report nonconformities to your customer. However, if the frequency of nonconformity exceeds process capability targets, the process has become unstable and requires corrective action.

In informing your customer when nonconforming product has been shipped you obviously need to do this immediately when you are certain that there is a nonconformity. If you are investigating a suspect nonconformity it only becomes a matter for reporting to your customer when the nonconformity remains suspect after you have concluded your investigations. Alerting your customer every time you think there is a problem will destroy confidence in your organization. Customers appreciate zeal but not paranoia!

Production permits or deviations are generally permitted for specific batches or a defined time period. This is to allow time for corrective action to be taken. It is therefore necessary to keep a log of the products and quantities produced that are subject to the production permit or authorized deviation. It is also necessary to ensure that when the batch or date when the corrective action becomes effective arrives, the production permit or deviation is withdrawn. Flags should be inserted into production schedules alerting planners to batches that are subject to authorized concession or production permit and when the date or batch beyond which authorization is invalid arrives.

When delivery subject to authorized concession or production permit commences the packaging should be duly annotated. Flags should be inserted into delivery schedules indicating which batches will be covered by authorized concession or production permit.

Action to preclude use or application (8.3c)

The standard requires the organization when appropriate to *deal with nonconforming product by taking action to preclude its original intended use or application*. The automotive additions require *customer to be informed promptly in the event that nonconforming product has been shipped*.

What does this mean?
Precluding intended use or application means either scrapping the product so no one can use it or regrading it so that it may be used in other applications. In some cases products and services are offered in several models, types or other designations but are basically of the same design. Those which meet the higher specification are graded as such and those which fail may meet a lower specification and can be *regraded*. The grading should be reflected in the product identity so that there is no confusion.

Why is this necessary?
This requirement responds to the Process Approach Principle.

If a nonconforming product cannot be made conforming or accepted as is, some other action is needed to prevent inadvertent use and this leaves the two options stated.

How is this implemented?
Regrading can be accomplished by assigning a new identity to the product. Scrapping an item should not be taken lightly – it could be an item of high value. Scrapping may be an economical decision with low-cost items, whereas the scrapping of high-value items may require prior authorization as salvage action may provide a possibility of yielding spares for alternative applications.

Records of nonconformity (8.3)

The standard requires *records of the nature of nonconformities and any subsequent actions taken, including concessions to be maintained*.

What does this mean?
The records of nonconformities are the documented details of the product (its identity), the specific deviations from requirements (what it is and what it should have been), the condition under which the nonconformity was detected (the environmental or operating conditions, i.e. what was happening at the time when the nonconformity was detected), the time and date of detection, the name of the person detecting it and the actions taken with reference to any instructions, revised requirements and decisions.

Why is this necessary?

This requirement responds to the Factual Approach Principle.

Records of nonconformities are needed for presentation to the authorities responsible for deciding on the action to be taken and for subsequent analysis. Without such records, decisions may be made on opinion resulting in the means for identifying opportunities for improvement being absent.

How is this implemented?

There are several ways in which you can document the presence of a nonconformity. You can record the condition:

- on a label attached to the item;
- on a form unique to the item such as a nonconformity report;
- of functional failures on a failure report and physical errors on a defect report;
- in a logbook for the item such as an inspection history record or snag sheet;
- in a logbook for the workshop or area.

The detail you record depends on the severity of the nonconformity and to whom it needs to be communicated. In some cases a patrol inspector or quality engineer can deal with minor snags on a daily basis as can an itinerant designer. Where the problem is severe and remedial action complicated, a panel of experts may need to meet. Rather than gather around the nonconforming item, it may be more practical to document the remedial action on a form. In some cases the details may need to be conveyed to the customer off site and in such cases a logbook or label would be inappropriate. It is important when documenting the nonconformity that you record as many details as you can, because they may be valuable to any subsequent investigation in order to help diagnose the cause and prevent its recurrence. An example of a Nonconformity Report is illustrated in Figure 8.4. Note that in this example provision is made to record the impact of the nonconformity on both the customer and the business. This provision enables personnel to demonstrate compliance with Clauses 6.2.2d and 6.2.2.3. Also the provision for recording the action taken with respect to suspect items responds to Clause 8.3.1 and if the decision is to Use-as-Is, the form also acts as a concession or waiver.

Re-verification of corrected product (8.3)

The standard requires *nonconforming product to be subject to re-verification after correction to demonstrate conformity to the requirements.*

What does this mean?

Any rework, repair, modification or other action taken to correct the nonconformity will change the product and therefore it needs to be subject to re-verification. This may involve verification against different requirements to the original requirements.

NONCONFORMITY REPORT					
Part/Process		Qty	Used on		**NCR/**
S/N	Location		Reported by		Date detected
DESCRIPTION					As relevant define: 1 What is nonconforming 2 Why it is nonconforming 3 The criteria that states it is nonconforming 4 The instruments used to detect the nonconformity 5 The conditions at the time the nonconformity was detected
IMPACT					
On Customer			On Business		As relevant define: The impact of the nonconformity on the customer and/or the business: (a) if no action is taken? (b) If the item is scrapped?
Decision			Acceptance authority approval	Date	Acceptance authority approval required if decision is to Use-as-Is
Date	REMEDIAL ACTION			Responsibility	
	Rework, repair or modification				As relevant define: 1 Action to correct the specific nonconformity 2 Action to seek out and correct other examples that might be nonconforming 3 Action to control the symptoms of the nonconformity until the root cause is located and a design or process change is completed 4 Verification action to be taken after remedial action is complete
	Action on suspect items				
	Containment action				
	Verification after remedial action				
EFFECTIVENESS OF ACTIONS					
All actions taken and effective? Yes No		Reviewed by:	Related NCRs		Date closed

Figure 8.4 *Nonconformity report (NCR)*

Why is this necessary?
This requirement responds to the Factual Approach Principle.

If a nonconforming product is accepted as is without correction, no re-verification is necessary, but if the product is changed the previous verification is no longer valid.

How is this implemented?
Any product that has had work done to it should be re-verified prior to it being released to ensure the work has been carried out as planned and has not affected features that were previously found conforming. There may be cases where the amount of re-verification is limited and this should be stated as part of the remedial action plan. However, after rework or repair the re-verification should verify that the product meets the original requirement, otherwise it is not the same product and must be identified differently.

The verification records should indicate the original rejection, the disposition and the results of the re-verification in order that there is traceability of the decisions that were made.

Consequences of nonconformity (8.3)

The standard requires the organization *to take action appropriate to the effects or potential effects of the nonconformity when nonconforming product is detected after delivery or use has started.*

What does this mean?
A nonconformity may be detected by a subsequent user of the product either within the organization or by the customer. Also a nonconformity might be detected prior to release and implicate products already in use such as when subsequent analysis reveals inaccurate measurements or when verification methods or acceptance criteria change. Such product may not have failed in service because it has not been used in a manner needed to cause failure but if part of the same batch or lot contains a common cause nonconformity all products are suspected. Action taken as a result of latent nonconformity may involve product recall, product alerts or the issue of instructions for correction.

Why is this necessary?
This requirement responds to the Customer Focus Principle.

The requirement acknowledges that problems may be detected after shipment or use that need action to prevent undesirable effects.

How is this implemented?
Nonconformities detected by internal or external users indicate that the controls in place are not effective and should give cause for concern. Details should be recorded

and an investigation conducted to establish why the planned verification did not detect the problem. Action should then be taken to improve the verification methods by changing procedures, acceptance criteria, equipment or retraining personnel.

When a nonconformity is detected by verification personnel in a product where products of the same type are in use, an analysis is needed to establish whether the nonconformity would previously have escaped detection. If not, there is no cause for alarm but if something has now changed to bring the nonconformity to light, an evaluation of the consequences needs to be conducted. It may only be a matter of time before the user detects the same nonconformity.

The procedures should cover:

- the method of receiving and identifying returned product;
- the method of logging reports of nonconformities from customers and other users;
- the process of responding to customer requests for assistance;
- the process of dispatching service personnel to the customers premises;
- a form on which to record details of the nonconformity, the date, customer name, etc.;
- a process for acknowledging the report in order that the customer knows you care;
- a process for investigating the nature of the nonconformity;
- a process for replacing or repairing nonconforming product and restoring customer equipment into service;
- a process for assessing all products in service that are nonconforming, determining and implementing recall action if necessary.

Analysis of data (8.4)

Collecting and analysing appropriate data (8.4)

The standard requires the organization to *determine, collect and analyse appropriate data to demonstrate the suitability and effectiveness of the quality management system including data generated as a result of monitoring and measurement and from other relevant sources.*

What does this mean?
The suitability of the management system as stated previously is whether it represents the best way of doing things rather than whether it is fit for purpose – this is expressed by the requirement for the management system to be adequate in Clause 5.6 (the omission of the word "adequacy" in this clause is perhaps an oversight). The effectiveness of the management system is judged by the extent it enables the organization to satisfy its stakeholders.

Appropriate data would be any data generated from the processes of the management system that assist in the determination of their suitability and effectiveness.

The standard does not indicate what data may be needed – the organization is required to determine the data needed, to collect it and then analyse it in order to provide a basis for determining the performance of the management system.

Why is this necessary?

This requirement responds to the Factual Approach Principle.

The requirements in Clause 5.6 indicate a need for data on which to judge whether the system is suitable, adequate and effective – this requirement addresses this need.

How is this implemented?

Determining what data to collect

Suitability is concerned with doing things in the most appropriate way, perhaps the best way. If the system enabled staff to waste resources or under-utilize physical and human resources and under-utilize knowledge and capability, the output may conform to requirements but productivity would be down, bottlenecks would occur in processes and there would be a struggle to meet delivery targets without certain staff working flat out. System suitability cannot be examined if one perceives the system to be a set of documents – it is only possible if one perceives the system as encompassing everything the organization needs to fulfil its mission.

Effectiveness is concerned with doing the right things rather than with doing things right. So if the system enables management to stop the development of products for which there was no requirement, discover a potential safety problem, anticipate customer needs ahead of the competition, cut waste by 50%, successfully defend a product liability claim, meet all the delivery targets agreed with the customer, you would probably say that the system was pretty effective. If on the other hand the system allows the shipment of defective products every day, loses one in three customers, allows the development of unsafe products to reach the market, or the failure of a revolutionary power plant, you would probably say that the system was pretty ineffective. So the first thing you need to do is establish what measures will be used to determine system effectiveness. Many systems are only designed to meet the standard with the result that you can deliver defective product providing you also deliver some that are not defective.

As with many of these requirements, the place to start is to assess each process and determine the data needed to establish that:

- the process is achieving its objectives;
- the process is being run in the best way;
- the process objectives remain relevant to the organizational goals.

Data on achievement of objectives might include statistics on percent conforming, throughput, response time, etc. Data on best practice may include measures of productivity, quality costs – the cost of prevention, appraisal and failure costs or benchmarking data from competitors or similar industries. Data on relevance of objectives may include contrasting the objectives and targets with the organization's objectives.

Data collection

Plan the data requirements carefully so that you only:

- collect data on events that you intend to analyse;
- analyse data with the purpose of discovering problems;
- provide solutions to real problems;
- implement solutions that will improve performance.

To analyse anything, you need data. Without data you cannot know for sure if your processes are under control and if your customers are satisfied. It is not sufficient to claim that you have had no problems unless you are confident that the processes in place will alert you to problems should they arise. You also need to take care to avoid the "garbage in/garbage out" syndrome. Your analysis will only be as good as the data with which you are provided. If you want to determine certain facts, you need to ensure that the means exist for the necessary information to be obtained. To do this you may need to change the input forms or provide new forms on which to collect the data. The data needed for corrective action is rarely of use to those providing it therefore design your forms with care. Reject any incomplete forms as a sign that you are serious about needing the data. A sure sign that forms have become obsolete is the number of blank boxes. It is also better to devise unique forms for specific uses rather than rely on general multipurpose forms because the latter have a tendency to degrade the reliability of the data.

> Even though you are not aware of any problems will the processes in place alert you to problems should they arise?

Methods are needed to collect the data from the sensors and transmit it to the analysis stations. This may involve not only collecting data locally for immediate analysis for control purposes, but also transmission to central analysis stations. At such stations the data on all factors affecting performance may be aggregated and information produced for use in measuring overall performance relative to objectives. For example, a common target may be set for all processes relative to utilization, nonconformities, response time, etc. Some processes may be better than others but when aggregated show that the corporate objectives are not being achieved. A computer network can aid data collection by enabling remote access and collection into interlinked databases. However, many organizations still rely on paper records and therefore you will need a means of enabling such records to be either submitted to the analysis points or collected from source. To achieve this you will need to insert submission or collection instructions in the relevant procedures that specify the records.

Many organizations use a Nonconformity Report to collect information on nonconformities. One report may deal with remedial action and another separate report may address corrective action in order to prevent the recurrence of one or more nonconformities. In this way you are not committed to taking action on every incident but on a group of incidents where the action and its cost can be more easily justified.

This requirement is similar to that in Clause 8.5.3 under the heading *Preventive action* as the data collected for preventive action serves a similar purpose. In one case an analysis of company-level data serves to identify overall trends and predict potential failures that will affect achievement of the goals. In the preventive action case, the data serves to identify local and overall trends, and predicts potential failures that will affect achievement of specified requirements for the product, process and the quality system. It would be sensible to develop a data collection and analysis system that serves all levels in the organization with criteria at each level for reporting data upwards as necessary. You should not treat this requirement separately from that for preventive action because the same data should be used.

A general plan of action would include the following actions:

- Identify the key parameters to be measured.
- Locate where in the process they are achieved.
- Install data collection methods in relevant procedures.
- Collect and analyse the data.
- Use suitable presentation techniques to draw attention to the results.

In collecting the data, care should be taken to avoid data paralysis. The various quality tools can be used to prioritize the identified problems and corresponding decisions. As with all data collection tasks, you should show a direct correlation between what you are collecting and the goals to be achieved. All conclusions should lead to positive action otherwise the effort has been futile.

Evaluating where improvements in effectiveness can be made (8.4)

The standard requires the organization to *evaluate where continual improvement of the effectiveness of the quality management system can be made.*

What does this mean?
Juran writes on improvement thus "Putting out fires is not improvement of the process". Neither is discovery and removal of a special cause detected by a point out of control. This only puts the process back to where it should have been in the first place.[4] Continual improvement in the effectiveness of the management system is concerned with identifying where the objectives, standards or targets established for activities,

tasks or processes are below those needed for the organization to accomplish its purpose or mission. At one level this means changing the targets so that they are harder to meet and at another level it means changing the objectives so that work is driven in a new direction.

Why is this necessary?

This requirement responds to the Continual Improvement Principle.

The analysis of the external environment (customers, markets, regulations, economy and society) will indicate whether the organization's objectives are relevant or whether they need to change. If the management system continues driving the organization against existing objectives, it will fail to satisfy customers and other interested parties, therefore improvement in the effectiveness of the management system is needed. There may be cases where no specific objectives are set for some aspects that affect organizational performance. For example, managers may be desensitized to the level of rejects, the level of waste, scrap, delays, absenteeism and illness believing there is nothing that can be done to change it but by default, objectives have been set to maintain the *status quo* – if no objectives are set, any level of performance is acceptable!

In evaluating the effectiveness of the management system, those areas where objectives have not been specified become opportunities for improvement.

How is this implemented?

Implementation of this requirement needs a two-pronged approach. A review of established objectives and a review of performance attributes that have no objectives.

Reviewing established objectives

An approach to take is to:

- identify the objectives and targets that have been established for each process and subprocess;
- establish whether these targets are being achieved;
- analyse the processes to determine the potential for raising the targets or eliminating targets on the basis that they no longer serve the organization's goals;
- assess the feasibility of meeting the raised targets or the impact of eliminating inappropriate targets;
- present the case to management for change.

Reviewing uncontrolled performance attributes

One approach is to undertake the following actions:

- Analyse each process and identify all process outcomes, i.e. tangible and intangible outputs and effects on all interested parties.

- Identify the outcomes for which no objectives and targets have been set. (Some outcomes may have objectives but no targets and are therefore not being measured.)
- Assess the significance of the outcome in terms of impact (both short and long term) on the organization's goals.
- Make proposals for setting objectives and plans for those outcomes that impact the organization's goals.

Analysis of customer satisfaction data (8.4a)

The standard requires *analysis of data to provide information relating to customer satisfaction*.

What does this mean?
This requirement seeks to take the data generated by the monitoring process addressed in Clause 8.2.1 and through analysis produce meaningful information on whether customers are in fact satisfied with the products and services offered by the organization.

Why is this necessary?
This requirement responds to the Factual Approach Principle.

Customer satisfaction is not something one can monitor directly by installing a sensor. One has to collect and analyse data to draw conclusions.

How is this implemented?
We can now look at that ways by which data can be collected relative to the different techniques of monitoring customer satisfaction:

- *Repeat orders*: This data can be collected from the order-processing process.
- *Competition*: This data is more subjective and results from market research.
- *Referrals*: This data can be captured from sales personnel during the transaction or later on follow-up calls.
- *Demand*: This data can be collected from sales trends.
- *Effects of product transition*: This data can be collected from sales trends following new product launch.
- *Surveys*: This data can be collected from survey reports.
- *Focus meetings*: This data can be collected from the meeting reports.
- *Complaints*: This data can be collected from complaints recorded by customers or by staff on speaking with customers.
- *Compliments*: This data can be collected from written compliments sent in by customers or by staff on speaking with customers.

As indicated above there are several sources of data, several ways in which it can be collected and several functions involved. Provisions need to be made for transmitting the data from the processes where it can be captured to the place where it is to be analysed. It is evident that sales and marketing personnel are involved and as information on customer perceptions is vital for these functions to manage their own operations effectively, it may be appropriate to locate the analysis process within one of these departments. In some organizations, customer support groups are formed to provide the post sales interface with customer and in such cases they would probably perform the analysis.

The process

The integrity of your process for determining customer satisfaction is paramount otherwise you could be fooling yourselves into believing all is well when it is far from reality. The process therefore needs to be free from bias, prejudice and political influence.

In defining the process you will need to:

- determine the sources from which information is to be gathered;
- determine the method of data collection: the forms, questionnaires and interview checklists to be used;
- determine the frequency of data collection;
- devise a method for synthesizing the data for analysis;
- analyse trends;
- determine the methods to be used for computing the customer satisfaction index;
- establish the records to be created and maintained;
- identify the reports to be issued and to whom they should be issued;
- determine the actions and decisions to be taken and those responsible for the actions and decisions.

Pareto analysis can be used to identify the key areas on which action is necessary. For example, it may turn out that 80% of the sales come from repeat orders indicating a slow down in the number of new customers. Also 80% of the complaints may be from one market sector with only 20% of the sales – an indication that 80% of customers may be satisfied. Alternatively, 80% of the compliments may come from 20% of the customers but as they represent 80% of the sales it may prove very significant. The important factor is to look for relationships that indicate major opportunities and not insignificant opportunities for improvement. Use the results to derive the business plans, product development and process development plans for current and future products and services.

Frequency of measurement

Frequency needs to be adjusted following changes in products and services, and major changes in organization structure, such as mergers, downsizing, plant closures, etc. Changes in fashion and public opinion should also not be discounted. Repeating the survey after the launch of new technology, new legislation or changes in world economics affecting the industry may also affect customer perception and consequently satisfaction.

Trends

To determine trends in customer perception you will need to make regular measurements and plot the results preferably by particular attributes or variables. The factors will need to include quality characteristics of the product or service as well as delivery performance and price. The surveys could be linked to your improvement programmes so that following a change, and allowing sufficient time for the effect to be observed by the customer, customer feedback data could be secured to indicate the effect of the improvement.

Customer dissatisfaction will be noticeable from the number and nature of customer complaints collected and analysed as part of your corrective action procedures. This data provides objective documentation or evidence and again can be reduced to indices to indicate trends.

By targeting the final customer using data provided by intermediate customers, you will be able to secure data from the users but it may not be very reliable. A nil return will not indicate complete satisfaction so you will need to decide whether the feedback is significant enough to warrant attention. Using statistics to make decisions in this case may not be a viable approach because you will not possess all the facts!

Customer satisfaction index

A customer satisfaction index that is derived from data from an independent source would indeed be more objective. Such schemes are in use in North America, Sweden and Germany. A method developed by a Prof. Claes Fornell has been in operation for 15 years in Sweden and is now being used at the National Quality Research Center of the University of Michigan Business School. Called the American Customer Satisfaction Index (ACSI) it covers seven sectors, 40 industries and some 189 companies and government agencies. It is sponsored by the ASQ and the University of Michigan Business School with corporate sponsorship from Federal Express, Sears Roebuck, Florida Power and Light and others. The index was started in 1994[5] and using data obtained from customer interviews, sector reports are published indicating a CSI for each listed organization thereby providing a quantitative and independent measure of performance useful to economists, investors and potential customers.[6]

Analysis of conformance to customer requirements (8.4b)

The standard requires *analysis of data to provide information relating to conformity to product requirements.*

What does this mean?

Data relating to conformity to product requirements is data generated from monitoring and measuring product characteristics. It comprises the results of all verification activities including those generated during design, purchasing, production, installation and operation. The data collected from customer feedback is also included where the cause of the complaint is product nonconformity. The requirement focuses on conformity and therefore the degree of variation is not relevant. A product either conforms or does not conform, and it is this that is required to be analysed. The next requirement deals with variation.

Why is this necessary?

This requirement responds to the Factual Approach Principle.

When setting your quality objectives, targets should be established for product conformity as a measure of achievement. Data needs to be collected and analysed for all products in order to determine whether these objectives are being achieved.

How is this implemented?

Data on conformity and nonconformity can be collected from the product verification points in each process but it is also important to collect data on the size of the population involved. In general terms it is important to know the overall ratio of conforming product to nonconforming product, i.e. of a quantity of products, how many were conforming and how many nonconforming. The data could also relate to specific product characteristics, such as reliability, safety, power output, strength, etc.

A concept that has become very popular is the sigma value – a measure of the capability of a process to produce conforming product (see Chapter 1 under the heading *Six sigma*).

Analysis of product and process characteristics (8.4c)

The standard requires *analysis of data to provide information relating to characteristics and trends of processes and product including opportunities for preventive action*.

What does this mean?

The data that can provide information relating to characteristics and trends of processes and product comprises all the product and process measurements taken. The measurements provide useful data for indicating variation in product conformity and process capability. Variation is measured when the characteristics are variable, such as dimensions, voltage, power output and strength.

Opportunities for preventive action may arise when the trend in a series of measured values indicates deterioration in performance and if the deterioration were allowed to continue, nonconformity would result.

Why is this necessary?

This requirement responds to the Factual Approach Principle.

The purpose behind the requirement is to generate information that can be used to bring about an improvement in product and process quality. It is only by analysing trends that the opportunity for improvement is revealed. A series of figures may appear random at first glance but on closer analysis proves the process to be out of control. The points on a graph may indicate measurements are within limits but further analysis may reveal that a change has occurred that if ignored, may result in the process generating nonconforming product in the future. Data analysis is therefore useful for assessing current performance and predicting future performance.

How is this implemented?

Measurements of product and process characteristics should be collected but remember to collect only data that is useful in improving conformity or capability. Just because it can be measured does not mean that it should be! In automated processes, the machine performs the analysis and adjusts the process. In operator controlled processes, the operator takes the measurements, conducts the analysis and makes the adjustments often using control charts, such as Average and Range charts, Median charts, p charts, np charts, etc. The characteristics measured are taken from the product specification. In management controlled processes, the data needs to be collected, reduced, analysed, interpreted and presented in a suitable form before any meaningful information can be transmitted to decision-makers. The data may come from many processes and locations. Analysing the Staff Development process for instance may require data from all managers on staff numbers, development needs, training programmes, etc. to be consolidated and interpreted before meaningful results will emerge. The output may be in the form of histograms showing the levels of competence in each department, the cost of training per employee, the proportion of turnover spent on training, etc. All information should relate to specific product or process characteristics to prevent the management being inundated with reports that add no value.

A methodical approach would be to generate a list for each product and process that includes:

- the product or process characteristics;
- the location(s) of the data capture point(s);
- the method of capturing measurement data (automation, operator or manager);

- the form in which the data is transmitted to the analysis point (raw data, reports, surveys, etc.);
- the unit responsible for data analysis;
- the methods used to extract meaningful information from the data;
- the form in which the resultant analysis is transmitted to decision-makers;
- the frequency of reporting;
- the levels through which the information must pass before reaching the decision-makers;
- the unit responsible for making decisions.

Data on product characteristics may be analysed and acted on by the operator, and therefore do not warrant separate indication of each characteristic. However, some characteristics such as reliability and availability are not measured by operators but from field data or service centres. For processes the characteristics will be key performance measures against the process objectives. Where a process involves many departments, extracting and analysing the data can be a complex process in itself. Juran uses the analogy of telephone transmission to illustrate the problems with communication:[7]

- *The amplifier*: It is restoring intensity of a signal that has become weak over long distances.
- *The filter*: A device for admitting a selected part of the spectrum and to reject all else.
- *Redundancy*: Multiple transmission channels to increase reliability.
- *Shielding*: A device to keep noise from invading the message, keeps the message secure and keeps out cross talk.

Depending on the significance of the information, all of these devices may be active in the channels transmitting information of product and process performance. Some of the devices may be installed deliberately to safeguard information from outsiders. Other devices may be used covertly to prevent information from reaching the top management. In the list described above, the penultimate bullet is inserted to bring such ploys out into the open. If the information is based on fact and is relevant to the product and process objectives, it should be allowed free passage. Redundancy may be warranted when the information is of vital importance to the business and some means of validating the facts is needed.

Analysis of supplier data (8.4d)

The standard requires *analysis of data to provide information relating to suppliers.*

What does this mean?
Information relating to suppliers includes that related to their performance regarding product and service quality, delivery and cost.

Why is this necessary?

This requirement responds to the Factual Approach Principle.

Suppliers are a key contributor to the performance of an organization and therefore information on the performance of suppliers is necessary to determine the adequacy, suitability and effectiveness of the management system.

How is this implemented?

There are several aspects of the purchasing process that can be analysed and used to reveal information about suppliers in order to determine their performance and opportunities for improvement. However, the resources allocated to the analysis need to be appropriate to the potential risks to the organization and its customers. It is therefore necessary to focus on those suppliers that indicate the greatest risk to the organization's performance.

Order value

One group of suppliers at risk are those that provide the highest value of products and services. High order value implies significant investment by the organization because such decisions are not taken lightly. Should the supplier fail, the organization may not be able to recover sufficiently to avoid dissatisfying its customers.

The supplier database should identify the value of orders with suppliers and from an examination of these data; a Pareto analysis may reveal the proportion of suppliers that receive the highest value of orders. The performance of those in the top 20% in the order value list will obviously have more effect on the organization than the performance of the other 80% and may warrant closer attention. If you establish the effort spent on developing these suppliers as opposed to the others, it may reveal that the priorities are wrong and need adjustment.

Order quantity

Another group at risk are those suppliers that process the greatest number of orders. If there is a systemic fault in their processes, many deliveries may contain the same fault. It is possible that some of the high order value suppliers are the same as the high order quantity suppliers such as those supplying consumables. The performance of the top 20% in the order quantity list may affect all your products especially if the product is a raw material, fasteners, adhesives or any item that forms the basis of the products physical nature.

Quality risk

The third group at risk comprises those suppliers that supply products or services that a product failure modes analysis has shown are mission critical regardless of value or quantity. You may only need a few of these and their cost may be trivial, but their failure may result in immediate customer dissatisfaction. The FMEA should show the probability of failure and therefore the Pareto analysis could reveal the top 20% of products that are critical to the organization in terms of quality.

Delivery risk
The fourth group at risk comprises those suppliers that must meet delivery targets. Some items are on a long lead time with plenty of slack, others are ordered when stocks are low and others are ordered against a schedule that is designed to place product on the production line just-in-time (JIT) to be used. It is the latter that are most critical although a JIT scheme does not have to be place for delivery to be critical. The top 20% of these suppliers deserve special attention, regardless of value, quantity or product quality risk. A late delivery may have ramifications throughout the supply chain.

Suppliers per item
The fifth group of suppliers is not necessarily a group at risk. Many organizations insist on having more than one qualified supplier for a given item or service just in case a supplier underperforms. As Deming points out "*A second source for protection in case of ill luck puts one vendor out of business temporarily or forever, is a costly policy. There is lower inventory and a lower total investment with a single supplier than with two.*"[8] Remember you don't have to be ordering from the different suppliers at the same time for a second source to be a costly policy. There are the costs associated with the evaluation and approval which are double for maintaining a viable second source. An analysis of purchased items by supplier will reveal how many items are sourced from more than one supplier. Those items sourced from the most number of suppliers are therefore candidates for a supplier reduction programme.

Costs
Cost is also a factor but often only measured when there is a target for suppliers to reduce costs year on year. An analysis of these suppliers may reveal the top 20% that miss the target by the greatest amount.

Once the top 20% have been identified in each group, further analysis should be carried out to establish how each of these suppliers performs on quality, cost and delivery, the amount of effort spent in developing these suppliers and the degree to which these suppliers respond to requests for action.

A common method of assessing suppliers was to send out questionnaires that gathered data about the supplier. These add little value apart from gathering data. A measure of how many of your suppliers have ISO 9000 certification does not reveal anything of value because it does not indicate their performance. Analysis of supplier data should only be performed to obtain facts from which decisions are to be made to develop the supplier or terminate supply.

Analysis and use of data (8.4.1)

The standard requires trends in quality and operational performance *to be compared with progress towards objectives and lead to action.*

What does this mean?
The data that needs to be analysed for meeting Clause 8.4.1 is that which will show performance against the customer requirements determined in Clause 5.2. The automotive customer will often specify what parameters are to be measured and reported on a regular basis. These parameters will often be generic, such as delivery performance, response time, nonconformities, returns, etc.

Why is this necessary?
Being customer focuses means measuring your performance in the same way as your customer. In order to do this.

How is this implemented?
In order to identify customer-related problems charts showing trends against customer expectations over time need to be produced and maintained. The target values should be indicated on the charts and where targets are not met, the charts need to be annotated with codes that relate to status reports and improvement plans.

A "Note" to this requirement suggests that the data be compared with that of competitors and/or appropriate benchmarks. If it is possible to obtain such data you need to validate the basis of measurement. What might appear superior performance might be result of measurements taken against a different standard to the one you are using.

Some automotive customers will require data reporting in specified formats. Templates are downloaded, filled in and uploaded to the customer server so that the customer can view performance from all suppliers.

Improvement (8.5)

Continual improvement (8.5.1)

The standard requires the organization to *continually improve the effectiveness of the quality management system through use of the quality policy, quality objectives, audit results, analysis of data, corrective and preventive actions and management review.*

What does this mean?

There are ten requirements for continual improvement in the standard including the one above:

1 The organization shall establish, document, implement and maintain a management system and continually improve its effectiveness.
2 The organization shall implement actions necessary to achieve continual improvement of the processes needed for the management system.
3 Top management shall provide evidence of its commitment to continually improving the effectiveness of the management system.
4 Top management shall ensure that the quality policy includes a commitment to continually improve the effectiveness of the management system.
5 The organization shall determine and provide the resources needed to continually improve the effectiveness of the management system.
6 The organization shall plan and implement the monitoring, measurement, analysis and improvement processes needed to continually improve the effectiveness of the management system.
7 The organization shall determine, collect and analyse appropriate data to evaluate where continual improvement of the management system can be made.
8 The organization shall continually improve the effectiveness of the management system through the use of the quality policy, quality objectives, audit results, analysis of data, corrective and preventive actions and management review.
9 The organization shall define a process for continual improvement.
10 Manufacturing process improvement shall continually focus on control and reduction of variation.

Requirement 1 is duplicated by requirement 8. Requirement 2 omits reference to *effectiveness* therefore improvement can be interpreted as applying to improved efficiency as well as effectiveness. Requirements 3 and 4 are linked in that they relate to policy. Requirement 6 serves to identify opportunities for improvement of which requirement 7 forms a part and requirement 5 provides the means by which improvement are to be made. Requirement 9 simply serves to apply the requirement for processes to be identified to continual improvement and requirement 10 goes beyond improvements in effectiveness to embrace improvement by better control. They are therefore not separate requirements but are all derivatives of the first requirement, each focusing on either the whole or a part of the improvement process.

Why is this necessary?

This requirement responds to the Continual Improvement Principle.

The policies, objectives and targets maybe those that are set on the basis of existing capability and performance – what was achieved last year. They may be those that were set last time customer needs and expectations were evaluated or the prevailing regulations determined. These may no longer be the policies, objectives and targets that are required to keep the organization focused on its purpose and mission. The *goal posts* may have moved! The competition may have pushed forward the frontiers of technology, innovation and performance. The market may have changed, the economic and social climate changed and in order to stay ahead of the competition, the organization has to aim for different objectives and targets, to set new policies that will form different behaviours. On the premise that for a system to be effective it has to fulfil its purpose and in the environment, in which the system operates, the route towards that fulfilment is always changing; the system will never be totally effective – there will always be a new goal to aim for. It is therefore necessary to continually improve the effectiveness of the management system even though by the time the system is effective the goal posts will have moved. There is nothing more certain in life than death, taxes and change!

> Check where the "Goal Posts" are located for they may have moved since the last time you looked.

How is this implemented?

This particular requirement duplicates that contained in the first paragraph of Clause 5.4.2 because Clause 5.4.1 required objectives consistent with policy and Clause 5.3 required policy to address continual improvement, therefore plans for continual improvement would be required. Planning for improvement is also addressed by Clause 8.1 and 8.5.1, so there is no shortage of clauses related to improvement planning. They all amount to the same thing – there is not a lot of difference between them but the most informative reference is that in Annex B of ISO 9004 where an improvement methodology is described.

Continual improvement was addressed in Chapter 4 but to understand the implications of this requirement it is necessary to understand:

- the composition of a management system;
- the types of improvement;
- the difference between random and continuous improvement.

The meaning of a management system was addressed in Chapter 1 so we have learnt that a management system is much more than a set of documents. As a management system consists of the processes required to deliver the organization's products and services as well as the resources, behaviours and environment on which they depend, it follows that in continually improving the management system, you need to continually improve the processes, resources, behaviours and the physical and human environment within the organization.

Improvement can be random and unstructured – arising from fire fighting measures, reactions to situations that have got out of control or to threats that appear on the horizon that must be dealt with in order to survive – but this is not strictly improvement, it merely restores performance to where it should have been in the first place. ISO 9000 requires a continual quest for improvement and this will only come about if you are continually measuring performance and acting on the results so as to improve conformity and improve capability. The frequency by which performance is measured should be appropriate to the parameter concerned. Some parameters change by the second, others by the year. The frequency of measurement should provide factual data that if acted on will move the organization forward incrementally or in great steps. What should not happen is that the measurements are taken too late to halt a significant decline in performance.

Improvement plans do not need to be consolidated into documents with the title *Continual Improvement Plan*. Separate plans may exist, focused on general or specific improvements. For example, there may be:

* new product development plans,
* new process development plans,
* new system development plans,
* corrective action plans,
* preventive action plans,
* staff development plans,
* plant development plans.

All these plans aim to provide the organization with an improved capability and for data management purposes, it may help to catalogue these plans under *Improvement* as well as product, process, department, division or corporate headings, etc.

Continual improvement of the organization (8.5.1.1)

The standard requires the organization *to define a process for continual improvement.*

What does this mean?

A process for continual improvement would be a series of interrelated activities, resources and behaviours that bring about an improvement in performance. Such a process would operate at several different levels in the organization rather than at one level. It would be impractical to have a single continual improvement process through which all improvements are channelled. It is more likely that continual improvement processes form part of other processes but the principles on which each is based would be the same.

Each managed process has three improvement mechanisms:

1 One that brings about *improvement by better control*: reducing variation about a mean.

2   One that brings about *improvement by greater efficiency*: reducing resources, doing more for less.
3   One that brings about *improvement by greater effectiveness*: raising standards, doing the right things, being ahead of stakeholder needs and expectations.

Why is this necessary?

The reasons for continual improvement have been given previously. Here we address the reason for a continual improvement process. Such a process is not a core process but a part of every process. In order for each process to operate effectively, continual improvement has to be embedded in the approach taken to its management. The process approach, if properly applied will not only bring every process under control but also subject those processes to continual improvement.

How is this implemented?

Quality improvement was addressed in Chapter 1 and little more can be added here without going into specific examples. The secret is to make continual improvement endemic in the organization, but don't go in for razzmatazz, treat every improvement as routine; an opportunity for improving control, finding better ways of doing things or finding different things to do that will delight customers and the way to do this is to install review and improvement cycles in each process, in each stage of each process and within each tasks of each stage of each process. Figure 8.5 shows three improvement cycles, each dealing with one type of improvement. Build these into your processes and you will bring about continual improvement.

Manufacturing process improvement (8.5.1.2)

The standard requires *manufacturing process improvement to continually focus on control and reduction of variation in product characteristics and manufacturing process parameters*.

What does this mean?

This requirement has to be taken in context of a manufacturing process that is already capable by virtue of the requirements of Clause 8.2.3.1. A process improvement that focuses on control and reduction of variation is therefore one that seeks to raise standards, achieve levels of variation beyond those in the specification.

Why is this necessary?

The less variation in product characteristics and process parameters, the less likelihood of product failure in the field.

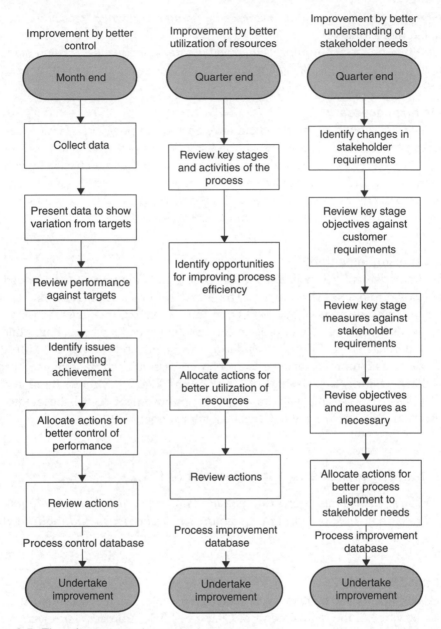

Figure 8.5 *Three improvement processes*

How is this implemented?

This might not be as difficult as it looks. It all depends on what the original specifications are. There may well be some latitude in the specification to reduce variation simply by better control of the process, i.e. investigating those deviations around the nominal values that although within limits have a common cause. Simply replacing the tools more often, changing the coolant, the oil viscosity or some other process parameter may bring greater stability to the process.

There may also be opportunities for changing the upper and lower limits so that they sit closer to the nominal. After monitoring performance over many batches, you might discover that the limits have been set far wider then necessary.

> ### Putting terms in the right order
>
> *Preventive action* serves to prevent the nonconformity from **occurring**.
>
> *Inspection* detects nonconformity.
>
> *Nonconformity control* identifies, segregates and rectifies the nonconforming item and any others that are potentially nonconforming.
>
> *Corrective action* serves to prevent the nonconformity from **recurring**.

Corrective action (8.5.2)

Eliminating the cause of actual nonconformities (8.5.2)

The standard requires the organization to *take action to eliminate the cause of nonconformities in order to prevent recurrence and requires the actions to be appropriate to the effects of the nonconformities encountered.* The automotive additions require a defined process *for problem-solving leading to root cause identification and elimination.*

What does this mean?

Corrective action is the pattern of activities that traces the symptoms of a problem to its cause, produces solutions for preventing the recurrence of the problem, implements the change and monitors that the change has been successful. Corrective action provides a feedback loop in the control cycle. Whilst the notion of *correction* implies that it could be as concerned with the nonconforming item as with the cause of nonconformity, correcting the nonconforming item is a remedial action. It doesn't stop it recurring. ISO 9000 does not use the term remedial action except in the context of a repair. The term in ISO 9000 used for correcting the nonconformity is *correction* – which is a little too similar to the term corrective action to avoid confusion. Preventing the recurrence of nonconformity is a corrective action. A problem has to exist for you to take corrective action. When actual problems don't exist but there is a possibility of failure, the action of preventing the occurrence of nonconformity (or any problem for that matter) is a *preventive action*. So we have Remedial Action, Corrective Action and Preventive Action, each with a different meaning. Remedial Action is covered by Clause 8.3, Corrective Action by 8.5.2 and Preventive Action by Clause 8.5.3 of the standard.

Although the natural inclination is to think of nonconformities in the context of product and manufacturing process, any departure from a requirement is a nonconformity.

A medical analogy

You have a head cold so you go to the doctor for advice. The doctor prescribes a remedy: "take two pills three time a day and the cold symptoms should subside". You do as the doctor advises and indeed the cold symptoms subside. *This is remedial action*.

You then return to the doctor year after year with the same problem and ask the doctor to prescribe a means of preventing the cold symptoms from recurring each year. The doctor prescribes a course of injections that will prevent the symptoms occurring. You take the course of injections and behold, you do not feel cold symptoms ever again! *This is corrective action* (as defined in ISO 9000).

You observe over the years that the course of injections work providing you continue with the regime. You have a daughter who has never had a head cold and, mindful of the treatment you are given, wish to preserve your daughter from the suffering you have had over the years. You seek the doctor's advice which is that you enter your daughter onto a course of injections to safeguard against the risk of attracting the cold virus. Your daughter undertakes the prescribed course of treatment and never experiences a head cold. *This is preventive action*.

It follows therefore that any failure to meet the organization's objectives is a nonconformity and its cause should be eliminated.

Action appropriate to the effects of the nonconformities encountered means that if the nonconformity is a random occurrence and of insignificant consequence, no corrective action might be needed; e.g. if a person makes a mistake and knows a mistake has been made – there is no pattern of behaviour that would suggest it would occur again. If there is frequent occurrence of nonconformity this indicates a systemic problem that can be prevented from recurrence by retraining, changing work practices or modifying product or process design. It is therefore not practical to attempt action to prevent recurrence for each nonconformity.

Why is this necessary?
This requirement responds to the Process Approach Principle.

Nonconformities are caused by factors that should not be present in a process. There will always be variation but variation is not nonconformity. Nonconformity arises when the variation exceeds the permitted limits. The factors that cause nonconformity on one occasion will (unless removed) cause nonconformity again and again. As the objective of any process must be to produce conforming output, it follows therefore that it is necessary to eliminate the causes of nonconformity. This does not simply apply to products of the production process but to products of all processes – mission management, demand creation and demand fulfilment.

How is this implemented?

The other requirements in this clause identify most of the steps needed to eliminate the cause of nonconformity. There are one or two stages that have been omitted. A more complete list of steps is shown in Table 8.2.

These will be addressed later in this chapter.

Table 8.2 *Steps in the corrective action process*

Step	Action	Clause 8.5.2 requirement
1 2	Collect the nonconformity data and classify Conduct Pareto analysis to identify the vital few and trivial many	Reviewing nonconformities
3 4 5	Organize a diagnostic team Postulate causes and test theories Determine the root cause of nonconformity	Determining the cause
6 7 8	Determine the effects of nonconformity and the need for action Determine the action needed to prevent nonconformity recurring Organize an implementation team	Evaluating the action needed
9 10	Create or choose the conditions which will ensure effective implementation Implement the agreed action	Implementing the action needed
11 12 13 14 15 16	Record the results of Pareto analysis Record the causes of nonconformity Record the criteria for determining severity or priority Record the proposed actions to be taken Record the actions actually taken Record the results of actions taken	Records of results
17 18 19 20 21	Assess the actions taken Determine whether the actions were those required to be taken Determine whether the actions were performed in the best possible way Determine whether the nonconformity has recurred If nonconformity has recurred repeat steps 1 to 20	Reviewing corrective actions

As the sources of nonconformity are so varied, it may not be practical to have a single corrective action procedure. It may be more practical to embody corrective action provisions in the following procedures:

- Failure investigation procedure.
- Nonconforming material review procedure.
- Customer complaints procedure.
- Document change procedure.
- Specification change procedure.
- Maintenance procedures.

Reviewing nonconformities (8.5.2a)

The standard requires *the documented procedure for corrective action to define requirements for reviewing nonconformities (including customer complaints).*

What does this mean?

As indicated above, the review of nonconformities means that nonconformity data should be collected, classified and analysed. The reference to customer complaints is that every customer complaint is a nonconformity with some requirement. They may not all be product requirements. Some may relate to delivery, to the attitude of staff or to false claims in advertising literature. Any complaint implies that a requirement (expectation, obligation or implied need) has not been met even if that requirement had not been determined previously. We have to accept that we could have overlooked something. Just because it was not written in the contract does not mean the customer is wrong.

Why is this necessary?

This requirement responds to the Continual Improvement Principle.

A review is another look at something therefore the first view of the nonconformity was when it was detected and recorded. The second view of it should aim to:

1 establish if the nonconformity had been predicted in the planning phase;
2 establish why the preventive action measures were not effective;
3 prevent it from happening again where possible.

How is this implemented?

Your corrective action procedures need to cover the collection and analysis of product nonconformity reports and the collection and analysis of process data to reveal process nonconformities. The standard does not require you to take corrective action on every nonconformity. Here it is suggested that the decision to act should be *appropriate to the effects of the nonconformities encountered.* It is therefore implying that you only need act on the vital few. To find the vital few nonconformities out of the total population that provide the bulk of improvement potential a Pareto analysis should be conducted. A Pareto analysis is a management tool that finds a few needles in a haystack of trivia, e.g. most of the wealth is concentrated in few hands or 10% of customers account for 60% of sales (for an explanation of the Pareto principle and its origins see Juran on *Quality by Design*). When dealing with nonconformity, the question we need to ask is "what are the few sources of nonconformities that comprise the bulk of all nonconformities?" If we can find these nonconformities and eliminate their cause, we will reduce variation significantly.

The first step is to assign a short description to the nonconformity, such as dry solder joint, hole not plated, broken track, incorrect part number, dirty terminal, etc. The next step is to sort the nonconformities by product and process. Then rank the nonconformities in order of occurrence so that the nonconformity having the most occurrences would appear at the top of the list. The result might be that for a particular product or process a few types of nonconformity would account for the greatest proportion of nonconformities. An example is given in Table 8.3. Here we see there are fifteen types

Table 8.3 *Pareto analysis*

Nonconformity type	Frequency	%	Cumulative (%)
Too much solder	400	34.19	34.19
Lifted tracks	230	19.66	53.85
Solder bridge	180	15.38	69.23
Dirty terminals	90	7.69	76.92
Component not flat on board	70	5.98	82.91
Cracked insulation	40	3.42	86.32
Uncropped component legs	30	2.56	88.89
Under spec plating thickness	30	2.56	91.45
Too little solder	20	1.71	93.16
Dry joints	20	1.71	94.87
Unplated holes	20	1.71	96.58
Broken tracks	10	0.85	97.44
Incorrect component fitted	10	0.85	98.29
Incorrect part number	10	0.85	99.15
Damage edge connector	10	0.85	100.00
Total nonconformities	**1170**		

of nonconformity with just four types accounting for 76% of the total. It follows therefore that if we eliminate the bottom four causes of nonconformity productivity would increase by a mere 3%. However, if we eliminate the four most dominant types of nonconformity, productivity would increase by staggering 76%.

Another way of ranking the nonconformities is by seriousness. Not all nonconformities will have the same effect on product quality. Some may be critical and others insignificant. By classification of nonconformities in terms of criticality a list of those most serious nonconformities can be revealed using the Pareto analysis. Even though the frequency of occurrence of a particular nonconformity may be high, it may not affect any characteristic that impacts customer requirements. This is not to say the cause should not be eliminated but there may be other more significant problems to eliminate first.

Before managers will take action, they need to know:

- What is the problem or potential problem?
- Has the problem been confirmed?
- What are the consequences of doing nothing, i.e. what effect is it having?
- What is the preferred solution?
- How much will the solution cost?
- How much will the solution save?
- What are the alternatives and their relative costs?
- If I need to act, how long have I got before the effects damage the business?

Whatever you do, don't act on suspicion; always confirm that a problem exists or that there is a certain chance that a problem will exist if the current trend continues. Validate causes before proclaiming action!

Customer complaints, rejected and returned product (8.5.2.4)

The customer can be mistaken, and customer complaints therefore need to be validated as genuine nonconformities before entering the corrective action process. Parts returned from dealers, customer manufacturing plants, etc. might not be nonconforming. They may be obsolete, surplus to requirements, have suffered damage in handling or have been used in trials, etc. Products may have failed under warranty and not be logged as a complaint but nonetheless they are nonconforming. Whatever the reason for return, you need to record all returns and perform an analysis to reveal opportunities for corrective action when appropriate. You should process these items as indicated previously but prior to expending effort on investigations, you should establish your liability and then investigate the cause of any nonconformities for which you are liable.

When parts are rejected subsequent to delivery it is indicative that your processes are not under control rejected parts analysis should be focused on determining the reason why the process failed to detect nonconformity because there could be some weakness in the process that if not corrected, further nonconforming parts might be shipped.

Nonconformity reduction

Previously it was suggested that action be taken on the vital few nonconformities that dominated the population. If this plan is successful these nonconformities will no longer appear in the list the next time the analysis is repeated. As the vital few nonconformities are tackled, the frequency of occurrence will begin to decline until there are no nonconformities left to deal with. This is nonconformity reduction (or special cause removal) and can be applied to specific products or processes. If you were to aggregate the nonconformities for all products and processes you would observe that it is quite possible to take corrective action continuously and still not reduce the number of nonconformities – no matter how hard you try you cannot seem to reduce the number. This is because the objectives and targets keep changing. They rarely remain constant long enough to make valid comparisons from year to year. There is always some new process, practice or technology being introduced than triggers the learning cycle all over again.

Determining the cause of nonconformities (8.5.2b)

The standard requires *the documented procedure for corrective action to define requirements for determining the causes of nonconformity.*

What does this mean?
The cause of nonconformity is the reason it occurred. What you observe when detecting nonconformity is the effect of the cause. A product is damaged, the immediate cause might be poor packaging or poor handling but these too have their causes and although they may appear unconnected there maybe a common cause which, if eliminated, resolves two problems of concern.

There are three types of corrective action, product related, process related and system related. Product-related nonconformities can be either internal or external and you will have nonconformity reports to analyse. Process-related nonconformities may arise out of product nonconformity but if you expect something less than 100% yield from the process, the reject items may not be considered *nonconformities*. They may be regarded as *waste*. Unlike products, process nonconformities are often not recorded in the same way and therefore the data is not as readily accessible. By analysing the process you can find the cause of low yield and improve performance of the process. Product and process nonconformities may be detected at planned verification stages and may also be detected during product and process audits. System-related nonconformities could arise out of internal and external systems audits but also arise as a result of tracing the root cause of a problem to a system inadequacy.

Why is this necessary?
This requirement responds to the Continual Improvement Principle.

All nonconformities are caused – all causes within your control can be avoided – all that is needed is concerted action to prevent recurrence. Nonconformity costs money and wastes resources. The fewer the nonconformities the more resources available for producing productive output.

How is this implemented?
Discovering the cause ("5 Why's")
To eliminate the cause of nonconformity the cause needs to be known and therefore the first step is to conduct an analysis of the symptoms to determine their cause. Simply asking why an event occurred might reveal a cause but don't accept the first reason given because there is usually a reason why this previous event occurred. Toyota discovered that asking *why* successively five times would invariably discover the root cause. There may be more or less than five steps to the root cause but it is critical to stop only when you can't go any further. The following example illustrates the technique.

A trainer arrives to conduct a training course to discover that the materials have not been delivered from head office as he expected them to be:

1 *Why were the materials not delivered? Answer: Because the administrators thought the trainer would bring his own materials.*
2 *Why did they think the trainer would bring his own materials? Answer: Because they had not been informed otherwise.*
3 *Why weren't the administrators given the correct information? Answer: Because the office manager had not communicated the agreed division of responsibility when setting up training courses.*
4 *Why had the office manager not communicated the agreed division of responsibility? Answer: Because the office manager had put other matters before internal communication in his order of priorities.*
5 *Why had the office manager not got his priorities right? Answer: Because he was not yet competent.*
6 *Why was the office manager not yet competent? Answer: Because the top management had made the appointment in haste.*
7 *Why had top management made the appointment in haste? Answer: Because they were not applying leadership.*

Therefore a lack of leadership (the second quality management principle) is the root cause. It took seven questions to get there but if we had stopped at question 3, and made the assumption that giving the administrators the correct instructions would prevent recurrence of the problem, we would be wrong. It might well prevent recurrence of the specific problem with that particular office manager but not similar problems with other managers. If the office manager forgot to issue the instructions, it indicates that he did not complete the process that commenced when the division of responsibility was agreed. This is quite typical. A meeting is held and agreements reached and when everyone departs they get on with what they were doing before the meeting, not realizing that a process has been initiated that needs to continue

> The root cause of most problems can be traced to lack of application of one or more of the eight quality management principles.

and be completed outside the meeting. If the staff were competent, they would complete the process before moving on.

Another technique is the cause-and-effect diagram (Aka Ishikawa Diagram or Fishbone Diagram). This is a graphical method of showing the relationship between cause and effect. Each type of nonconformity (an effect) would be analysed to postulate the causes; e.g. the question would be put to a diagnostic team: What could cause too much solder on a joint? The team would come up with a number of possibilities. Each one would be tested either by experiment or further examination of the soldering process and a root cause established.

Some nonconformities appear random but often have a common cause. In order to detect these causes, statistical analysis may need to be carried out. The causes of such nonconformities are generally due to non-compliance with (or inadequate) working methods and standards. Other nonconformities have a clearly defined special or unique cause that has to be corrected before the process can continue. Special cause problems generally require the changing of unsatisfactory designs or working methods. They may well be significant or even catastrophic. These rapidly result in unsatisfied customers and loss of profits. In order to investigate the cause of nonconformities you will need to:

1   identify the requirements which have not been achieved;
2   collect data on nonconforming items, the quantity, frequency and distribution;
3   identify when, where and under what conditions the nonconformities occurred;
4   identify what operations were being carried out at the time and by whom.

The investigation of parts rejected subsequent to delivery needs to probe not only into the root cause but also establish:

* Why the controls in place did not detect the nonconformity.
* Why the process FMEA failed to identify the potential failure mode.
* Why the process capability studies concluded the process was capable when clearly it wasn't.

It could be of course that the rejected parts are simply within the 3.4 ppm that were expected from a process with six-sigma capability but if the number of parts rejected already exceeds this limit process improvement action will be necessary.

Common problem-solving tools (8.5.2.1)

There are many tools you can use to help you determine the root cause of problems. These are known as *disciplined problem-solving methods*. A common method in the automotive industry is known as 8D meaning eight disciplined methods. Originally conceived by the Ford TOPS (Team-Oriented Problem-Solving) program in 1987[9] and upgraded and renamed "Prevent Recurrence" in 1992:

D1:   Establish the Team
D2:   Describe the Problem
D3:   Develop an Interim Containment Action
D4:   Define/Verify Root Cause
D5:   Choose/Verify Permanent Corrective Action
D6:   Implement/Validate Permanent Corrective Action
D7:   Prevent Recurrence
D8:   Recognize the Team

The 8D approach to problem-solving

1   *Use Team Approach*   Establish a small group of people with the knowledge, time, authority and skill to solve the problem and implement corrective actions. The group must select a team leader.

2   *Describe the Problem*   Describe the problem in measurable terms. Specify the internal or external customer problem by describing it in specific terms.

3   *Implement and Verify Short-Term Corrective Actions*   Define and implement those intermediate actions that will protect the customer from the problem until permanent corrective action is implemented. Verify with data the effectiveness of these actions.

4   *Define and Verify Root Causes*   Identify all potential causes which could explain why the problem occurred. Test each potential cause against the problem description and data. Identify alternative corrective actions to eliminate root cause.

5   *Verify Corrective Actions*   Confirm that the selected corrective actions will resolve the problem for the customer and will not cause undesirable side effects. Define other actions, if necessary, based on potential severity of problem.

6   *Implement Permanent Corrective Actions*   Define and implement the permanent corrective actions needed. Choose ongoing controls to insure the root cause is eliminated. Once in production, monitor the long-term effects and implement additional controls as necessary.

7   *Prevent Recurrence*   Modify specifications, update training, review workflow, improve practices and procedures to prevent recurrence of this and all similar problems.

8   *Congratulate Your Team*   Recognize the collective efforts of your team. Publicize your achievement. Share your knowledge and learning.

By courtesy of National Semiconductor, 2004.

Another version is shown in the boxed section indicating that the labels and interpretations vary. In both examples the terms do not quite align with those in ISO/TS 16949. D5, 6 and 7 are all part of the same action to prevent recurrence that ISO 9000 defines as "Corrective Action". The notion of a permanent corrective action implies there is a temporary corrective action. Corrective action either removes the root cause or it doesn't. If the "permanent corrective action" were effective it would also deal with the issues raised under D7 like modifying specifications but other interpretations[10] indicate this would form part of D6. It is not too important what the steps are called providing all steps are completed. However, the 8D technique deals with the control of nonconformity as well as corrective action because it addresses containment actions. Such actions do not prevent recurrence of nonconformity; they simply control nonconformity by stopping it occurring. This is why containment actions are specified in the Nonconformity Report of Figure 8.4. A corrective action report that simply deals with problem diagnosis and is illustrated in Figure 8.6.

Whilst 8D has a certain meaning, disciplined methods are simply those proven methods that employ fundamental principles to reveal information. There are two different approaches to problem-solving. The first is used when data are available as is the case when dealing with nonconformities. The second approach is when not all the data needed are available.

The seven quality tools in common use are as follows:

1　*Pareto diagrams* used to classify problems according to cause and phenomenon.
2　*Cause-and-effect diagrams* used to analyse the characteristics of a process or situation.
3　*Histograms* used to reveal the variation of characteristics or frequency distribution obtained from measurement.
4　*Control charts* used to detect abnormal trends around control limits.
5　*Scatter diagrams* used to illustrate the association between two pieces of corresponding data.
6　*Graphs* used to display data for comparative purposes.
7　*Check sheets* used to tabulate results through routine checks of a situation.

The further seven quality tools for use when not all data are available are as follows:

1　*Relations diagram* used to clarify interrelations in a complex situation.
2　*Affinity diagram* used to pull ideas from a group of people and group them into natural relationships.
3　*Tree diagram* used to show the interrelations among goals and measures.
4　*Matrix diagram* used to clarify the relations between two different factors (e.g. QFD).
5　*Matrix data analysis diagram* used when the matrix chart does not provide information in sufficient detail.

CORRECTIVE ACTION REPORT				
Problem name			CAR/	
Product/Process	Location	Reported by	Date observed	

PROBLEM DESCRIPTION	
	As relevant define: 1 Nature of problem 2 Acceptance criteria 3 Number of occurrences 4 Significance of problem (percent of total occurrences for this product)

DIAGNOSTIC TEAM			
Role	Name	Department	Location

IMPACT & PRIORITY				
On Quality	On Cost	On Delivery	Other impact	Priority Required Urgent ☐ Date High ☐ Medium ☐ Low ☐

ROOT CAUSE			
Cause	% Contribution	Verified	As relevant state the results of applying the 5 Why's test to the problem Reference relevant diagnostic reports
Cause 1			
Cause 2			
Cause 3			
Cause 4			
Cause 5			

RESOLUTION			
Solution for eliminating root cause	Responsibility	Date validated	As relevant define: 1 Product design changes and validation method 2 Process design changes and validation method 3 FMEA update 4 Control plan update

EFFECTIVENESS OF ACTIONS			
All actions taken Yes No	Number of potential recurrence opportunities since last action	Result	
Related NCRs/CARs	Reviewed by:		Date closed

Figure 8.6 *Corrective action report (CAR)*

6   *Process decision programme chart* used in operations research.
7   *Arrow diagram* used to show steps necessary to implement a plan (e.g. PERT).

The source of causes is not unlimited. Nonconformities are typically caused by one or more of the deficiencies in:

* communication,
* documentation,
* personnel training and motivation,
* materials,
* tools and equipment,
* the operating environment.

Each of these is probably caused by not applying one or more the eight quality management principles.

Once you have identified the root cause of the nonconformity you can propose corrective action to prevent its recurrence. Eliminating the cause of nonconformity and preventing the recurrence of nonconformity are essentially the same thing.

Evaluating the need for action (8.5.2c)

The standard requires *the documented procedure for corrective action to define requirements for evaluating the need for action to ensure that nonconformities do not recur.*

What does this mean?

All nonconformities are the result of something not going to plan no matter how insignificant the problem. Whether action is taken depends on the effects of the nonconformity. The plan may not be right, the deviation from plan may have no effect at all, it may be a one-off and unlikely to recur but on the other hand it may be disastrous (and likely to recur unless something is done about it). An evaluation of the need for action is therefore necessary to determine if action should be taken and if so when that action should be taken.

Why is this necessary?

This requirement responds to the Continual Improvement Principle.

With a multitude of problems to resolve, some method of determining priorities is needed and rather than require the recurrence of all nonconformities to be prevented, the standard requires quite sensibly an evaluation to determine the need for action.

How is this implemented?

In reviewing the nonconformity data, you can rank nonconformities by class or cost so that you reveal the most important problems to tackle. A simple classification is to classify nonconformities on the basis of affecting form, fit or function and this is sufficient for most purposes. However, there are various degrees of fit, form and function. You could use the same severity ranking that is used when conducting an FMEA (for more details refer to Table 8.5).

Although used in the design phase, these criteria can be applied in the production and operational phases to rank nonconformities – those detected before and after shipment. Clearly a nonconformity that falls into any category above low (6–10) is a candidate for immediate corrective action. However, many nonconformities may not have these effects and would therefore receive a ranking of 1 but this does not mean that such nonconformities are not important to other processes.

Determining and implementing actions (8.5.2d and 8.5.2.2)

The standard requires *the documented procedure for corrective action to define requirements for determining and implementing the corrective action needed.* The automotive additions require *the use of error-proofing in the corrective action process and for cycle times to be minimized when dealing with parts rejects subsequent to delivery.*

What does this mean?

The action needed is the action that will eliminate the cause of the nonconformity and therefore prevent its recurrence. You would think that after many years eliminating causes on nonconformity, there would be no nonconformities left, but you would be wrong, primarily because most of what purports to be corrective action is little more than remedial action. There are countless corrective action procedures being implemented that do not get close to eliminating the cause of nonconformity. They focus on the immediate cause and not the root cause.

Why is this necessary?

This requirement responds to the Continual Improvement Principle.

Getting at the root of the problem is crucial to corrective action. Action on the immediate cause is only a palliative – a temporary measure. Fixing the immediate cause will result in another nonconformity eventually appearing somewhere else. It is also important to minimize the time taken to isolate and eliminated the root cause in order to minimize the impact on the customer of further deliveries. While you are contemplating the cause of the nonconformity, batches of nonconforming product might be on their way to already dissatisfied customers.

How is this implemented?

It is important to distinguish between four separate actions when dealing with nonconformity:

1   Action to remove the specific nonconformity in the nonconforming item (this is addressed by Clause 8.3 covering the control of nonconforming product).
2   Action to discover other occurrences of the nonconformity (this should also be covered by the provisions to meet Clause 8.3 but is also addressed by Clause 8.5.2.3 under the heading *Corrective action impact*).
3   Action to prevent recurrence in the short term (this is the local action taken on the immediate cause and often referred to as "containment action").
4   Action to prevent recurrence in the long term (this is the action taken on the root cause).

At the time you review a nonconforming product you should consider whether other similar products could be affected. Some of these products might already exist; others may be in the process of being produced and others in the process of being designed. For example, a nonconformity might be that a component was fitted the wrong way round:

1   The first action is to remove the component and fit a new one the right way round.
2   The second action is to search for other assemblies and replace the component. Some regard this as remedial action because it is searching for like items. Others regard it as corrective action because it is preventing a recurrence of a problem in the same and similar products. There are also some people who regard it as preventive action because although the nonconformity has occurred in a specific product or process, it is only a potential nonconformity for other products or processes.
3   The third action is to display warning notices to alert operators and show them how to identify correct component orientation.
4   The fourth action is to install error-proofing measures and update the generic process FMEA.

When corrective actions require interdepartmental action, it may be necessary to set up a corrective action team to introduce the changes. Each target area should be designated to a person with responsibility in that area and who reports to a team leader. In this way the task becomes a project with a project manager equipped with the authority to make the changes through the department representatives.

Take care not to degrade other processes by your actions. The corrective action plan should detail the action to be taken to eliminate the cause and the date by which a specified reduction in nonconformity is to be achieved. You should also monitor the reduction therefore the appropriate data collection measures need to be in place to gather the data at a rate commensurate with the production schedule. Monthly analysis may be too infrequent. Analysis by shift may be more appropriate.

Your management system needs to accommodate various corrective action strategies, from simple intradepartmental analysis with solutions that affect only one area, procedure, process or product, to projects that involve many departments, occasionally including suppliers and customers. Your corrective action procedures need to address these situations in order that when the time comes you are adequately equipped to respond promptly.

Recording results of actions taken (8.5.2e)

The standard requires *the documented procedure for corrective action to define requirements for recording results of action taken.*

What does this mean?
If taken literally this requirement means that one need only record whether the action had the desired effect. But we are interested in not only doing what the standard prescribes but also doing what is necessary for effective process management. Records of all the intentions and actions relative to the elimination of nonconformity should be generated.

Why is this necessary?
This requirement responds to the Factual Approach Principle.

Records are necessary in this case to chart the trail from nonconformity to root cause and back again to effective action. This is so that you can check through the logic of the analysis and subsequent actions and verify that the planned action was implemented.

How is this implemented?
Through the corrective action process there are several things that should be recorded:

- the results of Pareto analysis,
- the likely causes,
- the root cause,
- the criteria for determining severity or priority,
- the tests conducted to validate the root cause,
- the actions proposed to eliminate the cause,
- the actions taken,
- the results of the actions taken.

In order for it to be possible to verify the actions taken, records need exist to provide traceability. For example, if your Corrective Action Report (e.g. CAR023) indicates that procedure XYZ requires a change, a reference to the Document Change Request (e.g. DCR134) initiating a change to procedure XYZ will provide the necessary link. The Change Request can reference the Corrective Action Report as the reason for change.

If you don't use formal change requests, the Amendment Instructions can cross-reference the Corrective Action Report. Alternatively, if your procedures carry a change record, the reason for change can be added. There are several methods to choose from, but whatever the method you will need some means of tracking the implementation of corrective actions. This use of forms illustrates one of the many advantages of form serial numbers.

Reviewing corrective action taken (8.5.2f)

The standard requires the documented procedure for corrective action *to define requirements for reviewing corrective action taken.*

What does this mean?
The review of corrective actions means establishing that the actions have been effective in eliminating the cause of the nonconformity.

Why is this necessary?
This requirement responds to the Factual Approach Principle.

Every process should include verification and review stages not only to confirm that the required actions have been taken but also that the desired results have been achieved. It is only after a reasonable time has elapsed without a recurrence of a particular nonconformity that you can be sure that the corrective action has been effective.

How is this implemented?
This requirement implies four separate actions:

1   A review to establish what actions were taken.
2   An assessment to determine whether the actions were those required to be taken.
3   An evaluation of whether the actions were performed in the best possible way.
4   An investigation to determine whether the nonconformity has recurred.

The effectiveness of some actions can be verified at the time they are taken but quite often the effectiveness can only be checked after a considerable lapse of time. Remember it took an analysis to detect the nonconformity therefore it may take further analysis to detect that the nonconformity has been eliminated. In such cases the Corrective Action Report should indicate when the checks for effectiveness are to be carried out and provision made for indicating that the corrective action has or has not been effective.

Some corrective actions may be multidimensional in that they may require training, changes to procedures, changes to specifications, changes in the organization, changes to equipment and processes; in fact so many changes that the corrective action becomes more like an improvement programme. Checking the effectiveness becomes a test of the system carried out over many months. Removing the old controls completely and committing yourselves to an untested solution may be disastrous therefore it is often prudent to leave the existing controls in place until your solution has been proven to be effective.

The nonconformity data should be collected and quantified using one of the seven quality tools, preferably the Pareto analysis. You can then devise a plan to reduce the 20% of causes that account for 80% of the nonconformities.

Preventive action (8.5.3)

Eliminating the cause of potential nonconformities (8.5.3)

The standard requires the organization to *determine action to eliminate the causes of potential nonconformities in order to prevent their occurrence and for such actions to be appropriate to the effects of the potential problems.*

What does this mean?

The difference between corrective and preventive action was previously addressed under corrective action. When actual problems don't exist but there is a possibility of failure, the action of preventing the occurrence of nonconformity (or any problem for that matter) is a preventive action. Potential nonconformities might arise due to the inherent nature of the product or the process or its design, production, installation or operation. The way a product has been designed or the way it is intended to produce, install or maintain a product might lead to a failure either during production, installation or maintenance or a failure when in service.

> One cannot plan for events about which one has no knowledge.

Action to eliminate a potential nonconformity could be any action taken to anticipate failure in products or processes and remove the cause. If one cannot conceive of a mode of failure no action could possibly be taken. This is often regarded as operating within the "state of the art". There may be a time in the future when the risk of failure is known but one cannot plan for events about which one has no knowledge. This means of course that if there is evidence in the organization or in the public domain of a mode of failure that could occur with the organization's products or processes, preventive action needs to be taken. The grey areas arise where such evidence remains unproven. Many organizations are reluctant to take action on the basis of opinion especially when the action may be costly. It is an area where the organization's values can be tested and where moral or ethical judgements may need to be made. In the environmental arena there is growing support for the precautionary principle (see panel) but you need to apply this principle with caution. A vivid imagination could result in very costly measures being taken to prevent an event that has never happened and is unlikely to happen;

but having said that, we didn't think that aluminium would burn until the Royal Navy's 3322 tonnes LSL, Sir Galahad was destroyed by fire in 1982 in the Falklands.

Even if one does not have a crystal ball to look into the future, the present can provide a wealth of information. Knowledge of current performance can provide the basis for predicting future performance (except for the value of investments of course!). Monitoring performance and observing an undesirable trend, then taking action before a failure occurs is taking *preventive action* even though there is no evidence that the occurrence could be imminent.

> **Precautionary principle**
>
> Not having the evidence that something might be a problem is not a reason for not taking action as if it were a problem.

Risk analysis, failure modes analysis, hazards analysis, stress analysis, reliability predictions or any similar type of analysis performed to identify design weaknesses is a preventive action *if* action is taken as a result of the analysis. Performing the analysis and doing nothing with the results is negligent. Preventive action can also be any action taken to ensure success. Research, planning, training, preparation, organizing and resourcing are all activities which if done well will prevent nonconformities arising.

Actions appropriate to the effects of the potential problems means that the probability of occurrence, the significance of the effects and the certainty of eliminating the cause should be taken into account when deciding on the action required.

It is clearly very odd that the authors of ISO/TS 16949 added no recommendations or requirements to this final clause of ISO 9001 when it is such an important one. Although FMEAs are addressed in other sections of ISO/TS 16949, the authors do not appear to hold the view that FMEA, planning, training and other provisions are *preventive actions*. This might stem from the perception that preventive action is the action taken to prevent the occurrence of a nonconformity that has already occurred elsewhere; i.e. having more to do with looking for other places nonconformity might occur than with original occurrences. But there is evidence within ISO/TS 16949 that contradicts this view. Clause 8.5.2.3 addresses corrective action impact or applying the controls to eliminate a nonconformity to other similar products and processes. So it remains an enigma.

Why is this necessary?
This requirement responds to the Continual Improvement Principle.

Regardless of the attitude of people towards quality, they all desire to be successful. No one really wants to fail but without foresight brought about by instinct, experience, diligence or luck, they will invariably fail. Organizations cannot put their results at risk and expect success to be achieved by chance. They have to take preventive action of the type described above if they are to succeed.

How is this implemented?

The other requirements in this clause identify most of the steps needed to eliminate the cause of a potential nonconformity. However, there are one or two stages that have been omitted. A more complete list of steps is given in Table 8.4.

These will be addressed later in this chapter.

It makes no sense to have a procedure with the title *Preventive Action Procedure* because the above sequence of actions does not occur in isolation. They are part of and embedded within other actions and processes. The saying "Look before you leap" is a preventive action but developed into a habit by the time we are teenagers. We sometimes forget or are distracted and don't look before we leap but it is not something where we stop and say to ourselves, "Now what's the next step – oh yes; I must now look before I leap" we just do it without a conscious thought. The first 12 actions in Table 8.4 might form the basis an FMEA Procedure with the other 12 actions being implemented by other processes. For example, the records will be generated from planned verification activities, not special preventive action activities. The diagnostic and implementation teams will not be special teams but the normal design and production staff acting in their normal roles. There may be a need on occasions for a special task force to resolve a particularly difficult problem but most of the time, preventive action will be performed as a routine part of your job.

Preventive action provisions may therefore be embodied in the following types of procedures:

- Business planning procedures
- Design planning procedures
- Production planning procedures
- Resource planning procedures
- Risk assessment procedures
- Hazard analysis procedures
- Training procedures
- Performance analysis procedures
- Design review procedures
- Design analysis procedures (reliability, safety, maintainability, etc.)
- Supplier/subcontractor performance review procedures
- Management review procedures.

The actions in Table 8.4 could be included in a general planning guide. It would be foolish to make them policies because they describe a general methodology and there will be occasions when not all apply.

Table 8.4 *Steps in the preventive action process*

Step	Action	Clause 8.5.3 requirement
1	Determine the objectives of the product, process, task or activity	Determining potential nonconformities and their causes
2	Organize a diagnostic team	
3	Perform an analysis to determine the factors critical to the achievement of these objectives	
4	Determine how the factors might act to adversely affect the product, process, task or activity (the mode of failure)	
5	Determine the potential effect of such condition on the achievement of the objectives	
6	Determine the severity of the effect on meeting the objectives	
7	Assess the probability of this condition occurring	
8	Postulate causes and test theories	
9	Determine the root cause of potential nonconformity	
10	Identify the provisions currently in place that will prevent this adverse condition occurring or detect it before it has a detrimental effect on performance	Evaluating the action needed
11	Assess the probability that these provisions will prevent the occurrence of this condition or of detecting it before it has a detrimental effect on performance	
12	Determine any additional action needed to prevent the occurrence of the potential nonconformity	
13	Organize an implementation team	Implementing the action needed
14	Create or choose the conditions which will ensure effective implementation	
15	Implement the agreed action	
16	Record the results of all the analysis	Records of results
17	Record the causes of potential nonconformity	
18	Record the criteria for determining severity or priority	
19	Record the proposed actions to be taken	
20	Record the actions actually taken	
21	Determine whether the actions were those required to be taken	Reviewing preventive actions
22	Determine whether the actions were performed in the best possible way	
23	Determine whether the nonconformity has occurred	
24	If nonconformity has occurred, undertake corrective action and review the preventive action methods	

Determining potential nonconformities (8.5.3a)

The standard requires *a documented procedure that defines requirements for determining potential nonconformities and their causes.*

What does this mean?

Potential nonconformities are those that have not occurred, therefore the determination of potential nonconformities is a quest to discover inherent characteristics of products and processes that if not changed will eventually result in actual nonconformities. As a nonconformity is non-fulfilment of a requirement, it follows that any requirement placed on or within the organization that may not be fulfilled provides the opportunity for a potential nonconformity. As indicated above, the determination of potential nonconformities involves the analysis and evaluation of risk. Once risks are known and the effect on the product, process or organization established, action can be taken to eliminate, reduce or control these risks.

Why is this necessary?

This requirement responds to the Continual Improvement Principle.

The prevention of potential nonconformities cannot be relied on to occur by chance – it has come about by systematic analysis of products and processes and their interrelationships both inside and outside the organization.

How is this implemented?

As indicated in Table 8.4, there are eight steps to determining potential nonconformities and their causes. Each step will now be addressed briefly.

Determining objectives, requirements and constraints

As nonconformity is non-fulfilment of a requirement, the first step is to determine what those requirements are. In some cases they will be product requirements, in others they will be process objectives or at a higher level the corporate objectives or quality objectives. They may also be requirements that act as constraints on other activities. For example, environmental requirements apply only if the operations you undertake impact the environment. If the operations do impact the environment, the requirements constrain what you can or cannot do.

It is necessary when considering the requirements for preventive action to avoid limiting your imagination to products because the potential for failure is just as present in the organization and its processes as in products – in fact it is these very weaknesses that are likely to be the cause of potential product nonconformities. Such objectives therefore will include the following:

- Corporate objectives (Clause 5.4.1).
- Product requirements (Clause 7.2.1).
- Process objectives (Clause 4.1c).

As the product will comprise units, components, materials, etc., each of these will be defined by requirements. As each process will comprise subprocesses, tasks and activities, each of these will be defined by objectives and input or entry requirements.

Organizing diagnostic teams
As indicated previously, specially organized diagnostic teams may not be needed because the prevention of nonconformity is the job of the product or process design team. It is usually a one-off activity. Once the provisions are put in place to prevent nonconformity, the team moves on to other things. At the corporate level, a team of senior managers or a research team may be needed to analysis the organization as a whole to discover risks to achievement of its objectives. At the subprocess and task level a team approach may also be necessary but at the activity level, it is the individual who should assess the risks, anticipate problems and either put in place provisions to prevent failure or take precautions.

Analysis of critical success factors
With every organization, process or product there are some critical factors on which their success depends. The success of electronic equipment is dependent on appearance, function and reliability. The success of an automobile depends on appearance, function, safety, reliability and maintainability. In most cases these factors will be defined in the product specification and will be the functions that the product is required to perform. With processes, success may depend on throughput, resource consumption, traceability and/or response. With the organization, success may depend on market intelligence, retaining competent people, short product development timescales, the quality of conformity or service standards – these will be the organization's objectives.

Answering the key question "What affects our ability to achieve our objectives?" reveals the critical success factors.

Failure modes
By taking each factor, one at a time and asking one or more of the following questions you develop a series of failure modes – the outward appearance of a specific failure effect:[11]

- How might this part or process fail to meet the requirements?
- What could happen which would adversely affect performance?
- What would an interested party consider to be unacceptable?

Previous experience is a good starting point for determining what could go wrong but often these failures only arise under certain conditions. When a product is stressed by being subject to extreme environmental conditions, it may fail. When a process is over-loaded or under-resourced or the operators are put under pressure, certain failures might occur. This can be presented as a Fault Tree diagram that describes the combination of events leading to a defined product/process/organization failure. For a description of this analysis see Smith (1997)[12].

Failure effects

For each failure mode, determine the effect of the failure through the hierarchy to the end product, the system and the external interfaces. The effect should be described in terms of what the customer might experience or the impact on overall organizational performance. A component may distort under load resulting in a failure of the assembly

Table 8.5 *Failure severity ranking*

Effect	Severity	Ranking
Hazardous without warning	Very-high-severity ranking when a potential failure mode affects safe vehicle operation and/or involves non-compliance with government regulation without warning.	10
Hazardous with warning	Very-high-severity ranking when a potential failure mode affects safe vehicle operation and/or involves non-compliance with government regulation with warning.	9
Very high	Vehicle/item inoperable, with loss of primary function.	8
High	Vehicle/item operable, but at reduced level of performance. Customer dissatisfied.	7
Moderate	Vehicle/item operable, but comfort/convenience item(s) inoperable. Customer experiences discomfort.	6
Low	Vehicle/item operable, but comfort/convenience item(s) operate at reduced level of performance. Customer experiences some dissatisfaction.	5
Very low	Fit and finish/squeak and rattle item does not conform. Defect noticed by most customers.	4
Minor	Fit and finish/squeak and rattle item does not conform. Defect noticed by *average* customers.	3
Very Minor	Fit and finish/squeak and rattle item does not conform. Defect noticed by discriminating customers.	2
None	No effect.	1

of which it is a part and consequently cause customer dissatisfaction. A process may deliver output late that in turn injects delays into other processes, which ultimately causes the organization to lose a customer. The omission of safety checks could endanger operators and subsequently put the line out of service or in breach of government regulations. The omission of records could leave a process downstream without necessary input data. This in turn could result in decisions being made without the full facts being available thus impacting work priorities that divert effort away from key projects.

Severity
Once the effects of failure have been identified, it does not follow that you have to act on all of them. Some will have more impact than others and the next step is therefore to assess the severity of the effect. A convention for ranking failure severity is shown in Table 8.5.

Probability of occurrence
A potential nonconformity is one that could occur but the probability of occurrence might be as low as 1 in 1,000,000 or as high as 1 in 2. With electronic components, standard failure rate data can be used to predict the probability of failure. With other failure effects you will have to rely on past experience and this is where the records from previous projects are useful. Simply asking the question, "When did we last experience this type of problem?", might bring forth useful information.

A convention for ranking the probability of occurrence is shown in Table 8.6.

Table 8.6 *Failure occurrence ranking*

Rank	C_{pk}	Failure rate	Probability of failure
10	0.33	>1 in 2	Very high: failure almost inevitable
9	0.33	1 in 3	
8	0.51	1 in 8	High: repeated failures
7	0.67	1 in 20	
6	0.83	1 in 80	Moderate: occasional failures
5	1.00	1 in 400	
4	1.17	1 in 2000	
3	1.33	1 in 15,000	Low: relatively few failures
2	1.50	1 in 150,000	
1	1.67	$\leq$1 in 1,500,000	Remote: failure is unlikely

Determining cause

The diagnostic effort needed to determine the potential cause of the identified mode of failure can be considerable and therefore it is prudent to determine the probability of occurrence first so that the diagnostic effort can be focused on the failures with highest probability of occurrence.

Here once again the "5 Why's" technique is useful to get at the root cause. In some cases it may be necessary to experiment using Design of Experiments (DOE) technique in order to determine which causes are the major contributors.

Evaluating the need for action (8.5.3b)

The standard requires *a documented procedure that defines requirements for evaluating the need for action to prevent occurrence of nonconformities.*

What does this mean?

The likelihood that an event might occur is not a command to take action for provisions may already be in place to prevent its occurrence, reduce or control its effects. The requirement therefore means that in evaluating the need for action, the existing provisions should be evaluated and on this basis, requirements for action to prevent the occurrence of nonconformity should be determined.

Why is this necessary?

This requirement responds to the Continual Improvement Principle.

The existing provisions may not adequately prevent the occurrence of nonconformity and therefore it is necessary to determine the additional actions that are needed. Events are not always mutually exclusive and therefore provisions designed to eliminate, reduce or control one failure mode may well eliminate, reduce or control others.

How is this implemented?

Adequacy of current provisions

All existing processes should include provisions for verifying output and monitoring progress even if they are informal. However, if the processes have evolved over a long period it is also likely that there will be many more controls in place. Every requirement of ISO/TS 16949 will, if implemented, impose control over the process. Such controls should have been installed to prevent failure but many were often installed just to meet the requirements of the standard and get the *badge on the wall.* If you now re-examine these controls from the perspective of establishing the failure modes they prevent, you will at least give justification to those that serve a useful purpose and provide an action list for those that don't.

Table 8.7 *Contrasting prevention and detection measures*

Prevention measures	Detection measures
Planning	Material testing
Training	Component inspection
Safety factors	Assembly testing
Component derating	Prototype testing
Component redundancy	Receipt inspection
Product identification	Design review
Error-proofing	Peer review
Warning notices	Alarms
	Alerts

The inherent characteristics of a product or process that will reduce the probability of occurrence should be identified. For a product these might be safety factors, component redundancy, error-proofing or specifying high-reliability components. For processes these might be error-proofing, inspections, tests, audits, identification techniques, skill training, peer reviews, etc.

Existing controls can be classified as those that prevent occurrence and those that detect occurrence. Table 8.7 illustrates the difference. There are two types of detection measures:

1 those that detect a failure and lead to a permanent change;
2 those that detect a failure and lead to a restoration action, such as to reset the product or process, or to replenish consumables.

Probability of detection
You can go a step further and rank the probability that the existing provisions will detect the failure. The ranking could range from "Certainty of detection" to "Absolute uncertainty of detection" with variations in between. Examine the existing provisions to see whether there are any alarms, alerts or reporting arrangements that would bring potential problems to the attention of management.

What may appear trivial on a case-by-case basis may well be significant when taken over a longer period or a larger population. Determining this deterioration requires some detective work that focuses on processes and not on specific products. Managers have a habit of reacting to events particularly if they are serious nonconformities in the

Table 8.8 *Ranking the probability of detection*

Detection	Criteria	Rank
Absolute uncertainty	Design control will not and/or cannot detect a potential cause/mechanism and subsequent failure mode; or there is no design control.	10
Very remote	Very remote chance the design control will detect a potential cause/mechanism and subsequent failure mode.	9
Remote	Remote chance the design control will detect a potential cause/mechanism and subsequent failure mode.	8
Very low	Very low chance the design control will detect a potential cause/mechanism and subsequent failure mode.	7
Low	Low chance the design control will detect a potential cause/mechanism and subsequent failure mode.	6
Moderate	Moderate chance the design control will detect a potential cause/mechanism and subsequent failure mode.	5
Moderately high	Moderately high chance the design control will detect a potential cause/mechanism and subsequent failure mode.	4
High	High chance the design control will detect a potential cause/mechanism and subsequent failure mode.	3
Very high	Very high chance the design control will detect a potential cause/mechanism and subsequent failure mode.	2
Almost certain	Design controls will almost certainly detect a potential cause/mechanism and subsequent failure mode.	1

form of a customer complaint. We are all poor at perceiving the underlying trends that occur daily and gradually eat away at our profits. If there is no means of alerting management to these trends clearly something is missing.

A convention for ranking the probability of detection is shown in Table 8.8.

There are some problems with this method as it is subjective. It comes down to one person's guess as to the likelihood of detection. Where a product is designed with sensors, the chances to detection can be calculated more precisely using component failure rates. But with business and work processes it is often up to the personnel running the process. For example, currently there is no process that will detect swarf inside an oil channel unless an endoscope can travel every channel and into every crevice. We rely on cleaning processes and manual observation of particles on a filter to determine whether the channels are free of swarf.

Risk priority number

The risk priority number (RPN) is the product of three numbers: Severity (S), Occurrence (O) and Detection (D). The ideal situation is where the product of $S \times O \times D = 1$. Any severity ranking of 10 should be eliminated by design. Any result above 50 should require action and therefore the priority needs to be given to those failures ranked above 50. However, RPN is a qualitative assessment of risk and takes no account of life cycle cost.

Financial impact

There is merit in extending the FMEA to include financial impact because there is not a linear relationship between RPN and life cycle cost. There might be some failures that are less costly to prevent or detect in process than to design out of the product. There might also be other failures that are more costly to prevent or detect in process due to the initial cost of equipment and recurring maintenance costs. A third scenario is where one solution to the reduction in RPN has a lower impact on product cost but higher impact on life cycle cost.[13] Traditional FMEA identifies a failure mode and its effect and postulates a cause and its probability of occurrence. The weakness with this approach is that it does not permit the investigation of several cause-and-effect chains each having a different probability of occurrence and thus a different solution in terms of cost.

Determining and implementing preventive action (8.5.3c)

The standard requires *a documented procedure that defines requirements for determining and implementing action needed to prevent occurrence of nonconformities.*

What does this mean?

Action to prevent a potential nonconformity may be in the form of a redesign of a product or process, the introduction of new routines, precautions, procedures, techniques or methods or involve changing the behaviour of personnel including management. The work environment may be the cause of some problems, and changing it should not be ruled out.

Why is this necessary?

This requirement responds to the Continual Improvement Principle.

The most cost effective action one can take in any organization is an action designed to prevent problems from occurring. It is always cheaper to do a job right the first time rather than do it over.[14] Preventive action therefore saves money even though there is a price to pay for the discovery of potential nonconformities. Once the analysis has been performed a few times, a pattern will begin to emerge where provisions made to handle one situation will prove suitable for other situations and so it will get easier until the analysis almost becomes a habit.

How is this implemented?

The action necessary to eliminate, reduce or control the effects of a potential nonconformity may be as simple as applying existing techniques or methods to a new product or process. In other cases it might involve designing new techniques and methods – something that may require additional resources and a development team. If the solution requires the involvement of more than one function, the formation of a multidisciplinary team may be necessary. This is not the same as a multifunctional team where representatives from each function meet to discuss a problem then go back to their departments and get on with what they were doing. A multidisciplinary team comprises people of different disciplines, brought together to pool their skills and knowledge. They may all be from the same function but the focus is on getting the problem solved, not playing departmental politics.

The steps you need to take to deal with specific problems will vary depending on the nature of the problem. The part that can be proceduralized is the planning process for determining the preventive action needed. A typical process may be as follows:

- Devise a strategy for eliminating the cause together with alternative strategies, their limitations and consequences.
- Gain agreement on the strategy.
- Prepare an improvement plan which if implemented would eliminate the potential problem and not cause any others.
- Prepare a timetable and estimate resources for implementing the plan.
- Gain agreement of the improvement plan, timetable and resources before going ahead.
- Calculate the Potential Risk Priority Number for the failure mode.

Some plans may be very simple and require no more than an instruction to implement an existing procedure. Others may be more involved and require additional resources. By incorporating the actions into a formal improvement plan you are seen to operate a coherent and co-ordinated improvement strategy rather than a random and unguided strategy. While those on the firing line are best equipped to notice the trends, any preventive action should be co-ordinated in order that the company's resources are targeted at the problems that are most significant.

Recalculating the RPN

When the improvement plan has been devised the RPN needs to be recalculated. Redesign of the product or the process may well reduce the probability of occurrence but the severity of failure might remain the same. By increasing the probability of detection the RPN may be reduced well below the original RPN indicating the planned action will bring about the necessary improvement. However, the important thing is that the planned action is taken and provision made with the associated documentation to indicate as much.

Error-proofing

Error-proofing is a means to prevent the manufacture or assembly of nonconforming product (see Appendix A). Errors are unintentional unplanned events in the design, planning or production of a product or delivery of a service. Machines don't run forever without attention. We occasionally forget things and we can either make machines and actions error-proof or we can provide signals to alert us of unintentional events that are about to happen, are happening or have happened.

Error-proofing can be accomplished by product design features in order that the possibility of incorrect assembly, operation or handling is avoided. In such cases the requirements for error-proofing need to form part of the design input requirements for the part. The Design FMEA should be analysed to reveal features that present a certain risk that can be contained by redesign with error-proofing features.

Error-proofing can also be accomplished by process design features such as sensors to check the set-up before processing, stop the machine if an abnormal situation arises (jammed mechanism or defect part), trigger audible signals to remind operators to do various things and stop the machine when the correct number of items have been produced (see Autonomation in Appendix A). However, signals to operators are not exactly error-proof, only a mechanism that prevents operations commencing until the right conditions have been set is proof against errors. In cases where computer data entry routines are used, error-proofing can be built into the software such that the operator cannot bypass a stage.

Recording the results of preventive action (8.5.3d)

The standard requires *a documented procedure that defines requirements for recoding of results of action taken to prevent occurrence of nonconformities.*

What does this mean?
If taken literally this requirement means that one need only record whether the action had the desired effect and as we are addressing potential nonconformities, a record of the absence of nonconformities would seem to address this issue. However, in addition (and not in place of), other records are necessary for effective process management. Records of all the intentions and actions relative to the elimination of potential nonconformity should be generated.

Why is this necessary?
This requirement responds to the Factual Approach Principle.

Records of the actions taken en route to discover and eliminate potential nonconformities serve a number of important uses:

- They show due diligence and consideration to matters that affect customer satisfaction and compliance with regulations.
- They can be used as evidence in any prosecution against the organization.
- They provide a basis for comparison of actual nonconformity against predicted nonconformity and therefore a means to improve product and process design techniques.

How is this implemented?

Through the preventive action process there are several things that should be recorded:

- the objectives or requirements of the product, process or organization;
- the critical success factors;
- the modes of failure, their cause and effect, their severity and probability of occurrence (an example of a Design FMEA is shown in Figure 8.7);
- the criteria for determining severity;
- current provisions to detect nonconformity;
- the probability that current provisions will detect the nonconformity;
- the actions proposed to eliminate the cause;
- the actions taken;
- the results of an analysis of nonconformity data showing the effectiveness of actions taken.

Reviewing preventive action (8.5.3e)

The standard requires *a documented procedure that defines requirements for reviewing actions taken to prevent occurrence of nonconformities.*

What does this mean?

The review of preventive actions means establishing that the actions have been effective in preventing the occurrence of the nonconformity.

Why is this necessary?

This requirement responds to the Process Approach Principle.

A lot of effort goes into preventing problems and therefore it is necessary to periodically review results to establish whether this effort is being effectively applied. Is the effort focusing on the right things or does it repeatedly fail to prevent any significant problems occurring?

FAILURE MODE and EFFECTS ANALYSIS – DESIGN/PROCESS

Subsystem passenger airbag system	Part or assembly process airbag	Specification or revision GL-50-265-1789 G	FMEA Ref./Issue DF2986-B
Function crash protection	Core team G Hunt, R Tuesday, J May, S Samadi	Design responsibility R Tuesday	Date 10/20/2000 Page 1 of 3
		Supplier ASSL Project Grampian	Prepared by G Hunt

No.	Item function or process stage or operation	Potential failure mode	Effect of failure	Cause of failure	Current controls	Current status O	S	D	RPN	Recommended action	Actionee	Action taken	Revised status O	S	D	RPN
1	Inflate airbag	Bag does not open on impact	Passenger injured	Malfunction of sensor	Sensor failure warning light	2	8	6	96	Add redundant sensor in parallel	Designer	Second sensor circuitry added 7/01/2001	1	8	2	16
2	Restrain passenger	Occupant unable to withstand inflation pressure	Injure lightweight occupant	Occupant not wearing seat belt	Seat belt warning light	4	8	10	320	Install switch to deactivate airbag unless seat belt is fastened	Designer	Switch and circuitry added 7/01/2001	1	8	3	24
			Bruise medium weight occupant	Force regulator malfunction	Regulator life testing	2	3	3	18	None	None	None				0

Figure 8.7 *Typical Design FMEA*

How is this implemented?
This requirement implies four separate actions:

1   A review to establish what actions were taken.
2   An assessment to determine whether the actions were those required to be taken.
3   An evaluation of whether the actions were performed in the best possible way.
4   An investigation to determine whether the nonconformity has occurred.

The first action is to trace forward from the failure analysis in order to locate evidence that the planned action was taken. The action taken may not be the same as planned simply because a better solution emerged and the FMEA was not updated (a common weakness but not drastic). The review should cause the FMEA to be updated but a more important issue is whether the *better solution* had the desired effect. An analysis of performance over a set time interval may reveal the evidence. It will show whether there are nonconformities occurring and if so, the preventive action has not been 100% effective. However, a study of actual nonconformities will reveal whether:

*   the root cause matches that in the FMEA;
*   the nonconformity is one that had been anticipated;
*   the solution merely created another problem.

The fact that nonconformities occur should not necessarily be cause for despair. If a process is very mature and the organization stable, there perhaps should be no nonconformities. But life is never thus. There is always change and some of it we can predict and some we cannot. A lot of what is needed to prevent potential nonconformity is to do with imagination, knowledge and commitment. You need imagination to postulate the modes of failure, knowledge to confirm your suspicions and isolate the causes and commitment to do something about it, especially where cost is involved and it is uncertain as to whether nonconformity will occur. There are costs versus benefits and often the benefits are external to the organization rather than internal such as protection of the environment and safety of people. The management has to balance the costs and make value judgements which if they share good values, the decisions will always fall in favour of the external parties (employees are external parties in this situation).

Summary

In this chapter we have examined the requirements contained in Section 8 of ISO/TS 16949. We have discovered that measurement is key to performance, for without measurement we have no idea how we are performing and where we need to focus our effort to improve performance. We have learnt that the measurement, analysis and improvement processes are not unique to Section 8 and have rationalized the duplication in the standard in this area. We have examined different ways of monitoring customer satisfaction and of performing internal audits and have described a method of conducting process audits that will reveal deficiencies and identify opportunities for improvement. We have learnt that the requirements for process measurement are quite challenging and may require significant change for many – extending process management into all facets of the organization, perhaps for the first time. We have examined the data analysis and continual improvement requirements, and discovered that the search for improvement opportunities will need to extend beyond product-related issues and into the boardroom – questioning policies and objectives, strategies, and goals. While the nonconformity control, corrective and preventive action requirements are not new, we have taken a fresh look at how these requirements can be implemented and described in some detail the failure modes and analysis technique for improving the robustness of both products and processes. However, once the improvement opportunities are identified, new levels of performance are within our reach because we have built change processes into the management system that ensure improvements are process oriented rather than function oriented and to ensure that the integrity of the system is maintained throughout any improvement programme.

Measurement, analysis and improvement requirements checklist

These are the topics that the requirements of ISO/TS 16949 address consecutively. Topics 112–309 appeared in Chapter 7.

Monitoring measurement, analysis and improvement process
General
310 Establishing processes necessary to demonstrate product conformity
311 Establishing processes necessary to ensure system conformity
312 Establishing processes necessary to improve system effectiveness
313 Determining monitoring, measurement and analysis methods

Identification of statistical tools
314 Determining statistical tools

Knowledge of basic statistical concepts
315 Utilizing statistical concepts

Monitoring and measurement
Customer satisfaction
316 Monitoring customer perceptions
317 Determining customer satisfaction monitoring methods

Customer satisfaction – Supplemental
318 Evaluating realization process performance
319 Determination of performance indicators
320 Demonstrating compliance with product quality and process efficiency requirements

Internal audit
321 Conducting internal audits for conformity with planned arrangements
322 Conducting internal audits for conformity with ISO 9001:2000
323 Conducting internal audits for conformity with the organization requirements
324 Determining effective implementation and maintenance of QMS
325 Planning the internal audit program
326 Defining audit criteria, scope, frequency and methods
327 Selecting auditors
328 Documenting audit procedures
329 Ensuring prompt action on audit findings
330 Following up audit actions

Quality management system audit
331 Auditing for compliance with ISO/TS 16949

Manufacturing process audit
332 Auditing manufacturing processes

Measurement, analysis and improvement – Food for thought

1 Do you manage the system as a series of processes or as a series of functions?
2 If no objectives are set for a process, will any level of performance be acceptable?
3 Only when you have put out all the fires can you improve the process.
4 Do you act on suspicion, or always confirm that a problem exists or might exist before taking action?
5 Have you discovered any root cause of a problem that cannot be traced to lack of application of one or more of the eight quality management principles?
6 Performing the analysis and doing nothing with the results is negligent.
7 Organizations cannot put their results at risk and expect success to be achieved by chance.
8 Have you re-examined existing controls from the perspective of establishing the failure modes they prevent?
9 Have you equipped every process with provisions for measuring its performance?
10 How often do you check that your objectives and targets are still relevant to the organization's goals?
11 If the questions in your customer satisfaction questionnaires were generated internally, how do you expect to obtain unbiased results?
12 When was the last time your internal audit programme found something that led to improved performance?
13 If you discontinued your internal audit programme, would anyone other than the internal or external auditors demand its reinstatement?
14 If your auditing approach has been to verify compliance with procedures, what approach are you intending to take now that the system has to enable the organization to achieve its objectives?
15 Do you consider that the system is effectively implemented if people are following the documented procedures or would you also verify that the planned results are being achieved?
16 Would you accept a box of 1000 components by simply checking one sample, if not why would you base your audit conclusions on a few unrepresentative samples?
17 Why shouldn't the manager perform the internal audits, and if you should think he or she is not competent to do so, why do you trust him or her to manage the function?
18 How do you know that each of the processes is achieving the planned results?
19 When was the last time you changed your operating methods in order to increase resource utilization?
20 Do you continue with the current level of product verification regardless of detecting no nonconformities?
21 Are you sure that those examining products or services for conformity apply the same criteria as those using them?
22 Is the data used by management to make decisions generated from the processes of the management system and if not why not?
23 Are there any data collection routines that are not triggered by a process in the management system?
24 How continuous is your continual improvement process?
25 When was the last time a problem recurred?
26 When did asking the question? What if … become a habit?

Appendix A – Glossary

The terms below are those used in this book for which alternative definitions may be provided in ISO 9000. For other terms the reader is directed to ISO 9000.

5 "Why"s These typically refers to the practice of asking, five times, why a failure has occurred in order to get to the root cause. For example:

1 Why did the machine stop? Because there was an overload and the fuse blew.
2 Why was there an overload? Because the bearing was not sufficiently lubricated.
3 Why was the bearing not lubricated sufficiently? Because the lubrication pump was not pumping.
4 Why was the lubrication pump not pumping? Because the pump shaft was worn.
5 Why was the pump shaft worn? Because there was no seal around the shaft and metal debris penetrated the gap.

6 "M"s These are the six words that are used to title the arms in a fishbone diagram. The words vary but the most commonly used "M"s are:

1 machines;
2 methods;
3 materials;
4 measurements;
5 milieu (surrounding environment);
6 manpower.

One could also add money and management.

8 "D"s A problem-solving method that is structured into eight disciplined steps. The eight basic steps are:

1 establish a team,
2 describe the problem,
3 develop interim containment,
4 define and verify root cause,
5 choose permanent corrective action,
6 implement corrective action,
7 prevent recurrence,
8 recognize and reward the contributors.

Note: Some of the terms are not consistent with ISO 9000 definitions for corrective and preventive actions.

Acceptance criteria The standard against which a comparison is made to judge conformance.

Accreditation A process by which organizations are authorized to conduct certification of conformity to prescribed standards. ·

Adequate Suitable for the purpose.

Appropriate Means suitable for its purpose or to the circumstances and required knowledge of this purpose or circumstances. Without criteria, an assessor is left to decide what *is* or *is not* appropriate based on personal experience.

Approved Something that has been confirmed as meeting the requirements.

Assessment The act of determining the extent of compliance with requirements.

Assurance Evidence (verbal or written) that gives confidence that something will or will not happen, or has or has not happened.

Audit An examination of results to verify their accuracy by someone other than the person responsible for producing them (see also ISO 9000 Clause 3.9.1).

Authority The right to take actions and make decisions.

Authorized A permit to do something or use something that may not necessarily be approved.

Autonomation Automation with the human touch. The purpose is to free equipment from the necessity of constant human attention, separate people from machines and allow workers to staff multiple operations. In Japanese the word is *Jidohka* (see also *Error-proofing*).

Benchmarking A technique for measuring an organization's products, services and operations against those of its competitors resulting in a search for the best practice that will lead to superior performance.

Calibrate To standardize the quantities of a measuring instrument.

Capability index, C_p The capability index for a stable process defined as the quotient of tolerance width and process capability where process capability is the 6σ range of a process's inherent variation.

Capability index, C_{pk} The capability index, which accounts for process centring for a stable process using the minimum upper or lower capability index.

Certification A process by which a product, process, person or organization is deemed to meet specified requirements.

Certification body See *Registrar.*

Class A group of entities having at least one attribute in common or a group of entities having the same generic purpose but different functional use.

Clause of the standard   A numbered paragraph or subsection of the standard containing one or more related requirements, such as Clause 7.2.2. *Note*: Each item in a list is also a clause.

Codes   A systematically arranged and comprehensive collection of rules, regulations or principles.

Commitment   An obligation a person or an organization undertakes to fulfil, i.e. doing what you say will do.

Common cause   Random variation caused by factors that are inherent in the system.

Competence   The ability to demonstrate *use* of education, skills and behaviours to achieve the results required for the job.

Competence-based assessment   A technique for collecting sufficient evidence that individuals can perform or behave to the specified standards in a specific role (Shirley Fletcher).

Competent   An assessment decision that confirms a person has achieved the prescribed standard of competence.

Concession   Permission granted by an acceptance authority to supply product or service that does not meet the prescribed requirements (see also ISO 9000 Clause 3.6.11).

Concurrent engineering   See *Simultaneous engineering*.

Continual improvement   A recurring activity to increase the ability to fulfil requirements (ISO 9000).

Contract   An agreement formally executed by both customer and supplier (enforceable by law) which requires performance of services or delivery of products at a cost to the customer in accordance with stated terms and conditions. Also agreed requirements between an organization and a customer transmitted by any means.

Contract loan   An item of customer-supplied property provided for use in connection with a contract that is subsequently returned to the customer.

Contractual requirements   Requirements specified in a contract.

Control   The act of preventing or regulating change in parameters, situations or conditions.

Control charts   A graphical comparison of process performance data to computed control limits drawn as limit lines on the chart.

Control methods   Particular ways of providing control which do not constrain the sequence of steps in which the methods are carried out.

Control procedure   A procedure that controls product or information as it passes through a process.

Controlled conditions Arrangements that provide control over all factors that influence the result.

Core competence A specific set of capabilities including knowledge, skills, behaviours and technology that generate performance differentials.

Corrective action Action planned or taken to stop something from recurring (see also ISO 9000 Clause 3.6.5).

Corrective maintenance Maintenance carried out after a failure has occurred and is intended to restore an item to a state in which it can perform its required function.

Criteria for workmanship Acceptance standards based on qualitative measures of performance.

Critical success factors Those factors upon which the achievement of specified objectives depend.

Cross-functional team See *Multidisciplinary team.*

Customer Organization that receives a product or service which includes purchaser, consumer, client, end user, retailer or beneficiary (ISO 9000).

Customer complaints Any adverse report (verbal or written) received by an organization from a customer.

Customer feedback Any comment on the organization's performance provided by a customer.

Customer-supplied product Hardware, software, documentation or information owned by the customer which is provided to an organization for use in connection with a contract and which is returned to the customer either incorporated in the supplies or at the end of the contract.

Data Information that is organized in a form suitable for manual or computer analysis.

Define and document To state in written form, the precise meaning, nature or characteristics of something.

Demonstrate To prove by reasoning, objective evidence, experiment or practical application.

Department A unit of an organization that may perform one or more functions. Units of organization regardless of their names are also referred to as functions (see *Function*).

Design A process of originating a conceptual solution to a requirement and expressing it in a form from which a product may be produced or a service delivered.

Design and development Design creates the conceptual solution and development transforms the solution into a fully working model (see also ISO 9000 Clause 3.4.4).

Design of experiments A technique for improving the quality of both processes and products by effectively investigating several sources of variation at the same time using statistically planned experiments.

Design review A formal documented and systematic critical study of a design by people other than the designer.

Disposition The act or manner of disposing of something.

Documented procedures Procedures that are formally laid down in a reproducible medium, such as paper or magnetic disc.

Effectiveness of the system The extent to which the system fulfils its purpose.

Embodiment loan An item of customer-supplied property provided for incorporation into product that is subsequently supplied back to the customer or a party designated by the customer.

Employee empowerment An environment in which employees are free (within defined limits) to take action to operate, maintain and improve the processes for which they are responsible using their own expertise and judgement.

EMS Environmental management system.

Ensure To make certain that something will happen.

Establish and maintain To set up an entity on a permanent basis and retain or restore it in a state in which it can fulfil its purpose or required function.

Evaluation To ascertain the relative goodness, quality or usefulness of an entity with respect to a specific purpose.

Evidence of conformity Documents which testify that an entity conforms to certain prescribed requirements.

Executive responsibility Responsibility vested in those personnel who are responsible for the whole organization's performance. Often referred to as top management.

Fagan inspection A software inspection technique in which someone other than the creator of a product examines it with the specific intent of finding errors. Software inspections were introduced in the 1970s at IBM, which pioneered their early adoption and later evolution. Michael Fagan helped develop the formal software inspection process at IBM, hence the term "Fagan inspection."

Failure mode and effects analysis (FMEA) A technique for identifying potential failure modes, and assessing existing and planned provisions to detect, contain or eliminate the occurrence of failure.

FIFO First in first out. A term used to describe a method of inventory control.

Final inspection and testing   The last inspection or test carried out by the organization before ownership passes to the customer.

Finite element analysis   A technique for modelling a complex structure.

First-party audits   Audits of a company or parts thereof by personnel employed by the company. These audits are also called internal audits.

Follow-up audit   An audit carried out following and as a direct consequence of a previous audit to determine whether agreed actions have been taken and are effective.

Force majeure   An event, circumstance or effect that cannot be reasonably anticipated or controlled.

Function   In the organizational sense, a function is a special or major activity (often unique in the organization) which is needed in order for the organization to fulfil its purpose and mission. Examples of functions are design, procurement, personnel, manufacture, marketing, maintenance, etc.

Geometric dimensioning and tolerancing   A method of dimensioning the shape of parts that provides appropriate limits and fits for their application, and facilitates manufacturability and interchangeability.

Grade   Category or rank given to entities having the same functional use but different requirements for quality; e.g. hotels are graded by star rating and automobiles are graded by model (see also ISO 9000 Clause 3.1.3).

Hoshin kanri   A Japanese term for a systems approach to goal achievement. *Hoshin* means a course, a policy, a plan or an aim. *Kanri* means administration, management, control, charge of or care for. Also known as policy deployment but it goes further than this.

IAF   International Accreditation Forum.

IAOB   International Automotive Oversight Body.

IATF   International Automotive Task Force.

Identification   The act of identifying an entity, i.e. giving it a set of characteristics by which it is recognizable as a member of a group.

Implement   To carry out a directive.

Implementation audit   An audit carried out to establish whether actual practices conform to the documented quality system. *Note*: Also referred to as a conformance audit or compliance audit.

Importance of activities in auditing   The relative importance of the contribution an activity makes to the fulfilment of an organization's objectives.

Indexing   A means of enabling information to be located.

In-process   Between the beginning and the end of a process.

Inspection   The examination of an entity to determine whether it conforms to prescribed requirements (see also ISO 9000 Clause 3.8.2).

Inspection authority   The person or organization that has been given the right to perform inspections.

Inspection, measuring and test equipment   Devices used to perform inspections, measurements and tests.

Installation   The process by which an entity is fitted into larger entity.

Intellectual property   Creations of the mind: inventions, literary and artistic works, and symbols, names, images, and designs used in commerce. Intellectual property is divided into two categories: industrial property and copyright.

Interested party   Person or group having an interest in the performance or success of an organization which includes customers, owners, employees, contractors, suppliers, investors, unions, partners or society.

ISO   International Organization for Standardization.

Issues of documents   The revision state of a document.

Jidohka   A Japanese word coined to describe machines designed to stop production whenever a defective part is produced (see also *Autonomation*).

Just-in-time   A method of lean production where the demand comes from the end of the process through to the beginning so that the only parts that are delivered are those that are needed at the time they are needed.

Kanban   A Japanese word for "tag" or "ticket" or "sign board" These tickets are used as a means of picking up and receiving the right quantity of parts required by a process thus ensuring parts are delivered just-in-time by preceding processes.

Kaizen   A Japanese word for improvement. It means continuing improvement in personal life, home life, social life and working life.

Lean production   A method of production that is demand driven (pull) rather than supply driven (push) as with mass production. There is zero waiting time, zero inventory, line balancing and reduction in process time with less space required for materials and finished product. This results in product being produced only to satisfy a demand. In lean production the person goes to the job and performs multiple tasks.

Line balance   Balancing the resources in a process or number of processes by optimizing speeds, feeds, batch size, number of workstations, operators, idle time, changeover time, cycle time and process yield.

Manage work   To manage work means to plan, organize and control the resources (personnel, financial and material), and the tasks required to achieve the objective for which the work is needed.

Management representative   The person management appoints to act on their behalf to manage the quality system.

Mass production   A method of production that is supply driven based on sales forecasts rather than firm orders. It produces large amounts of standardized products on parallel production lines that stretch from raw materials to finished product (vertical integration). In mass production the job comes to the worker who passes it on to the next worker to perform the next operation on the line.

Master list   An original list from which copies can be made.

Measurement capability   The ability of a measuring system (device, person and environment) to measure true values to the accuracy and precision required.

Measurement uncertainty   The variation observed when repeated measurements of the same parameter on the same specimen are taken with the same device.

Modifications   Entities altered or reworked to incorporate design changes.

Monitoring   To check periodically and systematically. It does not imply that any action will be taken.

Motivation   An inner mental state that prompts a direction, intensity and persistence in behaviour.

Multidisciplinary team   A team comprising representatives from various functions or departments in an organization, formed to execute a project on behalf of that organization.

Nationally recognized standards   Standards of measure that have been authenticated by a national body.

Nature of change   The intrinsic characteristics of the change (what has changed and why).

Objective   A result is to be achieved usually by a given time.

Objective evidence   Information that can be proven true based on facts obtained through observation, measurement, test or other means (see also ISO 9000 Clause 3.8.1).

Obsolete documents   Documents that are no longer required for operational use. They may be useful as historic documents.

OEM   Original equipment manufacturer.

Operating procedure   A procedure that describes how specific tasks are to be performed.

Organizational goals Where the organization desires to be, in markets, in innovation, in social and environmental matters, in competition and in financial health.

Organizational interfaces The boundary at which organizations meet and affect each other expressed by the passage of information, people, equipment, materials and the agreement to operational conditions.

Performance index, P_{pk} The performance index, which accounts for process centring and is defined as the minimum of the upper or lower specification limit minus the average value divided by 3σ.

Plan Provisions made to achieve an objective.

Planned arrangements All the arrangements made by the organization to achieve the customer's requirements. They include the documented policies and procedures, and the documents derived from such policies and procedures.

Poka-yoke Japanese term that means "mistake-proofing", a concept introduced by Shigeo Shingo to Toyota in 1961. It is a device that prevents incorrect parts from being made or assembled, or prevents correct parts being assembled incorrectly. Previously the term *baka-yoke* was used but as this means foolproofing and is rather offensive it was discontinued. Even mistake-proofing has evolved into "error proofing" to avoid the impolite implications. Error-proofing is one of the two pillars of the Toyota Production System (TPS).

Policy A guide to thinking, action and decision.

Positive recall A means of recovering an entity by giving it a unique identity.

Positively identified An identification given to an entity for a specific purpose which is both unique and readily visible.

Potential nonconformity A situation that if left alone will in time result in nonconformity.

Predictive maintenance Work scheduled to monitor machine condition, predict pending failure and make repairs on an as-needed basis.

Pre-launch A phase in the development of a product between design validation and full production (sometimes called pre-production) during which the production processes are validated.

Prevent To stop something from occurring by a deliberate planned action.

Preventive action Action proposed or taken to stop something from occurring (see also ISO 9000 Clause 3.6.4).

Preventive maintenance Maintenance carried out at predetermined intervals to reduce the probability of failure or performance degradation; e.g. replacing oil filters at defined intervals.

Procedure A sequence of steps to execute a routine activity (see also ISO 9000 Clause 3.4.5).

Process   A set of interrelated tasks, behaviours and resources that achieves a result (see also ISO 9000 Clause 3.4.1).

Process capability   The inherent ability of a process to reproduce its results consistently within preset limits during multiple cycles of operation.

Process description   A set of information that describes the characteristics of a process in terms of its purpose, objectives, design features, inputs, activities, resources, behaviours, outputs, constraints and the measurements undertaken to obtain data with which to manage the process.

Process parameters   Those variables, boundaries or constants of a process that restrict or determine the results.

Product realization   All those processes and resources necessary to transform a set of requirements into a product or service that fulfils the requirements.

Product   Anything produced by human effort, natural or man-made processes. Result of activities or processes (ISO 9000-2).

Production   The creation of products.

Proprietary designs   Designs exclusively owned by the organization and not sponsored by an external customer.

Prototype   A model of a design that is both physically and functionally representative of the design standard for production, and used to verify and validate the design.

Purchaser   One who buys from another.

Purchasing documents   Documents that contain the organization's purchasing requirements.

Qualification   Determination by a series of tests and examinations of a product, related documents and processes that the product meets all the specified performance capability requirements.

Qualified personnel   Personnel who have been judged by an accredited body as having the necessary ability to carry out particular tasks.

Quality   The degree to which a set of inherent characteristics fulfils a need or expectation that is stated, generally implied or obligatory (ISO 9000).

Quality assurance   Part of quality management focused on providing confidence that quality requirements will be fulfilled (ISO 9000).

Quality characteristics   Any characteristic of a product or service that is needed to satisfy customer needs or achieve fitness for use.

Quality of conformance   The extent to which the product or service conforms to the specified requirements.

Quality control   A process for maintaining standards of quality that prevents and corrects change in such standards so that the resultant output meets customer needs and expectations (*see also* ISO 9000 Clause 3.2.10).

Quality costs   Costs incurred because failure is possible. The actual cost of producing an entity is the no-failure cost plus the quality cost. The no-failure cost is the cost of doing the right things right first time. The quality costs are the prevention, appraisal and failure costs.

Quality function deployment (QFD)   A technique to deploy customer requirements (the true quality characteristics) into design characteristics (the substitute characteristics) and deploy them into subsystems, components, materials and production processes. The result is a grid or matrix that shows how and where customer requirements are met.

Quality improvement   Part of quality management focused on increasing the ability to fulfil quality requirements (ISO 9000).

Quality management system   The set of interconnected processes that enables the organization to achieve its objectives (*see also* ISO 9000 Clause 3.2.3).

Quality management system requirements   Requirements pertaining to the design, development, operation, maintenance and improvement of quality management systems.

Quality manual   A document specifying the quality management system of an organization (ISO 9000).

Quality objectives   Those results which the organization needs to achieve in order to improve its ability to meet needs and expectations of all the interested parties.

Quality planning   Provisions made to achieve the needs and expectations of organization's interested parties and prevent failure.

Quality plans   Plans produced to define how specified quality requirements will be achieved, controlled, assured and managed for specific contracts or projects.

Quality problems   The difference between the achieved quality and the required quality.

Quality requirements   Those requirements which pertain to the features and characteristics of a product or service which are required to be fulfilled in order to satisfy a given need.

Quarantine area   A secure space provided for containing product pending a decision on its disposal.

Registrar   An organization that is authorized to certify organizations. The body may be accredited or non-accredited.

Registration   A process of recording details of organizations of assessed capability that have satisfied prescribed standards.

Regulator   A legal body authorized to enforce compliance with the laws and statutes of a national government.

Regulatory requirements   Requirements established by law pertaining to products or services.

Remedial action   Action proposed or taken to remove a nonconformity (see also *Corrective action* and *Preventive action*).

Representative sample   A sample of product or service that possesses all the characteristics of the batch from which it was taken.

Requirement of the standard   A sentence containing the word *shall*. *Note*: Some sentences contain multiple requirements such as to establish, document and maintain. This is in fact an example with three requirements.

Responsibility   An area in which one is entitled to act on one's own accord.

Review   Another look at something.

Rework   Continuation of work on a product to make it conform to the specified requirements without additional procedures or techniques.

Scheduled maintenance   Work performed at a time specifically planned to minimize interruptions in machine availability; e.g. changing a gearbox when machine is not required for use (includes predictive and preventive maintenance).

Servicing   Action to restore or maintain an item in an operational condition.

Shall   A provision that is binding.

Should   A provision that is optional.

Simultaneous engineering   A method of reducing the time taken to achieve objectives by developing the resources needed to support and sustain the production of a product in parallel with the development of the product itself. It involves customers, suppliers and each of the organization's functions working together to achieve common objectives.

Six sigma   Six standard deviations.

SMS   Safety management system.

Special cause   A cause of variation that can be assigned to a specific or special condition that does not apply to other events.

Specified requirements   Requirements prescribed by the customer and agreed by the organization or requirements prescribed by the organization that are perceived as satisfying a market need.

Stakeholder   A person or an organization that has freedom to provide something to or withdraw something from an enterprise.

Statistical control   A condition of a process in which there is no indication of a special cause of variation.

Status   The relative condition, maturity or quality of something.

Status of an activity (in auditing)   The maturity or relative level of performance of an activity to be audited.

Subcontract requirements   Requirements placed on a subcontractor that are derived from requirements of the main contract.

Subcontractor   A person or company that enters into a subcontract and assumes some of the obligations of the prime contractor.

Supplier development   A technique for promoting continual improvement of suppliers by encouraging customer-supplier relationships and communication across all levels of the organizations involved.

System audit   An audit carried out to establish whether the quality system conforms to a prescribed standard in both its design and its implementation.

System effectiveness   The ability of a system to achieve its stated purpose and objectives.

Technical interfaces   The physical and functional boundary between products or services.

Tender   A written offer to supply products or services at a stated cost.

Theory of constraints   A thinking process optimizing system performance. It examines the system and focuses on the constraints that limit the overall system performance. It looks for the weakest link in the chain of processes that produce organizational performance and seeks to eliminate it and optimize system performance.

TQM   Total quality management.

Traceability   The ability to trace the history, application, use and location of an individual article or its characteristics through recorded identification numbers (see also ISO 9000 Clause 3.5.4).

Unique identification   An identification that has no equal.

Validation   A process for establishing whether an entity will fulfil the purpose for which it has been selected or designed (see also ISO 9000 Clause 3.8.5).

Value engineering   A technique for assessing the functions of a product and determining whether the same functions can be achieved with fewer types of components and materials and the product produced with less resources. Variety reduction is an element of value engineering.

Verification   The act of establishing the truth or correctness of a fact, theory, statement or condition (see also ISO 9000 Clause 3.8.4).

Verification activities   A special investigation, test, inspection, demonstration, analysis or comparison of data to verify that a product, service or process complies with prescribed requirements.

Verification requirements   Requirements for establishing conformance of a product or service with specified requirements by certain methods and techniques.

Waiver   *See Concession*.

Work environment   Is a set of conditions under which people operates and include physical, social and psychological environmental factors (ISO 9000:2000).

Work instructions   Instructions that prescribe work to be executed, who is to do it, when it is to start and be complete and if necessary how it is to be carried out.

Workflow system   A method of manufacture whereby value is added to the product in each process as it moves along a production line. Invented in 1910 by Charles Sorensen, first President of Ford Motor Company.

Workmanship criteria   Standards on which to base the acceptability of characteristics created by human manipulation of materials by hand or with the aid of hand tools.

Zero defects   The performance standard achieved when every task is performed right first time with no errors being detected downstream.

Appendix B – Related web sites

Articles and publications

An excellent range of articles and publications on ISO 9000:2000 at http://www.
transition-support.com
Articles from Quality World at http://www.iqa.org
Purchase past articles from *ASQ Journals* at http://qic.asq.org/infosearch.html
An excellent range of books on ISO 90000, ISO 14000 and related topics at
http://books.elsevier.com/
Information on the fathers of scientific management at http://www.accelteam.com/
scientific/index.html
Information about W. E. Deming at http://www.deming.org/
Articles from Dr. J. M. Juran at http://www.juran.com/
Quality magazines from http://www.qualitydigest.com/

Auditing

Transition arrangements for registered auditors at http://www.lrqa.co.uk/
Information on auditing practices at http://www.iqnet-certification.com
The official line from the International Accreditation Forum on the transition at
http://www.iaf.nu

Benchmarking

Information on benchmarking from the American Productivity Centre at http://www.
apqc.org/

Competence

The Institute of Personnel Development at http://www.ipd.co.uk/

Customer satisfaction

The American Customer Satisfaction Index at http://www.bus.umich.edu/
FacultyResearch/ResearchCenters/Centers/Acsi.htm
http://www.census.gov/mso/www/npr/acsi.htm

International organizations

International Automotive Task Force (IATF) web site http://www.iaob.org
International Automotive Oversight Body (IAOB) http://www.iaob.org
International Organization of Motor Vehicle Manufacturers (OICA) http://www.
oica.net
International Organization for Standardization (ISO) http://www.iso.ch

Process management

Articles on Process Management and ISO 9000 at http://www.dti.gov.uk/
Information from the Business Process Management Initiative at http://www.
bpmi.org/

Related initiatives

UK's manufacturing industry initiative at http://www.fitforthefuture.co.uk/
Tomorrows Company at http://www.tomorrowscompany.co.uk/whatis_tia.html
Information from the European Organization for Quality at http://www.
eoq.org/start.html

Six sigma

A good range of articles on Six Sigma at http://www.qualityamerica.com/six_
sigma.html

Standards

Mil Q 9858 can be obtained at http://www.daps.dla.mil
ISO Standards can be obtained at http://www.iso.ch
British Standards can be obtained at http://www.bsi-global.com/
Access news from the Technical Committee that created ISO 9000 at http://www.
tc176.org/
Standards on Investors in People at http://www.iipuk.co.uk/

Product recall

Recalls notified by UK Trading Standards at http://www.tradingstandards.net/
pages/recall.htm
Recall notices from the UK Department of Transport, Vehicle Inspectorate at
http://www.automotive.co.uk/fleetnews/start.htm
Recall notices from USA at http://fullcoverage.yahoo.com/Full_Coverage/US/
Consumer_News_and_Recall_Information/

Bibliography

Chapter 1

1 Figures from The International Organization of Motor Vehicle Manufacturers (OICA) (November 2004).
2 Ohno, T. (1978). *Toyota Production System*. Productivity Press, Portland Oregon.
3 Womack, J. P., Jones, D. T. and Roos, D. (1990). *The Machine that Changed the World*. Harper Perennial.
4 Maslow, A. H. (1954). *Motivation and Personality*. Harper & Row, New York.
5 Rollinson, D., Broadfield, A. and Edwards, D. J. (1998). *Organizational Behaviour and Analysis*, Addison Wesley Longmans. Based on Table 14.3.
6 Hoyle, D. and Thompson, J. (2001). *ISO 9000:2000 Auditor Questions*. Transition Support.
7 Hoyle, D. and Thompson, J. (2000). *Converting a Quality Management System using the Process Approach*, 2nd edition. Transition Support.
8 Boone, L. E. and Kurtz, D. L. (2001). *Contemporary Marketing*, 10th edition. Harcourt College Publishers, pp. 11–13, Chapter 1 Customer-driven marketing.
9 Juran, J. M. (1964). *Managerial Breakthrough*. McGraw-Hill.
10 Deming, W. E. (1982). *Out of the Crisis*. MITC.
11 *Ibid., Supra* 10. Juran as observed by Edwards Deming.
12 Pyzdek, T. (2001). *The Complete Guide to Six Sigma*. McGraw-Hill.
13 *Ibid., Supra* 9.
14 Watson, G. H. (1994). *Business Systems Engineering*. Wiley.
15 Shannon, R. E. (1975). *Systems Simulation*. Prentice-Hall.
16 Seddon, J. (2000). *The Case against ISO 9000*. Oak Tree Press.
17 Drucker, P. F. (1977). *Management: Tasks, Responsibilities, Practices*. Pan Business Management.
18 Hoyle, D. and Thompson, J. (2000). *Converting a Quality Management System using the Process Approach*. Transition Support.
19 Juran, J. M. (1992). *Juran on Quality by Design*. The Free Press.
20 Hammer, M. and Champy, J. (1993). *Reengineering the Corporation*. Harper Business.

21 Davenport, T. H. (1993). *Process Innovation: Reengineering Work through Information Technology*. Harvard Business School Press.
22 Hammer, M. and Champy, J. (1993). *Reengineering the Corporation,* Harper Business.
23 *Ibid., Supra* 19. Based on Figure 11.1.
24 Hoyle, D. and Thompson, J. (2005). *How to Satisfy Stakeholders – An Integrated Approach*. Transition Support.
25 *Ibid., Supra* 9.

Chapter 2

1 Department of Trade and Industry (1982). *White Paper on Quality, Standards and Competitiveness*. HMSO.
2 International Organization of Standardization (2000). *The ISO Survey of ISO 9000 and ISO 14000 Certificates Ninth Cycle 1999*, ISO.
3 Seddon, J. (2000). *The Case against ISO 9000*. Oak Tree Press.

Chapter 3

1 The ISO survey of ISO 9000 and ISO 14000 certificates, thirteenth cycle, December 2003.
2 *Standards, Quality and International Competitiveness*, HMSO, July 1993.
3 Adapted from an article in *Auto Express*, January 1999.
4 *Auto Express*, January 1999.
5 *ISO/TS 16949 Audit and Analysis Tool*, David Hoyle & John Thompson, 2004.

Chapter 4

1 Hammer, M. and Champy, J. (1993). *Reengineering the Corporation*. Harper Business.
2 Hoyle, D. and Thompson, J. (2004). *A Guide to Process Management*. Transition Support.
3 US Department of Defense (2001). www.idef.com All IDEF models can be downloaded from this web site.
4 Hoyle, D. and Thompson, J. (2000). *Converting a Quality Management System using the Process Approach*. Transition Support.
5 Juran, J. M. (1995). *Managerial Breakthrough Second Edition*. McGraw-Hill.
6 *Ibid., Supra* 4.
7 *Ibid., Supra* 5.
8 Imai, M. (1986). *KAIZEN, The key to Japanese Competitive Success*. McGraw-Hill.

9 *Ibid., Supra* 5.
10 Hoyle, D. (1996). *ISO 9000 Quality System Development Handbook.* Butterworth Heinemann.
11 International Organization of Standardization (2001). www.iso.ch
12 *Ibid., Supra* 10.

Chapter 5

1 Peters, T. and Waterman, R. (1995). *In Search of Excellence.* Harper Collins.
2 Drucker, P. F. (1977). *Management.* Harper's College Press.
3 Juran, J. M. (1995). *Managerial Breakthrough*, 2nd edition. McGraw-Hill.
4 Codling, S. (1998). *Benchmarking.* Gower.
5 *Ibid., Supra* 2.
6 Deming, W. E. (1982). *Out of the Crisis.* MITC.
7 Hoyle, D. (1996). *ISO 9000 Quality System Development Handbook.* Butterworth Heinemann.
8 Ohno, T. (1978). *Toyota Production System.* Productivity Press, Portland, Oregon.
9 Liker, J. K. (2004). *The Toyota Way.* McGraw-Hill.
10 *Ibid., Supra* 3, Chapter 17 Mobilizing for decision making.

Chapter 6

1 Fletcher, S. (2000). *Competence-based assessment techniques,* Kogan Page.
2 *Ibid., Supra* 1.
3 *Ibid., Supra* 1.
4 *Ibid., Supra* 1.
5 *Ibid., Supra* 1.
6 Deming W Edwards (1982). *Out of the Crisis,* MITC. Chapter 8 page 253.
7 Rollinson, Broadfield and Edwards (1998). *Organizational Behaviour and Analysis,* Addison Wesley Longmans.
8 Vroom, V. H. (1964). *Work and Motivation.* John Wiley, New York.
9 *Ibid., Supra* 7.

Chapter 7

1 Lyon, D.D. (2000). *Practical CM – Best Configuration Management Practices.* Butterworth Heinemann.
2 Sloan, Jr., A.P. (1963). My Years with General Motors. Doubleday.
3 Boone, L.E. and Kurtz, D.L. (2001). *Contemporary Marketing*, 10th edition. Harcourt College Publishers.

4 Measurement (1994–1999). *Britannica® CD 99 Multimedia Edition* ©. Encylo-paedia Britannica, Inc.
5 Metric system (1994–1999). *Britannica® CD 99 Multimedia Edition* ©. Encylo-paedia Britannica, Inc.

Chapter 8

1 Juran, J. M. (1995). *Managerial Breakthrough*, 2nd edition. McGraw-Hill.
2 Deming, W. E. (1982). *Out of the Crisis*. MITC.
3 Anonymous (1993). *Quality Network Action Strategy Summary.* General Motors.
4 *Ibid., Supra* 2.
5 American Society for Quality (2001). http://acsi.asq.org/.
6 US Government (2001). http://www.census.gov/mso/www/npr/acsi.htm.
7 *Ibid., Supra* 1.
8 *Ibid., Supra* 4.
9 Six sigma dictionary at http://www.isixsigma.com
10 From BNP Media, Troy, Michigan http://www.bnp.com/industrial_heating/archive/quality7-97.html
11 Smith, D. (1997). *Reliability, Maintainability and Risk*, 5th edition. Butterworth Heinemann.
12 *Ibid., Supra* 11.
13 Kemta and Ishii. (2000). *Scenario Based FMEA: Life Cycle Cost Perspective. ASME Design Engineering Conference.*
14 Crosby, P. (1979). *Quality is Free.* McGraw-Hill.

Index